Cambridge International A Level

Information Technology

Brian Gillinder

Series editor: Brian Sargent

Answers to practices questions and activities and all source files can be downloaded from www.hoddereducation.com/cambridgeextras.

Computer hardware and software brand names mentioned in this book are protected by their respective trademarks and are acknowledged.

Photo credits: page 3 © LDprod/Shutterstock.com; page 39 © Blackdiamond67/stock.adobe.com; page 40 © Sspopov/123RF.com; page 41 © Swapan/stock.adobe.com; page 43 © Insta_photos/stock.adobe.com; page 46 © Vantage/stock.adobe.com; page 50 © James/stock.adobe.com; page 51 © Franckito/123RF.com; page 64 © Julian Herzog/CC BY (https://creativecommons.org/licenses/by/4.0); page 65 *top* © Vadim/stock.adobe.com; page 65 *bottom* © Jim/stock.adobe.com; page 66 © Yingyaipumi/stock.adobe.com; page 68 © Monkey Business/stock.adobe.com; page 70 *left* © Jenson/Shutterstock.com, *right* © Andrea Danti/stock.adobe.com; page 72 © Tl6781/stock.adobe.com; page 74 © Prescott09/stock.adobe.com; page 76 © Monopoly919/stock.adobe.com; page 77 © Metamorworks/stock.adobe.com; page 120 © Scanrail/stock.adobe.com; page 123 © Vladimir Zhupanenko/stock.adobe.com; page 132 © Destina/stock.adobe.com; page 134 © Salita2010/stock.adobe.com; page 136 *top* © Ra3rn/stock.adobe.com, *bottom* © Zakhar Marunov/stock.adobe.com; page 138 © Sdecoret/stock.adobe.com; page 166 © Africa Studio/stock.adobe.com; page 208 both © Hodder & Stoughton Ltd; page 211 © Bonnontawat/stock.adobe.com; pages 367, 368, 388 © Brian Gillinder

Although every effort has been made to ensure that website addresses are correct at time of going to press, Hodder Education cannot be held responsible for the content of any website mentioned in this book. It is sometimes possible to find a relocated web page by typing in the address of the home page for a website in the URL window of your browser.

Hachette UK's policy is to use papers that are natural, renewable and recyclable products and made from wood grown in well-managed forests and other controlled sources. The logging and manufacturing processes are expected to conform to the environmental regulations of the country of origin.

To order, please visit www.hoddereducation.com or contact Customer Service at education@hachette.co.uk / +44 (0)1235 827827.

ISBN: 978 1 3983 0698 1

© Brian Gillinder 2023

First published in 2023 by
Hodder Education,
An Hachette UK Company
Carmelite House
50 Victoria Embankment
London EC4Y 0DZ

www.hoddereducation.com

The authorised representative in the EEA is Hachette Ireland, 8 Castlecourt Centre, Dublin 15, D15 XTP3, Ireland (email: info@hbgi.ie)

Impression number   10 9 8 7 6 5 4 3 2 1
Year   2027 2026 2025

Cover photo © Evgen3d - stock.adobe.com

Typeset by Aptara, Inc.

A catalogue record for this title is available from the British Library.

Printed in India

Contents

18 Mail merge   337

19 Graphics creation   364

20 Animation   394

21 Programming for the web   409

Answers can be found at www.hoddereducation.com/CambridgeExtras

Introduction

This textbook has been written to provide the knowledge, understanding and practical skills required by those studying the A Level content (Topics 12–21) of the Cambridge International AS & A Level Information Technology 9626 syllabus for examination from 2025. Students studying A Level will need to be familiar with the AS Level content (Topics 1–11) covered in Hodder Education's *Cambridge International AS Level Information Technology*.

How to use this book

This textbook, endorsed by Cambridge International Education, has been designed to make your study of Information Technology as successful and rewarding as possible.

Organisation

The book comprises 10 chapters, the titles of which correspond exactly with the topics in the syllabus. Each chapter is broken down into sections. While these sections largely reflect the subtopics in the syllabus, the order of the subtopics in a few chapters has been changed from that in the syllabus to make for a more logical approach to the content.

Features

Each chapter contains a number of features designed to help you effectively navigate the syllabus content.

At the start of each chapter, there is a blue box that provides a summary of the content to be covered in that topic.

> **In this chapter you will learn:**
> - ★ the types of new and emerging technologies
> - ★ the impacts of the new and emerging technologies.

There is also a box that lists the knowledge you should have before beginning to study the chapter.

> **Before starting this chapter, you should:**
> - ★ be familiar with the terms: project, collaborative working, scheduling and review.

Each chapter includes Activities to allow you to check your understanding of the concepts covered and practise the skills demonstrated in the text.

> **Activity 12b**
> 1 Describe a distributed ledger database.
> 2 Describe the contents of a blockchain created during a digital currency transaction.

The practical chapters also contain Tasks. The text demonstrates the techniques used to carry out the tasks. It provides easy-to-follow step-by-step instructions, so that practical skills are developed alongside knowledge and understanding.

> ### Task 18j
>
> The selection of recipients for an invitation to a conference could be made on the basis of where the recipient lives. Create a spreadsheet with the names of ten different recipients, five of whom live in one country, two in a different country, two in another different country and one in another. Create a master document of an invitation with mail-merge fields that selects the recipient depending on their country.

Finally, each chapter ends with Practice questions. These questions and their answers have been written by the author and the answers can be found at: www.hoddereducation.co.uk/cambridgeextras.

Practice questions

1 Layers are used when creating a computer graphic.
   a Explain why layers are used. (4)
   b Explain why layers are usually merged in the final image. (3)

2 a Describe how rulers, grids and guidelines can assist in the placing of objects when creating digital images. (6)
   b Describe the differences between the effects on an image of crop and cut out. (4)

3 Colour wheels are used in graphic image creation as an aid to choosing colours.
   a What is a colour wheel? (2)
   b Describe two limitations of using colour wheels. (2)

4 Why are colour-management systems (CMS) needed? (6)

Text colours

Some words or phrases within the text are printed in red. Definitions of these terms can be found in the glossary at the back of the book. In the practical section, words that appear in blue indicate an action or location found within the software package, for example 'Select the Home tab.' Words that appear in green indicate spreadsheet functions or formulas, for instance 'Pivot charts are created from the Insert, Pivot chart menu options.'

Problem-solving

When there are practical tasks that require problems to be solved, it is important to use your knowledge, logic and imagination. Firstly, it is necessary to understand the task so the problem can be identified. At A Level, you are expected to use your knowledge and practical skills, find the most efficient techniques and methods and then complete the practical tasks. The process of problem-solving in this context involves identifying and understanding the problem, gathering knowledge and using skills so the solution can be created, choosing the best method of creating the solution, and finally testing and reviewing the solution. By carefully following the instructions in the tasks and applying your knowledge and skills, you should be able to produce the required solution to the given problem and complete the tasks.

Assessment

The information in this section is taken from the Cambridge International Education syllabus. You should always refer to the appropriate syllabus document for the year of your examination to confirm the details and for more information. The syllabus document is available on the Cambridge International Education website at www.cambridgeinternational.org.

For the A Level part of the course, you will take two examination papers:

Paper 3 Advanced Theory (1 hour 45 minutes)

Paper 4 Advanced Practical (2 hours 30 minutes)

The Advanced Practical examination is based on Topics 17–21 of the subject content but may include practical tasks from AS Level Topics 8–10. Tasks are set in a problem-solving context. Students are expected to use the skills gained from studying the AS Level Topics 8–10 and apply them when solving the A Level practical tasks.

Command words

The table below, taken from the syllabus, includes command words used in the assessment for this syllabus. The use of the command word will relate to the subject context. Make sure you are familiar with these.

Command word	What it means
Analyse	Examine in detail to show meaning, identify elements and the relationship between them
Assess	Make an informed judgement
Compare	Identify/comment on similarities and/or differences
Contrast	Identify/comment on differences
Define	Give precise meaning
Describe	State the points of a topic/give characteristics and main features
Discuss	Write about issue(s) or topic(s) in depth in a structured way
Evaluate	Judge or calculate the quality, importance, amount or value of something
Explain	Set out purposes or reasons/make the relationships between things clear/say why and/or how and support with relevant evidence
Identify	Name/select/recognise
Justify	Support a case with evidence/argument
State	Express in clear terms
Suggest	Apply knowledge and understanding to situations where there are a range of valid responses in order to make proposals/put forward considerations

Notes for teachers

Key concepts

These are the essential ideas that help students to develop a deep understanding of the subject and to make links between the different topics. Although teachers are likely to have these in mind at all times when they are teaching the syllabus, the following icons are included in the textbook at points where the key concepts relate to the text.

Hardware and software 

Hardware and software interact with each other in an IT system. It is important to understand how these work and how they work together with each other, and with us in our environment.

Networks 

Computer systems can be connected together to form networks allowing them to share data and resources. The central role networks play in the internet, mobile and wireless applications and cloud computing has rapidly increased the demand for network capacity and performance.

The internet 

The internet is a global communications network. It uses standardised communications protocols to allow computers worldwide to connect and share information in many different forms. The impact of the internet on our lives is profound. While the services the internet supports can provide huge benefits to society, they have also introduced issues, for example security of data.

System life cycle 

Information systems are developed within a planned cycle of stages. They cover the initial development of the system through to its scheduled updating or redevelopment.

New technologies 

As the information industry changes so rapidly, it is important to keep track of new and emerging technologies and consider how they might affect everyday life.

12

IT in society

12.1 Digital currencies

Currency is money that can be exchanged for goods or services. Traditional currency is made up of physical coins and banknotes. The amount of physical money in circulation at any one time is controlled by the banks or financial institutions that issue the currency. Banks now hold a great deal of money in digital form on their computers, and not as physical currency in vaults. The economy of a country uses both physical and digital currency for the exchange of goods and services.

A **digital currency** is money that is not in a physical form (such as banknotes and coins) but is recorded electronically. It is also called electronic currency and can be used to buy services and physical items. Digital currency can be used within communities, such as in an online game, where it can be traded for virtual items or between members. Digital currency can be stored as an amount or balance on a **stored-value card** (SVC), where the funds are recorded on the card and not in a separate account.

Governments in most countries have central control over financial activities, such as those of banks and other financial institutions. The supply of physical and digital currency in the economy can be monitored and adjusted if the government thinks it necessary to do so. This is a centralised banking system. Currencies that are not controlled by a central banking system are called **decentralised** and are not monitored or administered by a government or regulatory body.

12.1.1 Centralised banking systems

A **centralised banking system** regulates and supervises the financial activities in a country or region. It controls the entire money supply, the reserves, interest rates and foreign exchange of the state. Debit and credit card systems are regulated by centralised banking systems but may be issued by many different banks.

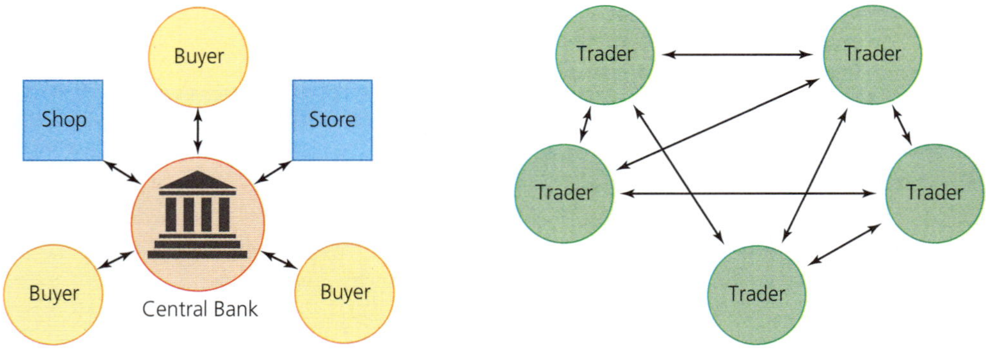

▲ **Figure 12.1** Centralised banking and decentralised banking

Central bank digital currency

Central bank digital currency (CBDC) is a form of digital currency, also called **digital base money**, that is set up by a government and regulated by laws. These currencies are part of the money system and the money supply and can be used as a means of payment and for storing value. They are the responsibility of the central bank of the government that set it up and the bank keeps a record of the amount of money in each account. The value of the currency is backed by the bank's reserves such as gold and foreign exchange. **Blockchains** are not used as they are not necessary for keeping the records secure. Each unit of currency in a CBDC system carries a unique serial number, which prevents counterfeiting. The use of central bank digital base money has both the convenience of cryptocurrencies and the protection from unpredictable fluctuations in value provided by traditional centralised banking.

Credit cards

A **credit card** is issued by a bank to a cardholder. An individual or business, called the 'merchant', will accept a credit card payment from the cardholder in return for services or goods and send the details to the merchant's bank – the acquiring bank. After a period of time, the issuing bank sends a bill to the cardholder requiring payment. Failure by the cardholder to pay the bill (or part of the bill) within an agreed time period will mean that the bank adds interest to the next bill.

The information and money flow between the bank, cardholder and merchant is called the interchange. The transactions that make up a credit card interchange include the cardholder presenting the card to the merchant as payment and the payment being authorised using a signature or a PIN (a Personal Identification Number known only to the user), or neither in the case of **contactless payments**. The transaction is deducted from the buyer's account and deposited into the bank's account and eventually ends up in the merchant's account. In effect, the bank that issued the credit card loans the cardholder the money to buy goods, until the holder repays the loan. The merchant is paid by the bank so the merchant does not know, or need to know, if the cardholder can repay the loan to the bank. This differs from the use of debit cards and **cryptocurrency** because the bank loans the money. No loans are made with debit cards and cryptocurrency.

Debit cards

With **debit cards**, the amount to be paid to the merchant accepting the debit card for payment is taken from the account of the cardholder with no bank lending the holder any money. If the cardholder does not have sufficient funds available, then the transaction will not proceed. A debit cardholder may not notice it when using the card, but debit card transactions are processed in several different ways.

Offline debit card systems are used in the same way as credit cards. There is no network connection from the merchants with the bank or account of the cardholder. The card is presented and authenticated, with a signature not a PIN, at the point of sale and the amount is taken from the holder's account within two or three days of the purchase. Offline debit cards are mostly used for paying for goods or services and cannot be used to withdraw or deposit money at ATMs or merchants. Often, a fee is payable by the cardholder when using an offline debit card and there is a maximum daily limit on use.

Online debit cards require every transaction to be electronically authorised using a network connection to the issuing bank, which allows the transaction to be processed immediately. The amount of the transaction is taken from the cardholder's account and transferred into the merchant's account at the moment of purchase. Authentication is by the cardholder entering a PIN. The merchant must have access to an electronic PIN pad for entry of the cardholder's PIN and for the transaction to be authorised. This type of debit card processing is far more secure and safe for the merchant and cardholder and is now the norm in many areas of the world. Online debit card systems allow the cardholder to withdraw cash at ATMs and at electronic point-of-sale checkouts at merchants, whereas offline debit card systems do not.

An ATM (automatic teller machine) enables users to insert their bank card, enter their PIN and withdraw cash, and also to look at their account balance and sometimes carry out other transactions, for example paying bills. ATMs can be located outside of buildings for use at any time or within restricted areas for more security.

Activity 12a

Explain the main difference between the use of a credit card and a debit card when paying for goods.

Electronic point-of-sale

Electronic points-of-sale (EPOS) used at checkouts or sales points in stores allow customers to pay using credit or debit cards. These are self-contained computer systems that carry out all of the tasks that an employee might do at a checkout to complete a sale. The amount of money that must be paid by the purchaser is calculated using devices such as barcode scanners, keypads, touchscreens and weighing scales that input data about the items being sold to the customer. At the time of checkout, the total amount to be paid is calculated and displayed to the customer. The customer uses a chip-and-PIN reader or a contactless card reader to pay with a credit or debit card. The amount paid is recorded and transferred digitally via the centralised banking systems.

▲ **Figure 12.2** EPOS terminal

Stored-value cards

Stored-value cards, such as gift cards, travel cards and pre-loaded debit cards, hold a fixed amount of money that can be withdrawn or used to buy goods. The store of value on the card is sometimes referred to as an electronic purse (ePurse) as the cardholder can take out money or add money to the purse. This is not the same as an electronic wallet, which is a web service or software that stores the user's logins, passwords and credit card details. Stored-value cards do not hold these details.

A value or amount of money is added to the card when it is first purchased. Once the amount is spent or used, the card is discarded. Gift cards from stores work in this way. Some cards allow a 'top-up' to be made by adding more value in a store or by an online transaction and the card is used as a purse.

An example of a stored-value card is the card issued by public transport or transit authorities. The card is meant to replace the use of cash in the transport payment system. A traveller is required to purchase a card pre-loaded with a fixed amount of money to pay for journeys. For each journey the system deducts the cost from the card when it is presented at the ticket barriers. The card may hold 'tokens' that represent set journeys and these are reduced in number on the card for each journey undertaken. The cards can only be used for travel and have no value for anything else. Travellers may be encouraged to use these travel cards through reduced fares per journey or discounts for frequent use of the cards. Reducing the amount of cash money used can reduce the risks of theft, fraud and errors in accounting for the cash by travellers and staff.

Stored-value cards and the ePurse system do not need a network connection for use. The value on the card is usually only available for viewing in the issuing store or when used, but some providers allow holders to check the stored value via their website.

A major risk of stored-value cards is the lack of security. Often the card requires no authentication for use and so, if lost, it can be used by anyone who finds it. They are untraceable, which allows easy transfer of money, for example as a gift, but, if lost, the value is gone. Also, if the card is damaged and unreadable, its value is gone.

Digital wallets

A **digital wallet** is an encrypted software application or service that stores user details on a computer device. These can include bank and credit card details, identification details, driving licence details or almost anything. Details stored in digital wallets can be used for making purchases or for verifying the user's identity or age.

Online wallets

An **online wallet** allows users to store their online shopping information such as logins, passwords, credit card details and shipping address in secure software. The wallet can be used to fill out online forms when purchasing goods, meaning all the user has to do is enter a password and the form is automatically completed.

Mobile electronic wallets

A **mobile electronic wallet** is an app or software that can be installed and used on a mobile device such as a **smartphone**. The wallet is used to store details of the user's credit or debit cards, loyalty cards and other payment options. When a purchase is made, the user opens the app on the smartphone, chooses the card to use and then taps or holds the phone against a terminal designed to accept digital payments. Since wireless **Near Field Communication (NFC)** is used to transfer the required data to and from the phone and terminal in the same way as contactless payments are made with physical cards, the smartphone has to be held quite close to the terminal.

The data in the mobile electronic wallet is encrypted and access is usually only gained by the use of two-step authentication, with PINs and fingerprints. This makes the system secure and helps to prevent fraudulent use of the smartphone if it is lost or stolen. Further, the ability to remotely wipe or disable a lost smartphone can help keep the wallet details safe and secure.

12.1.2 Decentralised banking systems

Currency systems that are not controlled by any central authority are said to be 'decentralised'. Virtual currencies and cryptocurrencies are decentralised forms of currency.

Virtual currency

A **virtual currency** represents a value, or amount of money, that is stored only in digital form. It can work in the same way as real money (coins and paper money) for buying goods or services, but it is not recognised as having any legal tender status by governments or financial institutions. Virtual currencies are not issued by central banks or authorities and are not regulated; instead they rely on trust between users and get their value from assets owned by the developers or from other mechanisms. They are mostly used between members of a specific virtual community, such as in an online virtual world or players in games on gaming networks. Virtual currencies are most often used for peer-to-peer transactions. The value of virtual currencies can fluctuate wildly in a very short time because the currencies are not regulated by any centralised authority.

Cryptocurrency

A cryptocurrency is a type of digital currency that uses cryptography technology to keep transactions authentic and secure. **Bitcoin** and **Litecoin** are two of the many cryptocurrencies. Control of a cryptocurrency is not centralised but works by using a **distributed ledger**. This means that the record of transactions is shared and replicated across many institutions, sites and countries. The records are synchronised between all of the sites. A blockchain system is usually used to do this.

Cryptocurrencies using blockchains make use of asymmetric encryption, with only the owner knowing the private key to access their records on the blockchain. If the private key is lost, then the records, transactions and value of the cryptocurrency are also lost as these cannot be accessed by anyone without the key.

Cryptocurrency data is stored across **peer-to-peer networks**. Devices on peer-to-peer networks are called **nodes**. Every node or user on the network has a copy of the blockchain. There is no main or official copy of the blockchain and no copy is trusted or referenced any more than any other. This decentralisation requires a large amount of computer resources and, as the system grows and grows over time, there is greater demand for computing resources. However, the distributed storage of the blockchains does ensure that all transactions can be securely authenticated.

When transactions are made using a cryptocurrency, they are checked to ensure that they are valid and can be allowed to occur. The checking requires computing power and, as more and more transactions occur, computing resources are often shared to accomplish this. During the checking of a transaction, a proof-of-work time-stamping system is used. The use of a user's computer resources to check transactions is rewarded with new cryptocurrency. The process of creating new currency in this way is known as mining. **Mining for cryptocurrency** has led to an increase in demand for more powerful, and cheaper, processing systems as well as increasing the demand for electricity.

Bitcoin

Bitcoins are created by mining and can be exchanged for goods or services or converted into other currencies. Transactions are verified by nodes using cryptography and recorded in a public ledger. The public ledger is distributed across all of the nodes and is called a blockchain.

Litecoin

Litecoin is derived from Bitcoin. It works in almost the same way as Bitcoin but processes transactions about four times as fast as Bitcoin. It also uses a more complex method of verifying transactions, which requires more processing power. This means that computing devices used for mining Litecoins are more expensive.

Activity 12b

1 Describe a distributed ledger database.
2 Describe the contents of a blockchain created during a digital currency transaction.

Characteristics of decentralised digital currencies

Digital currencies have a number of common characteristics, the main one being that they are all stored electronically with no physical presence.

The characteristics of decentralised digital currencies distinguish them from other types of currency. Decentralised digital currencies have no central controlling or regulatory authority, digital transactions are peer-to-peer over distributed and open networks, and transactions using blockchains are replicated by every node so the system is less susceptible to hacking attempts, software errors and system failures. Digital currency transactions are anonymous, with users having no need to identify themselves or be identified, and the transactions cannot be reversed or undone. Digital currency has a limited supply that depends on the algorithms used when it was created (Bitcoin is limited to 21 million coins and Litecoin to 84 million coins; some others have much higher limits) to try to prevent manipulation of its value. Also, there is no need for users to trust each other as all nodes verify, or not, the transactions.

Activity 12c

1 Describe the common characteristics of decentralised digital currencies.
2 Explain why users of decentralised digital currencies do not need to trust each other.

12.1.3 Advantages and disadvantages of digital currencies

Digital currencies enable global transactions to occur much more quickly and easily. The transactions are made over the internet and money can be transferred securely within minutes between accounts in different countries. Online shopping with sellers and purchasers around the world has been made possible by the use of digital currencies. Also, if the digital currency is a cryptocurrency, its value does not change due to exchange rate changes between different countries' currencies. One Bitcoin is worth one Bitcoin everywhere in the world, but one US dollar, for example, while still one dollar, would be exchanged into varying amounts of different currencies. Traditional bank money transfers can take several days to complete and involve more documentation. There is also the commission payable for exchanging currency. Digital currencies such as Bitcoin do not need exchanging, but the use of a credit card issued by a bank can incur high charges and poor exchange rates when purchasing goods from a different country or when used in a different country while on holiday.

However, digital currencies such as Bitcoin are not regulated or supported by central authorities. Their actual value can vary enormously over short periods of time. Users can be exposed to large losses as well as gains in their value. Currencies that are not regulated or supported can also be easily lost. Losing the passwords or keys to cryptocurrencies loses the entire value as the currency cannot be recovered. Exchanging and trading digital currencies can be difficult if there are few users of the currency. Security can be a problem because passwords or keys must be kept secure from **unauthorised access**. There is no central bank or credit card provider to protect or reimburse users of digital currencies if the currency is accessed and stolen. Digital currencies, apart from some stored-value cards issued by, for example retail stores, cannot at present be used in shops to buy goods such as a newspaper or sandwich, and usually cannot be exchanged for physical cash, for example at an ATM.

12.1.4 Impacts and risks of digital currencies

The impacts and risks of digital currencies, which include virtual currencies, cryptocurrencies and other types of currency, arise from the fact that the details of the values and transactions are stored electronically.

Individuals

Digital currencies are not regulated so there is little control over the value of each unit or coin, which means that all trading activities can have large effects on the value, for example the value of 'coins' can change rapidly and digital currencies are considered a high-risk investment. Cryptocurrencies are known for having unpredictable values and money can be lost very quickly. Virtual currency values can be whatever the users decide as they are not regulated.

Blockchain transactions, used in cryptocurrencies, are irreversible so care has to be taken to transmit payments to the correct address. Sending a payment to an incorrect address will result in a loss of the amount unless the receiver is good enough to return the payment. Given that transactions are anonymous, there is little that can be done to recover incorrectly addressed payments if the receiver decides to keep them.

Storing cryptocurrency private keys securely is essential. If the private keys are lost then the currency cannot be accessed and its value is lost forever. Hardware and software failures and **malware** can destroy or lose keys, so storing the keys in an electronic wallet and making multiple backups (and safely storing these) is essential.

However, digital currencies allow money to be moved around the world quickly and without many of the charges that banks make for foreign exchange transactions. Transactions using cryptocurrency take no account of international borders and money transfers are made securely and almost instantly.

People in poorer economies, in areas with high inflation or where having a bank account is rare can benefit from using digital currencies. In Kenya, anyone with a smartphone can store money and send money to anyone else using a system of mobile phone credits as money. Despite the high costs of cashing in these credits, their common and extensive use between people means that costs are avoided as there is often no need to actually convert these credits into cash. It is also physically safer than carrying cash or other high-value goods since the digital currency is encrypted and stored securely on the phone.

A concern with central bank digital currencies (CBDCs) is the privacy and rights of individuals. Current data-protection regulations and laws in the European Union, for example, allow individuals' data to be 'forgotten', which means that it has to be deleted if it is no longer required. With digital currencies, however, the data cannot be deleted or 'forgotten'. Individuals who wish to remain anonymous or do not wish their transactions to be tracked or remembered will continue to use physical cash.

Activity 12d

Explain why great care has to be taken by an individual when making a payment using a cryptocurrency.

Businesses

Digital currencies, especially the use of cryptocurrencies, have enabled the introduction of 'smart' contracts. Details of the contract are stored in a distributed ledger and this ensures that the money transactions are enforced according to the contract. Bitcoin allows smart contracts to be created in its blockchains. Because the details are built into the blockchain code when the contract is created, smart contracts cannot be altered, are trackable and cannot be reversed. One benefit is that the details can easily be seen at any time by all users involved in the contract. This attempts to ensure that money is not lost during transfer between organisations, internal divisions within organisations or contractors, or when crossing international borders.

Digital currency can make the purchase of high-value goods or property deals easier. A great deal of trust is required when people who do not know each other or who do not meet make transactions involving large sums of money. By placing the money in an account held by a third party, buyers and sellers have more security and less need to rely on trust. The third-party account, called an 'escrow' account, only pays out to the seller when the goods are transferred to the buyer. The seller knows that the money is available because the buyer cannot stop the payment once the goods are safely transferred, and the seller does not get the money until this has occurred. With digital currency, this process can be automated and expanded to involve multiple individuals or organisations. The currency can only be moved when all the participants have authenticated the transaction.

In order to attract investments in cryptocurrencies, some companies, due to being unregulated, have offered unrealistic interest rates. This has fooled investors who are new to cryptocurrencies into purchasing large amounts of a currency, which has then lost most of its value. Such schemes have become a common trap for unwary investors.

Businesses can benefit from central bank digital currencies (CBDCs) because payments to and from them are easier and quicker. Sellers can be assured that money will arrive in their accounts as agreed. The conditions are encoded into the digital transactions and cannot be changed without everyone's agreement.

The drawbacks of using central bank digital currencies are few. One of the main concerns is that a bank issuing a CBDC may be vulnerable to a 'bank run'. This is when many investors or customers try to remove their money all at once or within a short time from a bank or a group of banks. With physical cash, the banks can simply refuse to give out any more and close their doors while they try to deal with the crisis. With digital money, there is little control and transactions occur very quickly. To combat this risk, some governments and banks restrict the use of their digital currency by financial institutions and do not allow individuals to access it.

In a similar way, a central bank digital currency made available to businesses and individuals may have a damaging effect on other banks. Businesses and individuals may decide to move their accounts and money to the CBDC and start a 'bank run' on the other banks. This risk has meant that central banks have been reluctant to create digital currencies that are available to everyone.

Governments

Digital currencies, especially cryptocurrencies, are seen by many governments as a threat simply because they are decentralised and unregulated. The usual role of central banks in lending money to financial institutions or purchasing financial assets to create more money in an economy is undermined by cryptocurrency. Unregulated and excessive growth of money supplies can lead to an increase in the rate of inflation, rapid price increases and the loss in value of the country's money compared to that of other countries. For this reason, many governments are seeking ways to regulate and control cryptocurrencies. One way is to try to amend existing laws, but this is complex and may have unexpected effects on other currency regulations. Another way is for a government to issue its own cryptocurrency and ban all others. Some financial experts say that this approach is to be resisted as it would give total control of all money supply and transactions to governments. This would remove the right of citizens to keep their transactions transparent and anonymous.

Some Asian countries have banned the use of cryptocurrencies, while others have accepted their use. Bans have been imposed because of the risk of too much currency leaving the country, which might make the economy unstable. In other areas of the world, cryptocurrencies are becoming accepted.

The European Parliament is starting to use blockchain systems to create its own decentralised ledger for its own finances. This will attempt to ensure that their workings are transparent and resistant to fraud. Switzerland is incorporating cryptocurrency into its economy and passengers on its public transport systems will be able to use Bitcoin to pay for travel, avoiding the use of cash. In Germany, cryptocurrency is accepted and can be traded freely but is taxed if traded for euros. These, and other countries, have recognised that cryptocurrencies are becoming more popular but have not recognised them as 'real' currencies as yet. This avoids creating new regulations but risks destabilising the economy of the country. There is no control over the amount of money in circulation, the rates of exchange or who is allowed to create, use and trade the currencies.

CBDCs are being introduced as governments recognise that, while digital currencies can make transactions easier and add to economic growth, they have some risks. The benefits of CBDCs to governments are not only the almost instant transfer of money without the need for lengthy processes involving banks or financial institutions, but also the ability to provide low-cost bank accounts to almost everyone. In addition, more taxes can be collected because, with a CBDC, there is no easy way to hide income or move it to other countries without the issuing bank knowing. All transactions are recorded in the distributed ledger. For the same reason, criminal activity by individuals, organisations or businesses is much more difficult to hide as transactions can be tracked more easily. Unlike some cryptocurrencies, the use of a central bank digital currency is regulated by the issuing bank.

Activity 12e

1 Describe what is meant by a 'smart' contract.
2 Describe two advantages of smart contracts compared to traditional ones.

Global economy

The use of digital currencies has led to increases in business activities around the world. Money can be moved much more easily using digital currency than with real money. There is no need to physically ship paper money or valuable materials, such as gold, silver or precious gems, between countries to pay for goods.

E-commerce has been made possible by digital currencies and has increased business between countries. Customers can buy goods from anywhere in the world. E-businesses in underdeveloped countries can easily trade worldwide.

Digital currencies can be linked to global stock markets, where money can be paid out depending on movements of share prices into accounts anywhere in the world. This can happen in noticeably short timescales, allowing more global transactions to occur.

12.2 Data mining

Data mining is described as the practice of extracting useful information or knowledge from very large repositories of data; it is also called **knowledge discovery** in databases. Data mining uses tools with self-learning algorithms and techniques from artificial intelligence, statistics, neural networks and **machine learning** to analyse massive collections of data called **data sets**. Data sets are made up of collections of numbers, images, values that describe features (such as a person's date of birth), documents and/or files.

Data mining can use artificial intelligence and machine learning with self-learning algorithms to learn about the data sets to improve its performance. It can monitor and determine relationships within the data and use the relationships to learn and re-examine the data and provide results in real time. If new data sets become available, machine learning can use what has been learned to analyse the new data. Artificial neural networks are also used because they can automatically create new relationships in the data as they learn from the data being processed.

Vast amounts of data are collected by organisations such as businesses that need to try and make sense of it. Looking for patterns, trends and associations within the data will help the organisations make decisions. The organisations can then take actions based on those decisions. Many industries, such as manufacturing, marketing, scientific research and the space industry, continuously collect raw data so their data-mining processes are ongoing, repetitive and iterative to ensure that their analyses and reports are regularly updated.

Other uses of data mining are in national security and surveillance, where very large amounts of data are analysed to help with national interests and assist with law enforcement.

12.2.1 The data-mining process

The **cross-industry standard process for data mining (CRISP-DM) reference model** is the standard method for carrying out data mining. It has six phases, each of which has its own defined tasks and outputs. The phases that make up the CRISP-DM model are business understanding, data understanding, data preparation, data modelling, evaluation and deployment.

1: Business understanding

In this phase, the business sets out its goals and decides what it needs from the data-mining process. A business goal might be to increase sales from social-media advertising. A data-mining goal might be to find out how many, and what type, of its customers use the different types of **social media**. It will then be able to target these customers with customised advertising.

The factors that will affect the data-mining goals are then determined. The risks, constraints and contingencies are assessed along with the costs and benefits of carrying out the data mining. The success criteria for the whole process are set in this process, for instance to be able to identify which customers use the different types of social media and where advertising should be targeted to increase sales.

A detailed plan is produced that includes the tools and the techniques to be used. The plan should include a list of tasks that have to be carried out, how long these will take, the resources required and all the inputs and outputs of each process.

2: Data understanding

Data is collected from the sources available to the business. The amount of data collected must be large enough to ensure that it does actually contain any patterns and trends that would be useful to the business. On the other hand, the amount of data must be concise and manageable so that the results of the data mining are available in a reasonable amount of time to be useful.

A description of the data is produced in a report. The description details the format, the number of records and fields and any other information that has been discovered during collection. The data is then explored to ensure that it will provide the information set out in the data-mining goals.

In this phase, the data is also assessed and verified for its quality. The characteristics of high-quality data include being accurate, complete, valid, consistent and fit for purpose. The results of the assessment are given in a data-quality report.

3: Data preparation

The data to be used is selected according to its relevance to the data mining, its quality and any technical constraints such as restrictions on the **data types** or the quantity of data that can be handled. The data is then 'cleaned' by detecting and removing corrupt or inaccurate records and removing irrelevant parts of the data. This is to ensure that the data is of high quality.

The next task is the construction of any new records that need to be derived from existing records in the data, for example new records may have to be created for customers who have made no purchases for several years.

Combinations of data can be created by merging data, such as common features of different social-media types, or by aggregating data into summaries, such as how many purchases or the method of payment used by customers, for example how many customers use a debit card, a credit card or contactless payments.

4: Data modelling

In this phase, the tools decided upon in the business-understanding process are used to carry out the mining. A number of data-mining techniques can be used alone or in combination with each other.

Data-mining techniques include looking for relationships, or associations, between data and classifying data into predefined groups using decision trees, neural networks, linear programming and statistics. Data is clustered by grouping together data with similar characteristics that are used to define the groupings and to predict relationships between the variables. Regular events or patterns that recur during business periods are discovered using decision trees, or a set of decision trees is used to test data against conditions until a final decision is made about the trends.

The mining is documented so that the process can be repeated. A test procedure is created and documented so that the data-mining model can be properly evaluated. Often the data to be used is divided into a number of subsets that are used for testing, validation and training purposes. From the testing, the performance of the model is assessed and any necessary alterations to the settings of its parameters are made along with any amendments to the data structure that were created during the data-preparation process. This is repeated until the model is deemed to be performing as required.

5: Evaluation

This phase evaluates the model against the business-success criteria from the business-understanding process. All the reports from the previous processes are taken into account. If the model does not meet the criteria, then the phase will have to go back to the previous phases and any amendments will have to be made. Changes at the data-preparation and modelling phase will have to be made to ensure that the success criteria are met. If the model does meet these criteria then it is approved and ready to be deployed.

6: Deployment

The data-mining model is deployed (made available to the business to use) when it is found to be satisfactory and to meet the requirements of the business. Plans and documents for its use, maintenance and monitoring are created so that the process can be repeated and amended if necessary. Finally, the results of the data mining, the knowledge about the trends and patterns, are made available to the business in a format that it can use.

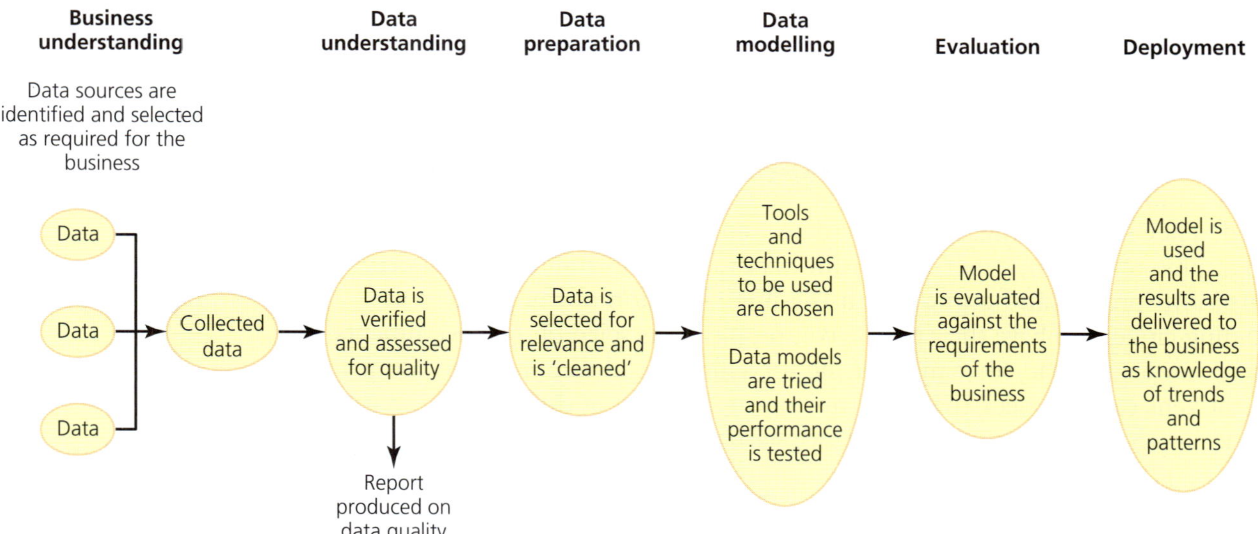

▲ **Figure 12.3** A summary of the CRISP-DM data-mining process

Activity 12f

1 Describe what is meant by a 'data set'.
2 Explain why data mining uses artificial neural networks.

12.2.2 Uses of data mining

Data mining has many uses and we will now describe how and why some of those uses occur.

National security

Data mining can be used to assess crime patterns and allow crime-prevention resources to be allocated accordingly. Patterns can be detected in financial transactions, such as money transfers, and in communications, and these can be used to track organisations and individuals that could be a threat to national security or may have carried out crimes.

Attempts have been made to use data-mining techniques to try and identify possible extremist groups or individuals before they can act. Data mining can be used in the identification of suspect communications by analysing telephone and email communications for language patterns. Patterns can be identified in the movements across the world of suspect groups of people and these patterns can be used by security services to try and identify possible threats.

Law enforcement agencies use data mining to look for patterns in travel, financial transactions and criminal activity in the records of shared databases within their own countries and those of other countries. The sharing of data between countries is subject to domestic laws and international agreements but can include billions of records from thousands of resources, making the results from data mining more reliable. It can help detect and combat people and drug trafficking, cybercrime and terrorism-related activities.

Surveillance

The ability to collect and analyse large amounts of data about online communications allows governments to monitor the activities of their citizens at home and abroad. Communications are intercepted and collected from internet service providers and the major internet companies. The type of communications intercepted and collected include files transferred between users, social-network information such as posts, comments and photographs, video and voice messages, VoIP chats and calls, emails, text messages and chats. From the massive amount of data collected and shared between government agencies and between governments around the world, patterns, trends and groupings can be discovered using data-mining techniques. This surveillance is justified by the governments involved. They maintain that it can help detect and combat crimes such as child pornography, terrorism, cyber-terrorism, tax evasion and fraud. Civil liberty organisations protest and argue that it is a huge invasion of the privacy of law-abiding citizens to have their private communications spied on by governments. Further, they argue that, since most of the surveillance and data mining is done in secret, citizens cannot check that governments are respecting their rights and that the information is being used in accordance with data-protection laws.

Businesses

Retail companies collect data from customers in many ways. Vast quantities of data are provided by records of sales in stores and online, loyalty card schemes, surveys and data bought from other companies or from social media. This can be mined to analyse shopping habits to determine what people buy, when, and in what combinations. Knowing what items people buy at the same time can enable supermarkets to display these items closer together or to send out coupons offering incentives on them. Knowing when items are bought and what sequence they may be bought in can help in advertising campaigns. Delivery companies use data mining to help determine loading patterns for delivery trucks and for distribution schedules between warehouses and retail outlets.

Financial services

Stock market trading patterns and trends can be analysed using data mining of the records of the trading of stocks and shares over many previous years. A trend is a movement in values in either overall upward or downward directions over a period of time. Trend analysis looks at many current trends and uses them to try to predict future trends. While there can be no guarantee of 100 per cent accuracy, using vast amounts of data can uncover, for example, which stocks and shares are growing in value more than others. This analysis can be in the short-term for making a quick profit, or over the medium- and long-term for more certainty of profits over many years. Data mining can help financial institutions to decide which stocks to buy and when to sell them.

Credit card companies use data mining to try to discover how customers use their credit cards and how often customers change from one bank card provider to a different one. The results are used to offer customers incentives to use their cards more or in different ways, by offering one-off or time-limited deals to dissuade them from changing cards. Credit card companies can also use the results of data mining to combat fraud by detecting unusual patterns in customers' transactions or buying of goods. Data mining can be used to assign customers a credit score based on how their details compare to the trends and patterns discovered by analysing millions of other customers' financial activities. A credit score is a number based on an analysis of a person's finances and financial history and represents how worthy they are of being allowed to have more credit from the bank.

Scientific research

There are many scientific research papers on a variety of topics published every year. Data mining can be used to examine and analyse these publications in order to group together research on similar topics. This can then be used by scientists to discover patterns in the discoveries, which in turn could be used as a basis for more research or for the production of summaries of research. Scientists from different areas of research would then be able to quickly and easily follow the research of others, which, in turn, aids their own research.

Scientists use data mining in astronomy to analyse the huge amounts of data collected from visual, radio and other telescopes around the world. They are looking for patterns in the data that might lead to new discoveries about the universe.

Research into new drugs for the treatment of diseases can include the mining of data from different trials of drug use. This helps in the development of new drugs, medicines and treatments.

Health care

The introduction of electronic health records (EHR) and the sharing of these among health care facilities has allowed a huge amount of data about patients and illnesses to be collected. Mining this data has enabled health care providers around the world to improve the quality and efficiency of their organisations. Health care professionals are able to identify best practices and the most effective treatments by studying these across many patients. These are then shared with other health care professionals and used to save the lives of more patients and to improve their quality of life. By analysing and comparing the effects of different medicines and drug treatments on large numbers of patients, the results can be used to determine a standard treatment for illnesses that will save lives. Medical errors may be reduced if the results of data mining are used to improve diagnoses, treatments and patient care.

An important use of data mining in health care is detecting and reducing fraud. Fraudulent insurance claims by patients, doctors and health care providers can cost insurance companies or government health providers millions of dollars that would be better spent on genuine patients. Data mining can show the usual patterns of claims by patients, doctors and health care providers and help to establish a 'normal' baseline against which the claims can be compared. A 'normal' baseline is the number and type of claim that would be expected if there was no fraud occurring. For example, doctors and health care providers would normally be expected to make claims that reflect the type of patients and illnesses that they deal with. A doctor treating only kidney ailments, for example, would be expected to claim for drugs, medicines and treatments related to those ailments, but not for other diseases. Unusual patterns of claims, referrals for treatment or prescriptions can be spotted and investigated by insurance companies before the claim is paid.

The outbreak of diseases and the spread of pathogens (micro-organisms that cause diseases) can be studied by using data mining. By discovering patterns in changes in the pathogens or by identifying any clusters of patients infected with the pathogens, it is possible to detect outbreaks of diseases earlier and attempt to prevent their spread.

The analysis of social and economic trends

Government policymakers rely on the data collected about their citizens to make better decisions for their country or region. The data can include information about migration, a growing or ageing population and a change in wealth by different groups of citizens. Data mining allows trends to be discovered in the data and allows governments to make preparations for the future.

Social-media data mining collects and analyses information from various social-media platforms. Posts, comments, tweets, images and the items clicked on are analysed to create social profiles and to predict how groups will react to marketing strategies and social pressures. The use of artificial intelligence and powerful computer systems allows this analysis to occur in real time. Companies can exploit the opportunity to adapt their marketing quickly according to the conversations and communications of social-media users. Political parties can also use such data mining to try to influence voters in elections by quickly analysing chats and posts and targeting their campaigns accordingly.

An economic trend indicates how well, or not, a country or region is managing financially. The economy of a country or region includes all the financial transactions that occur. Again, there is a massive amount of data that has to be analysed. Data mining is used to discover the underlying trends in the economy. These trends can be used to make forecasts of stock market prices or the effects of changes in taxes and employment rates.

12.2.3 Advantages of data mining

During the data-understanding phase, all the data sets to be mined are gathered together. This has the advantage of collecting the information together so it can be more easily managed, documented and assessed prior to be used.

The main advantage of data mining is that it helps in decision-making. By providing information about trends and patterns in sales or marketing history, retail businesses can make decisions about future sales campaigns or what goods to sell or not to sell. By analysing sales data, data mining can help to identify the need for, and lead to the development of, new products that could increase the profitability of a business. Forecasts of who is going to respond to marketing campaigns can allow more targeted advertising to increase sales and profits. Data mining can identify areas where potential new customers can be acquired by targeted advertising or marketing materials. Data mining can also identify the existing customers who can be encouraged to buy more goods from the company. These customers can be retained and this will add to the company income and increase the competitive advantage of the company.

Financial institutions and banks can examine historical records and discover patterns and trends that help to detect fraudulent transactions and protect their customers from financial loss.

Data mining can also assist in detecting potential security risks by showing typical patterns in user activities and revealing unusual activity compared to the norm. The knowledge that patterns and trends could be discovered by security agencies may act as a deterrent and help prevent criminal activities. Criminals who know that their activities can be discovered may not, as a result, actually carry out the crimes.

12.2.4 Disadvantages of data mining

Data mining can reveal patterns and trends but does not show the significance or value of the findings. Skilled technical and analytical specialists are needed to interpret the output created by the data-mining tools. A disadvantage of data mining is that, although it can identify relationships and connections between the data, it cannot pinpoint their causes. For example, there is a link between increased sales of airline tickets just before a flight is due to take off and the ticket purchase being online. This does not necessarily mean that the closeness of the take-off time causes the purchases to be made online. There are many other factors that only a skilled analyst would identify once the trend has been discovered.

Data mining cannot be used to predict the activities of an individual. Just because millions of credit card users show a particular pattern of spending, for instance buying a drink at the same time as some food, does not mean that a particular individual will always do the same. Data mining only shows up the overall trends and patterns in large numbers of spending activities.

Data mining also requires a large amount of relevant data to be available. Where insufficient data is available, data mining does not always provide a reliable or meaningful set of trends or patterns. National security agencies have attempted to use data mining to help discover possible trends in terrorist activities, but the number of incidents is, fortunately, too few to provide specific patterns. Also, where the results of data mining are used to assess individuals based on how well they compare to overall trends, errors and anomalies can occur. For example, when assessing a customer's financial prospects, banks may use data-mining results to allocate individual credit scores. Sometimes this can lead to a customer not being offered loans because their profile and scores do not match what the bank sees as acceptable when making loans.

On the other hand, the amount of data may be overwhelming. National security agencies have reported that the amount of data they are expected to mine and analyse in the very short timescales required of them presents a considerable challenge. Expensive computer systems, sophisticated data mining and reporting software, and highly skilled personnel are required to deal with the continuous stream of live data.

The quality of the data being used for data mining has an influence on the results. The presence of duplicate records, the age of the data, and human error in the input can all affect the outcomes in ways that sometimes cannot be recognised. A large amount of time and effort has to go into cleaning the data. Unwanted or inaccurate data has to be removed. The same coding must be used throughout (for example **No** is always shown as **0** and not sometimes as **N, no** or **n**) and dates must always be recorded in the same format.

Databases may not be compatible or be able to exchange data effectively. It is important that all the computer systems and databases to be used for data mining can work together so that several databases can be searched and mined simultaneously to make use of the vast amounts of stored data.

Mining of data acquired from indirect sources, for instance data purchased from a collecting organisation by another business, may suffer from inaccuracies that are not apparent to a new user. An example could be the use of a shared bank account. It would be obvious to the bank, which would easily recognise transactions for the individual customers using the account, but if the data was combined with other similar accounts and anonymously mined by others looking for patterns in individual transactions, there would be errors in the overall results.

The wide range of specialist data-mining tools available makes it difficult and expensive for businesses to choose the most suitable for their needs. Data-mining tools can be complex and difficult to use, requiring personnel who are skilled in their use. Extracting the required information can be daunting and may be beyond the skills of the employees in some businesses.

Privacy and ethical issues are of major concern to many people. Massive amounts of data are collected about customers of businesses and banks and about patients using health care facilities. These people are concerned that the information being collected and shared about them may fall into the hands of unscrupulous people who will use it for fraud, blackmail and other criminal activities.

Health care experts argue that the risk of data falling into the wrong hands is worth taking to save lives. A solution is to allow patients to decide whether or not they wish their data to be used for data mining. Doctors argue that this may mean the loss of important data sources and reduce the reliability of the results of data mining if many patients decide to withhold their data.

In banking, concerns about the use of personal data and the consequences of it being lost or viewed by unauthorised people has led to extensive precautions being taken to protect the data. However, data mining requires that large amounts of personal data be shared, combined and analysed. While the data can be anonymised, customers are still concerned that their details could be exposed to fraudsters. Banks are being encouraged to use strong authentication techniques such as **two-factor authentication (two-step verification**) for customer logins. They could also use masking techniques to hide credit card details, encrypt stored data and remove data about customers whose accounts are no longer active. They could also allow customers to withdraw their consent to the use of the data or require an annual renewal of consent by the customer.

There is concern that much data mining is carried out by individuals and organisations using their own software algorithms, which are not scrutinised or checked by public regulators. The public do not really know what data mining is being done on their data and to what use it is being put.

Another concern is that the data collected by private and public organisations for their data-mining purposes is later used for another purpose. This has been called 'mission creep' and can occur whether the data was collected with the specific consent of customers, for example, or was acquired by other means. Using data for purposes other than the purpose for which it was originally collected can lead to consequences that affect people in ways that were not intended and can be harmful.

Activity 12g

1 Explain why small companies do not often use data mining to analyse their sales.
2 Explain why data mining is very costly.

12.3 Social-networking services/platforms

Social-networking services provide an online platform for people to share ideas, views, digital videos and images with others. Various types of social-networking services exist to connect people with existing friends, connect professional people according to their jobs or employment, and to connect people looking for information about family history or for carrying out other research. Online social-networking platforms share common features. Most allow and encourage the creation of profiles by the user with the facility for the user to add or create content linked to that profile. The profile can be linked to similar profiles on the platform to create a social network. Social-networking services use Web 2.0 technology and standards because these allow the active collaboration of users who can interact with each other on websites in ways that were not previously possible. Web 2.0 technology places the emphasis in websites on user-generated content, **cloud computing** and **social networking**. Today, the term 'Web 2.0' is not often used because nearly all websites use its standards and features but, without it, social networking as it is known today would not have been possible. Profiles created on one social-networking platform are specific to that platform and not usually shared with other platforms. Users have to create a profile for each platform they use.

The facility for users to create content on websites is used by businesses, which encourage customers to review their purchases and purchasing experience, their trips and visits to places. Reviewers can interact with each other in a form of social networking. This can lead to increased sales and more profit for the businesses.

> **Activity 12h**
> Explain how Web 2.0 technology has enabled social networking to be developed.

12.3.1 Chat rooms

A **chat room** is an online space that is used to share information with other users. While instant messaging is only between two people, chat rooms involve multiple users at the same time. The participants in a chat room are connected together via the internet (or other network) and are usually, but not always, set up to discuss a specific topic. Topics can be about almost anything. Users of chat rooms often have a nickname or user ID that allows them to hide their real name. It is important to choose a nickname or ID that does not reveal who you are and not to disclose personal information about yourself when using chat rooms. However, this means that participants can be anonymous, enabling them to express views or behave in ways that would not be tolerated in real, live conversations or situations. Young and vulnerable people can be placed in danger when using chat rooms as other users may not be who they pretend to be.

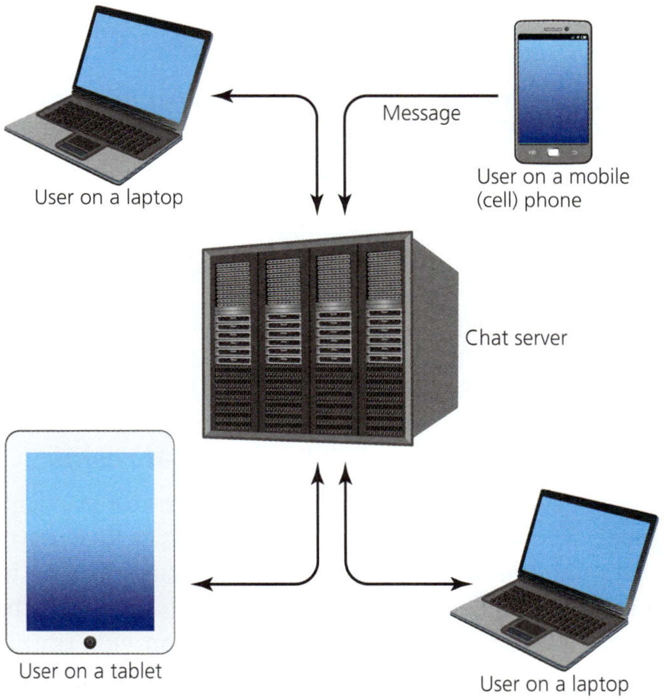

▲ **Figure 12.4** A chat server with clients

Chat rooms use a protocol called **Internet Relay Chat (IRC)** that breaks the data into packets that are sent between the user's device and the chat room server. When a user logs into a chat server, the server can show all the chat rooms that are available and give the user a choice of room to join.

The server sends all the messages from the user to all the other users.

Some chat rooms allow private messages to be sent between users and these messages appear separately from the group messages.

Types of uses

Use by individuals

Chat rooms are set up with rules that govern how the users should behave while participating. This is to prevent, for example, the use of bad or inappropriate language, the spreading of hate or racist ideas, or advertising. Other restrictions could prevent the use of repeated messages to overwhelm others. These rules are important when the chat room is open to the public but would not really be necessary for those set up for employees of businesses or government departments.

Chat rooms are usually set up for specific groups of individuals, for example a chat room for those interested in railways, or those interested in hill walking, or for any other activity or topic. Sometimes, participation is restricted to only those who are invited, but others are advertised and open to anyone.

Use by businesses and organisations

Business owners and employees set up chat rooms to communicate with other business owners or employees. This is to 'network' with them, which means to make contact, to discuss business ideas and to encourage trade. Businesspeople who chat in chat rooms can increase their awareness of other businesses and make other businesses aware of them. This can result in an increase in commerce between the businesses. By hosting a 'special event' chat room, for example a discussion on new trade regulations, business owners can hold discussions that advertise their goods and services to encourage others to trade with them.

Use by governments

Governments can use chat rooms to hold discussions with interested groups of people. While some chat rooms are open to anyone, most government chat rooms are restricted because they are set up to allow discussion of policy between departments or for discussion of issues that are not made public. Governments' chat rooms allow employees who are in remote locations, for example in different cities, to hold discussions.

Government departments, for example a revenue department, can set up a chat room to have a private discussion of taxes with an individual or with a business. This can be used instead of a face-to-face meeting.

Advantages and disadvantages of chat rooms

Chat rooms have a number of advantages for participants. Most providers make access to their systems free or at very little cost. Communication is near instantaneous so conversations and discussions can be in real time. This is a major advantage to businesses, which can set up chat rooms for employees to gather ideas and views on products, provide training for employees without having to gather them all together in one place, or hold meetings without requiring employees to leave their primary place of work. Businesses can set up chat rooms for individual customer support where a customer makes an initial enquiry and is then connected to a specialist at the company for detailed support. The use of chat rooms can save businesses considerable amounts of money by reducing the costs of employee meetings and of customer support. Individuals benefit by not having to move from their current task as they can chat while doing other things on the computer.

There are disadvantages of using and communicating in chat rooms compared to other forms of social networking. An obvious disadvantage is that the participants need access to the internet and have to be online at the same time as other users of the chat room.

A more important disadvantage is that users can choose names for their profiles or accounts that do not reveal who they really are. This can make using chat rooms risky, especially for young people because they may reveal details that allow other users to learn private and personal information or even trick them in to face-to-face meetings. The content that users type into or post in chat rooms is not filtered, moderated or controlled, so inappropriate language, intimate or indecent photographs or images may appear unexpectedly and to those for whom this is unsuitable.

Text-only chat rooms do not provide any information about, for example, body language, and misunderstandings may occur. Security is also an issue as chat rooms are not designed to deal with malicious software.

12.3.2 Instant messaging

In contrast to chat rooms, **instant messaging (IM)** is real-time communication over the internet between two users who are known to each other. The people who can communicate are listed in a contacts list that allows users to be alerted when others are online. Some IM services use push technology that transmits a message character by character so recipients can see messages as they are being composed. By combining communication features into one software messaging application or app, many IM systems now include the sending of images and videos and the ability to make audio and video calls. The inclusion of group chats into some of these apps has blurred the distinctions between all of these services. Instant messaging is text-based and between two people. If it is expanded to include groups of people and multi-media, it becomes a messaging system.

Most IM apps lack the facility to communicate with other companies' IM apps or software because there is no agreed standard protocol for the sending of instant messages. The **Short Message Service (SMS)** that forms the basis of text messaging commonly used on mobile (cell) phones uses well-established standard communication protocols to allow devices to exchange text messages over communication networks. Instant messaging services have no such standard protocols so users of one service cannot message to another service using IM. Each IM service provider uses its own proprietary protocols, although there are some third-party applications that can use several protocols to allow users to message across different services. The creation of open standard protocols like Jabber, which became the **Extensible Messaging and Presence Protocol (XMPP)**, enabled modern messaging systems to be developed and the use of IM to decline in popularity.

To use an IM service, a user must download and install IM software or an app from a provider, create an account user profile and log in. The app or software usually acts as a client connected to the IM server, providing the server with the user's connection information and contact lists. The server checks to determine if any users on the contact list are also logged in and, if so, sends a message back to the client app, which changes the status of that user to 'online'. The user can then choose another user to message and the two are connected together, with windows opening up to allow the users to type. The apps can use the connection information provided by the server to make a peer-to-peer connection directly between the users so that the server is not involved in the communications between users. More often, the apps act as clients of the server, which stores and forwards messages between clients. When the communication session ends, the server should delete the connection information provided by the app and change the user status to offline.

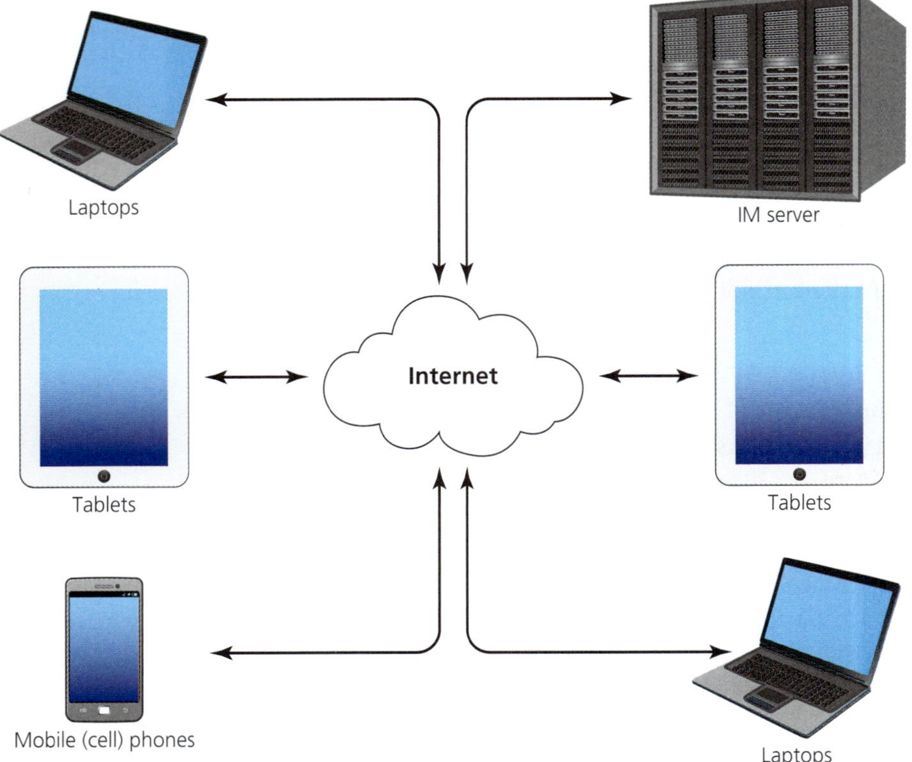

▲ **Figure 12.5** Client–server connections of online users

IM is not as secure as the messaging services that are more commonly in use today. IM service providers store the connection information, and often the messages, on their servers. Some encryption may be used but often it is not, so confidential or private information should not be sent by instant message as it may be intercepted and viewed by unauthorised persons.

Types of uses

Use by individuals

Individuals use IM to send short, concise messages to another user. The message arrives almost instantly and the other user can reply immediately. This allows a real-time conversation to occur between the users. IM is often used to inform the other person of, for example, the user's location, their time of arrival at a pre-arranged meeting or to exchange thoughts or comments.

Use by businesses

IM is used by businesses to exchange information between employees about their activities, but it can create security risks. Phishing attempts, the delivery of viruses and the theft of employees' contact lists are examples of the security risks linked to IM. Instant messages may not be encrypted so can be vulnerable to eavesdropping.

Businesses are also required to comply with data-protection and other laws governing business information so instant messages often have to be kept and archived along with all other data. For this reason and due to the security issues, many businesses have strict policies regulating the use of IM for business purposes.

Use by governments

Government employees use IM in the same way as anybody else. However, governments have to keep all electronic data exchanges secure and have to retain them for records of their activity and for archiving, and the use of IM is included in this. Archiving instant messages requires the use of IM scanning services, content storage services and other systems to store the large amount of data generated. This, and the lack of privacy and security issues with IM, means that government employees are restricted in their use of IM.

To enable the use of IM by government employees but to try to avoid the security issues, some governments, for example the French government, have created their own IM app for mobile devices. The French IM app, called Tchap, provides end-to-end encryption to allow informal communications between government employees and agencies. Its compulsory use will replace the use of all other IM apps by French government employees.

> **Activity 12i**
>
> Explain why confidential information should not be sent via instant messaging.

Messaging systems

Modern messaging systems that can be used on smartphones have the ability to make audio and video calls as well as send images, videos and texts between users over the internet. Some examples of features that can be included are:

- group chats that allow more than two people to have an online conversation in real time
- file transfer allowing images and documents to be exchanged
- mobile payments using a mobile device to authorise a transaction instead of using cash or credit card
- games
- **push notifications** (messages, notifications, alerts or requests for a transaction) are sent from a server to a client device instead of being initiated by the client; an example is a message from a courier stating that a parcel is to be delivered at a certain time
- status updates, which are notifications with information about the progress or status of a task, process or transaction
- **emoticons**, which are a pictorial representation of facial expressions, for example :-), created out of letters, numbers or punctuation marks
- virtual assistants, which are software applications that can respond to questions or requests; online chat virtual assistants are called 'chatbots'.

Each messaging system uses its own software app to connect users but still within its own network as systems still do not easily connect with each other or are not permitted to do so by the provider. Web browser-based clients allow other computing devices to use the messaging systems.

Examples of modern messaging systems include Apple's iMessage®, Facebook's Messenger®, WhatsApp®, Snapchat®, Viber™, Telegram®, Line® and Google Chat™. All offer text messaging but many also offer voice calls, video calls and multimedia exchange options.

Some messaging systems use end-to-end encryption to try to prevent messages being read or modified by unauthorised persons. Encryption is explained in *Cambridge International AS Level Information Technology* and is referred to in the sections on communications technology in Chapter 14. The message is encrypted by the sender and can only be decrypted by the recipient. Any messages stored on servers are stored in encrypted form so cannot be understood by third parties. Sometimes the messages are only encrypted when in transit across networks but are stored as plain text. This is not as secure as end-to-end encryption. Companies that use end-to-end encryption are unable to give security or law-enforcement authorities access to messages as they do not have the means to decrypt them.

Text messages

Text messages are mostly sent between users of mobile (cell) phones. Text messages use the Short Message Service (SMS) or the Multimedia Messaging Service (MMS). SMS allows only text to be sent, but with MMS digital images, video, audio and emojis can be added. Emojis are graphic symbols that represent words, ideas and concepts in a drawing. They are designed to be understood regardless of the language of the reader.

Types of uses

Use by individuals

Personal and business text messages are often used when a short, simple or informal message is needed as they are a quick and easy way to communicate. Instead of an email or phone call, a text message can be used when it would be inappropriate or impolite to interrupt the recipient, who may be busy or even asleep.

Use by businesses

Text messages are used by businesses to confirm orders, and to alert to and check delivery times with customers, and to send texts advertising their goods or services.

Banks and other organisations that deal with confidential details or materials use text messaging to send a **one-time code, token or password (OTP)** to users when they log in to their online services. The use of two-factor authentication (two-step verification) adds an extra layer of security because the OTP is sent to the user by a different communication method than the login ID and password. As many people always carry their mobile phone with them and have easy access to it, this system can work well when the access to the phone is also controlled by a PIN or fingerprint. Authentication techniques are explained in Chapter 14 'Communications technology'. The drawbacks are that the user must be in possession of the mobile phone that receives the text, and the text message is in plain text so can be read by anybody with access to the phone.

> ### Activity 12j
> Describe how a bank could use a text message to check that the person logging in to an online account is authorised to do so.

Use by governments

Text messages are used by governments to keep citizens informed. Alerts can be sent to all citizens with mobile phones when an emergency or crisis happens.

Advantages and disadvantages of instant messaging

Some of the advantages of instant messaging are that the communication is in real time, can be considerably cheaper than text messaging as the messages pass over the internet, and is less intrusive than a telephone or video call. Messages are short compared to emails and can be composed and read quickly. Instant messaging can be carried out at the same time as doing other tasks without interfering too much with those tasks. Composing or reading an email interrupts other tasks more than an instant message does.

Instant messages can, like those in chat rooms, be misinterpreted by the recipients. The advantage of IM is that messages are composed and sent very quickly so errors or confusing sentences can be sent. As there are no visual clues, only the text in the message is seen, so unclear messages that are not double-checked before they are sent can be misleading or misunderstood.

IM is not secure. Messages are not usually encrypted unless it is part of a modern messaging app where end-to-end encryption may be used. In a business environment, monitoring the IM of employees can be difficult. Employees can send unencrypted sensitive information by instant message or can inadvertently download malware. In the same ways that other forms of social networking are insecure, IM can compromise the data of a business. Another drawback is that employees can be distracted from their main task by IM. Most businesses require each employee to have their own corporate login provided by the company for IM so their activities can be tracked, and personal messaging is forbidden.

12.3.3 Forums

A website, or part of a website, that allows users to communicate with other users by posting messages is called an **online forum**, a **web forum** or an **internet forum**. Other names for web forums include **online bulletin boards** and **discussion boards**. Most forums can be viewed by visitors to the site, but most allow posts to be made only by users with an account. Unlike the users of chat rooms and IM services, members of a forum do not have to be online at the same time as other members. Postings in forums are visible for considerable periods of time, unlike those on chat rooms and IM, which are lost once the session is closed.

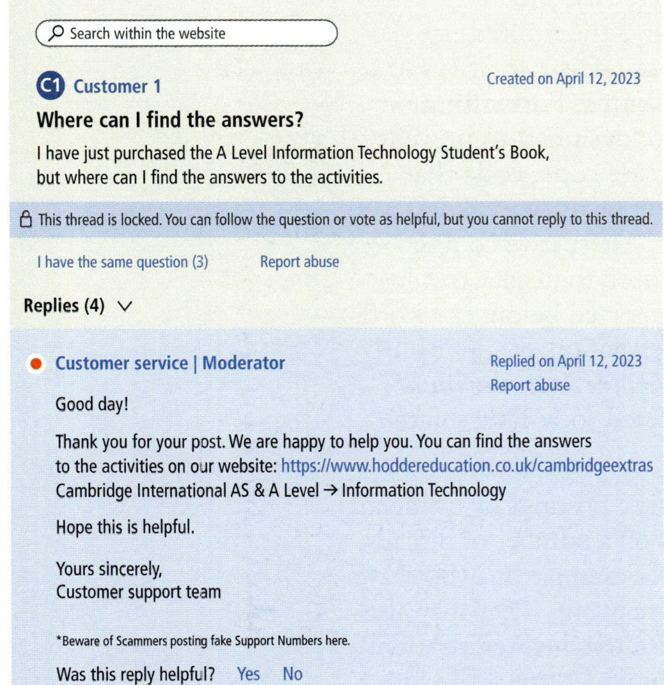

▲ **Figure 12.6** A Help section in a web forum

A user with an account is called a member of the forum. Many forums have an administrator and a moderator. The administrator, who can also be a moderator, is responsible for the maintenance of the site and controls who can have an account and who can be a moderator. Moderators have access to all of the posts of members and regulate the posts to ensure that they meet any rules of the forum. Postings by members have to be moderated before they become visible to all. A forum is usually set up to discuss a particular topic and the rules of the forum are designed to prevent inappropriate postings and to ensure that postings relate to the purpose of the forum.

Some online forums are 'flat' forums, where each post or reply is added to the previous one and there is no organisation of the relationship other than the order of the posting. Most forums are 'threaded', so that the relationships and relevance of the postings are linked to a particular topic. A collection of posts to a forum is then called a 'thread', and this has a title, a description of the thread and an original post. Members can add their comments and postings to the thread until it is closed by a moderator. The number and order of threads can be changed or fixed by administrators and moderators depending on the importance or usefulness of the thread. Postings on a forum are usually shown in chronological order. A typical forum will show threads in order of newest first, but the posts and replies to the posts for that thread will be in order of oldest first. Members can create 'ignore lists' that hide or minimise threads and posts that they do not want to see. Threads often end with members stopping posting through boredom, the topic being exhausted or members going off-topic.

Online forums can use **cookies** to record members' use of the forum, topics and to store login details to automate repeat logins and show whether a thread has new posts for the member to view. These cookies can be disabled or removed at any time. Regulations in some parts of the world require forums, like other websites, to seek permission from the user before using cookies.

Other features of online forums are being able to upload file attachments, some private messaging between members, embedding videos from other social-networking platforms, the use of avatars (an image that shows up when a member makes a post and represents the member) and the limited use of formatting and other coding in posts or replies to posts. The use of HTML in actual posts is discouraged or forbidden by most forums. Forbidding or disabling HTML coding in posts is a security measure designed to prevent client-side code from being inserted into posts. Client-side code, code that runs in the member's web browser, can be used to exploit security weaknesses and bypass access controls such as logins and passwords. Alternative coding systems such as Bulletin Board Code (BBCode) can be used to display text in bold, italic or different colours. In this system, a tag is inserted before the word, for example [b]This[/b] is [i]"Cambridge[/i] [b][i]International[/i][/b] [u] A Level"[/u], meaning it will appear as **This** is *Cambridge **International** A Level*" when the post is viewed and the code is rendered to HTML by the web browser.

Most forums have automatic mechanisms to prevent multiple postings by members who may try to get answers or make the same comments on a number of threads, and to prevent duplication of threads by members who, for example, may not have bothered to search for a relevant thread but just create a new one. They may also make use of mechanisms such as CAPTCHA to prevent automated registrations, automatic deletions of the accounts of unused memberships, and the facility to suspend members who break the rules. Users who regularly break forum rules or abuse others on forums can be blacklisted on a database that web forms can check automatically when a user applies to become a member. This can help to ensure that 'forum spammers' do not disrupt the genuine activities of forum members.

Types of uses

Use by individuals

Individuals use forums for discussion and research. Almost any topic can be discussed. Research, or the search for information, can be carried out by posting questions. Family history, car repairs, appliance servicing and other topics can all be found on web forums with members willing to help others in the search for knowledge or advice.

Use by businesses and organisations

The main reason that businesses use web forums is to connect with customers and to enable customers to connect with each other. By connecting with them, businesses can allow customers to interact with the business and increase sales and customer loyalty. Customers can ask questions about products, how to use products, how to repair or maintain products, or how to make use of after-sales or customer service. Other customers can assist, or enhance, the traditional support role of a business by answering questions posed by other customers. This feature can save the business money as their customer support staff may not be so busy answering telephone calls or may not have as many emails to respond to. The 'outsourcing' of customer support in this way has the benefit of being able to advise or help many customers at once.

Market or consumer research can be carried out using web forums. Customers can provide feedback about goods or services and help generate new ideas for products. This method can be cheaper than traditional surveys or focus groups. One benefit to the business is that, while the business controls and supervises the postings, customer feedback is usually reliable and honest as it is generally not anonymous and can be commented upon by other customers. Unwarranted or unfair criticism of products is usually quickly countered by other customers so the business will gain customer support and loyalty.

Use by governments

Government agencies and departments use websites to distribute information and to promote government initiatives and ideas. This is one-way communication to the public. Web forums can be used, along with other social-networking and social-media systems, to enable citizens to interact with and to participate in government. Governments can get feedback from citizens on their policies, ideas and performance, making communication between governments and citizens a two-way process. Governments can therefore become more transparent and accountable in their decision-making. Many governments are using a range of social-networking platforms to do this, not just web forums.

Advantages and disadvantages of forums

Web forums are a good way for businesses to connect with customers. Contact with customers is direct and immediate. Businesses that encourage customers to use their web forums to discuss issues and ideas about their products will attract visitors to their sites, which in turn will mean more customers and more sales. It will also encourage customers to return to the business, so increasing sales.

One of the drawbacks to using web forums is that the discussion has to be started in a way that will attract and retain participants. This is not an issue when a forum is set up to discuss a specific topic or product with invited participants, but setting up a discussion in the hope of attracting visitors can be challenging. Moderation of forums can be time-consuming and requires skills and knowledge in order that inappropriate comments or questions can be blocked. Blocking too many posts will reduce the attractiveness of the forum and users will go elsewhere. This can have a negative effect on a business. Similarly, having too few posts because of lack of interest will reduce the usefulness of the forum. A lack of new or returning customers will reduce the business's sales and profits.

In some countries, the business or individual that provides a web forum on their website is held responsible for the content of the forum. This means that any offensive or illegal language, the use of materials that are copyrighted by others or the defamation of people or businesses could lead to the prosecution of the website owner.

12.3.4 Email

Electronic mail, or email, is a method of exchanging messages between users on computer networks and over the internet. Email uses the **store-and-forward method** of message exchange. A message is sent by a user to an email, or mail, server, which forwards the message to another user. Both the sender and recipient use email client software applications to log in to the **mail server** but they do not have to do so at the same time. The server will receive, store and forward the email message as and when the users log in. Email protocols and mail servers are also explained in Chapter 14 Communications technology.

Email client software is available for almost all computing devices so emails can be sent and received from desktop computers, laptops, tablets and smartphones. To use email, a user has to have an email address, an email client (which is usually a software application or app but can be web-based), and a connection to a network or the internet. Email systems allow users to check emails while mobile by using cellular networks or **Wi-Fi** connections to the internet.

Every user has a unique email address that identifies the email box from which, and to which, emails are sent and received. An email address is made up of two parts; the local part, which is the unique username, and a domain part, which identifies the organisation that issued the email address. The two parts are separated by an @ character.

The process of sending and receiving emails is:

» The sender composes the email in an email client and addresses it to the recipient. An email can contain plain text and HTML codes to format the text. Originally emails were only plain text, but the adoption of the **Multipurpose Internet Mail Extensions (MIME)** standards allows the inclusion of multimedia such as audio and video files and digital images. Attachments can also be added to email messages. The use of HTML has enabled emails to be displayed with all manner of text formatting, to carry inline graphics and have weblinks embedded in the message. A plain text email is usually sent as well as an HTML version because some email clients cannot use HTML. The use of HTML in email messages raises security issues because it can allow malware to be carried in emails as well as be used in phishing attempts.

» The email client prepares the message and readies it for sending using the **Simple Mail Transfer Protocol (SMTP)**. All internet email servers use this protocol for sending emails. Some propriety systems such a Gmail™ and Microsoft® Outlook® use their own protocols for sending email within their systems, but they have to use SMTP when sending to other email providers. Every email message has a header, which contains the email address of the sender, the date it was sent, the email address of the recipient, the email addresses to which the message is to be copied (if the sender has entered these), an address to which to reply (this may or may not be the same as the address of the sender), and the type of content, which is usually MIME. A message identity number is assigned to the message and this is included in the header. This information can be used to track the email message's passage through the systems and to log its progress. If the email message is invalid because the recipient email address is invalid or does not exist, the sender is alerted almost immediately. Alerts can also be made to the sender when the message is delivered or if the message cannot be delivered.

» The email client places the email to be sent in the outbox of the sender and sends it to the email server of the sender's email provider.
» The server determines, using SMTP, the address to which it should be sent. The domain part of the email address is resolved from the Domain Name System (DNS) and the server looks up the address of the email server of the recipient and sends it. The message may be passed using SMTP to several other email servers before it reaches the server of the recipient.
» The receiving email server looks up the local part of the email address and, if valid, places the email message in the inbox of the recipient. The message is stored on the server until the recipient collects it.
» The recipient uses an email client, which may be web-based, a desktop application or an app on a smartphone, to log in to the email server.
» The email message is downloaded to the recipient's client software using Post Office Protocol (POP3) or Internet Message Access Protocol (IMAP). It can then be read by the recipient.

The diagram in Figure 12.7 summarises the process of sending and receiving an email message.

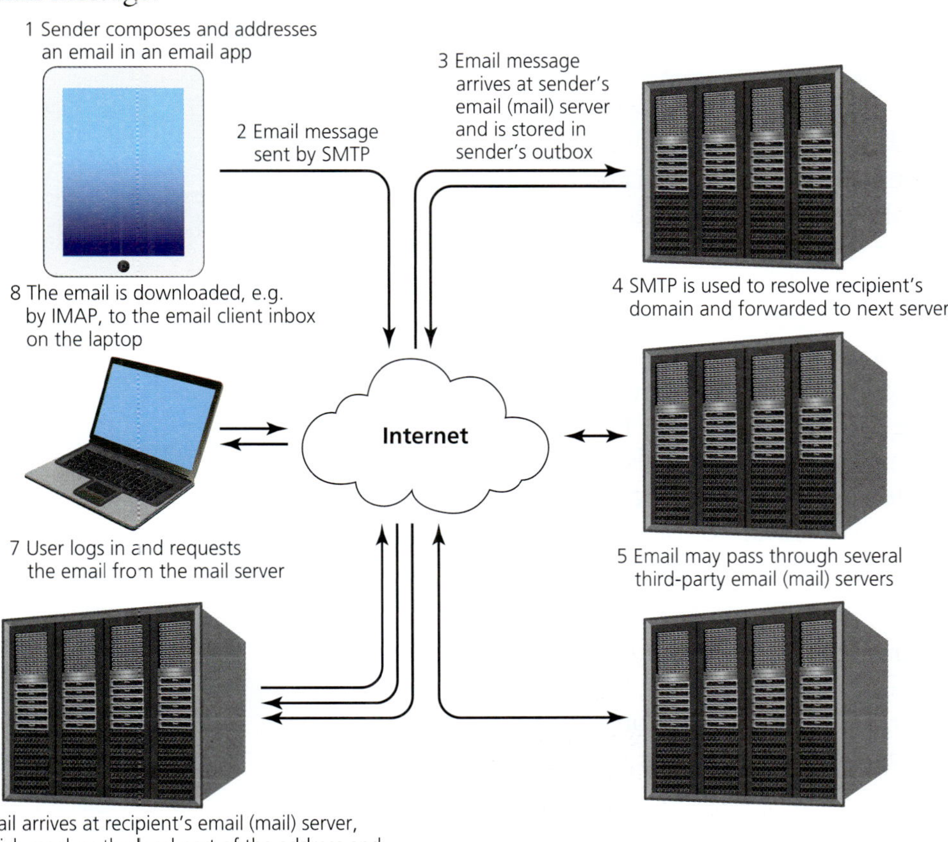

▲ **Figure 12.7** Summary of the process of sending and receiving an email message

Post Office Protocol version 3 (POP3) is used by email clients to download messages from an email server. By default, email clients using POP3 will delete the messages from the server after retrieving them. This has the drawback to the recipient of only being able to read the email on one device. **Internet Message Access Protocol (IMAP)** allows several email clients to access and retrieve email messages from servers. The messages are not automatically deleted but remain on the server until the recipient chooses to delete them. The recipient can download and read messages using any number of devices.

Once an email message is downloaded using POP3 or IMAP, it can be read when the recipient is offline. If webmail is used, the recipient has to be online to read messages. The use of webmail has the advantage that there is no need for a specific email client as any compatible web browser can be used to send and receive webmail. However, webmail has the disadvantage that, if the connection is insecure, it can be freely read by third parties. Users of webmail are usually required to connect to servers using HTTPS, which uses encryption.

While email is a convenient and easy method of communicating, email messages are not encrypted during transmission or when stored on email servers. Users can encrypt their messages, but this is not the default. Confidential and sensitive information, documents or other secret data must be exchanged using secure systems that use encryption. For this reason, again, users of webmail are usually required to connect to servers using HTTPS.

Activity 12k

1 Explain why email is described as being a store-and-forward system.
2 Describe the make-up of a typical email address.
3 Explain why users might choose to use webmail rather than a dedicated email client to send and receive their emails.
4 Explain why some users prefer not to use webmail.

Advantages and disadvantages of email

Email has some advantages over other forms of social networking. Senders and recipients do not have to be online at the same time. Emails are sent and are stored on email servers ready for the recipient to log in and download them at a convenient time. The feature of emails that allows them to be sent or copied to many recipients at once can be used by businesses for keeping customers informed or for advertising. Automating the sending of emails in response to an email enquiry and automatic notifications of delivery or the reading of emails can save large amounts of time for busy employees. Email can reach recipients around the world in shorter times than traditional letters, is not subject to the problems of time-zone differences that can affect the use of chat rooms or instant messaging, and is relatively cheap to use. Also, emails can carry images, links and file attachments as well as text, which makes them ideal for communicating information.

Email has, along with other forms of social networking, the disadvantage of being insecure. Emails are not encrypted by default, have multiple copies stored on many email servers around the world, and are stored for indeterminate amounts of time. Once an email has been sent, the sender usually has no knowledge of the route it takes or where it is stored and under whose jurisdiction it comes on the way to the recipient. Emails can be intercepted, retrieved and read after the sender and recipient have finished with them. For this reason, confidential, sensitive and personal information is not safe or secure in emails.

Large quantities of unwanted emails, called spam emails or just spam, can distract recipients, fill up inboxes and email servers and slow down network traffic. Although anti-spam software and filtering by email providers has reduced the amount of spam emails, they are still a nuisance to email users. Very large volumes of emails sent on an email server may cause it to crash. This is known as email bombing.

Email messages, as well as instant messages, are used by fraudsters attempting to obtain personal and confidential information by phishing techniques. By using false email addresses that appear to come from reputable organisations such as banks, phishing tries to direct the recipient to a fake website by placing instructions, warnings or alerts and links in email messages.

12.3.5 Blogs and microblogs

Weblogs or **blogs** are social-network platforms that use software tools to post content. They are web pages that can be used by authors as diaries or journals, as a platform on which to display their work, or to present information and opinions on a specific topic, in an engaging way by using images, fonts and headings.

Most blogs have a similar page structure, with a header, a main content area, and sidebars that carry information about the authors and their profiles. There is usually an area for links and privacy policies, contact lists and other information about the page. Static web pages often do not have these **elements**. Unlike blogs, static web pages are not updated regularly and some are never updated.

Types of uses

Use by individuals

Individuals create and use blogs for many reasons. With access to the internet, bloggers can communicate with a very large audience and 'talk' about any topic they wish. Bloggers can attempt to make money from blogs that become very popular. The blog can be used as a marketing tool through the addition of advertisements, collaborating with advertising networks, selling of digital products or even memberships.

Microblogging is the addition of small snippets of content to a blog. A short piece of text, a single image or a link to a video or other content is a microblog.

Use by businesses and organisations

Businesses use blogs for advertising purposes and as way to increase customer satisfaction. Customers can be kept up to date about the activities of the business. News services use blogs to inform readers and express their opinions.

Use by governments

Governments and government agencies use blogs to attempt to keep citizens informed of their activities. An example of a US government agency blog is NASA's blog, which is usually updated daily with information about its space activities. Another example is the Indian government's use of blogs on its official portal to spread news and information and to request feedback from its citizens.

Advantages and disadvantages of blogs and microblogs

Blogging and the use of blogs to share content with others has enabled people to communicate with audiences around the world. Many people have made blogging a full-time occupation, taking advantage of working from home. By having a lively and interesting blog about themselves, their activities or their chosen topics, they can attract audiences and visitors to their sites. Selling advertising spaces and links for redirections to retailers' own sites can generate income for the blogger. The disadvantages are that bloggers work alone and need to be self-motivated because they may have to work long hours to generate enough income to make a living and bloggers cannot work and earn money when they are ill.

Bloggers are responsible for any postings on their blog and may be liable to prosecution or litigation. Bloggers who make derogatory comments, defame their employers or blog about company business without authorisation from the business may be required by the employers to remove the blog, be subject to disciplinary action or be fired by the employers, or sometimes all of these. There is the possibility of online insults, attacks and threats against a blogger. Internet 'trolls' can use the anonymity of postings to target bloggers. As with other social-networking methods, these attacks can have severe detrimental effects on the targeted person.

Many businesses now make use of blogs to advertise their services and products. Sponsored posts on bloggers' sites add to the income of bloggers, and raise awareness of the products and bring new customers to the businesses' own sites. Search engines will automatically scan blogs along with other websites for keywords and show the results to customers looking for product information. Having a large number of mentions will increase the number of times that a customer sees the business information.

The disadvantages of using blogs are that very few customers actually take part in blogs run by businesses, so the feedback and comments do not always produce reliable data. Some postings may be untrue, inflammatory or derogatory, so careful consideration by moderators has to be made about whether these can be publicly shown. Postings to company blogs must be constantly monitored and it is important for businesses to respond quickly to postings in case customers lose interest and do not return. This can make the job of company moderators difficult and time-consuming, distracting them from more important company business.

Businesses using blogs must have a company policy about the content and use of postings, and must make sure all employees are aware of the policy and take steps to enforce it.

12.3.6 Social media

The term 'social media' is used to describe applications and websites that have facilities for people to post and share content in real time. There may be some restrictions put in place, but usually as soon as a user adds or posts content, other users of the social-media platform can view it. Social-media sites can be accessed using apps on smartphones and tablets as well as by using software applications or web browsers on desktop and laptop computers. Dedicated social-media websites, forums, blogs and microblogs, chat rooms, photograph- and video-sharing sites, virtual worlds, some forms of online gaming and other ways of sharing content can all be included under the broad term of 'social media'.

Social-media sites have a number of features in common with each other, for example they all require users to create a profile specific to that social-media platform before the user can make use of the platform. A typical profile on popular social-media sites will include the name and a photograph of the user, a collection of personal information about the user, a list of 'friends' or people who are connected to the user, and sections for the user to post their content and for others to add their comments. The content can be made available to any other member or can have restrictions as to who can view it. All enable and encourage the connection of user profiles with other profiles, all depend upon content generated by users and all use Web 2.0 technologies.

Joining and creating a profile on a dedicated social-media platform is usually free but there are often restrictions on who can join. An email address is essential. Age restrictions are the most common, with persons under 13 years old being banned. This is to attempt to protect young people from exploitation or exposure to inappropriate materials. A profile contains information about the user and is shown on a web page unique to that user. Links and connections to other users can be automatically created by comparing existing user profiles with the list of email contacts found in the user's smartphone or webmail account, by an automatic search of the site's profiles of existing members or by a manual search by the user of the existing member profiles. Searches can be based on any information that could be found in a profile. Most sites will allow the details of a user profile to be hidden from public view by adjustments to privacy settings by the user. Keeping personal details hidden is a method of preventing the details from being stolen and used for malicious or fraudulent purposes.

Types of uses

Use by individuals

Individuals make use of social-media platforms to create and maintain personal relationships, and to interact with others by sharing content such as photos, videos, messages, views and comments about their everyday activities, holidays and other experiences.

Social media can be a useful way of keeping in touch with other people since there is a great deal of personal information posted in profiles. Applications for employment can include references to a user's social-media profile for employers to see. Some employers will explicitly check the social-media profiles of job applicants and their postings to try and find out more about the applicant.

There are digital divides among social-media users. While the apps or software to use social media are usually available for free, some populations lack easy and cheap access to computer devices and to the internet. Poor or remote areas also lack the infrastructure to enable the population to access social media. Where social media is readily available, young people tend to access it more often than older generations.

Activity 12l

Why do some employers check the social-media profiles of prospective and existing employees?

Use by businesses

Businesses use social media for marketing purposes to connect with or contact existing and potential customers, to track and analyse social-media conversations and trends, and to allow their employees to keep in contact with the business and each other on an informal basis. Businesses use data mining to look for trends in the large amounts of social-media data that they can collect or buy from the owners of social-media platforms. Of particular interest to businesses are social-media conversations that go 'viral' or spread rapidly, because these are the conversations or posts that will be seen by the most users in a short time.

One of the features of social media is that individuals can link the profiles of others to their own so that they are automatically shown or alerted to any new postings. This is known as 'following' and more individuals follow businesses than follow celebrities on social media. Businesses can exploit this to draw attention to their products and services by making regular updates to their profiles or postings so that these are automatically shared with their followers and raise the brand awareness.

The postings can also contain links to the business website, which increases the number of visitors and can increase sales. Customers who are aware of a brand and of who runs the business are more likely to buy new or replacement products from the business because of what they can see on social media. In this way, the business uses social media to 'humanise its brand', a strategy also known to increase customer loyalty and sales.

News organisations make considerable use of social media to distribute news information and opinions. The organisations have taken advantage of the fact that many people use social media very often. News items posted on social media are quickly distributed, so the opinions of journalists and news commentators can spread across the world very quickly. Drawbacks of using social media as a news source are that it is difficult to verify the accuracy or honesty of a news item or opinion without cross-checking with other sources, and that the views and opinions of the writer may be biased, not appropriate or extreme, but there is no means of countering them. Because social-media postings can spread unchecked so quickly and be added to and reinforced by replies or additional postings, extreme opinions and fake news can easily become accepted as the norm. So, while social media is an important medium for alerting people, it has to be used alongside more traditional methods of getting information to the public.

Use by governments

Governments use social media to communicate with their citizens. Information can be disseminated and opinions, views and comments can be returned to the government on social media by its citizens. Governments can try to give their citizens more involvement in the decisions that affect them. Some governments use social media as a way to reach more people than they would through traditional methods such as newspapers or radio and TV.

People can use social media to try to influence governments and raise awareness of issues in their region. People can also use social media to show support for or argue against ideas, policies and political leaders. In democracies, governments use social media to encourage citizens to participate in government, provide feedback to the government and to hold government to account for its actions.

Use in education

Social networking is used in education for collaboration and communication between students, teachers and parents. The use of mobile devices to connect to other students to ask questions, assist others with homework or share documents can improve learning. Teachers can be asked questions and send replies without the student having to physically find the teacher. Parents can be kept informed of their children's activities and school events via social networking.

Information can be discovered and discussed on social-media platforms, for example blogs or wikis, that students can use and share in their studies. Learning can be enhanced in areas such as reading and writing. Students who dislike reading books often readily read web pages and social-networking posts. Students are often more motivated to read online messages, news, information and articles than to read books. Social networking can provide distance-learning opportunities for students who are unable to attend school.

Use in finance

Financial institutions use social networking to advertise their products and services. In most countries, financial institutions, for example banks, are closely regulated but are allowed to advertise for new customers. Using social networking enables the institutions to bring themselves to the attention of millions of potential customers. Advertisements can be placed into news feeds, onto web pages that host blogs, onto video-sharing sites and sent to potential customers by email. Advertisements can also be placed into online shopping sites.

Online reviews and comments from customers are a useful way of monitoring the success, or otherwise, of a financial product or service. These can be made visible across different social-media platforms and can act as advertisements.

Customer support is another use of social networking. Monitoring social networks can provide quick responses to problems and inform customers of attempts to resolve them. For example, a response to the reporting on social-media platforms that a bank's online services are not working can be very quick, with an update as to when the problem is fixed.

General financial advice and explanations have been posted on social networks by major banks to educate the public. Podcasts of information have been made available. These uses also serve to enhance the reputation of the institution providing the free advice.

Financial institutions also gather and analyse large amounts of data from the use of social networks. Data mining can help to discover possible new markets for their products and services.

Use in health care

Social networking is used in health care by patients and by health care providers. Patients can access information about health conditions and discuss this with others. Self-diagnosis and treatments can be discovered without reference to professional health care providers. While suitable for minor afflictions, it is not recommended as a replacement for consulting a proper doctor as information exchanged on social networks about medical conditions and treatments can be misleading, inaccurate and possibly dangerous. However, a discussion with others who have the same medical conditions and are willing to share their experiences can be helpful in understanding and treating illness. This is peer support and can help considerably in, for example, giving up smoking, as sharing the experience with others using social networking can increase motivation.

Public health and outbreaks of disease can be monitored using social networking. Contact tracing of individuals that may have come into contact with disease carriers can be accomplished using, for example, instant messaging, contact apps on mobile devices or posts on social-media platforms. In times of public health crisis, social-media platforms can very quickly spread information and alerts.

Health care providers and professionals use social networking to share information between themselves and with patients. Reminders and alerts about scheduled medical treatments can be sent by instant message or email. Reminders and advice on how to avoid infections or injuries can be posted to social-media platforms. Updates and news about health care provision can also be posted.

People training to be health care providers can access information and can engage in group discussions to support their learning and training.

Advantages and disadvantages of social media

Social media has enabled people from all around the world to connect and communicate with each other. People can connect to family and friends, to others with similar thoughts, religious views and interests, and share content such as text, images and videos. The advantages of using social media for this are that it requires few skills, can be low cost and can be seen very quickly by others regardless of the location of the users.

The disadvantages of using social media for communication with others include the fact that an internet or mobile connection is required and this may be expensive in some areas of the world. Personal or confidential information can be seen by others, and can be stolen or lost if users do not take precautions to keep data safe and secure. If unauthorised people obtain and use the personal data of others, they can use it to commit fraud, or embarrass or humiliate users of social media.

Other disadvantages are that adding posts to social-media conversations without carefully considering the previous posts or feelings of others can produce anxiety and stress in others and create a negative perception of the person making the post. Cyberbullying can lead to low self-esteem, depression and other mental health issues. Cyber-stalking can produce online and physical threats to users. The spread of false or misleading information is difficult to control. It is also difficult to distinguish real information from fake.

12.3.7 Impact of social networking

On individuals

Using social media can have positive effects for individuals. For example, it can provide help and support for people with health problems. There are social-media groups for people living with various diseases and those with disabilities. By sharing issues and problems on social media, individuals can be supported and helped.

There are concerns, however, that the use of social media can have adverse effects on individuals. Users are more likely to be extrovert and open on social media but are less likely to be afraid of hurting the feelings of others, compared to when talking face-to-face. This can often lead to open and frank discussions about topics. On the other hand, as posts and comments are permanently displayed and available, the uncontrolled and hurtful views of others can have damaging effects on the psychological health and mental well-being of individuals. Individuals who frequently check on the activities of friends can develop a fear of missing something and being left out of activities, or of not being part of the group. This has been linked to mood swings and depression in some people. There are also the fears of being hacked, of personal details being used by unscrupulous people, or of employers, school and college officials reading the social-media posts, and these can lead to increased levels of stress and anxiety in individuals.

The ability to check up on others and to find out personal information has drawbacks. Users can 'stalk' other users by looking at status updates, posts, images and comments and using these to track the movements and activities of individuals. New friends or acquaintances can be investigated, and ex-partners or colleagues can be 'watched' or followed online. As well as upsetting and causing fear and anxiety in the person being stalked or watched, the stalker can also develop mental health issues that can have a lasting effect on their behaviour.

Most social-media platforms do not carry out background checks on users when they join and set up a profile. Often all that is required is an email address and the answers to a few questions. Provided the email address is valid then the answers to the questions can be fake. This makes it easy for people to create false profiles and to hide their real identity. People seeking new relationships can set up a fake social-media account with a profile designed to attract and interest others.

Fake profiles can be used by adults who attempt to befriend young people with the aim of influencing or abusing them. By pretending to be much younger, using false photographs and creating a fake profile containing the type of personal information of a teenager, an adult can pretend to be of the same age group as an intended target. Young people are always advised to keep the amount of personal information they show in their profile to a minimum, never to disclose these details to strangers, never to post compromising photographs of themselves or others, and never to befriend or connect with strangers on social media. Young people are also advised to report any inappropriate or unusual requests to make contact with strangers. Parents are advised to check who their children are in contact with on social media and what their children are posting. However, checking a child's several hundred online connections can be very difficult.

When individuals are not approved of or not accepted by their peers on social media, they can become emotionally distressed. This effect is seen in teenagers who try to emulate the appearance and personality of others on social media to try and be accepted or become part of a peer group. If they fail, or are mocked online, then they can develop health issues. On occasions, retaliation can occur and the person being mocked becomes a bully by mocking others in turn. Bullying online is called 'cyberbullying'. Studies have shown that occurrences of cyberbullying are increasing on social-media platforms. The intensity, duration and frequency of being cyberbullied can all have a negative impact and significantly lower the self-esteem of an individual.

There are numerous cases of people losing their jobs after posting inappropriate or silly images or comments to their social-media profiles. Employers can check what is being posted to business accounts and to private accounts and may not approve of the postings. This can become an issue even after many years as social-media postings remain visible for a long time.

Activity 12m

1 Explain how some social-media platforms allow users to create fake accounts.
2 Explain why parents find it difficult to protect their children from fake users on social media.

On businesses

Business profiles on social media allow companies to have a global audience with little cost to themselves. Most social-media content can be produced quickly and cheaply, often with little formal training. This compares to the high-quality content that needs to be created by advertising agencies for use in traditional advertising on television or in cinemas or on posters. Further, while traditional advertising takes considerable time to develop and appear, social-media advertising can be almost immediate and reach many potential customers very quickly. Another benefit of using social media to advertise is that the postings are more permanent so they remain visible far longer than advertising on billboards and television and in cinemas. However, when there are many posts being added, advertising can quickly be displaced by new posts and disappear from sight.

Businesses can offer their followers on social media the possibility of customised discounts or offers in order to increase sales. An airline company in the USA offered free food and a free flight to the first customers to use its new service out of San Francisco or Los Angeles. The customers were required to post their location and time on a specific social-media platform when checking in for the flight.

The use of social-media platforms or sites to advertise products and to attract new customers can be subject to abuse. There is evidence that some customers are paid for posting good reviews for products, that some reviews are fake because the customer making the posting has not purchased the product, that some reviews are posted by employees or owners of the business, or that some postings are generated automatically by computers.

On governments

Social media has made citizens more aware of the activities of their governments and allowed more insight into what governments are doing. Governments can use social media to distribute their ideas and views and become more transparent in their activities. Citizens can more easily hold their governments to account because of the free use of social media to quickly spread news and ideas.

Social media is difficult for governments to control as it is freely available on the internet. In some areas of the world, social media is controlled and policed by governments that are concerned that their citizens might use it to cause dissent and unrest in the country or to publicise activities that the government might not want people to know about. Methods of control by governments include passing laws that make it illegal to use some or all social media, blocking access to social-media platforms, controlling internet access or restricting access only to government-approved social-media platforms. Other attempts at censorship of social media include making it illegal to post links to certain websites or to share information from certain sites.

A positive use of censorship of social media is the banning of links and information about child pornography, child and people trafficking, drug trafficking, terrorism or similar threats to society. It is illegal in most countries to post or share indecent images of children, and those who do so, even if they are underage themselves, can face prosecution and punishment. The easy accessibility and freedom of social media in most countries, however, makes it very difficult to monitor, police and control everything that is posted.

Activity 12n

Explain why using social media may cause anxiety and stress.

12.4.1 Society

A society is a grouping together of people. It can be a small group of people with common interests, a community or a nation of people. Information technology and its tools for storing and manipulating data and for communicating information have had impacts on the common traditions, institutions and group activities of a society and on the way that people work in the society. The use of computing devices, mobile technologies, multimedia technologies and the internet all have an impact.

12.4.2 Sport

Athletes can use wearable technology to monitor and record their heart and breathing rates, speed, distance covered, location and other performance information. Small electronic devices can easily be attached to clothing, footwear and body areas to communicate data collected from sensors to central computer systems where it can be analysed. The use of data mining and artificial intelligence systems to track performances can add to the information that coaches and trainers have at their disposal. Coaches and trainers can use the analyses to enhance the athlete's or team's performance. Decisions that affect the team strategy can be made during games by using IT to track and monitor players on the field.

Sensors and video recording and replay can be used to enhance the experience of spectators and to assist in making judgements during games and tournaments. Sensors placed in balls, on goal posts or on courts can ensure that scoring decisions are accurate, fair and unbiased, and the paths of balls can be tracked and analysed by computers almost immediately for the production of game statistics or to show replays overlaid on live action. This can add to spectator involvement and enjoyment and increase the popularity of the sport, including supporters being able to view and enjoy watching sports from around the world in real time. The use of modern telecommunication links and high-speed internet connections allow live viewings of sporting events anywhere in the world. Viewers can use digital television systems, smart television sets with internet connections, mobile devices such as tablets or smartphones and other computing devices to view sports fixtures and tournaments as they happen. Social-media platforms via apps and web-based systems can also be used. Despite the high costs to the end user for access, the impact is that sport has been made available to more viewers than ever before by using information technology. By being able to view more sport, more people may be encouraged to take up sport themselves.

Decisions by on-field referees or umpires can be reviewed by **video-assistant referees (VAR)**, the **television match officials (TMOs)** or third umpires who watch the game unfold in real time. A VAR or TMO has the same status as other match officials and, by using video technology and wireless communications, can assist the referee on the field in making decisions that affect the play. Critics have suggested that the systems are intrusive and spoil the flow of the game. Supporters maintain that they enhance the fairness of decisions.

▲ **Figure 12.8** VAR in football

Software packages can be used to monitor and analyse data about nutrition, training and fitness. Software can calculate body fat percentage, body mass index and fitness compared to others. Sharing the data on social media with other sports people can provide an element of competition but may have detrimental effects on the individual. For example, comparing their own body shape and weight with those of others can be psychologically damaging or result in lower self-esteem if the others appear to be doing better.

Prosthetic limbs and better sports equipment can be designed using computer-aided design software and created with **3D printing** technology. This can ensure that the fit and feel of the limbs and equipment is customised to individual athletes, which can lead to enhanced performance.

Activity 12o

Explain why football fans attending a match complain that the use of video-assistant referees (VARs) spoils their experience of the game.

12.4.3 Manufacturing

Manufacturing has become safer for workers through the use of information technology. Workers can be trained using **simulations** of dangerous or complex machinery or situations and using online training courses that workers can carry out in their own time and receive instant feedback on.

Virtual reality and augmented reality are used to train employees to use machinery. These are also used during the manufacture of goods to assist in the process. Precise placement of parts, instant feedback on the manufacturing process for quality control, monitoring by supervisors and error reporting can increase the consistency and speed of manufacturing.

Industries are using **robotics** to:

» speed up manufacturing processes
» make them safer, as humans no longer do as many dangerous tasks
» carry out tasks in hazardous environments
» work continuously.

Manufacturing companies do not have to employ as many workers as would be used on a manual production line, but they have to employ technicians to service and maintain the **robots**. The cost of employing manual workers is saved, but the cost of skilled technicians can be high. Overall, the use of robots in manufacturing can decrease the labour costs and, because robots can work faster, more consistently and continuously, the costs of manufacturing have been reduced.

▲ **Figure 12.9** Robots assembling solar panels

12.4.4 Health care

Information technology has improved the monitoring and treatment of patients. Advanced machines controlled by microprocessors can be used to continually monitor patients, sending data to nurses and doctors and providing alerts when the condition of a patient changes or is of concern. This can enable the nurses and doctors to monitor and treat more patients simultaneously, can reduce the number of nurses and doctors needed to treat patients, can automatically display information in easily understandable formats such as graphs and charts, and can record data for analysis. Patient data can be more easily shared between doctors and health care providers to add to the amount of available data that can be used for data mining. This enables trends and patterns in diseases to be discovered.

Doctors and nurses can post their observations to medical communities using social media or cloud computing systems. This enables more discussion and the sharing of research into diseases and treatments. Treatments can be improved through the sharing of information.

Monitoring of patients can be carried out remotely. Data from sensors placed on patients at home (for example sensors that monitor blood pressure, heart rate and electrocardiographic [ECG] data) can be sent in real time over the internet to health care providers. Patients can be continually monitored and doctors alerted if their condition changes. This can improve patient care and support. If necessary, emergency services can be called more quickly and patients treated faster so they are more likely to survive and recover.

Wearable robotic devices can be used to monitor patients' health and report to health care providers. These devices are still being developed but will enable doctors to remotely monitor and collect patient data.

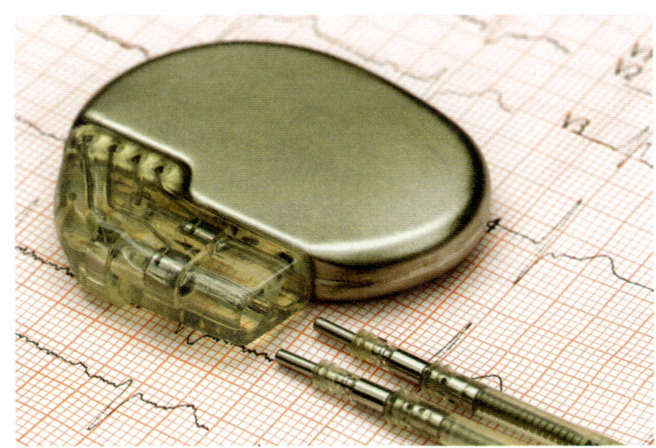

▲ **Figure 12.10** Wearable monitoring system

The use of artificial intelligence and robotics has enhanced the development of new prescription drugs and medicines for treatment of patients. AI can analyse more documentation and research papers than humans and, combined with other data-mining techniques, can more quickly analyse the effects of drugs and medicines. Robotic devices can test more compounds more reliably and consistently, with repeatable results, than human technicians and can keep more in-depth and searchable records for later analysis. For example, a common compound in toothpaste, triclosan, has been found by AI robotic testing to be useful in treating malaria caused by parasites resistant to other anti-malarial drugs.

3D printing can be used to create and dispense prescription drugs to patients. Drugs can be customised to a specific patient and 'printed' whenever required. There is no need for large quantities of the drug to be produced in a factory or stored in a pharmacy. In the future, drugs could even be printed out at the home of the patient as and when required.

Health care providers can be contacted online to make appointments, ask for the resupply of drugs on prescription and seek advice. The use of VoIP and video calls between patients at home and doctors can save the time of travelling to the health centre and help the doctor to see more patients in a day. These calls can be made using a smartphone.

The use of artificial intelligence, robotics and wearable computing has improved the treatment and care of patients. Expert systems, with the use of artificial intelligence, can assist doctors in the diagnosis of illness and the application of the most suitable drugs and treatments to improve patient recovery times and survival rates. Robotics and wearable computing enable surgeons to carry out surgery and medical procedures more safely, precisely and accurately. Using these computer-controlled methods, patients have surgery that is less invasive, often experience less pain during recovery, have reduced scarring and recover more quickly.

Robotic devices including facilities for **telepresence** are used in health care. Telepresence means enabling a person to appear to be present in a location when they are physically elsewhere. In health care, a doctor or nurse can appear to the patient and discuss their condition as if they were physically next to each other.

Telesurgery is a type of telepresence where a surgeon performs a surgical procedure while not being at the same location. The surgeon controls the robotic arms and tools on a robotic device to carry out the surgery. The surgeon may be next to the patient in the same room, in another room or at a different location altogether. However, telesurgery relies not only on the robotic device but on reliable, high-speed communication systems to exchange the required data between the robotic surgical tools, the control system and the surgeon.

Prosthetic limbs can be controlled by monitoring the electrical signals sent to muscles along nerves and using microprocessors to translate these into signals to control actuators on the prosthetic limbs. The microprocessors can receive data from sensors in advanced prosthetic limbs to measure the angles and forces when the wearer walks or moves. The microprocessor can analyse how the person walks and change the stiffness of the limb joints in real time according to the requirements of the movements. This enables the wearer to move and walk more easily and safely.

> ### Activity 12p
> Describe how wearable computing systems can assist a surgeon in operating on a patient's eye.

12.4.5 Education

Education is partly about how to discover information, how to access information and how to exchange and distribute information with and to others. Using information technology also requires people to be able to determine the truthfulness and validity of the information they find. The vast amounts of data and information means students need help selecting, analysing and evaluating information.

Using information technology improves the educational experience of students because lessons and courses can make use of multimedia, images, text and links to materials from resources stored almost anywhere in the world. Physical libraries of books and resources are being replaced by digital libraries and resources that can be accessed using the internet.

Classroom-based teaching takes place in a room where students sit at desks and a teacher leads them through a topic. Worksheets, visual aids and other teaching materials are used to support the students' learning. A virtual classroom uses information technology to enable students and teachers to communicate, interact and discuss topics. It is online and often uses **video-conferencing** techniques to allow students and teachers to be connected at the same time. Instructor-led learning is where a teacher, or instructor, guides and assists students through a learning process. It can be for groups or individuals, online or face-to-face.

▲ **Figure 12.11** An online classroom

Teachers can enhance lessons with IT and can, perhaps more importantly, customise lessons for individual students. Teachers have had to learn new skills to be able to use IT effectively both in the classroom and remotely. Spaces and areas, for example rooms in homes, where students can access lessons remotely, are now also classrooms.

12.4.6 Finance

Information technology has changed banking from a system where customer accounts were kept at branches and transactions were done in person, to a system where banking can be carried out from almost anywhere that has an internet connection. Electronic banking uses automated teller machines (ATMs), telephone banking and internet banking.

Information and computing technology have enabled banks to store accounts and financial information in large databases for ease of retrieval and manipulation, to provide more flexibility for their customers and to save the costs of having any bank branches or many staff. Services such as credit and debit cards, automatic teller machines (ATMs), electronic funds transfer (EFT), electronic points-of-sale (EPOS), mobile banking, telephone banking, internet banking and electronic data exchange (EDI) have all been made possible by IT. IT has made banking more available to individuals and businesses and so has increased commerce.

Internet banking, online banking or e-banking means that anyone with a suitable computing device and an internet connection can perform almost any banking activity. A person still needs to be physically present at a bank or ATM when the transaction requires physical proof of identity or involves cash money.

Enquiries about bank account balances, transfer of money between accounts, paying of bills, converting money between foreign currencies, managing accounts and reporting fraud can all be carried out using internet banking. Most, if not all, of these tasks and the opening of new accounts can be carried out at any time of the day and from mobile devices such as smartphones.

The introduction of internet banking has had an impact on the banking activities of people with disabilities. Internet banking has enabled these people to carry out their banking activities without travelling to a branch or finding a branch with disabled access, and means they can access the services at their own pace without stress or pressure to proceed faster because others are waiting in line. There are some difficulties in that people with sight problems may not be able to read the screen, or disabled people may not be able to have access to their own or a suitable computer. The use of speech recognition and text-to-speech software has given people with poor sight access to internet banking. People with motor disabilities could find using a keyboard or mouse difficult, and those with cognitive disabilities could find understanding how the system works difficult.

> ### Activity 12q
> Describe how the use of online banking has made life easier for people with disabilities.

Information technology has made **e-business** possible. The electronic transfer of funds, the use of debit and credit cards, online banking and online commerce use the computer systems, networks and public communications systems that make up the internet. Individuals can purchase goods from sellers anywhere in the world, at a time that suits them and without leaving home. The main impacts include increases in global commerce and global wealth, and that commerce can be at almost any time of the day and night, and be between people and businesses based anywhere in the world.

Expert systems can be used by banks in financial planning and giving advice to their customers. The use of expert systems to provide financial advice has impacted on customers in that advice is more consistent and can be given faster than by using human advisers. Expert systems do not forget details about the customer's financial information, but human advisers may do so. Training humans to give advice takes time and may not be the same for each adviser. An expert system can be more efficient than humans when dealing with many customers and may give more consistent advice than humans.

Banks have made contactless payments available using cards and mobile phones. The use of contactless payments has made cashless payments more common and the use of cash money has decreased. In some countries, payments by mobile phone are more common than payments by cash.

Financial trading has been enhanced by the use of information technology. Computer systems can be programmed to automatically track the value of stocks and shares, and to buy and sell them. Human traders have instant access to the prices of stocks, shares and goods, and can use IT to look for trends and patterns in the prices so they can make better decisions about whether to buy or sell.

Business reports make use of **Extensible Business Reporting Language (XBRL)**. XBRL is based on XML, which is the extensible mark-up language used to transfer information over the internet, and has standardised the financial information that appears in the public reports of companies so that it can be sorted and searched more easily by traders.

Computer systems can calculate and display financial data in various formats, for example graphs and charts, that assist owners of businesses in making decisions. If the owners of a business are taking out a loan to expand, the interest and the costs can be shown in a way that makes it easier to understand and predict the possible effects on the finances of the business.

Business data can be securely transferred online and payments can be received and sent electronically. Financial transactions are faster and more secure than by traditional methods such as the movement of cash.

Personal finance has been made easier and simpler by using information technology. Personal accounting, online banking, purchasing and selling from home and by small businesses have all been made possible by IT and by reliable, high-speed communications systems. The internet has enabled e-business to flourish among small and large companies. Software for personal finance can display data in the form of graphs and charts that assist in the understanding and planning of financial matters.

> **Activity 12r**
> 1 Describe how information technology can assist business owners in understanding their finances.
> 2 Describe how information technology has enabled financial traders to make more profits when trading goods, stocks and shares.

12.4.7 News sources 

Information technology has increased the speed with which news is collected, collated, researched, checked and distributed by journalists. The development of high-capacity communications systems using **fibre-optic** and **satellite communications systems**, data storage systems and mobile devices has enabled journalists to find, research and report on news topics from anywhere in the world at a speed not possible using traditional methods. News items are almost permanently stored for later reference and research. Experts on the news topics, and others such as local residents, can be interviewed using mobile devices with video calls.

Traditional newspapers use IT to create and print their editions. The newspaper stories are collated and set out by editors using IT systems and sent for printing. Editions of the newspaper or magazine are sent around the world to be printed locally to where they will be sold. IT has cut the cost of distributing newspapers and magazines as there is now no need to transport them.

News services are making increasing use of social media. Because social media is available for anybody to post news and stories, postings must be checked for accuracy and reliability by journalists who intend to use them in their own stories. Social media often posts news stories more quickly than traditional journals, newspapers or television news reports. News journalism is changing from long, fully-checked, in-depth articles to short posts, videos or tweets that can be corrected later if necessary.

12.4.8 Family and home

Information technology in the home has enabled better communications with friends and relatives, access to vast amounts of information for research and education, control of devices in the home environment (including voice-controlled devices) and monitoring and surveillance for security purposes.

Home networks allow access to the internet using computing devices such as laptops and smartphones. A wireless network can easily be set up in a home so that each member of the household can have individual access to shared resources such as printers. More usefully, Wi-Fi connections provide access to control networked devices and appliances, to view outside surveillance cameras or baby monitors, and to access the internet.

Access to the internet enables home users to keep in touch with friends and relatives and to search for information to help with school or college studies. The use of VoIP, video calls and messaging systems allow the exchange of chat, home videos and photographs between family members who are far apart.

Video and audio editing applications on home computing devices such as laptops and tablets have enabled home users to create their own videos for uploading to social media.

Home networks allow users to use apps on their smartphone to control central heating or air conditioning systems, lighting and televisions. Smart devices connected to the home Wi-Fi can be controlled remotely using the internet. Users can turn on and alter heating, air conditioning and lighting ready for when they arrive home. Internet-enabled devices and appliances can be configured to monitor themselves and send usage data, for example electricity use, to suppliers or to re-order goods, for example a smart refrigerator can reorder perishable foods when stocks on its shelves are low.

▲ **Figure 12.12** Google Home Hub™

Smart television sets allow voice control by the user. Infrared remote controls are replaced by the use of voice commands. This requires the TV set to 'learn' the voice and dialect of the user and to store these for reference. Security concerns have been raised over the use of voice control of TV sets and other devices because these devices must always be listening out for commands. Whatever is spoken in the vicinity of the devices can be recorded and often this is on the servers of the supplier of the device. Personal and confidential conversations can be overheard with the resulting loss of privacy.

12.4.9 Entertainment

The creation and distribution of TV programmes has been changed by the use of IT. Television programmes, from news to entertainment and movies, are recorded, edited and distributed using computing technology. Programmes are recorded using high-definition television cameras, and the video and audio files are stored on high-capacity storage systems. Editing is carried out in the digital domain where special effects can be added, unwanted materials removed and titles added with ease. Digital television programmes are distributed to viewers using satellite communications systems and the internet, as well as by more traditional terrestrial broadcast systems. Broadcasters can reach a wider audience that crosses regional boundaries using satellite transmissions, but this makes applying copyright restriction and censorship more difficult for regional governments.

Streaming video over the internet has changed the way people view television programmes and movie content. The content is available as and when the viewer wants to watch and not when the broadcaster decides to show it. Viewers have more control over what they watch and when they watch it.

Computer technology has enabled the creation and streaming of home-made movies, or short videos, to be uploaded to internet streaming websites. This has allowed people to reach audiences around the world to share their ideas and views and to entertain. By using subscriber services, entertainers can make a living from their productions, whereas this would be difficult using traditional methods.

The streaming of digital audio has enabled recording artists to make their music available very cheaply to the listening public. The costs of recording a song with a computer and distributing it using a digital streaming website or on social media are very low compared to hiring a recording studio and making the song available on CD. Well-known recording artists' songs are also available via digital streaming, made possible by information technology. The digital streams can be accessed from almost any computing or mobile device. Smartphones with **Bluetooth®**-connected headphones are a popular choice for listening when moving around.

The internet has enabled radio stations to reach global audiences. Listeners can access thousands of radio stations from almost anywhere in the world because the radio station streams its output over the internet. This means content can be shared between different countries.

The creation, storage and distribution of computer games is dependent on computing technology. Computer gaming consoles are specialised computing devices designed specifically for playing games. The impact of computer gaming has been to create a new business and commerce area in the global economy and to drive the development of better computing devices. For example, faster central processors and processors for graphics rendering, the development of high-speed RAM and solid-state hard disk data-storage systems, and the development of very large capacity optical and holographic storage devices have been driven by the need for more powerful computing systems and devices for gaming and entertainment.

As the need for more computing resources in devices for playing games increases, computer consoles may be replaced by a client–server arrangement whereby a low-specification device is used for accessing high-performance game servers over high-bandwidth communication networks. This arrangement makes the games more available, with more profits for the game suppliers, to more users because there is no need for them to buy an expensive console.

Computer games have been shown to improve visual attention skills, problem-solving, hand–eye coordination and fine motor and spatial skills in players. Playing strategy-based computer games can increase planning, resource management and logistical skills because resources are limited in these games and decisions must be made on how to use resources to the player's best advantage. Other, more complex, skills such as multi-tasking, tracking variables and the management of multiple objectives have been shown to be improved by the playing of strategy games.

12.4.10 Government

Governments use information technology to:

» gather information from their citizens
» distribute information to their citizens

» enable citizens to apply for government jobs

» enable citizens to apply for documentation such as identity cards, driving licences, passports and other legal documentation.

The use of IT has made the payment of government funds to citizens faster and more secure. Pensions, tax repayments and other monies due to citizens can be paid directly into the bank accounts of citizens.

The use of IT and the internet has enabled citizens to have access to more government information and take part in government activities. Governments can use websites and social media to allow citizens to interact with and hold it to account. Information can be made available for viewing on websites but also sent directly to citizens' social-media accounts.

In emergency situations, text messages can be sent to the smartphones of citizens to warn of approaching events or to give advice. Live streams of virtual meetings have enabled governments to continue to function in times when real meetings are not possible. In the coronavirus pandemic that began in 2019, government ministers and officials, along with many other citizens around the world, could not meet in person so held virtual meetings and sessions to enable government to continue with its business and inform the people. For example, in the UK, Parliament held its first-ever virtual session and broke a 700-year-old tradition that all its government sessions must be held in person. Without IT, using video conferencing and calls, this session would not have been possible, and the government members would have had to attend in person, which was a dangerous practice at the time, or the government would have had to be suspended. Similar use of IT occurred across the USA and around the world. The use of information technology helps governments to inform, safeguard and assist their citizens and to govern the country in times of extreme peril.

One impact of the increasing use of IT by government departments is on the environment. In the UK, the government is the country's largest purchaser of IT equipment, which uses a large amount of power and resources. This adds to the release of carbon dioxide in to the atmosphere and the UK government has pledged to reduce this impact by the simple act of turning off the computers at night.

12.4.11 Politics

Politicians use IT to keep in contact with their electors. They hold web forums that can be accessed from anywhere with a suitable web-enabled device and can reach a wide audience. These can collect a wide range of views and suggestions from the electorate and allow the electors to question the politicians. A weblog can be produced by or for politicians to give the perception that the electors have greater access to their elected representatives. It provides a chronological record of postings and may include multimedia elements. Emails, with attachments such as posters or leaflets, from politicians can be sent to registered subscribers to keep them updated with information about the activities of the representatives. Politicians usually have a website with information about themselves and their activities. Online questionnaires can be used to collect views and suggestions from citizens to help the politicians better serve the people. Citizens have greater access to their representatives and can influence the decisions that they make in government through the use of IT.

Data-mining techniques are used to analyse large amounts of data to try to discover voting trends, to anticipate the demands of voters, and to predict how political candidates will perform in elections in different regions. This allows election campaign managers to target areas with advertising, to target specific groups of people on social-networking platforms, to raise awareness of their political candidates, and to influence voters to try and get their candidates elected. Data mining can also reveal spending patterns, budget trends and other patterns in government that can be used by politicians to inform their plans and show citizens how the government is performing. This affects how citizens view their government, gives them more information and enables them to be more involved with government.

The development and deployment of apps for smartphones made specifically to promote a politician are available. Voters can ask questions, search for information and engage in peer-to-peer discussions with politicians, and this can help or influence their decision on which candidate to vote for.

Social-networking platforms allow many millions of citizens to learn about the policies, views and ambitions of their politicians and their governments. Social-networking platforms enable discussions involving many people and can increase the numbers included in politics. However, social networks can be manipulated, for instance by carefully and methodically promoting only certain points of view and specifically excluding opposing ideas. Extremist and generally unpopular views can be spread with little opposition.

Websites have been set up to find and add to online petitions. For instance, www.change.org alerts subscribers by email to petitions and surveys set up to influence politics. Users can easily post their agreement or disagreement and add their name to petitions. Local, regional and countrywide petitions can be set up to make people aware of issues that are of concern.

IT has made possible the use of electronic voting systems. Voters can cast their votes in person by using a voting machine at a polling station. Results of elections can be announced much more quickly than with manual counting. Some electronic voting systems enable voters to cast their votes remotely. Web-based voting systems using a web browser can be accessed anywhere using the internet, meaning voters do not need to travel to a polling station and can vote while away from home, for example if they are on holiday or living abroad. Electronic voting has increased the turnout of voters as it provides greater access to voting.

Electronic voting also enables disabled people to participate more in elections. For example, visually impaired people cannot easily use punched cards or place their mark on voting papers without assistance. This means that the secrecy of their votes is compromised when using manual methods of voting. The use of voice recognition or eye movements to control computing devices can enable votes to be cast on websites. Online voting can therefore encourage voting among people who usually do not go to vote because they cannot undertake, or dislike, the experience of voting in person.

There is a digital divide between those who can access these platforms using IT and those who, for a variety of reasons, cannot. Consequently, one of the impacts of the increase in the use of IT in politics has been to disenfranchise those people. People who cannot access social networks are deprived of their right to take part in the process of government. The use of online voting systems may therefore make it easier for some people to cast their vote, but it excludes many others because they cannot access the IT systems.

Electronic voting also has drawbacks and hidden biases. It is subject to interference and errors caused by unauthorised access by those wishing to alter the outcome of the voting, by votes not being transmitted to the central systems due to communication or machine failures, or by deliberate 'rigging' of the votes, for instance the electronic alteration of the numbers of votes cast.

It is difficult to verify the identity of a voter if they vote remotely, so end-to-end auditable voting systems are available. These provide a voter with a receipt that proves that they voted, but not who they voted for. Other methods of authenticating the voter include providing each voter with a unique code, often as a barcode, that can be scanned as they enter a polling station or entered into the web page when voting.

12.4.12 Individuals and organisations

The greatest positive impact of IT is to enable individuals and organisations to communicate, share ideas and work together regardless of their geographical location. It allows, for example, flexibility in working hours and in locations so employees can work from home and employers can have employees working anywhere in the world. IT allows individuals to access education and resources (for example using websites, email and video conferencing) and to become part of a global society. IT enables organisations to have greater communication with employees and customers, to analyse customer information, for example by data mining, to increase business and to manage their finances.

Negative impacts on individuals can include reduced interactions with others, job losses and reduced physical activity. Negative impacts on organisations can include the increased costs of using IT and the security issues associated with the storage and use of data.

12.4.13 Monitoring and surveillance in society 

Information technology is used for the monitoring and surveillance of property and people in many ways and for a variety of purposes. Property and people can be monitored remotely using cameras wired into closed-circuit television systems or accessed using private computer networks or the internet. Individuals and companies use surveillance to protect their homes and businesses from intruders and other unauthorised persons. The security systems use small cameras placed around the exterior and interior of houses, and the images and sometimes audio can be accessed by the use of apps on smartphones or other computing devices.

Security systems for businesses use cameras placed to allow security guards to view certain areas in an office from the location or remotely. A few security personnel can monitor large areas or complex buildings. If a situation arises that needs attention, security guards can be sent to the location of the situation and their progress and safety can be monitored. With the use of night-vision cameras, the monitoring can occur all day, every day, without interruption.

Traffic movements and driver behaviour are monitored by using surveillance cameras placed at road junctions, over roads and motorways, and at accident blackspots. Many countries have centralised traffic control systems that use surveillance cameras to watch traffic conditions and enable traffic control officers to alter traffic lights and send emergency services to accidents. **Automatic number plate recognition (ANPR)** uses optical character recognition to read licence plates and create data that can show a vehicle's location every time it is photographed by an ANPR camera. ANPR is often used in speed-limit enforcement. By accessing the central records of vehicle registrations, the vehicle's owners can be discovered and traced and fines automatically issued.

▲ **Figure 12.13** ANPR cameras can be positioned over the road and used in police vehicles

Law enforcement agencies use surveillance cameras to monitor areas to try and prevent crime. Streets and open areas can be monitored and watched for people committing crimes and for known criminals. The use of CCTV and remote monitoring by law enforcement in public areas, along with the use of **facial recognition** software, is subject to intense debate in some countries. While it is accepted in some countries as a normal activity of law or security forces, in many countries it is not. There is great concern that watching citizens going about their activities in public areas is a massive invasion of their privacy and civil rights. In the UK, most of the monitoring of areas and people is carried out by private companies trying to protect their property and employees. This has raised concerns from regulatory bodies when the companies monitor people without telling them that it is happening. In an attempt to protect the individual's right to privacy, any surveillance has to be notified to the people being watched by, for example, a notice on a wall or post. This may appear to defeat the point of the surveillance, but has been shown to be a strong deterrent to crime.

Internet-connected cameras (IP cameras) are small and can be placed almost anywhere that they can connect to a network or the internet. Law enforcement departments, companies and individuals can place them well out of reach of the public, by hiding them in rooms or in doorbells so they can watch over areas and properties.

In China, surveillance cameras are used to ensure that the population abides by laws and regulations. This proved very successful in monitoring how its citizens obeyed the lockdown rules imposed as a result of the coronavirus pandemic. Close monitoring of people to ensure that they did not go out from their homes saved many lives.

In South Korea, its 8 to 10 million surveillance cameras were used to track people who may have come into contact with people infected with Covid-19. The tracking of credit card transactions was also used as it shows where people have been. This was combined with data taken from monitoring the locations of cell phones, which are recorded automatically and with great accuracy. Cell phones are connected to several masts at once so their location can easily be calculated. With all this data on individuals, the authorities were able to retrieve names and addresses from service providers for all the people who came close to an infected person and locate them. People could then be checked by medical professionals. In situations where the privacy of individuals and their safety, for example from a disease, are in conflict, there are difficult choices to be made by governments. For instance, is the privacy and the right of an individual not to be watched all the time more important than their safety?

Facial recognition software can be used to scan the images from surveillance cameras to identify known criminals. This has led to catching criminals without arrest warrants being in place. Large gatherings of people, such as at sporting events, markets and airport arrival and departure lounges, can be scanned for those suspected of plotting terrorist attacks, for example. Information technology can carry out this scanning and identification faster, and possibly more accurately, than humans can. Again, this raises the question of whether the infringement of the privacy and rights of law-abiding individuals can be tolerated in order to catch

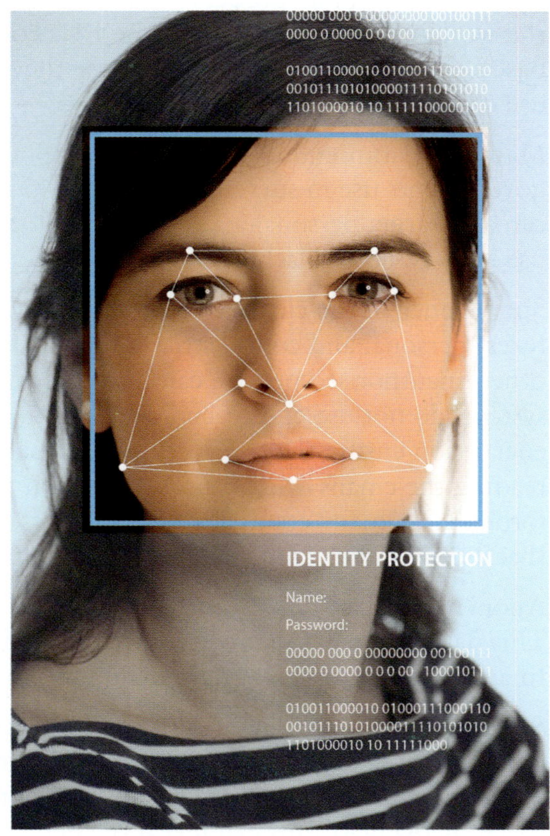

▲ **Figure 12.14** Facial recognition technology

criminals or prevent terrorist attacks. Different governments and countries have differing views on this matter.

Aerial surveillance issued by law enforcement and the military is carried out by the deployment of remotely controlled devices such as drones. These are equipped with cameras and listening devices and can be monitored from locations very far from their base. Drones can go where larger aircraft cannot. However, the use of drones can cause safety issues as they can fail and fall on people or interrupt the flights of aircraft if used carelessly.

Satellite technology to monitor people from space is made possible by the use of high-definition cameras, precision aiming, high-speed communications systems and remote-control technology. Events from around the world can be watched in real time by observers based anywhere. Military commanders, and their political masters, can watch their troops in action from the safety of their command posts.

Surveillance of computing devices and their use has an impact on society. Cell phones can be tracked to determine their location and where they have been. Although in reality this only tracks the phone itself, individuals can be tracked by the assumption that the owner is with the phone. Computing devices can be monitored and their use reported to businesses and governments using websites' cookies (small files placed in the computer). In many countries, websites have to explicitly request permission from the user to place cookies that report usage data. This has given the user control over much of the data that is reported back to the website owner. However, the interruption of the web browsing experience has become a nuisance to users because many websites cannot be viewed without a lengthy check of what cookies are to be allowed. A refusal to allow cookies often results in the user being unable to view the site at all. The European Union (EU) has data-protection laws that prevent data being exported beyond its borders without checks, so some websites from around the world cannot be viewed in the EU. For example, some news websites based in the USA are blocked because they refuse to place restrictions on their use of data.

Malware that monitors and reports computer usage to others is a tool that some governments use in surveillance of their citizens. Malware, for example spyware, can also target personal and confidential data and report it to others who can use it for fraud or blackmail.

Surveillance of the electronic communications between computing devices can reveal personal and confidential information about users. Governments, and other unauthorised people, can intercept and read emails because most email systems do not use end-to-end encryption. Apps that make use of end-to-end encryption ensure that only authorised users can read the messages. While reading the emails of citizens is often seen as an invasion of privacy and is forbidden in many countries, governments and law enforcement agencies argue that it assists in tracking and apprehending criminals.

Anonymised data can also be collected and sold to third parties for analysis. This data can be mined to reveal trends and patterns, for example in the financial activities of a population in a country.

Surveillance of social-media use and postings can be carried out by monitoring and tracking individual postings to determine the personal and political views of individuals. Data mining and analysis of the vast amounts of data from social-media platforms can reveal the personal interests, political views and beliefs, friendships and associations, activities and buying habits of individuals, groups and populations. Companies and governments can use this information to discover trends and patterns that assist in government planning, security arrangements and marketing. It can be argued that by using social media and displaying information for all to see and read, users are allowing themselves to be monitored by companies and governments, but others argue that the use of data gathered from social media should be carefully regulated to protect users because these users did not intend their postings to be used in this way.

Data from credit cards and other electronic payment systems can be collected and analysed to track people's purchasing habits or movements. By analysing and combining the findings from this data with data from cell phone usage or other electronic communications, it is possible to create an 'electronic trail' that shows where and when individuals have travelled. This can be used by law enforcement or other government agencies to monitor and locate individuals. In many countries this monitoring has to be approved by a judge or by a warrant in order to access the data, but in others it does not. In countries where approval has to be sought, any unauthorised collection, access and use of the data to monitor individuals is forbidden and punishable under the laws of the country. It is argued that this restricts the activities of law enforcement and may allow some criminals to escape justice, but civil-liberty organisations argue that this is a price that people should be prepared to pay to keep their freedoms and rights. However, the lawful monitoring of electronic communications has proved to be a valuable tool in reducing illegal immigration, preventing the trafficking of children, and identifying and catching those who create, trade in and view child pornography.

12.5 Technology-enhanced learning

12.5.1 Computer-based training

Computer-based training (CBT) is an interactive learning method that can also be called computer-assisted learning or computer-based instruction. It involves the use of a personal computer or mobile device that is, usually, connected to a network. The aim of computer-based training is to increase the performance of learners through the production, management and use of appropriate digital resources and educational technology. Learners can access the training materials at any time and can work at their own pace because a computer system takes the place of the teacher or instructor. A drawback of having no teacher present all the time, although teachers can be contacted, is that there is no one on hand to answer any urgent questions the learner might have, as there would be in a traditional classroom situation. This is known as 'asynchronous' study, where the student and teacher communications are separated in time.

Computer-based training can be made available on almost any topic. While the theory of swimming, bicycling or flying aircraft can be taught with CBT, at some point the actual real activity must be carried out.

CBT teaching techniques

CBT uses different teaching techniques to keep learners motivated and interested in the materials to try to ensure that learners complete the training course properly. A CBT training course set up by a business to train its employees will be designed to take into account what the staff need to be taught, what skills they already have, and the level of technical expertise they have for accessing the course. Colleges also develop CBT courses with specific objectives for their students while considering the technical aspects required to give students access. The costs of collating and writing materials and constructing the courses and tests can be high. CBT can help reduce overall costs as fewer teaching staff may be required when it is used to train large numbers of employees or students, but it is expensive to develop courses for just a few employees or students.

In CBT, the role of the teacher is taken over by the computer. The CBT system also monitors the process of teaching and the progress of the students, provides assessments for testing the students and provides feedback assessing student results. CBT is available to students in three different ways: CD-ROMs, web-based training or as **networked courses**.

Activity 12s

Explain the drawbacks of asynchronous study with computer-based training.

Optical discs

Materials and course resources can be provided on CD-ROMs, DVD-ROMs or Blu-ray discs to individual students to use whenever they wish. Students use their own computing devices to access the materials and to store any test results. There is often no need for the computing device to be network or internet capable, but it must have the use of a suitable CD, DVD or Blu-ray drive. This type of CBT is useful for learning a foreign language at home or for learning a specific work-based skill such as how to repair or maintain a machine tool.

Web-based training

Web-based training is provided over the internet using any appropriate web browser. Sometimes this is called cloud-based training. Students can access the materials from anywhere and at any time, provided they have a suitable internet connection. Activities and tests can make use of resources found on the internet and students may be able to be in contact, in real time in live 'chats', with other students and supporting teachers. Web-conferencing can be used for discussions.

Courses available on the local area network

Schools, colleges or businesses may use local area network-based (LAN-based) systems within the school or organisation for their own students, and not the wider internet. Many students can access the materials at the same time using networked courses. These allow more control over what resources and materials are made available to students and ensure that the distractions found on the internet are not available. Also, the materials are not available outside the organisation (networked courses are explained in detail later in this section).

Features of computer-based training

Most CBT systems provide the same features to their users. After the initial logging-in procedure, a student will find instructions on how to use the training, glossaries of the most important terms used in the training, interactive tutorials on topics, graphics to illustrate topics, video and audio learning materials, examples taken from real-life scenarios on the topics if appropriate, and sets of tests and assessment materials to allow the students to monitor their own progress, get feedback and report to the course supervisors. The ability to repeat the course or sections of the course is a major benefit to employees and students using CBT. If a section is not understood or needs practising, or if the assessment requires it, a student can repeat the section as many times as is necessary.

Computer-based training uses a combination of teaching materials. Tutorials are used to explain techniques and concepts. Some tutorials are in the form of recorded lectures given by teachers who use them to explain and show demonstrations of how to carry out physical activities such as fitting replacement parts, or carry out an experiment in science, or explain concepts in mathematics or literature. The lectures are presented in a logical manner to enable students to follow the teacher's explanation.

Audio and visual materials can be incorporated into the tutorials used in CBT. These can be part of the linear flow of the course where they are inserted and follow on from exactly how the original lecturer or instructor would have used them in a lesson, or they can be accessed by links that take the student away from the main course.

Learning theoretical concepts is difficult for everyone and there is evidence that most people remember only about 20 per cent of what they see or read first time. CBT provides the means for students to repeat topics as many times as they wish or need to. Repeating and practising topics improves learning. CBT should always provide a 'drill and practice' session after every tutorial, chapter or module.

Games and simulations included in CBT improve the motivation and concentration of students and help improve problem-solving skills. Entertaining games can improve the chances of students completing the training rather than giving up through boredom. Simulations can provide opportunities for students to try out and enhance their newly acquired skills. Simulations that provide realistic scenarios can prepare students for dealing with real-life events without the risk of being endangered. They also allow the scenarios to be repeated, recorded and analysed.

Assessment of students and feedback to students is an important feature of most computer-based training systems. Standard or customised assessment tests can be created by the teachers or supervisors. Testing at the end of each topic gives the student valuable feedback and the possibility of repeating topics and sections to improve performance, and provides teachers and supervisors with monitoring data. The data can be used to provide customised assistance and guidance to students and count towards final gradings. Some CBT systems provide certification at the end of the courses.

Advantages and disadvantages of CBT

CBT of employees has benefits for employers. High-quality, consistent and focused training using the same course materials is provided to employees without the cost or time needed to set up classrooms with teachers for the training. Also, differences in the individual styles of teachers are eliminated, travel costs for employees or visiting teachers are much reduced or eliminated altogether, and employee progress and understanding can be more easily tracked and recorded. For situations where employee skills need regular updating or testing, when new processes and procedures are introduced, or if the business has many locations in different countries, CBT provides a good method of regularly training or retraining employees; it is especially useful where fast, intensive training is required to deal with important new product updates or issues.

On the other hand, drawbacks such as the lack of physical practice or interaction with other employees, or the lack of an instructor to show how tasks should be done, can make CBT courses less suitable for some businesses. Also, some employees may not be very motivated, may skip sections, or may not study in sufficient depth to fully comprehend all the topics, but instead may just do the minimum to complete the course.

Benefits of CBT include that courses can be undertaken at the student's own pace and at a time and place convenient to the student; the use of multi-media, tests and quizzes can be stimulating and motivating; and students can easily stop and restart from where they left off. Progress can be tracked, recorded and reported both to the student as feedback and to the teachers for monitoring purposes. Students who live too far away from traditional schools and learning centres can take part in courses and can take courses that would not normally be available to them due to budget restrictions in their school. For example, learning a minority language or specialised technical skills may not be possible in some areas, but courses can be made available by CBT.

Drawbacks include that students must be self-motivated and able to work unsupervised. This may be difficult for some students, causing them to fall behind or fail to complete the course. Important dates may be missed due to lack of supervision or pressure from teachers. Some students may fail to meet deadlines for the submission of work if not under constant pressure from their teachers. Students may skip sections in order to complete the course quickly. Some sections may be quickly looked at but not fully understood. If a section has no associated test, assessment or reporting then there is no obvious need for the student to complete that section. Important or crucial knowledge may be missed. The lack of interaction with other students or teachers may make the course boring to take. The level of a CBT course may not be properly matched to individual students as it uses the same materials for all students. In practical learning, CBT can support the learning of techniques but cannot fully replace the 'hands-on' experience.

12.5.2 Online tutorials

Online tutorials allow students to take part in a learning environment where they interact with other students and with teachers without having to physically attend a class. All online tutorials share the common features that they are studied online as an activity, are carried out by the students as self-study and have a specific outcome for their learning. Students and teachers can be online at the same time or students can study and contact teachers as and when the need arises.

A group of students can log in and be online with a single teacher, or a group of students can be online together and learn together without a teacher, or a single student can be online with a single teacher or tutor for one-to-one teaching or mentoring. This is known as 'synchronous' study, where communications are available between the student and the teacher and they are in contact in real time.

The tutorials can be one of two types. A recorded tutorial provides video and audio in a linear sequence that has only one route through the tutorial, while an interactive tutorial has a number of different routes through the work. The material in interactive tutorials can be studied in any order. A mixture of both types is used in many tutorials.

Online tutorials are used in universities and colleges for teaching and monitoring academic studies, and in business organisations for industrial and vocational training. They enable students and other learners to access teachers and tutors that they would not otherwise be able to because they live too far away, cannot travel or do not have the resources, time or money to get to the classrooms.

Online tutorials do require a stable, adequate-speed internet connection and a suitable computing device in order to participate. The tutorials are usually restricted to students or employees of the organisation; some may require payment to join and some may be free to all. Free online tutorials usually have minimal, if any, contact with teachers or tutors and may consist of little more than a few video demonstrations. These cannot properly be classified as tutorials.

Online tutorials use a common interface shared by the students and teacher, with all online at the same time. Often, this is accomplished by using web browser-based software so the tutorial can be accessed using any web-enabled device. Modern browsers can implement audio and video connections so there is no need to download or install specialised software on to the student's computing devices.

Sometimes, an online tutorial will allow the student to work individually and alone and communicate with a teacher who responds later. The student and teacher are not online at the same time. This type of tutorial allows students to be in complete control of their learning in terms of timings and progress.

Most online tutorials will try to emulate the experience of being in a real, face-to-face classroom with other students and a teacher. The facility to share documents by uploading and downloading to and from a centralised storage area, the simultaneous editing of documents, sharing screens and live discussion of topics all enhance the feeling of being in a real classroom.

The recording of an online tutorial for later review by students and teachers is a feature that increases learning as well as accountability. The student can review the work covered to ensure that there is understanding, and the teacher can ensure that the work covered was appropriate and met the topic objectives. This is important because the experience of an online tutorial can be different for each of the students and for the teachers, whereas in a real classroom the experience is mostly the same for all.

A good online tutorial will provide almost instant feedback to students at any time of the day, will include explanations and discussions aimed at the level appropriate for the student, being not too easy nor too difficult, and will provide regular assessments with feedback so that students can monitor their progress.

Advantages and disadvantages of online tutorials

Online tutorials have all the benefits and drawbacks associated with computer-based training. Students can join online tutorials with their fellow students and with teachers, or they can learn when they choose. For the providers of the tutorials, costs are lower than running traditional classes or lectures as fewer teachers are needed, and experts can create the tutorials from wherever they live and do not have to travel to the school or college to deliver them. The teachers have to spend time creating and updating the materials and carefully select the right media to ensure that the audio, video, graphics and web pages are appropriate and suitable for the students. They also have to decide on the length of the tutorial to ensure that it covers the topics properly but does not go on so long that students' attention wanders.

Activity 12t

Describe how online tutorials allow students to have greater control over their learning.

12.5.3 Networked courses

Networked courses are a type of computer-based training used in schools and colleges as well as in businesses. Networked courses are taken online with access to course materials and assessments provided by a school or business. Usually, the materials are stored on an intranet with access restricted to members of the school or organisation, but some can be accessed over the internet by creating an account and logging in.

Networked courses can enable students to learn with others by providing a support community for students to ask for help or get advice from other students. Materials provided by the course can be regularly updated quite easily as they are stored centrally. Assignments and tests can be set and submitted to teachers for assessment, but some may be automatically marked by computer to give instant feedback. Teachers can monitor the progress of students and be involved with the learning as little or as much as is required by each student. Networked courses may allow the students to set and monitor their own targets throughout the course as a way of motivating them to achieve more.

As a result of a test or assessment at points during the course, the student may be directed along different paths through the course compared to other students. Students may also choose to go through the course along different routes. These courses are called 'non-linear' and have branched sets of paths through the work.

Another, simpler, type of course presents all the materials and assessments in a linear fashion so that all students work their way through in exactly the same way, just like in a traditional classroom teaching environment. This type of course is easy to write and set up.

Some courses allow teachers to assess students before they start the course and set up customised routes through the course for the student before they start. This type of course takes considerable time and effort to set up and manage as the individual needs of many students have to be accommodated.

Linear courses do not take the experience or ability of individual students into account when presenting materials, but non-linear courses can be customised to each student.

Advantages and disadvantages of networked courses

Networked courses have similar benefits and drawbacks. An advantage of these over other types of course is that they are usually taken with other students, and peer-to-peer help and collaboration is available between students. The course materials are available on intranets, so access is more reliable and no internet access is required. Navigation around the course and student interaction may be greater than in other courses as networked courses are usually more specifically aimed at particular groups of students or learners. A disadvantage is that the courses may not be available from outside the intranet of the school or organisation, which restricts the times and locations from which the courses can be accessed. Online tutorials and massive open online courses do not have these restrictions.

Activity 12u

1   Explain why a car mechanic learning how to service a new type of car may prefer to learn from a course supplied on CD-ROM rather than a web-based course.
2   Explain why companies may prefer to use networked courses for training their employees rather than a web-based course.

12.5.4 Massive open online courses

A **massive open online course (MOOC)** is named as such because there can be almost an unlimited number of students enrolled on the course at the same time. These are 'open' because almost anyone can join up to the course and there are no formal admission or enrolment processes or academic requirements. They are 'online' because the course is delivered using the internet, and they are a 'course' because their purpose is to teach a specific topic. MOOCs are produced by universities and colleges such as Harvard, businesses such as Microsoft, and organisations such as the British Council or the United Nations to reach large audiences around the world. Many universities and organisations do not produce MOOCs because they are reluctant to give away their resources online for free.

MOOCs include all the methods of imparting knowledge and skills found in CBT courses: video recordings of lectures and demonstrations, games, problem-solving activities, simulations, tests, quizzes and assessments, assignments and online tutorials. Also, social-media discussions and interactive forums can be used to allow students to communicate with others and to gain feedback or answers to questions about the course topics. However, some MOOCs do require a payment for full access to all of their features. Teacher feedback and certification at the end of the course usually require a fee as a condition of full enrolment. MOOCs can be used as part of a wider and more extensive learning programme for students. A student may be required, or wish, to take a MOOC while studying a related topic.

Advantages and disadvantages of MOOCs

MOOCs have a number of benefits for students. They are available for many topics or areas of study and are mostly free. Students can be in any part of the world where internet access is available and join a MOOC that is published and made available from anywhere in the world. Students and teachers can make contacts all over the world and become known in their fields of study to a global audience. Students can work through the course at their own pace and at times convenient to themselves while performance and achievement can still be easily and regularly monitored throughout the course.

There are, however, some drawbacks to the use of MOOCs for learning. A number of barriers to participation exist. Students must have a reliable, stable and adequate-bandwidth internet connection. Materials such as video, graphics and often large files will need to be accessed without the problems caused by a poor or unreliable internet connection. Websites for MOOCs do not adequately provide for students with accessibility problems such as visual, auditory and motor disabilities and these students are effectively excluded. Most MOOCs are available only in English so there is a language barrier that prevents students who do not speak or understand English from taking part. Lastly, while most MOOCs are currently free, more and more are asking for or demanding fees for full access. Students who cannot afford to pay these fees will not be able to participate. There is also the problem that, while some universities and colleges use and provide MOOCs, most MOOCs cannot be used as credits for university or college qualifications.

12.5.5 Video conferencing

Video conferencing enables users to meet with others without having to travel to conference venues. There are many benefits for businesses of using video conferencing, including the reduction in the time taken to organise a meeting, the reduced disruption to employee work schedules of having to attend a meeting away from their usual workplace, the reduction in the environmental impact of travelling employees, the reduction of the costs and times of travelling by employees to a meeting venue, and the avoidance of problems caused by time-zone differences.

A video conference requires multiple cameras and screens so that the participants can all see each other, a set of microphones and speakers to exchange the spoken words, and video-conferencing software to exchange the video and audio over the internet. As there is a reduction in the social interaction between participants, the software also ensures the orderly communication between the participants.

Most video-conferencing services not only allow participants to engage in meetings where they can see and hear each other but can also share screens and remotely access each other's desktops. The conference can also be broadcast to viewers not actively involved in the actual conference using the conference software, a web interface or social media. Chat rooms and text messages, VoIP and digital whiteboards can be integrated into video conferencing. These features can be put to use in teaching and learning.

Students and teachers can participate in video conferences between schools that are geographically distant from each other. This is useful for connecting students with others from around the world and enabling teachers to teach students from other countries. Students can experience other cultures from around the world, which can encourage tolerance and understanding of others.

Lessons or lectures can be recorded and viewed later for revision or for clarification of the topics, or if students are unable to participate in person.

Advantages and disadvantages of video conferencing

Students who cannot physically attend a school because they live in remote areas of a region can take part in classes by video conferencing. Costs of providing education can be reduced by combining classes from different schools by video conferencing. This type of distance learning gives students access to learning materials, teachers and discussions that might not otherwise be available to them.

There are a number of drawbacks to using video conferencing in education. The costs of the additional hardware and the appropriate software can be prohibitive in some regions. The technical problems of latency of the audio or its lack of synchronisation with the video, interruptions due to power outages, dropped internet connections or equipment malfunctions can make a video-conferencing session unworkable. The lack of the physical presence of a teacher, or the lack of complete coverage of the classrooms by the cameras, may allow some students to be easily distracted or avoid taking part in the lesson. It may also make the monitoring by the teacher of individual students' progress during the lesson more difficult and make the teacher's job of encouraging all students to take part more difficult.

> ### Activity 12v
> Describe **two** drawbacks of using video conferencing in teaching.

All computer-based training systems and their components, such as online tutorials, networked courses, MOOCs and video conferencing, require access to a suitable computing device or devices and appropriate software or web browsers.

A network connection and reliable access to the internet is also a prerequisite. The exception is the use of CD-ROM- or DVD-ROM-based material or courses, when a standalone computer may be used. However, as many modern laptops, tablets and smartphones do not have optical devices, these types of courses are becoming less common.

12.5.6 Impact of delivery methods

Computer-based training, online tutorials, networked courses, MOOCs and video conferencing enable students to take part in and follow courses of study from which they would otherwise be excluded. This has allowed students to meet other students from around the world, gain a global perspective and increase their knowledge, understanding and tolerance of other cultures.

The use of computer-based training methods presents challenges to students and teachers. Online courses and tutorials often present the teaching in much the same way as traditional classroom teaching. Some teachers simply transfer their teaching materials into digital format and expect students to work their own way through it. Good online courses that embrace information technology and use a well-integrated mixture of multimedia and interactive materials will motivate and interest students. This has been shown to increase their performance and raise their grades. However, this increase in performance and grades relies on students being well-prepared and carrying out the required course activities thoroughly. Research has shown that under-prepared students perform less well when using online courses than if they were in a traditional classroom. Under-prepared students are also more likely to drop out of online courses and this slows their progress through college. Where colleges have moved to using many more computer-based and online courses over traditional face-to-face lessons and lectures, this has had the impact of reducing course completion among less well-prepared students.

However, the involvement and use of online tutorials increases the likelihood that students who are well-motivated and prepared will complete the main course. Students who are highly motivated do well when using online courses and tutorials. It increases their independence and willingness to take responsibility for their own education, and encourages them to use email, social media and video calls to contact their teachers to ensure that they understand their course targets, goals and assignments.

Activity 12w

Describe **two** benefits of using video conferencing in teaching.

Practice questions

1 Explain why governments view cryptocurrencies as a threat to their economy. (6)

2 Explain how using a digital currency can assist an art dealer when paying for a high-value painting. (6)

3 Explain why individuals should be careful when using and investing in cryptocurrencies. (6)

4 Describe why the use of data mining may be of concern for the security of personal data. (6)

5 Explain why a student taking a holiday walking along the
   Great Wall of China might prefer to use a blog rather than
   a traditional hardcopy diary to record the trip.   (8)

6 Discuss the benefits and drawbacks to businesses of using
   chat rooms during the development of a new product.   (8)

7 Describe the processes that an email message goes through
   to pass from a sender's email client on a laptop to the inbox
   of a recipient on smartphone.   (8)

8 Describe how the role of a teacher has changed with the
   introduction of information technology.   (6)

9 Discuss the benefits and drawbacks of using voice-control
   systems to control devices and access information in the home.   (8)

10 Describe how the use of computer-based training can affect
   the motivation of students to learn.   (8)

11 Schools and colleges may try to reduce the costs of providing
   lessons by employing fewer teachers and using computer-based
   training (CBT) methods instead. Explain why the use of CBT to
   teach students may not reduce the overall costs.   (6)

13 New and emerging technologies

In this chapter you will learn:
- ★ the types of new and emerging technologies
- ★ the impacts of the new and emerging technologies.

Before starting this chapter, you should:
- ★ be familiar with the terms: radio transmission, smartphone, compression, recycling.

13.1 Types of new and emerging technologies

13.1.1 Artificial intelligence

Artificial intelligence (AI) or **machine intelligence** is defined as the ability of computer systems, or computer-based machines, to carry out tasks that would normally only be possible by humans. AI mimics or copies human thinking processes, for example problem-solving and learning from experience, and is the ability of computing devices or software to interact with the environment in a manner similar to that of humans. AI involves being aware of the environment (through using sensors) and carrying out actions depending on what is sensed. AI would, like a human, examine and analyse its environment and then use algorithms to take actions. The actions taken by AI are designed to increase the chances of achieving a set target or goal.

Early AI algorithms had a step-by-step approach with precise sets of instructions and choices for problem-solving, but humans do not simply follow a set of pre-defined instructions to achieve a goal. While some problems can be solved in this way, most humans learn by experience and apply their knowledge to new situations. Artificial intelligence definitions now take this into account and, to be deemed AI, a system must be able analyse its surroundings or external data, carry out actions, learn from any new data gathered as a result of the actions and carry out further actions to achieve a goal. The concept of learning is important because it ensures that AI can deal with new, unexpected or changing situations without human intervention. AI can make decisions on its own.

Humans also solve problems and make decisions quickly by intuitive means or just by guesses. Humans make assumptions based on their knowledge of the world around them, use what is known as common sense and take into account abstract feelings or thoughts when making decisions. The challenge of developing artificial intelligence is to recreate all of these in machines. Developments in computing technology, such as supercomputers, advances in

▲ **Figure 13.1** A supercomputer built for artificial intelligence

computational sciences, game theory and machine learning involving the study of algorithms and statistics, contribute to the advances in AI.

Impacts of artificial intelligence

Almost any task that requires thoughts and decisions can be carried out by artificial intelligence systems. Rules-based expert systems, using if–then rules to go through large bodies of knowledge, make use of AI to solve complex problems by reasoning. This is used in air-traffic control, speech recognition, diagnoses of illness and disease, diagnosing faults in machinery and engines, identifying organic materials, and many other sectors.

Self-driving cars and trucks use AI to analyse and control the movement of the vehicle. Self-driving vehicles must be able to cope and navigate safely in unknown or high-risk environmental conditions, such as in areas not in their pre-installed maps or in poor weather. They must be able to make instant, difficult decisions in unexpected or potentially dangerous situations. However, moral and ethical issues are a challenge to AI systems.

Online trading of stocks and shares makes use of AI to analyse financial markets and make decisions to trade or not to trade. Forecasts and predictions can be made on future trading prices by data mining to discern trends in the large amounts of data. AI can operate without breaks and can react to changes in financial markets much faster than humans, resulting in greater profits. It can also combat fraud by analysing patterns and spotting unusual financial activity. Financial records can also be kept by AI.

AI has an impact on health care. It is enhancing the care that is given to patients, and is helping around the world with the diagnosis of illnesses and diseases and reducing the number of trained professionals required. AI is used in data mining and in expert systems based on the analysis of large numbers of incidences of diseases and their symptoms. Doctors can use these systems to access the knowledge of experts and increase the speed of diagnosis of an individual. AI and expert systems may also be used by individuals to self-diagnose and seek treatment sooner.

AI has assisted in the creation of prescription drugs to combat disease by reducing the costs and time taken to develop new drugs. AI identifies diseases and conditions that could be successfully targeted by new drugs, or by existing drugs that could be used in new treatments. AI and data mining can be used to predict the effects of existing and new drugs against diseases. The time taken for clinical trials of drugs has been reduced by using AI to identify suitable patients for the trials, to analyse the data from the trials and to raise alerts if the trials do not produce useful or helpful results. Artificial intelligence systems can also reduce the amount of manual work required by technicians by automating testing, classifying and statistical analysis of the tests and results, helping to allocate specific drugs depending upon patients' likely response to the treatment. Patients are more likely to receive drugs that will improve their condition and speed up their recovery rate.

In the care of patients, AI is used in chatbots that provide patients with access to advice on their health, in wearable devices, for instance **fitness trackers**, that record, monitor and alert the user to health issues, and in analysing the health of a patient. Virtual nursing assistants use AI to monitor patients and answer questions more promptly. Personalised genetic analysis by AI can reveal genetic conditions and provide suggestions for possible courses of treatment. Lifestyle and disease management can be researched and suggested by AI systems. The use of AI has increased survival rates from diseases and improved patient care.

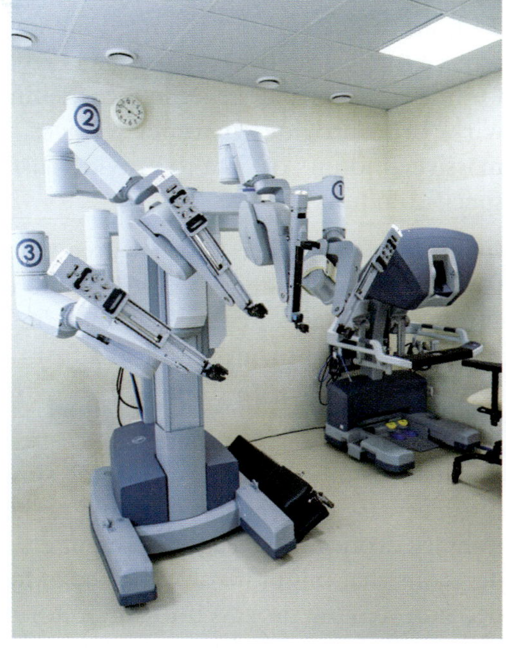

▲ **Figure 13.2** An AI-controlled surgical robot

Virtual assistants using AI are found in computer operating systems that power desktop computers, laptops and smartphones to provide smart assistants and enhance the user experience. Voice control of devices, controlling smartphone camera effects and monitoring battery levels all make use of artificial intelligence. Devices in the home linked to online AI systems that analyse and interpret the spoken commands and behaviour of users can control heating, air conditioning, lighting or other household systems, or they can be used to answer questions, play music as smart speakers or order goods online. While these devices can make life easier and more pleasant for the users, there are serious concerns about the privacy and security of the large amounts of data that need to be collected so that the AI can learn and make sensible decisions that reflect the individual user's needs.

Notifications and customised advertising in social-media feeds rely upon AI to provide data to the user when they log in to their social-media applications depending on their usage patterns. Smart email apps can use AI to scan emails, reject spam emails and categorise emails for easier access by the user. Smart reply systems suggest replies in contexts decided by artificial intelligence.

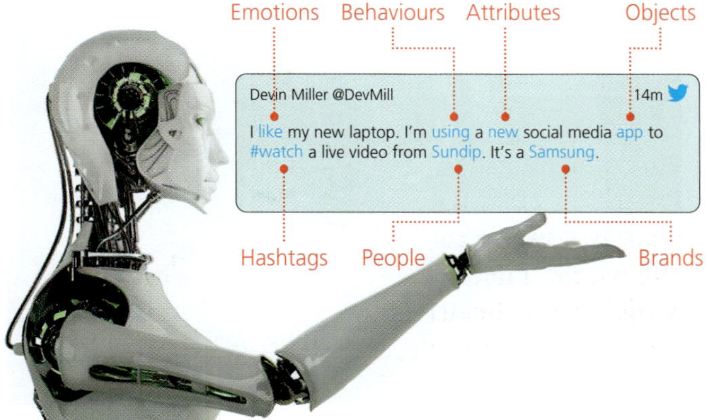

▲ **Figure 13.3** AI analysing a social-media post

In business, chatbots can be used to interact with users online. Chatbots, or chatter robots, provide a chat interface that allows a human user to interact with a computer program using AI. The AI uses natural language processing, video and audio analysis to provide customer care by answering questions and enquiries at any time of the day or night. Customer service can be provided 24 hours a day without the need for staff. Costs can be reduced as chatbots can replace many employees. Customers receive instant responses to their queries. Online retailers and travel services selling flights and hotels can use chatbots with AI to provide customer services. The AI uses data mining to analyse massive amounts of data to provide customers with answers customised to their specific enquiry.

Computer games are another area where AI enhances the user experience. Gamers can play against AI-generated opponents called bots. The game bots can take the part of other cars in racing games or opponents in action games. AI-powered opponents can learn from the actions of the player and adjust and evolve their play accordingly.

Military uses of artificial intelligence are mostly concerned with the fast analysis of data in combat scenarios, training simulations and decision-making. The aim is to decrease the reaction time and make more informed decisions. AI-controlled military drones and weapons present ethical concerns about who or what makes decisions that might involve the use of force. The removal of humans from the decision-making process into a monitoring role has been opposed by many people because they say that AI cannot be trusted to distinguish properly between friend and foe or to decide on a proportional response to a military threat.

Artificial intelligence requires high-powered computing to run, which means more electricity has to be generated; depending on how the electricity is produced, this can be damaging to the environment. These systems consume large quantities of rare metals, plastics and other materials when manufactured. The extraction and production of the raw materials causes damage to the environment through mining and the disposal of the waste products. Pollution of waterways with waste water and excess heat from the extraction and refining processes may become a problem.

13.1.2 Augmented reality

Augmented reality (AR) is the enhancement of a real-world experience by overlaying computer-generated objects onto real objects. The computer-generated objects can be obvious overlays giving information or directions, or can be made to appear to the user's visual, auditory and other senses in much the same way as the real objects in a scene.

Augmented reality changes the user's perception of the real world, whereas virtual reality (VR) completely replaces the real world with a simulation. In contrast to virtual reality, augmented reality scenarios are real, for example a doctor cannot perform a real operation using VR but can use augmented reality to assist in a live surgical procedure.

Augmented reality systems use similar technology to virtual reality. While the hardware is similar, however, there are a number of important differences. AR is in real time so AR headsets must include a camera to capture live scenes, there must be microphones to

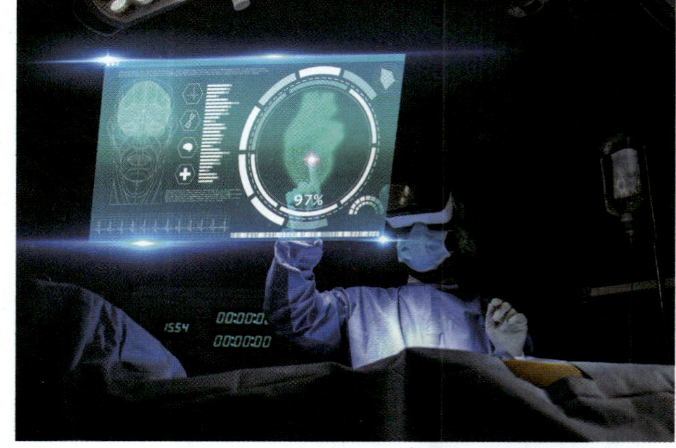

▲ **Figure 13.4** AR being used in a medical procedure

capture any external sounds, and there must be processors that can integrate these in real time with exterior scenes. Output devices such as headsets or eyeglasses must be able to overlay the images onto the scene that the viewer sees around them. Cameras can be built into eyeglasses and the images can be displayed on screens that replace the glass or can be projected into the eye of the viewer. Sometimes, people with reduced visual capabilities have been shown to be able to make out objects better in this way. Attempts have been made to create AR-capable contact lenses using microprocessors, LEDs and wireless antennae, but these have not appeared commercially. A head-up display (HUD), where the AR images are shown on a clear display in front of the viewer's line of vision, can also be used.

Impacts of augmented reality

Emergency services can use AR to help train first responders and doctors. First responders wearing AR equipment can be alerted to danger areas and assisted in finding victims needing rescue. First responders can also be helped by the overlay of information and of geographical and navigational guidance on a real scene. Safe routes, dangerous objects and local information can be added to the real surroundings seen by the viewer.

Computer games make use of AR to immerse players in the game, placing computer game characters and features into real-life situations. Gamers can search for and interact with characters and objects placed in geographical locations using GPS on their smartphones to run the game software app.

The impact of augmented reality on businesses has been to increase product awareness and sales. Customers have an enhanced experience of buying because they can interact with a product before buying it, can visualise objects better and make more informed decisions. The fashion industry and real estate businesses use AR to help sell their products to customers anywhere in the world and at any time. However, it does depend upon the potential purchaser having access to a smartphone or mobile device and to the internet. Tourism has benefitted from the use of AR; places, buildings and objects can have their information displayed to visitors without the need for printed posters or expensive displays. Museums can overlay details and videos of artefacts as visitors move around the displays.

The benefits of using augmented reality in education include students being more motivated and engaged in learning. This increases their participation in lessons and lectures and enhances their learning of topics. Individuals can access relevant information in real time as it applies to their immediate situation. This allows better and more accurate navigation, more informed decisions about what to buy and greater awareness of their surroundings. Individuals with reduced cognitive or intellectual abilities can be helped in their activities.

The use of augmented reality in medicine means that diseases can be diagnosed more accurately and surgeons can perform more accurate procedures. This increases the chances of safe recovery for patients.

Augmented reality raises some issues about the privacy of users and others. People using AR can deliberately or inadvertently invade other people's privacy. A group of people may be taking part in an AR experience and be quite content and in agreement with the sharing of their experiences and actions with others. However, bystanders and property owners may not be happy to have their actions and property appear in AR situations as they have not given permission and have no control of the use and storage of the information.

Further drawbacks include expensive hardware and software to implement augmented reality, which may be too costly for many educational and business organisations. An individual using augmented reality experiences has less contact with reality, potentially increasing the chance of social isolation. As with other computer-based activities, there may be some adverse effects on eyesight as well as other health issues resulting from prolonged use.

13.1.3 Virtual reality

Virtual reality (VR) is a simulation of the real world using computing technology. Sensors and actuators are used to provide the inputs and outputs to give the necessary sensory feedback for the simulations. Virtual reality headsets are used to provide video and audio to the user so as to immerse them in the simulation and exclude the real world. Forces, vibrations and other sensory feedback can be given to the user by 3D touch systems, also called haptic technology, or by kinaesthetic communication devices, such as joysticks, 3D mouses, gaming steering wheels and game controllers. These are connected wirelessly and are used to move objects within the virtual reality world and to provide movement, positional and touch feedback to the user.

▲ **Figure 13.5** VR headset in use

Impacts of virtual reality

Users of virtual reality can experience scenarios such as trekking in the mountains, touring an archaeological site, exploring a museum or art gallery, or become immersed in a computer game without leaving home. More serious uses include meditation and health therapies, occupational health, physical rehabilitation and professional training scenarios. Data about the environment can be captured and combined with video, still imagery and audio to create 3D virtual environments. These can be especially useful for teaching in schools, for viewing houses virtually to rent or buy, or for providing low-risk virtual scenarios for people in potentially high-risk professions, such as doctors, aircraft pilots or firefighters. Procedures like surgeries can be virtually practised, refined and shared with other surgeons before carrying them out in reality. VR can even be used to assist people to overcome fears and anxieties such as fear of spiders or of flying. Exposing people to virtual situations where there is no risk of being hurt or injured can treat anxieties, phobias and psychoses. This is called virtual reality exposure therapy (VRET). It has been shown to be highly effective in treating some fears.

Computer video games are currently the main application for VR, where gamers can reach out and touch or turn their heads to see objects around them to enhance the gaming experience. Often a complex controller is not required and the situation closely simulates real life. In entertainment, VR can be used to stream live theatrical and concert performances in real time, allowing very large numbers of viewers to experience the event. Drawbacks of using VR include the high cost of the equipment and software and the limited number of scenarios available, say in education settings. It has also been suggested that excessive use of VR may create an over-reliance on the virtual world and impede communication between people in the real world.

13.1.4 Robotics

The study and science of robotics includes the branches of engineering and computer and other sciences that deal with the design and creation of machines that replace or copy human actions. Robotic machines or robots all have features in common: a computer system, a mechanical structure or frame, and a power supply.

A computer system is always included. Computer programs control the robot and determine what tasks are carried out and when the tasks are done. The programs may have user input of commands or may have a degree of independence. A robot designed to paint cars on an assembly line will be controlled by user inputs of command instructions, while a robot designed to explore new or dangerous areas will be equipped with artificial intelligence to make its own decisions. The robots use sensors to react to their environment and actuators carry out appropriate tasks.

The mechanical structure is customised to the purpose of the robotic device. A robot designed for cleaning floors has a structure to support the cleaning tools, power supply, navigation devices and the computing hardware. Night-vision cameras would be found on security robots but not on robotic vacuum cleaners.

The machinery needs a power supply to provide an electrical supply to the components. This is often a portable supply such as a battery or solar energy supply.

Activity 13a

1 Explain why the mechanical structure of a robotic device that freely moves around cleaning floors is different from the structure of a robot that cleans windows by hanging from wires down the side of a building.
2 Describe the information that artificial intelligence needs to determine from the data it receives from sensors fitted to a floor-cleaning robot.

The computer programming of a robotic device depends on its function. Robotic devices can enhance the movements, the accuracy of movement and the strength of humans. Humans have complete control over this type of robot by programming them to do specific, often repetitive, tasks or by remote control over wireless telecommunications systems.

Some robotic devices require human operators to command their tasks, and the robot will 'work out' how to complete the task. The human operator supervises the robotic device in carrying out the tasks. Sometimes, the human operator has to approve or specify any changes of operations, such as a change in actuator during a task.

Fully autonomous robots carry out tasks without human interaction. Tasks range from very complex explorations of areas, where the robot has to interact with the environment and make decisions itself, to relatively simple, repetitive tasks such as fitting parts to cars on factory assembly lines. The use of artificial intelligence to control robots allows the robot to learn from its actions and apply the learning to similar situations. This enables robots to have some autonomy in their actions where human interaction is limited or not feasible. A robotic explorer on a distant planet cannot, with the limitations and delays in direct communications, easily be controlled by a human operator so it must have the ability to make some operational decisions itself.

Activity 13b

Some scientists think that humanoid robots are not necessary.
1 Explain why a robot that looks human in shape would not be suitable for assembling the doors of a car.
2 Explain why humanoid robots are not popular with the public when used for greeting and serving guests in hotels.

Impacts of robotics

Robots can be built to operate effectively in places that are too dangerous or where it is impossible for humans to go. In space exploration, autonomous or semi-autonomous robotic machines can be sent to explore space or other planets without expensive life-support machines or concerns for their safety, and no consideration needs to be given for its return to Earth. Robots can also be used instead of staff in places like hospitals to carry equipment between areas, thereby reducing exposure to disease for staff and patients, as was done during the Covid-19 pandemic.

Robots with actuators can be used in artificial limbs (prosthetics) that enhance the quality of life for people who have missing limbs. Prosthetics enable athletes with missing limbs to take part in sports that were previously closed off to them. Prosthetics that include computing devices to analyse and control the shape and movement of the limb can make movements and appearance more natural.

The manufacturing industry has been impacted the most by the introduction of robotics. Robots increase productivity because they can carry out tasks very quickly and accurately, such as the assembly and testing of components. They are also more efficient than human workers because robots can work continuously without breaks or vacations. The numbers of workers required to produce goods is therefore reduced, so a manufacturer will reduce the costs of employment. However, companies may have to employ specialised workers or technicians to maintain and repair the robots. These technicians will be expensive as they need to be trained and qualified in the use of the robots. Also, workers who work with the robots, for example ensuring that the raw materials are in place, have to be trained and skilled in the use of robots to make goods. While using robots can make working environments safer for humans, robots cannot carry out tasks that they are not programmed to do. They cannot be creative, cannot respond to unexpected conditions and can create hazards that were not present before, for example fast-moving heavy robot arms can be hazardous in factories.

▲ **Figure 13.6** Robots in car manufacturing

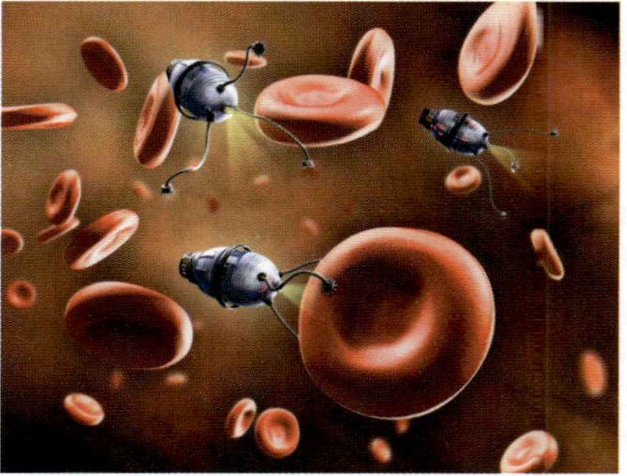

▲ **Figure 13.7** Nano robot with camera for passing through blood vessels

In health care, robots have a wide range of uses. Autonomous robots carry out tasks normally done by health care workers such as nurses, like taking temperatures, and can move between patients. This was especially useful during the Covid-19 pandemic, where these autonomous robots supplemented patient care when health care workers were in short supply or needed elsewhere. Surgeons use robotic tools for operations all over the body and can even control them remotely from another part of the world. Using small but accurate robotic tools means surgery is less invasive, therefore reducing the recovery time of patients. Health care professionals can even use extremely small robotic devices, micro- or nanorobots, inside the human body to carry cameras and tools to explore and carry out tasks within the body.

> ## Activity 13c
> 1 Explain why autonomous delivery drones may present hazards.
> 2 Explain how autonomous delivery vehicles can assist in health care facilities.

13.1.5 Computer-assisted translation

Computer-assisted translation (CAT) translates one human language into another. **Machine translation** attempts to translate languages without human input and is used as part of computer-assisted translation. It makes use of large, customised dictionaries and tools.

Translation of human languages using only human translators always has the risk of mistranslating words and sentences or altering the context. Computer-assisted translation is less likely to do this since it relies on vast dictionaries and a variety of software tools, has no mechanism or programming to do so, and is overseen by human translators.

Computer-assisted translation uses software tools to convert one language into another. Translation memory uses a large database of words and segments of the source language along with the equivalents in the target language. Translated sections of documents are stored in the database and are reused for subsequent documents. Software **alignment tools** are used to divide the documentation into sections so that the source text and the target text can be kept together and stored for use in future translations.

> ## Activity 13d
> 1 What is the main difference between machine translation and computer-assisted translation?
> 2 State two reasons why a research student would use computer-assisted translation, instead of relying on machine translation, for translating an ancient religious document.

Impacts of computer-assisted translation

Computer-assisted translation is used by companies to adapt user and technical documentation for various devices and products that are sold in different countries, from printers to computer games and simulations. Games can be adapted to suit different languages and cultures by localisation of menus and currencies or removal of culturally inappropriate phrases. The use of CAT improves the consistency of terminology and instructions in the translations, and reduces the need for human input, which reduces costs.

Creative documents such as novels are not easily translated by CAT because CAT is not good with context and expressions in language. For these, humans are required to proofread and correct errors.

CAT is used by global organisations, businesspeople, tourists, diplomats and embassy personnel to overcome the language barrier. Apps for smartphones or other devices can translate words and documents in real time with no need for interpreters.

In health care, CAT has enabled medical records, documentation, data and research information to be made available to health care providers in different countries to enhance the treatment of patients who may be travelling between countries. Research can be shared between doctors and surgeons around the world. Co-operation between doctors and surgeons who do not speak each other's language is made easier with CAT. A drawback is that care must be taken to ensure the accuracy of the translation of technical words and terms.

CAT can be used by law enforcement or in judicial courts when foreign-language speakers are present. However, there is always the danger that an inaccurate translation may lead to misunderstandings and miscarriages of justice.

Using computer-assisted translation has drawbacks. CAT software produces a literal word-for-word translation without any understanding of the context or nuances shown by the languages, and therefore all text translated by CAT software has to be reviewed and edited by a skilled human translator to take account of context, dialect and words with several meanings. Any use of slang or colloquialisms in the source material needs to be translated properly by humans. High degrees of accuracy usually require expert knowledge of the systems by translators. Also, as the CAT software may store translations for future use in memory translation systems, there are security concerns if the documents being translated are confidential.

Activity 13e

1 Why are automatic translations of text for globally available computer games checked by humans?
2 Identify one other aspect that must be checked for global audiences.

13.1.6 Holographic imaging

Holographic imaging uses computer science, electrical engineering and optics to create three-dimensional images. Holograms are recordings of the interference and diffraction patterns in light fields that describe how light moves in every direction through points in space. Holographic images can be viewed from different angles whereas photographs cannot. **Holograms** are recorded using **laser** beams, but photographs can be recorded using normal light.

Creating a holographic image uses two laser beams made by splitting a main beam. One laser beam, the reference beam, is used to illuminate the recording medium, and the second beam, the object beam, illuminates the object to be recorded and is reflected or scattered onto the recording medium. The interference pattern created by the two beams is recorded. Computer-based holography overcomes some of the problems with photographic holography because specialised photographic recording media are not required and computers can create the optical wave patterns from digital data.

Impacts of holographic imaging

Because holograms are very difficult to copy and forge, they are printed onto labels on goods, bank notes and bank cards, for example credit and debit cards. Counterfeiters making copies of original goods cannot

▲ **Figure 13.8** A hologram on a bank note

copy the holograms properly, so holograms are a useful way to tackle fraud. The use of holographic imaging on labels reassures buyers that they are purchasing original goods and not cheap copies.

Holographic imaging is used in many branches of medicine and health care. X-ray holography produces images of internal body structures that can be viewed in 3D, such as the eyes or skeleton, and anatomy can be taught using such images. Holographic images can also be created to show 3D views of healthy and diseased organs. Doctors and surgeons can use them to diagnose, plan and carry out treatments and surgical procedures with greater accuracy.

> **Activity 13f**
>
> Explain why holographic labels are not often counterfeited.

13.1.7 Holographic and fourth-generation optical data storage

Optical data storage systems were first introduced to store music and computer software. The compact disc (CD) and MiniDisc were the first generation; the digital versatile disc (DVD) and a high-capacity MiniDisc (Hi-MD) were the second generation. These replaced vinyl records and tape systems. Third-generation optical storage is used for high-definition movies, which are distributed as Blu-ray discs.

Fourth-generation optical discs can store much more data because they use holographic imaging. **Holographic data storage**, or three-dimensional data storage, can store up to hundreds of gigabytes and has faster data-transfer times because it uses the depth of the recording medium, unlike CD and DVD optical disc storage systems that use only the surfaces. A CD can hold about 783 megabytes and a DVD can hold up to almost 16 gigabytes of data, but holographic data storage offers capacities of 300 gigabytes or more per disc. Capacities of several terabytes per disc are hoped for in the future.

Holographic data storage uses two laser beams: a signal beam and a reference beam. The signal beam is shone through a liquid crystal display to produce the binary information, which is then shone onto the recording medium. The second beam (the reference beam) is guided along a separate path onto the recording medium, and where the two beams meet they create an interference pattern. The interference pattern is recorded as a hologram of the data at different depths in the layers of the medium.

Impacts of holographic and fourth-generation storage

The massive amount of storage available on holographic and fourth-generation optical storage systems can provide the capacity required for storing things like holographic images and the long-term safety of the data. This has some advantages for storing medical images, computer game data and business data.

Holographic and fourth-generation optical storage systems have a projected life of over 50 years, compared to a typical life of 5 years for magnetic storage devices, so medical data can be stored for a very long time. They have very large capacities and very fast access rates, with transfers approaching 40 times that of DVD storage. Low power consumption of these storage devices means that they can be used in small, portable devices suitable for use in health care facilities. These features make this form of storage highly suitable for the archiving of medical data.

The cost per gigabyte of storage is much higher than that of other systems such as magnetic hard disks, solid state devices (SSDs) or external USB storage devices. Moreover, the development of reliable, high-bandwidth internet connections into homes, enabling video streaming and online gaming, has made the need for expensive high-capacity data storage systems for movies or computer games unnecessary.

A more important drawback to the use of holographic systems is that external ultraviolet light sources can erase the data on some CD-RW and DVD-RW discs if exposed to sunlight. The UV light acts on the media in the same way as the recording laser, meaning data can be corrupted and lost.

Activity 13g

Explain why users are advised not to place rewritable CDs, DVDs and Blu-ray discs in direct sunlight but the same advice is not given for commercially produced CDs, DVDs and Blu-ray discs.

13.1.8 3D printing

Three-dimensional (3D) printing uses sets of instructions sent to a 3D printer to add successive layers of materials to create three-dimensional objects. 3D printing may use polymers, metals and ceramics to create the objects. It is also called **additive manufacturing**, in contrast to **subtractive manufacturing** where materials are cut away to create objects.

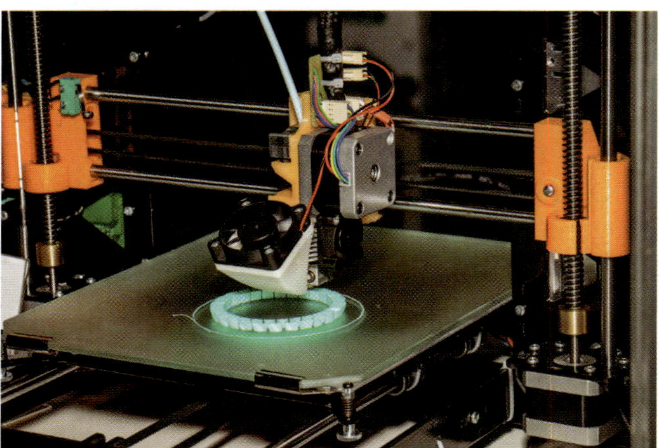

▲ **Figure 13.9** 3D printer

To create an object by 3D printing, the instructions for the printer have to be written. Computer-aided design (CAD) software is used to create a model and export the required set of instructions. Also, a digital camera can be used and software using photogrammetry techniques analyses the images and extracts the data needed to create the instructions.

3D scanners can be used to scan original objects from many different angles and create digital representations. The instructions for creating the digital object can be exported to CAD to produce a copy of the original object.

Activity 13h

Explain the difference between additive manufacturing and subtractive manufacturing.

Impacts of 3D printing

3D printing has a wide range of applications. Manufacturing companies use 3D printing to produce both the tools and the parts needed for manufacturing goods, such as cars or domestic appliances, as well as spare parts for manufactured items. In the car manufacturing industry, underbody chassis structures and fuselages can be printed and made as one continuous structure to avoid the need for complex joins and welding of several parts. The aviation industry manufactures parts for aircraft using 3D printing because it produces less waste and reduces the manufacturing costs; 3D-printing can produce the parts to exact requirements as and when required anywhere in the world without the need for keeping vast numbers in stock. In construction, concrete can be laid in layers by 3D printing, as well as 3D printing being used to create walls, doors and windows.

In health care, 3D printing can have extensive uses. Patients can now have customised, and robotic, computer-controlled prosthetics that fit perfectly and vastly improve their quality of life. Hip and knee replacements can now be 3D printed to fit exactly to existing bone structures and created out of materials that are inert in the body, increasing comfort and reducing pain. Body movements and functions can be almost completely restored. Additionally, 3D printing has enabled health care professionals to create specialised surgical and medical instruments with great precision, including recreating body parts to plan, and to practise and train for surgical procedures without touching the actual patient.

3D printing can be used in tissue engineering. Tissue engineering combines a base material, the 'scaffold', with cells and other biological materials to make working biological tissues. Some tissues have been engineered successfully, but fully working biological organs, such as hearts and kidneys, have not yet reached the stage where they can be implanted into bodies. 3D-printed very thin layers produce the scaffold for cells to grow on, for example to grow artificial skin for grafts to treat skin burns. The thin layers are 3D printed using a 'biological ink' made from blood plasma and skin cells. In a similar way to making artificial skin and bone, artificial blood vessels can be 3D printed. Using biological polymers, a base can be printed with small holes and channels along which cells can be arranged and grow. 3D printing may overcome the problems that exist in making extremely small blood capillaries, because polymer strands are not fine enough and the live cells do not arrange themselves as in real live tissue.

Hospitals, drugstores and pharmacies will be able to 3D print customised drugs for individual patients. Individuals may one day even be able to print their prescription drugs at home, making customised drugs on-demand, using data files provided by their doctor or pharmacist. However, there are drawbacks. The home user must have access to a specialised 3D printer, have access to the required files to instruct the printer, and have a supply of the required chemical constituent of the drugs. The drug supply to the individual is not under the control of the medical professionals and mistakes in dosage or use may have an adverse impact on the individual's health.

Beyond health care, sculptors and prop-makers in the film and theatre industry use 3D printing to make quick, cheap props. In law enforcement and forensic pathology, evidence can be recreated for study by scientists. In a similar way, archaeological artefacts can be reproduced for study purposes. Scanning fragile artefacts and making 3D-printed copies allows scientists to study them while protecting the artefacts themselves from damage or loss.

3D printing removes the need for the skilled, complicated and time-consuming chiselling, milling and grinding of shapes of new parts and items during development stages. New shapes and objects can be formed quickly and relatively cheaply by printing compared to hand- or machine-tooling methods. Iterations are quicker and cheaper to make. There is no need to wait for the creation of new expensive tools or moulds that are then discarded.

Activity 13i

Explain why governments do not want prescription drugs to be 3D printed at home by patients.

13.1.9 Vision enhancement

Vision enhancement technology and devices aim to improve the quality of life of people with vision impairment. Vision enhancement technology can also assist other groups of people, for example those in health care, and assist drivers of vehicles and the military in their operations.

Impacts of vision enhancement

Combining technology can provide remote assistance to people who are completely blind. Headsets connect blind people to an operative who sits at a computer screen that displays the view taken by cameras in the headset and guides the wearer around their environment. The details are then communicated to the wearer with a small earbud that speaks out the details in real time. Using AI operatives would enable the visually impaired user to be more independent.

▲ **Figure 13.10** Vision enhancement glasses for macular degeneration

For people who have some vision, visors (**smartglasses**) are available that cover the eyes and display images and information on small screens. These devices use augmented reality to add to the visual experience of the wearer. With 5G or Wi-Fi connections, smartglasses can enable visually impaired people to take part in video calls and video conferencing.

Implanting devices directly into the human eye can help to enhance vision. A small part of the retina can be replaced by a small electronic array that sends signals to the optic nerve and then to the brain, which 'sees' the objects being viewed. The device in the retina is wirelessly connected to a small video camera that is worn by the wearer as eyeglasses. A computer chip artificial retina that replaces part of the retina with light-sensitive sensors is also used in some patients in an attempt to restore or improve some sight.

Vision enhancement can be useful for people who are not visually impaired too. For example, the use of augmented reality and head-up displays in cars provide drivers with more information, such as GPS navigation, as they are driving. Head-up display warnings are detected and processed more quickly by drivers, resulting in a safer journey. Similarly, head-up displays in the cockpits of aircraft assist pilots by displaying information in their line of view. Pilots of aircraft that fly very fast and are highly manoeuvrable, for example fighter jets, have to gather and process a huge amount information very quickly. The use of 3D head-up displays can provide all the data that the pilots need within their field of view. Additionally, as humans have poor night vision, visual enhancement can improve the ability to see in low light conditions. Night-vision goggles or glasses that use sensors to capture low levels of light and display the results on screens are used by the military and by many others who wish to see in the dark, including workers in the mining sector.

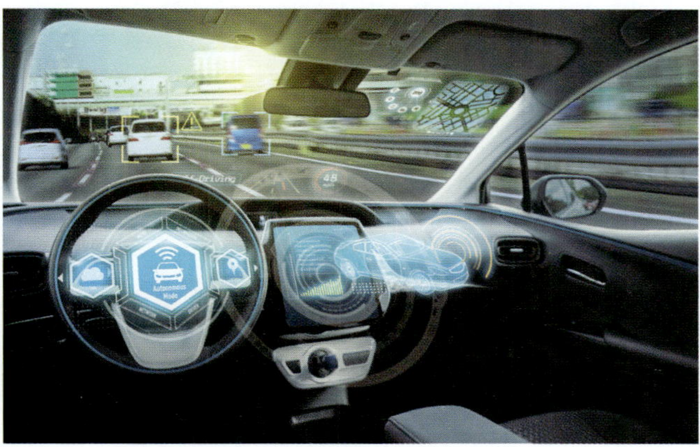

▲ **Figure 13.11** Head-up display in a car

13.1.10 Wearable computing

Wearable computing, or body-worn computing, refers to the wearing or attaching of computing devices onto the body. The devices may be small, for example fitness trackers, an Apple watch and head-mounted displays like Google Glass™, or larger electronic systems that help in manufacturing environments. Wearable devices are usually worn on the wrist, hung around the neck, worn as eyeglasses or strapped to the body in some other way. Referring to smartphones and personal digital assistants (PDAs) as wearable devices is not accurate because smartphones are carried rather than worn.

Impacts of wearable computing

Wearable computing provides useful information or services to the wearer. For example, fitness trackers can sense movement and activities and monitor heart rate and distance travelled; smartwatches often combine fitness trackers with functions from other devices like smartphones and media players. Smartwatches are more portable than a tablet or smartphone and can be worn on the wrist. However, they are usually too small to allow easy use with fingers, meaning a smartphone is often still required. Moreover, the battery life can be short and unpredictable and the connectivity raises security issues. The personal data stored on the smartwatch may become accessible to unauthorised users.

Activity 13k

1 Explain why a smartphone or tablet is not considered to be a wearable device despite being carried by users.
2 Describe how a smartwatch can assist in preparing an athlete for a race.

Smartglasses or head-mounted devices have operating systems built into them and work almost independently of other mobile devices. Video screens, sensors, wireless communications systems and internet capability may be included, but some features may depend on the smartglasses being connected to a smartphone. A modern smartglass system consists of a wearable frame with a pair of video screens that can display images but can be seen through. An embedded computer system provides the functions, the wireless connectivity and in some cases access to the mobile (cell) telephone networks. Smartglasses are designed to add data and information to the views seen by the wearer.

Wearers can interact with and control their smartglasses using a connected smartphone, physical touch by buttons or touchpads fitted onto the frame of the device, gestures that are recognised by the glasses, or voice recognition systems that respond to spoken commands. Eye tracking can also be used where movements of the eye, the focus or direction of a gaze can control the glasses' functions. Most systems aim to enable a completely hands-free operation of the smartglasses.

Smartglasses can use augmented reality systems to enhance the wearer's experience of the real world or to assist in carrying out tasks. Navigation systems enable wearers to find their way to locations, identify locations and have additional information shown to them. Information is superimposed on to the view seen by the wearer and shows details of the objects in view, identifies objects, adds information about landmarks and applies facial recognition to viewed people to identify them to the wearer. Law enforcement officers can use smartglasses as body cameras to view and record incidents and people. Images of faces can be sent to a central database and compared to facial recognition software to identify suspects, retrieve their details and track the movements of individuals.

Wearable devices can be worn by ill or elderly people to monitor them continuously. Data can be collected and sent to health care providers or to emergency services by wireless or mobile networks. Responses to changes in the patient's condition can be very fast and, in the case of an emergency, because the device will also show the location of the patient, first responders can be sent very quickly. Devices can allow patients to monitor body parameters, such as blood sugar level, to easily decide if they need to eat or take medication. Similarly, devices worn by children who experience seizures at night can quickly alert parents if the child is in distress. Smartglasses are used in health care to augment the view of surgeons during procedures and to assist patients in their recovery.

Activity 13l

Explain how wearable computing devices communicate with other devices.

13.1.11 Blockchain technology

A blockchain consists of a list of records, known as 'blocks', that grows with every transaction. The blocks are linked to create a chain.

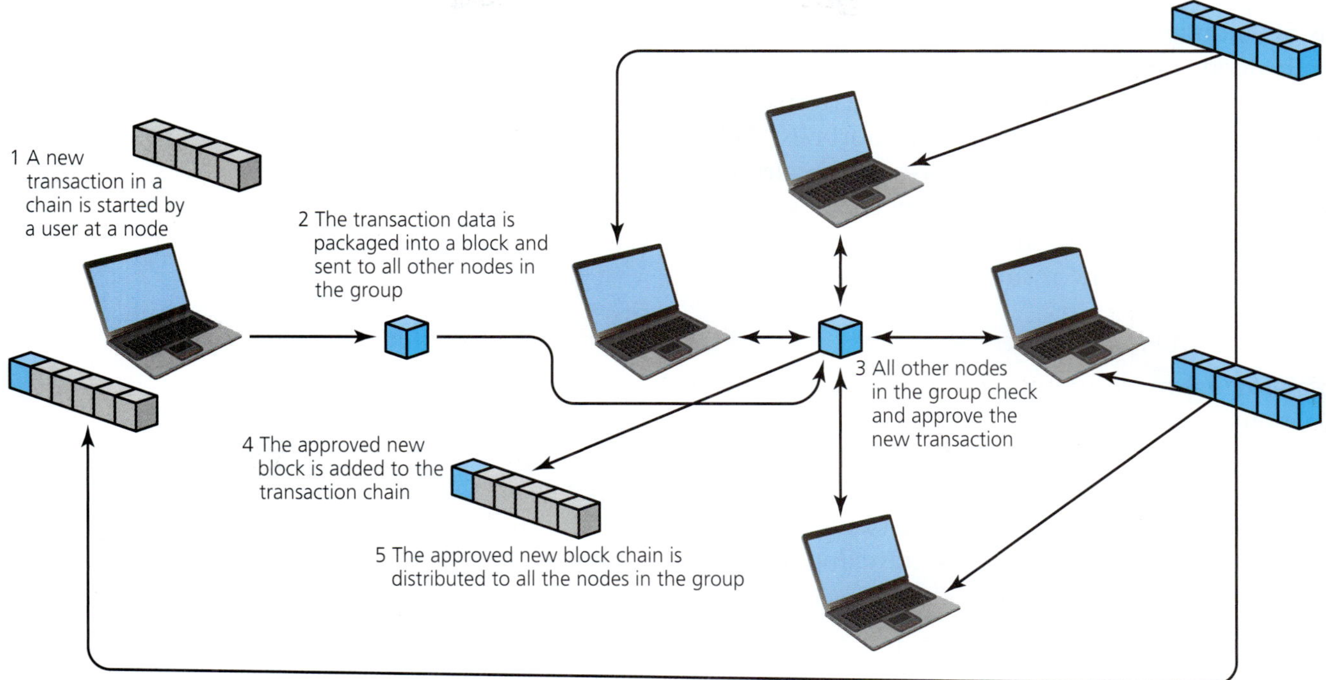

▲ **Figure 13.12** How a blockchain is made

Blockchains are created using cryptography and, because each block contains details of the previous block, a timestamp and a record of the transactions, they cannot easily be altered once they are made. This makes blockchain systems very secure since any alterations mean that all the other blocks also have to be changed. Altering blockchains after they are created is difficult as it requires all subsequent blocks to be amended and needs the consent of all other creators of the blocks. A blockchain system allows the integrity of the previous blocks to be confirmed all the way back to the original block in the chain.

Blockchain technology uses a decentralised record or ledger of transactions on a peer-to-peer network. It is used in cryptocurrency systems and smart contracts, and could be used in digital voting systems and for securing medical records. A smart contract allows for agreements to be made and carried out between two parties without the involvement of a third party to check that each side has fulfilled their obligations. The transactions are set up and processed using blockchains, and contracts cannot be altered without both parties agreeing. Transactions are automatically carried out only when pre-set conditions in the contracts are met. Using blockchains to record votes could stop voter fraud but, while votes could not be altered and would be verified by everyone, individual votes could become visible to all voters. Storing medical records, or other records such as digital identities, using blockchains ensures that the records cannot be tampered with or used by unauthorised persons.

13.1.12 Internet of Things

The **Internet of Things (IoT)** describes the interconnection of devices to exchange data over any communications network. The network does not have to be the 'internet' itself but can be any network, such as a home WLAN or company LAN or WAN, so the term is slightly inaccurate. Devices can be any object that can connect and exchange data, such as a laptop, smartphone, smart watch, smart television set, home appliance, camera and many more. The development of new technologies has allowed the IoT to emerge and develop. Technologies include high-speed data communications systems such as 5G mobile (cell) phone networks, high-bandwidth broadband connections, embedded computing systems in home appliances such as robotic cleaners, room thermostats, lighting systems, external doors, smart speakers and home security systems like smart doorbells. The use of remote-controlled or autonomous robotics in industry or health care is part of the IoT, where high-speed communications allow the devices to exchange data in almost real-time situations so an instruction sent from far away is carried out almost immediately and the result seen almost at once.

Impact of the IoT

The IoT has made an impact in many ways. Individuals can use a smartphone to monitor and control devices in their home. The introduction of everyday devices with embedded computing systems and the use of high-speed **mobile communications** systems linking the devices into the IoT has created smart homes, where individuals make use of their smartphone as a controller at home, from remote locations and when travelling. For example, a smartphone can be used to:

» monitor home security sensors and cameras
» monitor room temperatures
» monitor the contents of a smart refrigerator
» control home entertainment devices
» open and shut doors
» switch lights on and off
» turn on kettles and ovens
» control heating and climate control systems.

Wearable computing devices, such as smart watches and fitness trackers, can connect into the IoT to collect and present data, including text messages from smartphones or health information from sensors worn on the body, to the individual. Smartglasses using augmented reality can provide a continuously updated head-up display overlaying information onto the view of the real world.

Organisations use devices connected into an IoT to:

» automatically load and unload goods, for example from ships and trucks into warehouses
» monitor their products while they are in operation, for example aircraft engines while in flight
» monitor and control remote gas and oil extraction in remote or dangerous areas.

The IoT has enabled devices to be connected together and reduces the costs of operating workplaces with the use of embedded sensors that detect room occupancy. The data from the sensors can be sent to remotely connected microprocessors to control the lighting and climate in the room. There is no need for physical wiring, which also cuts costs. In medicine and health care, devices connected to the IoT allow patients to be monitored remotely, medical and delivery robots to function, and surgeons to remotely carry out or oversee procedures.

13.1.13 Molecular data storage

Molecular memory attempts to store data using chemical molecules as the storage medium instead of using electronic circuitry. While currently in the experimental and development stages, **molecular data storage** promises to eventually be able to store vast amounts of data in extremely small spaces. It may soon be possible to store terabytes of data in only a few cubic millimetres of space. Data is stored in mixtures of customised molecules. These molecules could be long-chain polymers, DNA or much smaller molecules such as aldehydes or carboxylic acids. The data is encoded in several ways. The molecule may hold an electric charge, change its capacitance or change its behaviour towards light beams. In the same way that HHDs use magnetic forces and SSDs use electrical charges, these are reversible so can be used repeatedly to store, retrieve and rewrite data. The technology is in its infancy, very expensive and currently confined to research laboratories.

13.1.14 Autonomous transport systems

The term **autonomous transport system** includes any vehicle or system that moves people, resources such as construction materials, or baggage such as goods, suitcases or luggage, from one place to another with little or no input or control by a human operator or driver. Autonomous vehicles are usually used for human transport or goods delivery and are single, solitary vehicles such as a driverless car or truck, but an autonomous transport system consists of numerous vehicles that are interconnected and can communicate with each other and with a central control system. Autonomous robot vehicles can deliver goods to homes or deliver materials around, for example, a factory or hospital with very little human intervention.

Autonomous vehicles, such as driverless cars, are equipped with sensors to provide data that can be used by the computer system to 'build' a 3D model of the environment and surroundings in which it is operating. The sensors include cameras and detectors for the two systems that are found in driverless cars:

» millimetre-wave radar
» light detection and ranging (LiDAR).

The autonomous vehicle uses computer systems, often including AI, to analyse and act upon the input from the sensors and learn how to react to the data, and uses a navigation system, which may be GPS, to travel between pre-determined locations. The system must be able to react extremely quickly to previously unknown occurrences and to learn from them, so the use of AI will enable the vehicle to avoid people, objects and dangerous situations.

The use of autonomous vehicles can significantly reduce the number of collisions between vehicles so increasing passenger safety, enable people who are unable to drive to travel independently of others and so reduce their social isolation, and enable passengers to do other tasks while travelling.

Autonomous transport systems reduce traffic congestion while increasing traffic throughput, so there are fewer traffic jams but more cars, buses or trains passing through, and may reduce the use of land for more roads or parking facilities. Robotic vehicles may increase the speed and reliability of delivery of goods. The impact on the environment may be positive as there will be less pollution and less power may be needed for the newer generations of vehicles.

13.1.15 Energy from wireless signals

Electromagnetic waves, including those that carry Wi-Fi signals, have energy that can be captured, turned into DC (direct current) electricity and used to power devices. **Wireless transmission of power** is not new; it was first demonstrated by Nikola Tesla in 1899, but it has not often been used. Cordless toothbrushes and mobile phones can be charged using wireless methods, but few other devices use it. The charging methods used for these require the devices to be in close contact with the charger. The benefit of wireless charging here is really only safety and convenience. Charging from wireless signals enables charging at greater distances. As the IoT grows and more devices, all needing to be powered, are connected, wireless charging with energy extracted from 4G, 5G, Wi-Fi and other wireless telecommunication signals is becoming practicable. The location of the device may mean physical cables are difficult to install, solar energy may not be available, or batteries are difficult to replace, so the devices will be made to power themselves via wireless. Also, devices inserted inside the human body, such as heart pacemakers or medicine-release systems, may be powered by the wireless signals that pass though human bodies all the time.

13.1.16 The impact of new and emerging technologies on scientific research

New and emerging technologies have changed and enhanced the way scientific research is carried out. Computing technology can be used in observation, experimentation and analysis of data to produce hypotheses that are then tested in scientific research.

Observation using augmented reality systems, artificial intelligence and robotic devices can add to the data collected in traditional ways. Image enhancement can remove imperfections and compensate for missing pixels in the data, and the addition of false colours to images helps humans to understand the images. Artificial intelligence is used in data mining. It can analyse and collate massive amounts of data found in scientific research publications in short time periods. It can discover trends and patterns that help to discover many things, such as new materials for construction and engineering, new chemicals for industry to make products such as cleaning materials, and new drugs for combatting disease. AI can also analyse images, like those of stars and galaxies from telescopes collected by space observations, to discover patterns to aid the research. AI requires high-performance computers, cloud computing, vast amounts of data storage and high-speed communications systems to mine and analyse data from all over the world and from space-based research tools.

In research for new medicines, robots have been used to mix together chemicals to find out what new compounds can be created. Robots can speed up the work, work repetitively and more consistently and produce results quicker. New drugs can become available sooner. Remote control robotic devices or robot arms can be used to handle hazardous or dangerous chemicals and materials without putting the scientists at risk.

Scientific research documents placed on the internet or accessible cloud storage are automatically translated using computer-assisted translation into different languages, making them available to others all over the world. Human corrections may have to be made as CAT is not ideally suited to technical documentation, but the time taken to translate the documents is much reduced. This makes research available more quickly to other scientists for review, comment and use.

Holographic imaging is used to visualise objects in three dimensions. When used with virtual reality, it can enable scientists to move around and inside objects in ways that are not physically possible. Chemical molecules, for example new drugs, can be examined and researched. Parts of the molecules can be moved to find out the effect of doing so. Biological structures can be digitised and the images studied.

Scientists use holographic and fourth-generation storage because their research can generate huge amounts of data and this must be kept safe. Current holographic and fourth-generation storage is expensive, but it may be the only choice for saving and archiving massive quantities of data for future study. The reduced data-access times of holographic and fourth-generation storage are useful for accessing and analysing the large amounts of data.

3D printing of scanned and digitised objects, for example fossils, enables the study of objects that are too fragile to touch. The printed models can be handled and moved without fear of damage to the originals. Multiple copies can easily be made available to different scientists or the 3D printer files sent across the world for printing.

Vision enhancement, with smartglasses, headsets or display screens, can allow scientists to study very small or very distant objects, or objects that reflect light outside the range of human vision. Robotic devices with cameras can enhance the images in various ways to study the environment in which the robotic device is operating.

> **Activity 13m**
>
> Explain why scientific research often requires the use of high-performance computer systems.

13.1.17 The impact of new and emerging technologies on the environment

E-waste and recycling

E-waste, or electronic waste, is discarded electronic devices. The discarded devices can be recycled, reused or dismantled. **Recycling** of complete devices often means they are reused by others without being dismantled for parts. Dismantling and recovering parts and components to be used in new devices is true recycling. Gold, silver, nickel and aluminium, used in circuitry and connections, are not usually directly hazardous to health, but high levels of them are to be avoided. Cadmium, lithium, manganese, copper and lead, and many others, are used in electronic components, circuit boards and solder joints in varying quantities, and all can cause environmental damage. Components and parts that cannot be reused or recycled, either because it is not economic to do so or because they are no longer suitable, have to be disposed of. Rare, dangerous and other metals are often left in the discarded waste. Eventually these metals will leach into the environment and water supplies and may cause damage to crops, animals and human health.

Polymers and plastics used to make cases, supports, covers, wiring and switches are difficult to reuse and recycle. These remain in the environment for many years without degrading. When plastics are burnt or do eventually degrade, they may release noxious compounds or elements into the environment. Chlorine is used in the manufacture of, and is found in, some plastics and its release is harmful to life.

The disposal of e-waste can be by mechanically stripping the material from the device, or by burning the materials and attempting to recover some materials from the ashes. Burning destroys the components and releases noxious fumes into the atmosphere unless carried out safely and carefully. Industrial containment and safe plants are required to prevent contamination of the air, water supplies and surrounding lands.

Vast quantities of components and parts that cannot be recycled or recovered are shipped from one country to another for disposal and may end up buried in landfill. Sometimes, this may be because other countries have better disposal facilities, lower costs or fewer regulations regarding waste disposal. However, the shipping of e-waste around the world can add to the environmental damage.

People living near recycling facilities, or near to landfill areas where e-waste is dumped, may feel the effects of hazardous materials, and may become ill or suffer other health problems. Exposure to hazardous e-waste has been known to affect unborn children.

Power consumption

As previously discussed, all new and emerging technology requires power to work. The increase in computing power demanded by new technologies used in manufacturing and by businesses, organisations and individuals increases the demand for electrical power. The incorporation of microprocessors, internet capability and communications systems into new areas, for instance home appliances, telecommunications with mobile devices, health care, scientific research and many other areas, has led to further increases in demand for electrical power.

Manufacturing processes

The use of new and emerging technology in manufacturing has reduced costs, increased productivity and increased the consistency and quality of goods. However, there has been an impact on the environment. New technologies in manufacturing often require more electrical power as they run continuously through the day and night. They use raw materials that are already scarce, for example rare metals, and that require extensive mining to extract. New technologies in manufacturing also create waste materials that are difficult to dispose of without considerable cost, and create goods that do not degrade so their components remain in the environment for many years. Plastics are a major component of many new technologies, for example robotic devices that are used in many manufacturing processes.

Beneficial impacts are that manufacturing waste has been reduced, goods last longer so do not need to be replaced so often, working environments are safer for workers, and workers do not need to work very long hours to produce as many goods so they have a healthier lifestyle. For example, using 3D printers in a manufacturing process can reduce overall production times because there is no need for complex milling or drilling of complex shapes, but only the removal of the support materials and final clean-up. This produces longer-lasting goods in a safer workplace. Waste is reduced and there is less power consumption. Robotics also reduces waste as there are fewer errors in production and goods are more consistently assembled. Using robotics with AI can help to detect production problems, for example by more quickly recognising and comparing the production items with pre-set values of perfect objects, so fewer imperfect goods are made and then rejected.

Practice questions

1 Describe how analysing social-media posts can be used to target users with advertisements. (5)

2 Describe how shoppers could make use of augmented reality (AR) when buying clothes in a store. (6)

3 Explain why surgery from remote locations using robotic devices is more risky than surgery carried out by a surgeon in person. (6)

4 Discuss the impact of wearable computing devices in health care. (6)

5 Discuss the benefits and drawbacks of using robotic devices to manufacture automobiles. (8)

6 Discuss the benefits and drawbacks of the use of computer-aided translation by a company that produces TV sets for the global market. (6)

7 Describe the impact of holographic imaging on commerce. (6)

8 Describe how the disposal of computing devices that are old and no longer usable can pose a danger to the environment. (6)

14 Communications technology

Before starting this chapter, you should:

★ be familiar with the terms: component, peripheral, internet and internet-based services, hardware, software, types of computing device (for example desktop and laptop computers), tablets and smartphones, networks and associated types and components, malware and backup.

14.1 Networks

14.1.1 Types of network

A computer network connects computing devices, called 'network nodes', together to enable the exchange of data and access to shared peripherals, for example printers or storage devices, the world wide web, the use of internet-based services such as email, messaging, cloud storage systems and video streaming, and many more.

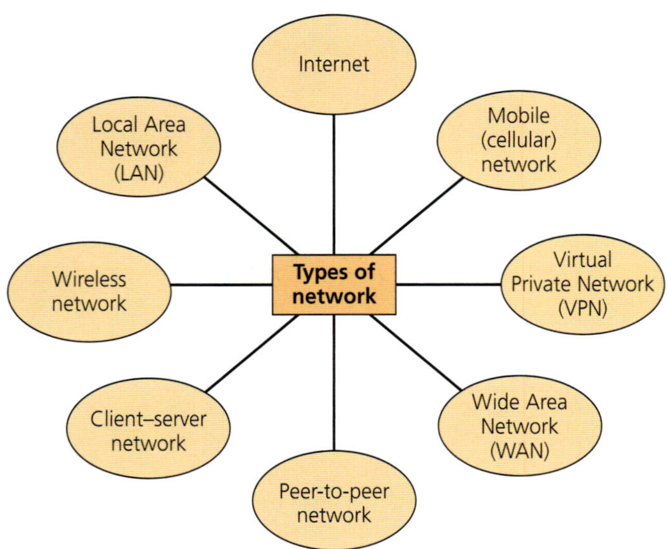

▲ **Figure 14.1** Types of network

Local area network

A **local area network (LAN)** is an internal network that is restricted to a specific area or organisation, such as a school or college campus, a home or a business, and can only be accessed by users belonging to the organisation.

LANs communicate at **layers** 1 and 2 in the **Open Systems Interconnection (OSI) model**, which is a conceptual model that describes networking in seven layers. LANs use, for example, **ethernet** technology, which is designed for communication over relatively short distances compared to WANs. LANs do not normally use public telecommunications systems but may be interconnected into a metropolitan area network (MAN) to create a larger LAN that covers, for example, a city for use by city officials in different buildings.

Characteristics of LANs

LANs can have various network topologies connecting nodes together using physical cables, wireless connections and optical connections, and are characterised by covering a relatively small geographical area of usually only a few kilometres. This can be extended, for example by using VPNs, between buildings or campuses to cover much larger areas.

Data transfer speeds over LANs can be very high, often in excess of hundreds of gigabits per second. The maximum possible data transfer rates are rarely achieved because some capacity is taken up by the processing of the protocol headers, error corrections, packet collisions and media imperfections.

Installation of LANs is relatively easy and has a lower cost than WANs because the area to be covered is smaller, and maintenance is easier because the connection media and devices are similar and compatible. The scalability of LANs is good because they can be extended quickly and easily by adding new devices and transmission media.

Uses of LANs

LANs are used by schools, colleges and other educational organisations to connect together computing devices, such as desktop computers, laptops, tablets and smartphones to enable the exchange of data such as images and videos, files and applications and other digital resources. Peripherals such as printers, scanners and networked storage systems can be shared by network users, which reduces the number, and the running costs, of peripherals.

Networked resources, courses and teaching materials can be shared and made available on the network. Teachers can be sent work by students to assess and can send back the results. Businesses, organisations and governments use LANs to connect computing devices to share and exchange data, information and resources. Company files, house-style documents and templates can be centrally stored and accessed by employees so that all company documentation has the same style, layout and appearance.

Protocols used by LANs

Using protocols specified by the Institute of Electrical and Electronics Engineers (IEEE) ensures that networked devices can connect to others anywhere in the world. Ethernet is a networking technology that divides data into frames (packets) that carry the source and destination addresses and basic error-checking data, for transmission over wired connections. Wi-Fi technology does the same for wireless connections. Ethernet and Wi-Fi are the standard technologies for use in LANs and WANs. They work at the lower layers of the TCP/IP and OSI models, and their frames carry the TCP/IP packet information. The TCP/IP model is a conceptual model that describes networking in four layers. The **network protocols** used on LANs are discussed later in this chapter in Section 14.6 'Network protocols'.

Advantages and disadvantages of LANs

▼ **Table 14.1** Advantages of LANs

Advantages of LANs
• LANs reduce the number of peripherals required so there is a reduction in the initial costs of purchase and the running costs of peripherals as there are fewer to maintain. • Users may have access to a wider range of peripherals and can choose which printer (inkjet, laser, larger paper size) to use for a particular job. • Data, files and documents can easily be shared between users, and company documentation and templates, for example house styles for documentation, can be stored centrally for access by employees. Centrally stored data can be kept safer and more secure by automatic scheduling of backups and archives, and authentication techniques can control access to it. • Central storage can provide an application server that is easier to maintain and update than having multiple copies of software applications installed on individual computers. • LANs allow communication, for example by email, so that messages, documentation and notifications can be sent to many users, saving the time and cost of duplicating individual messages. Employers can control and monitor what employees are sending and receiving over LANs. • Users on a LAN can share a single access point to the internet, where the data leaving and entering the LAN can be controlled and monitored. • Shared internet connections from LANs reduce the number of Internet Protocol (IP) addresses that are needed, so one **IP address** is shown to the internet and hides the IP addresses of the LAN devices from public view. This increases privacy and security. • A LAN reduces the time and costs of the maintenance of computers, other devices and software used on the LAN. New software, software updates and other administrative tasks can be centrally organised and distributed. • LANs in a home allow members of a household to share peripherals, share a single internet access and set up a smart home controlling and accessing household goods and appliances. The use of Wi-Fi to connect tablets and smartphones, and other mobile devices, allows household members to move around the house with their devices and to use the devices from any location in the house.

▼ **Table 14.2** Disadvantages of LANs

Disadvantages of LANs
• The smaller geographic areas covered by LANs can be a disadvantage compared to greater coverage by WANs and mobile networks. Using WANs, mobile technologies and VPNs to extend a LAN adds to the complexity, cost and difficulty of use. • A network administrator, or technician support, is usually needed to diagnose and solve network problems and to service and maintain a LAN. This can add to the costs of running an organisation. It can also create privacy concerns for users as a LAN administrator usually has access to all stored data and information on the LAN. Users should encrypt their data and files, but this adds to the complexity of use and to the time taken to process the data. If users store their data on removable storage to protect it, then this undermines the purpose of having a LAN. Also, they may introduce malware via the removable devices. Internet browsing history and personal emails are also available to a LAN administrator when used on a company LAN. • Malware can easily spread from device to device over a LAN. Computer viruses, ransomware, Trojans, spyware, adware and other malware can be passed from one device to another over a LAN. Data and files can be damaged, lost or stolen more easily if used and stored on a LAN.

Wide area network

A **wide area network (WAN)** is a computer network that covers a much larger area than a LAN. It makes use of telecommunications systems to connect nodes from locations distant from each other, such as between cities, regions and countries. Leased, private data lines between geographically separated offices create circuits on public telecommunications systems to become a WAN for global organisations.

WANs operate at layers 1, 2 and 3 in the OSI model, which specify the structuring, management and control of network traffic and allow the exchange of data over long distances. WANs use technologies such as Multiprotocol Label Switching (MPLS), Asynchronous Transfer Mode (ATM), frame relay and X.25 for data exchange over public communications systems.

Characteristics of WANs

A WAN is distinguished from a LAN by its use of communication media such as public or leased telecommunications systems and network infrastructure that are not owned or run by a single organisation. Typically, a WAN will cover a greater geographical area than a LAN because the connected LANs are far enough apart from each other to require the use of telecommunication lines. The internet may be regarded as a wide area network that uses public telecommunications systems.

WANs make use of different transmission media including the various types of copper cabling, wireless, fibre-optic, microwave and satellite communications systems, often for very long-distance connections.

A WAN uses switching and **routing** technologies to direct the data from the sending device to the receiver. Data travels across the WAN in packets, and is stored on intermediate devices and then forwarded to the next device (called the 'store-and-forward' method of transmission). This is one of the reasons that sending data across WANs may not be as secure as on LANs. Encrypting the data is a solution to this problem.

More **data transmission** errors occur on WANs compared to LANs because the longer distances involve many interconnections and differing technologies. These add to the background noise that interferes with the data transmission and can cause delays in the propagation of data across the WAN. Video and audio streaming may be seriously affected by propagation delays.

WANs have lower data transfer speeds than LANs because the different transmission media used for the connections require conversions between different protocols; the requirement for **routing protocols**, the different transmission media and the number of intermediate devices reduces data transfer speeds. WANs can transfer data at about 150 to 200 megabits per second, whereas LANs can reach rates of hundreds of gigabits per second.

Uses of WANs

Wide area networks are used to connect computing devices into a network covering a large geographical area, for example several countries. Businesses, organisations, educational establishments and governments use WANs to exchange data between computers in their remote offices around the world.

The internet can be considered to be a WAN, and is used by people all around the world to exchange data, share files, use email and other social-networking platforms, and search for information.

Protocols used by WANs

Wide area networks use High-level Data Link Control (HDLC), Point-to-Point Protocol (PPP) and frame relay to create the frames (packets) that are transmitted across a WAN, in the same way as Ethernet or Wi-Fi protocols create the frames for LANs.

HDLC requires little configuration and is used by **routers** to connect LANs into a WAN using leased telecommunications lines. A router on a LAN uses HDLC to repackage packets from the LAN and send them along the line to a router on another LAN. The packets are converted back for use on the other LAN. HDLC provides error correction for data sent over WANs.

PPP is based on HDLC and is also used to connect LANs via leased lines. It is customised for use on dial-up connections into WANs or into the internet. It authenticates and checks the quality of a dial-up connection.

Frame relay is used in **packet-switching** networks such as an Integrated Services Digital Network (ISDN) and in permanent virtual circuits (PVN) and enables data to be transferred across public telecommunications systems. Frame relay can be used to set up a PVN across public telecommunications systems so that a user appears to have a private connection to the WAN or the internet. It can also be used to connect a home LAN, via a router, to the internet via broadband.

Frame relay works at the layers 2 and 3 – the Data Link and **Network layers** – of the OSI model and puts data into variable-sized frames (packets). Frame relay does not provide error correction, so it can be unreliable. Devices using frame relay are expected to provide any required error-correction procedures. It can be set up to allow a quality of service (QoS) configuration to prioritise, for example, video streams.

Multiprotocol Label Switching (MPLS) is a routing system used on WANs to carry packets of data using other protocols, such as frame relay. It uses short labels instead of long network addresses to identify nodes and passes the packets quickly between nodes. Its use can increase data transfer speeds.

Advantages and disadvantages of WANs

▼ **Table 14.3** Advantages and disadvantages of WANs

Advantages	Disadvantages
• For a company with locations in different areas, connecting LANs into a WAN means that resources can be centralised at head office or in a **data centre**, making the maintenance, updates, backup and archiving of resources simpler. Users can access all of the company resources regardless of their physical location. • The use of private, leased lines or **virtual private networks** means that company data does not pass openly over the public telecommunications networks so is kept private and secure. • WANs can also be used to carry voice calls, using VoIP, so the costs of telephone calls between office locations can be reduced. • Conference calls, and video and web-conferencing, can be made using WANs and the discussions can be kept entirely within the company.	• WANs have high costs of initial set-up because locations are far apart. Technicians setting up WANs may have to travel long distances several times to ensure the correct operation of the WAN. • The gathering of data and resources into centralised locations, such as data centres, may increase the security of the data at the data centre, but can reduce the security resources and personnel available at remote locations, due to costs. This may make the data at these locations less secure and more vulnerable. • WANs may have increased maintenance difficulties and costs. • A centralised data centre must run continuously with little or no downtime. Administrators must ensure that the data and resources are available around the clock because company locations using the data centres on a WAN may be in different time zones and work at different times. • The maintenance and management of the links and of leased telecommunication lines increase the costs to companies. This cost may be higher when compared to the use of VPNs over the internet where the costs of the telecoms lines are borne by the telecoms operator and, in the case of problems, traffic is automatically redirected.

Activity 14a

1 One of the main differences between a LAN and a WAN is the geographical area that they each cover. Describe three other differences.
2 List the main protocols used on LANs.
3 List the main protocols used on WANs.

Client–server network

A **client–server network** system consists of devices (the clients) that connect to servers providing the services requested by the clients. Clients can be personal computers, laptops or smartphones that request data, files, applications and websites from servers.

A client makes a request to the server, which responds to the client by sending the results of the request back to the client. The client and the server must use the same protocol, operating at the **Application layer** (layer 7) of the OSI model. The data transferred between clients and servers may be encrypted for security. Servers have a scheduling system to prioritise requests and to limit the number of requests from clients. However, very large numbers of requests may result in a **denial of service** (DoS).

Characteristics of client–server networks

Client–server networks use the request–response model. The client requests a service from the server, which gives a response. Clients do not share any of their resources with the servers and rely on the servers to provide services and content as requested. The server does not initiate client–server sessions but waits for a client to request its services. Servers can support many clients at once. Low-powered clients use, over a network, the higher processing power and greater primary and secondary storage of servers. The client has only to be able to act as an interface for the user and to be able to initiate requests to the server.

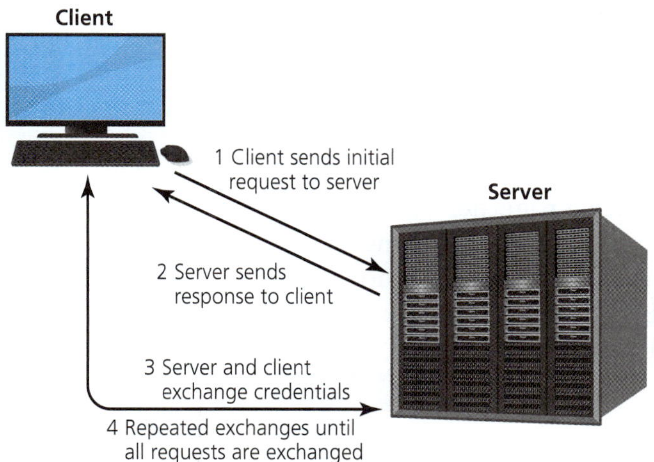

▲ **Figure 14.2** Client–server model of networking

Server time is used efficiently in the client–server model because connections between clients and servers are made only when needed. Client–server networking is not the same as using a terminal to access a mainframe computer. Connections into a mainframe computer from a terminal maintain the connection all the time that the user is logged in at that terminal.

Uses of client–server networks

Web servers store web pages and only send them to web browsers (the clients) when the browsers request them. A series of requests and responses may be required to complete the transfer of all the components of a web page. Similarly, an email client will request a message from an email server when the user logs into the email account, for example from a smartphone. Printing from a client to a networked printer, the upload or download of files between clients and **file servers**, and the posting of photos to social-media websites are further examples of client–server networking.

Protocols used by client–server networks

Client–server networks use the communication protocols of the network (LAN or WAN) upon which they run. Moving files between a laptop and a file server using the **File Transfer Protocol (FTP)** is an example of a client–server network. The exchanges between client and server use protocols that work in the Application layer of the OSI model, making use of application programming interfaces (API) provided by software creators for their applications.

Advantages and disadvantages of client–server networks

▼ **Table 14.4** Advantages and disadvantages of client–server networks

Advantages	Disadvantages
• Client–server networks have central administration where access rights and allocation of resources are managed by the servers, allowing much greater control over who can or cannot access and use the resources compared to peer-to-peer networks. • Files are stored in one central area and can be found and administered more easily by all users than if they are stored on local hard disks. • A backup and recovery schedule of the files on the server is administered centrally and not by individual users, who do not need to take responsibility for this.	• The files or services cannot be accessed by clients or users if the server or network fails. • Managing the servers and **access rights and permissions** can be more complex and usually needs skilled IT technicians, adding to the costs of maintaining the client–server network. • There is a great deal of network traffic when client–server networks are used and this can result in congestion on the network and a perceived reduction in performance by users.

Peer-to-peer networks

Peer-to-peer networking connects nodes (physical devices) together on an equal basis. Nodes may share some of their resources, such as processing power, data storage systems or **bandwidth** with other peers, but there is no central server to control or co-ordinate the sharing. Because resources are shared between nodes, peer-to-peer networking can be seen as a form of distributed computing.

Characteristics of peer-to-peer networks

Most peer-to-peer networks are unstructured and connections are made as and when, and where, they are required. It is easy to quickly create a local peer-to-peer network by this ad hoc approach. Peer-to-peer networks are very robust and resistant to changes in the network, so nodes can join and leave the network without causing significant errors.

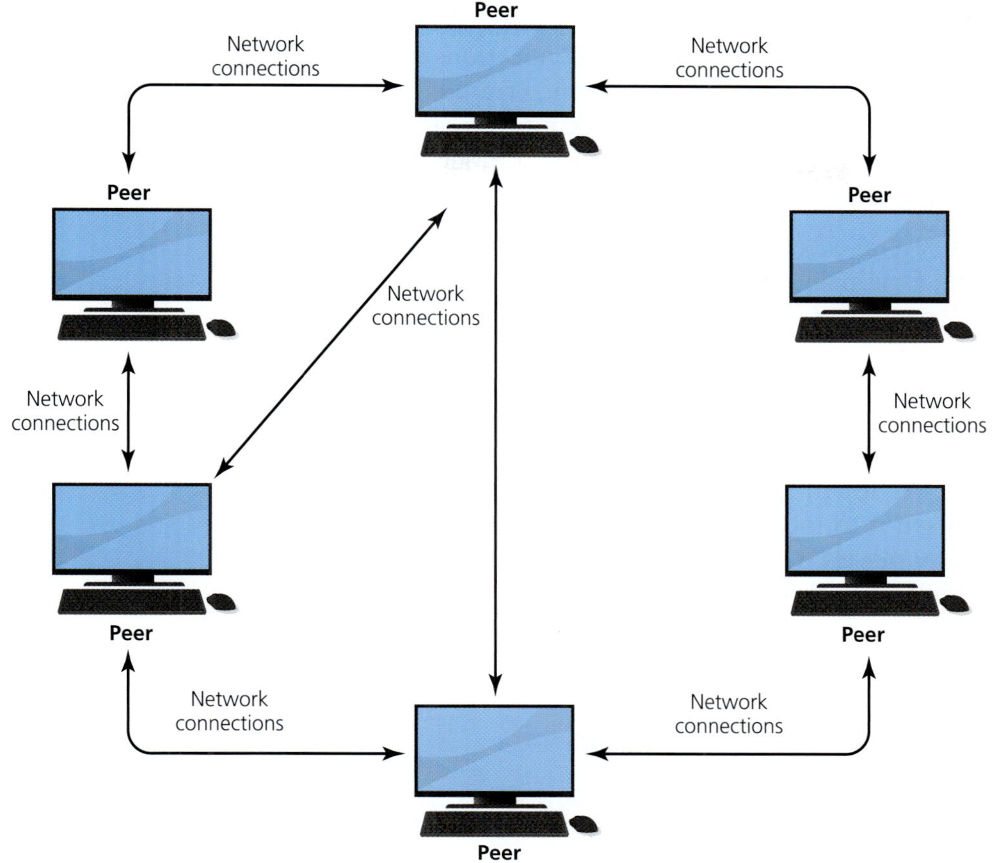

▲ **Figure 14.3** Peer-to-peer model of networking

Some peer-to-peer networks have a structure imposed on them to organise the nodes into a specific **topology**. Each node has to maintain a database of the other nodes around it and how data should be directed across the network. A distributed hash table (DHT) provides a lookup table of all the nodes in the network. Structuring a peer-to-peer network makes searching for resources more efficient in locating other nodes and reducing network traffic. Removing and adding nodes in a structured peer-to-peer network can cause severe disruption to the network until the tables are updated on every node.

Uses of peer-to-peer networks

Peer-to-peer networks are used to distribute and share digital content, such as digital books, music, computer games, videos and images, to users. Users can search, locate and download digital media from several nodes at once. As the number of nodes storing and sharing the content grows, the capacity of the network increases.

File sharing between nodes is a popular use of peer-to-peer networking. Files can be searched for and downloaded from many nodes at once, which increases the speed of download to a node. Once downloaded, the node can make the file available to other nodes. BitTorrent is an example of file sharing using peer-to-peer methods. Large files can be quickly downloaded using BitTorrent. Software companies can use their own proprietary systems to distribute operating systems or applications.

Computer files shared by peer-to-peer networking can be altered by any node and can carry malware. Complex systems have been developed for checking the integrity of shared files.

Protocols used in peer-to-peer networks

Peer-to-peer networking protocols all provide the means for connections to be made between nodes without a central server. Digital currency transactions, for example Bitcoin and Litecoin, use peer-to-peer networking. Bitcoin uses a protocol that encrypts transactions and allows nodes to join and leave the network at any time.

BitTorrent's protocol is used by several client applications to prepare, transmit and request files over a network. Originally, a small text file would be prepared. The text file, called a 'torrent', contained information, for example file names and sizes, about the files to be shared and the IP addresses of the 'trackers' that stored data about the location of peers on the network that had the complete requested file. Computers that held the complete requested file were called 'seeds'. The trackers were computers that stored lists of seeds and together formed 'swarms'. To download a file, a BitTorrent client first obtained the torrent file and used it to locate trackers and seeds. Multiple requests for torrents from many peers could overwhelm a peer-to-peer network and lead to 'query flooding' of the network. BitTorrent now uses a distributed hash table instead of seeds. BitTorrent downloads require less bandwidth, and are often completed in far less time, than downloads using FTP or HTTP.

Advantages and disadvantages of peer-to-peer networks

▼ **Table 14.5** Advantages and disadvantages of peer-to-peer networks

Advantages	Disadvantages
• Using a peer-to-peer network removes the costs and administrative time of having a central server. Devices can be quickly and easily added to or removed from peer-to-peer networks, so a peer-to-peer network can be created or shut down quickly with little configuration needed. • Removal or addition of nodes does not affect the operation of the other nodes. • Exchanging files between mobile devices is quick and convenient for users over peer-to-peer networks, such as Apple AirDrop® for iPads® and iPhones®. • The user's privacy is increased as there is no central record of who is using the network and what it is being used for.	• Peer-to-peer networks can be insecure as each device has to manage its own security. • There is no central storage of data so archiving and backups of data and files have to be carried out by individual nodes. • No central server means that there is no central record of the users or of what data and files are being exchanged. Illegal activities are difficult to discover and prevent when users use peer-to-peer networks. • Multiple requests from and to nodes, or searching for resources, create a high amount of network traffic. • Multiple requests also increase the processing required by the nodes, with the result that requests often fail and nodes do not find what they are looking for.

Activity 14b

Explain why the lack of central control in peer-to-peer networking has drawbacks.

Virtual private networks

Virtual private networks (VPNs) allow nodes to connect to private LANs using public telecommunications systems, so users can use them to exchange data as if they were connected directly to the private LAN. It appears to the user that the private LAN is being extended across a public telecoms network to wherever they may be located. Because VPNs can connect across the internet, an employee can be anywhere in the world and still connect to the employer's private LAN. VPNs can be used to connect several LANs into a WAN.

VPNs are available to individuals and business users. Network traffic is directed to a VPN server where the user's IP address, data and other details are changed so they appear to be coming from the VPN server.

VPNs use **tunneling** protocols that operate at layers 1 and 2 of the OSI model. A tunneling protocol using authentication and encryption provides a secure connection that allows the exchange of sensitive and confidential information between the users and their companies or organisations.

Characteristics of VPNs

A virtual private network creates a point-to-point connection between two nodes over a public telecommunications system. A VPN can connect two nodes, for example two laptops on different networks, to create one logical network. Connecting a single node to a network over a VPN makes it a logical part of the network regardless of its distance from the network. Most basic VPNs do not support all LAN protocols, but commercial installations of VPNs can extend the LAN without any reduction in its capabilities.

VPNs require users to authenticate so that their data is transferred securely. Usually, a username and password are used, but businesses and organisations may require their employees to use biometric methods, a dongle or two-factor authentication. Digital certificates can be used where VPNs connect two or more networks. The certificates can be stored to allow the networks to connect automatically. Encryption can be used to ensure that any data captured or intercepted is unintelligible to unauthorised viewers. Integrity checks on the data detect attempts to interfere with the data.

Uses of VPNs

Businesses, organisations and governments use VPNs to create a secure connection. Extending a LAN over public telecommunications networks between different buildings or areas allows employees to connect into their LANs from remote locations. Employees working at home can securely access the company resources as if they were physically present in the office. Voice over Internet Protocol (VoIP) calls can be made more secure by using a VPN.

Individuals can use a VPN to create a secure connection when connecting mobile devices, such as laptops or smartphones, to public unencrypted Wi-Fi to keep user data and information secure. Using a VPN when shopping online keeps shipping and credit card details encrypted and secure.

VPNs can also be used to avoid regional restrictions (geo-restrictions) when accessing media and information on the internet. A VPN connection can provide an internet address that reflects where the VPN connects into a LAN, instead of where the user is physically located. This is useful when connecting into a corporate LAN as the user is seen by the network services as being local so there are no regional restrictions. However, this feature of VPNs can be used to avoid restrictions on the use of copyright materials. Most video-streaming services have mechanisms in place to recognise when a user is connecting via a VPN and to prevent the illegal use of the streaming service.

Accessing social media from countries where social-networking platforms are censored or banned is made possible by the use of VPN connections. Using VPNs can attempt to get around restrictions, so in some countries the use of VPNs is banned.

Protocols used by VPNs

The protocols are the **Layer 2 Tunneling Protocol, Internet Protocol Security (IPsec)** and **Transport Layer Security** (SSL/TLS). Layer 2 Tunneling Protocol creates the data packets, providing headers and packet information, for transmission over a VPN. Layer 2 Tunneling Protocol itself does not encrypt data but creates the connection – the tunnel – and leaves the encryption to IPsec. IPsec encrypts an IP packet and places it inside an IPsec packet at one end of a VPN connection, and decrypts it back to the original IP packet at the other end.

Transport Layer Security (TLS) has replaced the Secure Sockets Layer protocol (SSL), and provides the authentication and encryption for data transmission over VPNs. Symmetric cryptography is used, with the keys being exchanged by TLS using a 'handshake' at the start of a connection session. The connection is secure from unauthorised users because the keys cannot be discovered and any interference with the transmitted data can be detected. IPsec may be unreliable because **firewalls** may block IPsec data packets due to an inability to read the encrypted IP addresses. TLS connections are not affected in this way so are reliable.

Point-to-Point Tunneling Protocols (PPTP) can also be used to create a tunnel and encapsulate the data packet. An additional protocol, for example **Secure Shell (SSH)**, handles the encryption. SSH creates the tunnel and carries out the encryption of the tunnel but not of the data.

Advantages and disadvantages of VPNs

▼ **Table 14.6** Advantages and disadvantages of VPNs

Advantages	Disadvantages
<ul><li>The main advantages of VPNs compared to other means of communication over public telecommunications systems are encryption and privacy. The encryption of data transmitted over a VPN ensures that the data is kept secure.</li><li>VPNs can help keep the user's internet use, search history and web browsing private. Some internet service providers analyse the type of data, for example email, web use or video streaming, and adjust the available download speeds or bandwidth. This adjustment may restrict the bandwidth available to a user; using a VPN makes this data analysis almost impossible.</li><li>Using a VPN to connect geographically distant company LANs over public telecommunications systems is cheaper than renting leased lines as it uses the internet with all its connections already in place. It allows the centralising of resources into data centres so maintenance costs are reduced and security is enhanced.</li></ul>	<ul><li>VPNs can have an adverse impact on the performance of a network. The network traffic of a VPN is directed via the servers of the VPN provider and these add a performance overhead to the traffic. VPNs take time to process and transmit data and requests for websites, so the overall network performance is reduced.</li><li>VPNs can be difficult and complex to set up and maintain. Free VPN services offer little if any technical support and often change their settings or shut down completely to avoid legal issues.</li><li>There may be compatibility problems when using VPNs. While most operating systems and devices are compatible with most VPNs, companies have to ensure that their employees and workers are using the same type of device to connect to the corporate LAN or WAN.</li><li>The privacy and anonymity of users is not completely secure. VPN providers may monitor the activity of their users and sell the IP addresses and search data.</li><li>The use of a VPN may be illegal in some countries and users may be subject to criminal penalties.</li></ul>

Activity 14c

Explain why company employees working from home use VPNs to connect to the company network over the internet.

Mobile networks

Mobile, or cellular (cell), networks are wireless networks. Mobile networks work from base stations, located on poles, tall buildings or towers, which cover a small geographic area called a 'cell'. Base stations provide the mobile network coverage using radio signals that can be received and used by mobile devices such as cell phones. Cells in mobile networks are often drawn as hexagonal in shape because a hexagon is the best theoretical shape for placing cells together. However, the actual, real shape of a cell depends on the terrain of the location.

Each base station uses a range of frequencies, but this is not the same range as used by a station next to it. If the same range of frequencies were to be used, they would interfere with each other and the stations and mobile devices would not be able to communicate properly. One example of this interference is co-channel interference, where two transmitters work on the same frequency and both transmissions are received. The transmissions can cancel each other out or pass data between them, meaning the connection is lost or there is crosstalk, for example where calls from both transmissions are heard by all callers.

There is a limited range of frequencies available for mobile networks, so frequencies have to be reused. Base stations that are next to each other are not assigned the same frequencies. Frequencies can be reused by stations that are separated from each other by at least one other cell. The frequencies used by mobile networks do not travel great distances, so cells that are separated by several kilometres will not usually interfere with each other.

Mobile network designs include a switching system because there is a possibility of the available frequencies being in use, or of some interference of frequencies. Mobile devices can automatically switch to free frequencies or ones with less interference, or between base stations to achieve a good connection. Users can move around without interruptions to calls or data access because their mobile devices automatically and seamlessly switch frequencies without the user noticing.

The availability in an area of up to three, or possibly more, base stations makes the switching possible. This is used in cities where base station coverage is smaller than in open areas. Base stations and devices 'hop' between frequencies using spread-spectrum techniques during calls and data exchange, which increases the capacities of mobile networks. Multiplexing of **data streams** is used to increase the capacity of the networks.

The size of the cell covered by a base station depends on the nature of the area. In rural areas, with few buildings and low hills, the area could cover several square kilometres, but in cities with tall buildings cell coverage can be just a few square metres. City coverage with mobile networks requires many more base stations and these can be seen attached to existing fixtures such as telegraph or power poles, road signals and the sides of buildings in an attempt to provide complete coverage. Some base stations use directional antennae to direct their signals in a specific direction. Stations at the edge of a provider's coverage or covering a particular area in a city can aim their signals directly at that area.

In theory, a specific area should be covered by three base stations so as to provide seamless coverage and continuous access to voice calls, data exchange and other communication requirements. Base stations are connected to each other and to the central service provider to allow connections to the global public telecommunications systems. Each station requires its own power supply and a high-capacity connection to the central system. The expense of providing full coverage means that there are some areas, usually poorer regions, remote areas such as mountains or hills, or along major roads between cities, that have little or no coverage.

The size of the area covered also depends on the radio frequencies used for the mobile network. Low frequencies can cover larger areas than higher frequencies. The use of the 450 MHz frequency can reach up to 50 km from the base station, but the 2100 MHz frequency reaches less than 15 km. However, the lower frequencies can carry far less data and so can have fewer simultaneous calls than the higher frequencies. Lower frequencies are more suitable for rural areas due to fewer potential users. Higher frequencies have greater capacity and can be used in areas such as large buildings to provide high-capacity mobile coverage that is restricted to very small areas with many users, due to the structure of walls, for example. Also, because the lower frequencies can reach further than the higher frequencies, they are more prone to interference from base stations using the same frequencies.

Most mobile network providers now use the higher frequencies to avoid the technical problems of interference between stations, to increase their capacity and enable higher data transmission rates. Mobile devices such as smartphones can use many different frequencies and can switch between frequencies depending on their location.

Mobile networks use multiplexing methods to allow several streams of information to interleave and use a single communications channel. This means that many more calls and data streams can be carried on fewer frequencies, which increases the capacity of the network. Codes included in the data streams enable base stations and connected devices to change frequencies automatically as the user moves about between cells or if the network gets congested. 'Handoffs' from one base station to another are usually seamless and not noticed by callers.

Characteristics of mobile networks

Mobile networks have a number of radio base stations providing wireless connections for mobile devices. A switched network enables voice calls to be made between users of one mobile network and users on other networks. The base stations are connected into the public telecommunications network so that users can connect to other users of the system with voice and data calls. There is also a packet-switched network to transfer mobile data between users and the internet.

Uses of mobile networks

Individuals use mobile networks to enable them to use their computing devices away from a fixed point. Users can make voice calls, exchange data, use social media, connect into their employer networks and access the internet while moving around.

A mobile network allows for the seamless movement of connected devices from one cell to another. A user can walk around cities or travel in cars, trains or buses between cities and still be able to access internet services and voice calls because base stations and devices automatically hand over connections from one station to the next.

Customers can receive updates on delivery times from drivers of delivery vehicles and the vehicles can be tracked by the owners of the delivery company. By tracking the position of the driver's smartphone or dedicated device using GPS, the location can be sent back to the company using the mobile network. The satellite GPS system only sends out the data that can be used to calculate the location of a device and does not allow for two-way communications, so mobile networks are used to send location and tracking data back to a company.

Satellite navigation systems in vehicles that use GPS to locate their position and to plan routes to destinations can also receive traffic updates and weather reports using mobile networks. The traffic information is transmitted over data connections on the mobile network and is received by the satellite navigation device either by using its own connection (SIM) card or by a Bluetooth connection to a smartphone. By combining the GPS travel route information with the traffic and weather reports, alternative routes can be calculated automatically or the driver can be shown a choice of routes.

Protocols used by mobile networks

Mobile networks enable devices to connect into public telecommunications systems for voice calls, exchanging text messages, transferring data and accessing internet services. Second (2G) and **third (3G) generation mobile networks** use general packet radio services (GPRS), a wireless technology, to send data in packets. GPRS can provide data transfer speeds of up to 114 kbps, but the speeds are dependent on the amount of network traffic. If the amount of traffic on the network increases, speeds drop considerably due to congestion. **Fourth (4G)** and **fifth (5G) generation mobile networks** use **Internet Protocols (IP)** to transfer data.

Advantages and disadvantages of mobile networks

▼ **Table 14.7** Advantages and disadvantages of mobile networks

Advantages	Disadvantages
• Mobile networks allow individuals to make voice calls, exchange data and access internet services, such as email, while on the move. There is no fixed connection and devices can be used away from home, while travelling and in public places like hotels, cafes and shops. • Wireless contactless payments may be made over mobile networks, avoiding the need for dedicated connections. Small businesses and individual traders can make use of contactless payments and avoid the need for cash transactions. • Employees can work from home, while travelling and while in hotels because connections can be made into company networks. Voice calls, text messages and emails can be sent and received over mobile networks, allowing employees and employers to keep in contact.	• The use of mobile networks can increase the security risks to data and information. Wireless transmissions can be intercepted so it is essential that personal and confidential data is encrypted. Not all applications that can use mobile connections provide for the encryption of data. Social-media applications and email are often sent unencrypted. • The use of mobile networks can be expensive. While a number of pricing plans are available to users, the costs of calls, text messages and data exchange can be high. • Coverage of areas by mobile networks can be limited. High buildings or those built with materials that block radio signals or tunnels can cause loss of signal. In rural areas and areas of low population density, coverage can be poor or of low quality, not provide all the features of mobile technology or even be non-existent. This can be caused by the terrain, where hills or other obstructions interfere with the signals, or by the lack of base stations because it is not seen as economical by the network providers.

14.1.2 Tunneling

Tunneling is the use of protocols (tunneling protocols) that wrap, or encapsulate, data so that it can be moved between private networks over public telecommunications systems. The complete data packet with the private network protocols and its payload is encapsulated in a larger packet headed by the tunneling protocol information so that it can travel across public networks. Encrypting the tunneling packet contents helps to keep the data from unauthorised users. Tunneling enables packets using 'foreign' protocols to travel over networks that do not use those protocols. For example, when an IP packet has to travel from one IP LAN over a public telecommunications network to a destination IP LAN, it is:

» encapsulated inside a tunneling packet
» carried over the public telecommunications network
» deposited at the destination IP LAN
» unwrapped from the tunneling packet
» sent on its way on the destination IP LAN.

Another example of tunneling is encapsulating IPv6 packets inside IPv4 packets so they can travel on IPv4 networks. In a similar way, tunneling enables IPv4 packets to travel on IPv6 networks. VPNs are created using tunneling protocols to establish private networks over the internet.

14.1.3 Network architecture and topologies

Network architecture refers to the design of the network and is the way network devices and services are arranged and connected together along with the rules by which they can communicate. Peer-to-peer and client–server are two examples of different network architectures. The architecture of a network includes the hardware components such as routers, **gateways** and user devices (for example laptops, smartphones and printers), the transmission media in the network (such as physical copper wires or fibre-optic cables), and wireless communications systems over microwave or radio (for example Wi-Fi or mobile/cellular connections). It also includes the protocols that are used to set the rules for the data transmission and the topology of the network.

Network topology refers to how the nodes (devices such as computers and routers) and the links (the connections) in a network are arranged. The **physical topology** of a network shows the physical layout and connections of the transmission medium, such as the copper wires or fibre-optic cabling, and shows how they connect to each node in the network. The **logical topology** shows the ways that data travels in the network.

> ### Activity 14d
> Research, draw and make brief notes on the types of physical network topologies that are commonly used in LANs.

14.2 Components in a network

14.2.1 The role and operations of components in a network

Network components are the hardware items required to create a network and make a network function. Every component has a role to play in the transmitting, sending and receiving of data by devices, and in the connection and control of the flow of data across and between networks.

Network interface cards, repeaters, **hubs** and switches are used to connect computing devices together into a network. Gateways, routers and firewalls are used to connect networks together.

Network interface card

A **network interface card (NIC)** (a network interface controller or network
adapter) connects computing devices to a wired or wireless transmission
medium. The electronic components of a NIC may be separate components
or be in a single chip. Network interfaces may be integrated into the main
computer hardware so a separate NIC may not be required.

The role of a NIC

The electrical components of a NIC provide the access to the data transmission
medium. They send and receive electrical signals to and from copper wires,
modulate and demodulate the light beams that travel along fibre-optic cables,
and exchange data to and from the radio signals that carry data via wireless
connections. This operates at the **Physical layer** of the OSI model. A NIC
provides the protocols used to allow communications on LANs and WANs.

Each NIC has, in its hardware, a unique identifying code called a **media
access control (MAC) address** that is assigned during its manufacture. A
MAC address is usually six groups of two hexadecimal digits. It contains the
manufacturer code and its own unique identifying codes. MAC addresses of
NICs are used at the **Data Link layer** of the OSI model of network protocols.

The operations of a NIC

Network interface cards are responsible, under the control of the operating
system on a device, for exchanging data between the central processor and the
transmission medium. Internal data buses carry the data between the NIC and
the CPU. The parallel streams of data that the CPU is sending out over the
network are converted into a serial data stream and placed into data packets
(frames). Data packets from the network are received serially, converted and, if
necessary, re-arranged into the correct order, from the frames into parallel data
streams. The streams are placed on the internal data buses and sent to the CPU
for processing.

When data is to be sent across a network from the device, the NIC is notified
that a frame is being created in a **buffer** by the operating system. The buffer is
read by the NIC using direct memory access (DMA) to determine the address
to which it is to be sent. The NIC creates a data frame, which is sent to the
transmission medium and transmitted on the network. When the frame has
been sent, the NIC notifies the operating system that it has gone. The process is
repeated until all the data has been sent in frames.

NICs monitor the network's transmission medium for frames. NICs examine
all the frames to read the address and, if the address in the frame matches
the NIC address, the frame is written into a buffer using DMA. Frames that
are not addressed to the NIC are discarded and ignored. The contents of the
frame are checked for integrity using a checksum calculation. If the contents of
the frame are acceptable, the NIC notifies the OS that a frame has arrived and is
available for processing.

Wireless network interface card

A **wireless network interface card (WNIC)** is designed to connect to, and
exchange data over, a radio-based network instead of a wired network. WNICs
are available to connect to Wi-Fi networks, Bluetooth and mobile networks.
Wireless NICs designed to work on Wi-Fi networks, and to connect to each
other, provide a similar networking capability as wired NICs. WNICs designed
for Bluetooth do not provide these capabilities.

The role of a WNIC

WNICs are responsible, in a similar way to wired NICs, for exchanging data between the central processor, under the control of the operating system of the device, and the wireless medium.

The operations of a WNIC

WNICs exchange data from the internal data buses of the computing device in the same way as wired NICs, and work at layers 1 and 2 of the OSI model. A WNIC adds additional information to a data frame to allow for the sharing of radio frequencies, dealing with errors and the higher number of retries needed when using wireless connections. Ethernet on wired connections has few errors but wireless connections are prone to interference and loss of signals, so a WNIC has to be able to manage more retries when sending frames.

Wireless network cards can work in two ways. In **ad hoc mode**, they can connect stations together into a peer-to-peer network. In **infrastructure mode**, connection is to a **wireless access point** (WAP) acting as a central point of the network and providing access to the LAN.

WAPs provide a network name (its **service set identifier** or **SSID**) and require authentication from the WNIC when a connection is started. The authentication techniques are used to encrypt the data during the exchange.

WNICs have, just like wired NICs, a unique media access control (MAC) address used when specifying where the data has to go and where it is being sent from. Data to be sent is divided into frames (packets), which contain the encrypted data, information about the protocols to be used when processing the data, the source and destination addresses and the data needed for dealing with errors and retries.

When sending a frame, the WNIC modulates the radio carrier wave of the wireless connection with the data and transmits it. The latest versions of Wi-Fi use more than one radio carrier wave, on different frequencies, at the same time and make use of frequency-division techniques to increase the rate of data transfer. The WNIC is responsible for packaging the data, modulating it on to the radio carrier waves and sending it out on the frequencies used in the Wi-Fi network. WNICs demodulate the radio carrier waves that they receive to extract the data frames. As with wired NICs, WNICs examine all frames being transmitted by others and discard frames that are not addressed to them. When a frame addressed to the specific WNIC is received, the CPU is interrupted and the data is processed.

Network repeater

Repeaters are used to extend networks by connecting several network sections together into a bigger network to increase the network range. They are needed because the power of the signal that can be carried on copper and fibre-optic cabling, and by wireless, is limited. In the case of wireless networks, obstructions can also limit the range of a network.

The role of a repeater

A repeater is placed at, or preferably just before, the point at which the original signal starts to be degraded or lost. In a wireless network, a repeater can be positioned so that its signals bypass an obstruction. The role of a repeater is only to pass signals along from one network section to the next, so they receive the transmissions on one network section and re-transmit them on the next.

The operations of a repeater

In computer networks, repeaters work only at the Physical layer of the TCP/IP or OSI model because they do not process the data packets.

The repeater receives the original data frame and its electronic circuitry sends an identical frame onward. The frame is not processed at all other than ensuring that the new frame is physically and electrically identical to the received frame. Repeaters consist of a receiver to take the data frame from the original network, electronic circuity to recreate the frame, and a transmitter to send out the new copy of the data frame. This corrects the loss of signal strength, the decrease in signal compared to noise on the medium and any errors on the timings of frames. The frames are regenerated, so repeaters are sometimes called 're-generators'.

While light beams can travel very long distances along fibre-optic cables, the light is absorbed by the physical material of fibre-optic cabling, so repeaters are used to extend the distance even further. Repeaters in fibre-optic network cables receive the light signals along the cable, convert the light signals to electrical signals and pass them to the transmitter in the repeater. The transmitter converts the electrical signals back into light signals and sends them along the next fibre cable.

The conversion of the light into electrical signals and back into light again takes time and reduces the rate of data transmission along fibre-optic cabling over long distances. Because many different wavelengths of light are sent down a single fibre to provide very high rates of data transfer and one repeater can only regenerate one wavelength of light, many repeaters are required for each fibre. This makes the manufacture, installation and maintenance of these repeaters expensive and complex to deal with the required high data rates. The use of these repeaters also reduces the bandwidth of fibre-optic networks. Newer forms of optical repeaters have been developed that amplify light signals without converting them into electrical signals and back again. These repeaters can deal with all the wavelengths of the light, require less power to operate and do not reduce the bandwidth as much.

Wireless repeaters are used to extend wireless networks. Wi-Fi signals are restricted in their range by their low power, by obstructions and by interference from other radio signals. Security cameras, wireless doorbells and microwave ovens can emit **radio waves** that interfere with Wi-Fi signals, especially in the 2.4 GHz band. Wi-Fi repeaters, called 'Wi-Fi extenders', receive data from the main network and re-transmit it. A wireless repeater contains a radio receiver and a radio transmitter. Several antennae may be fitted to allow incoming and outgoing transmissions to occur at the same time. The transmissions between the repeater and its connected devices use different frequencies from the original.

Wi-Fi extenders are used in homes and small businesses to extend Wi-Fi coverage. However, large companies and other organisations usually install additional wireless access points instead. The reason for this is that Wi-Fi extenders, while creating a stable connection, can reduce the network bandwidth by as much as **50 per cent**, add to the wireless interference in the area and reduce the security of the Wi-Fi network.

Wi-Fi extenders add to the interference on Wi-Fi networks because they have to use the same radio bands as all other Wi-Fi devices and their signals may overlap. Wi-Fi extenders reduce the security of the network because they can extend the Wi-Fi coverage outside of the home, are often poorly secured by the homeowner and can be attacked. The Wi-Fi security is provided by the extender and not by the main router, so the security has to be carefully managed by users during installation. Wireless access points are usually physically wired to the main network so do not reduce the bandwidth and are more secure because the security is managed by the central network.

A multi-port repeater, or network hub, in a wired network has more connections, or ports, to enable the data frames to be repeated onto several network sections at once. Frames arriving at one port are replicated at all other ports. The sections all appear as the same network to users. Repeaters for connecting and extending wired networks using Ethernet have been superseded by hubs and switches.

> ## Activity 14e
> Describe the difference between a Wi-Fi extender and a WAP.

Network hubs

Hubs are repeaters with many connections or ports. Like repeaters, hubs work at the Physical layer of the OSI model because there is no processing of the data inside frames.

The role of hubs

The role of a hub is to replicate incoming data frames. Frames are received, regenerated and sent out on all ports except the port at which the frame arrived. A hub is not able to determine which data frames should go to which recipient, so all sections of the network carry all the frames with no management of the network traffic. When several nodes attempt to transmit frames at the same time, collisions occur and the rate of data transfer reduces. Because all frames are replicated to all ports, there is a higher number of collisions on the network. The detection of collisions on the network is by the nodes and not the hub. Hubs do not make efficient use of a network and their use can make a network appear to users to have a low rate of data transfer.

Connecting sections of a network together with a hub makes all the sections appear to be the same network. A connection from the main network through a hub to multiple sections enables more than one network device, or node, to share a single connection into the main network. Hubs act as **bridges** between network sections and the type of connection does not have to be the same at each port, so a hub can be used to convert between different types of wiring.

Because hubs regenerate all the frames passing through them, they can capture frames for network analysis and monitoring. This can create a security issue because the contents of the frames can also be discovered.

The operations of hubs

As with repeaters, hubs are used to extend a network over a greater distance. Hubs are also used to add multiple connections to an existing network, in a star topology, so that several nodes can use and share a single connection into the main network. All the sections connected by a hub act as if they were the same network.

A frame received at one port is regenerated and transmitted to all other ports apart from the port at which it arrived. The regeneration ensures that all the copies of the data frames sent out from the hub are 'as new' to avoid the problems associated with long cable runs. This is also a store-and-forward process because the frames are received, stored and then forwarded to their destination.

The use of hubs has been superseded by the use of switches and routers, and they are no longer part of the modern networking standards.

Network switch

A **network switch** has the same role as a network hub, with the added ability to manage the network traffic by examining the contents of frames.

The role of a switch

The role of a network switch is, like a hub, to connect sections of a network together, but it can also manage the network traffic. A network switch acts as a bridge between network segments. A switch examines all received frames to determine their destination, regenerate the frames and send them only to the port connecting to that destination. This significantly reduces the network traffic on the network. However, if the destination cannot be determined, the frames are sent to all ports other than the receiving port.

A switch can be used to extend a network, to enable the connection of several nodes or devices into a network by sharing a single connection, or to deliberately divide a large LAN into segments. Switches can be used to compensate for different rates of data transfer on the connected segments because they use the store-and-forward transmission method.

Switches can be strategically placed and configured in a large network so that network traffic between nodes on a segment is confined only to that segment, while maintaining the ability of the nodes to communicate with all of the other nodes on the LAN. Switches work at layer 2 (the Data Link layer) of the OSI model and, by using the unique address of nodes on the network, they can direct network traffic accordingly.

While some switches can work at layer 3 (the Network layer), and above, most do not. Switches are usually restricted to bridging network sections of the same type, for example all Ethernet or all wired. Switches are not normally used to connect, for instance, an Ethernet wired network to a leased line from a telecommunications system because it is the role of routers to make connections between different types of network. However, some multi-layer switches can be configured as routers because they work at the other layers of the OSI model and have additional network traffic management features allowing them to exchange network traffic between different types of network.

Switches can be used to monitor network traffic. A process called 'port mirroring' copies all the frames passing through a specified port to another port, which can be connected to a computing device to analyse the frames. Network administrators use this to detect unauthorised data frames on the network in a process called 'intruder detection', and to capture frames for analysis, which is called 'packet sniffing'. Packet sniffing can also allow unauthorised users to analyse frames to extract the data.

Switches can be categorised into unmanaged, smart and managed:

» An unmanaged switch requires no configuration by the user because it has inbuilt settings that cannot be changed. Unmanaged switches are used to connect additional devices or network segments. These are used in small LANs, for instance in small businesses or homes, where network traffic is light. They can also be used for temporary connections as no setup or configuration is required.

» A smart switch allows an administrator to make some limited configuration changes to its basic settings via a web interface, for example the bandwidth to each port can be set or a virtual LAN (VLAN) can be created.

» A managed switch is configured by the network administrator according to the needs of the network. The administrator uses a web interface or a command-line interface via a **telnet** connection or a serial port (not found on most smart switches) directly into the switch. The switch can be fully configured to prioritise different types of network traffic, disable or enable ports as required, manage the bandwidth available to different ports and create VLANs.

The operations of a switch

Switches can work at all layers of the OSI model. While it is possible to configure switches to work at layer 1 (the Physical layer), those components that can work only at layer 1 are repeaters or hubs and not switches because there is no management and they can do no more than replicate the frames.

Switches that work at layer 2 (the Data Link layer) in the OSI model are able to manage network traffic. Using the hardware MAC address of nodes, a layer 2 switch can direct frames to the port on which the destination node is located. Frames are not sent to other ports, so the traffic in a section is restricted to only necessary traffic. Collisions on each section are reduced as there are no unnecessary frames present. By directing traffic between specific ports, the rate of data transfer can be increased.

To be able to direct network traffic, switches can be configured by the network administrator and can 'learn' to direct frames to specific ports by monitoring frequent traffic between ports. By examining the contents of a frame during the store-and-forward process, the frame can be sent to the specific destination port.

Working at layer 3 (the Network layer) and at the higher layers, modern switches can be made to function as routers directing network traffic based on hardware addresses, IP addresses and URLs. Port monitoring and traffic analysis can examine the data packets to create VLANs, block specific ports to certain traffic or prioritise traffic between specific ports. Switches that can work at the highest layers can be configured for prioritising web use or for content delivery. This can ensure, for example, that high-definition video is transferred without **buffering** issues.

Wireless access point

A wireless access point (WAP) is used to connect devices with wireless capability into a wired network. Usually, a WAP provides Wi-Fi for the wireless connection and an Ethernet connection on the wired network.

The role of a WAP

A mobile device, for example a laptop, tablet or smartphone, can use its Wi-Fi capability to connect to a WAP and then use the wired network to access services available on the LAN. The wired connection of a WAP can be direct to a router that connects to the internet. Devices connected to the WAP can then access the internet via the router. A WAP allows several devices to connect at once and to share the wired connection to the router.

A WAP may be built into a router that provides the Wi-Fi connection for several devices and the connection to the internet. This is the typical physical configuration found in home routers for connection to a broadband service.

The home user has a single physical box that acts as a Wi-Fi access point for devices in the home and as a router connecting to the internet along public telecommunications lines. Smart television sets, home appliances, home assistants and other home-based Wi-Fi-enabled devices often do not have a wired connection so require a WAP to connect to each other or to the internet. Usually, they connect to the WAP in the home broadband router supplied by the internet service provider.

On a wired LAN all devices are trusted at the Physical layer using their hardware address and usually have their location fixed, but devices trying to connect to a WAP are often mobile and so are authenticated for security. WAPs allow any device within range to connect and become part of the LAN. Anyone can approach a WAP and attempt to connect their device. WAPs can be configured to use encryption to protect the data, so a password or passphrase is required when a device attempts to connect. Users wishing to connect their device to a WAP are required to enter the passphrase. The passphrase can be published so that anyone can use it or a WAP can be configured not to require a passphrase at all. For security, the passphrase is usually kept secret and revealed only to trusted users.

Confidential passphrases prevent access by unauthorised users to company LANs. Visitors will not know the passphrase so cannot connect their devices. Employees who need to move around company buildings can be given the passphrase to allow them to connect their mobile devices from wherever they are working in the buildings.

A WAP can be used to create a wireless hotspot in hotels, cafes and other public places where users can connect their devices and have internet access. Published or displayed passphrases are provided in hotels, cafes and other public areas to allow guests and customers to use the Wi-Fi connection.

A 'strong' passphrase used with the latest versions of the security protocols for wireless connections (WPA, WPA2 and WPA3) provides good encryption levels and data is considered secure. Earlier security versions (for instance WEP) are not so secure. WAPs without a passphrase, often used for public **Wi-Fi hotspots**, which allow any device to connect, are not secure. Insecure hotspots should be avoided unless additional security measures are taken by users. The provider of a public hotpot may have access to the data and content of web pages visited by users. The data in frames exchanged over hotpots may be examined by others who intercept the Wi-Fi signals. Users should make use of SSH/TLS (Secure Shell and Transport Layer Security) and HTTPS (**Hypertext Transfer Protocol Secure**), which use end-to-end encryption and help keep data secure. The use of a VPN that has an encrypted connection can ensure that data exchanged over a public Wi-Fi hotspot is secure.

The operations of a WAP

WAPs connect into a LAN using wired transmission media to provide wireless connectivity, usually Wi-Fi, for devices that can then use the wired connection into the LAN. The devices connected to the WAP share the bandwidth available on the single connection. WAPs built into home broadband routers provide Wi-Fi connections, via the router technology, to the internet.

WAPs contain a wireless network card (WNIC) to provide Wi-Fi connectivity. WAPs transmit **beacon packets**, in beacon frames, using the Wi-Fi protocols to inform WNICs of the presence of Wi-Fi connections. Beacon packets carry the name of the wireless network, called the 'service set identifier (SSID)', so that devices scanning for Wi-Fi networks can display the name to the user. The SSID can be set to be blank or any set of characters up to 32 bytes long. Usually, the SSID describes the Wi-Fi connection, for example **HotelGuest**, so it can be easily identified in a list of Wi-Fi networks. Beacon frames also carry a time stamp and details of the types (for instance infrastructure or ad hoc) and formats (for example the frequencies in use) of the available Wi-Fi connections. Beacon frames are sent out at set intervals from all WAPs.

Attempting to hide a Wi-Fi network by leaving the SSID blank, or disabling its broadcast over a Wi-Fi, is not very effective because every WNIC has its own unique hardware address, similar to the MAC address of a wired card, called the 'basic service set identifier (BSSID)'. This is either set during manufacture or calculated from its serial number and cannot be changed. While the SSID may be hidden, the BSSID is still visible because it is needed when addressing frames and packets during data exchange.

When a device is attempting to connect to a WAP, it scans the Wi-Fi radio bands for any access points in range. The device is listening for the beacon frames that are sent out at set intervals. When the scan is complete, the device displays the list of Wi-Fi networks to the user to choose from. Alternatively, if a Wi-Fi network has been used before and is remembered by the device, it will connect automatically with no user intervention. Even after a connection is established, devices listen for beacon frames in case, as the user moves, there is a stronger radio signal available. Most devices can change connection from one known Wi-Fi network to another without user intervention.

When a connection from a device is chosen, the WAP requests the passphrase and proceeds with the connection only if the passphrase is entered correctly. Entering an incorrect passphrase results in the WAP rejecting the connection. There is no limit on the number of times that a passphrase may be entered incorrectly. Entering a correct passphrase allows the connection to be established. The details of the connection are exchanged and the method of encryption is set up. All devices connecting to a wireless access point must have the same SSID and must use the same security system as the WAP.

WAPs provide the security method for connection using wireless protected access (WPA). WPA2 and WPA3 are now the standard security for Wi-Fi and require much longer passphrases, to generate longer keys, of at least 192 bits. They also carry out integrity checks on the data packets. WPA2 and WPA3 are available on all new devices. When a Wi-Fi connection is set up, the WAP uses the passphrase from the user to generate keys for the end-to-end encryption of the data during its transmission between WAP and the device. Early encryption methods, such as wireless enabled protection (WEP), were not secure enough so were superseded by more secure methods and are no longer used. One of the reasons for the lack of security was WEP's use of 64- or 128-bit keys that, once entered by the user, did not change and were used for every packet. The low number of bits for the key, its repeated use and its poor checks on the integrity of packets made it possible for hackers to extract the data from the packets.

Network gateway

A gateway connects two separate networks together. Unlike switches and routers, a gateway can use the different protocols on each network to allow network traffic to travel between the networks. Gateways can be implemented as hardware components, software applications or a combination of both.

The role of a gateway

A gateway enables devices on one network to access services on another network. In practice, gateways operate with switches and routers to exchange data between networks. A gateway used on the network of a company, school or organisation to connect to the internet may also carry out firewall or proxy server functions. On a network, a particular node can be specified as a 'default gateway'. Data packets that have an address that is not recognised on the LAN will be sent to the default gateway, which is configured to pass them to the next network.

A gateway can also carry out network address translation (NAT), which changes the IP addresses of data packets as they are exchanged between networks. For example, each device on a home LAN will have its own unique IP address but will share the single broadband router's IP address on the internet. The gateway will translate the internal LAN IP address to the external internet IP address, and back again, as packets travel between the LAN and the internet.

The operations of a gateway

A gateway translates the set of protocols used on one network into another set used on another network. This can be carried out by hardware, by software applications or by a combination of both. A gateway, often implemented as part of a router, is placed at the entry and exit points of a LAN. It is used to exchange IP data, carried in Ethernet or Wi-Fi frames, with a WAN or the internet, which use different transmission protocols.

A data frame that is destined for the other network arrives at a gateway and is examined to determine its destination and source addresses. It is repackaged into a frame appropriate for the external network. Additional information is added to the frame by the gateway. This includes the internet address of the gateway sending it out. The gateway address now appears as the new source address and this information is required by the protocols on the external network. The frame is then passed along the external network connection to be delivered to its destination. When a gateway receives a frame from the external network or internet this process is reversed. Working with routers, the frames are sent by the gateway on to the LAN to be picked up by the destination node.

Network bridges and routers

A bridge connects two or more networks together. Frames are transferred to and from each network on to the others. A router is a component that connects sections of networks, and different networks, together. A router may connect networks with different transmission media by containing the appropriate network interface cards for each type of medium.

The role of a bridge

Multi-port bridges connect several network sections together.

The operation of a bridge

Bridges use the store-and-forward method of transfer. A complete frame has to be received and stored before it can be processed by the bridge. Each frame is regenerated and sent onward. Bridges can create an internal database, or table, that stores the MAC addresses of nodes obtained from the frames passing through. This feature is also used in switches and routers. Frames can be examined for destinations and the bridge can determine from its table whether or not to send the frame out of specific ports. If a destination MAC address is not recognised, the frame is sent out on all ports.

The operations of switches and routers are based on the functions of a bridge.

The role of a router

Routers use the store-and-forward method of transmission of data frames and are responsible for sending, and directing, data frames across a network. When a frame is received, it is stored and examined to determine the destination address. The frame is then forwarded to the destination. Routers send frames from one router to another router until the frames reach their destination. If the destination is not known to a particular router, that router forwards the frame to another router. This continues until the frame either reaches its destination or is discarded.

Routers are used, together with bridges and gateways, to connect home and business LANs to the internet. A LAN will have a gateway to connect to a router, which will connect to other routers on the internet. Typically, the gateway and router will be built into the same physical 'box' and be seen as a single unit by the home user. The unit has connections for the home LAN, usually both wired and wireless, and a connection to the cable, leased line or telecommunications system provided by the internet service provider (ISP).

Business or company routers are used to connect the routers on the LANs or WANs of large organisations, colleges, universities and governments, as well as ISPs, into the main internet network of fibre-optic and satellite links.

The operation of a router

Routers have two or more connections to different networks. By using the appropriate NIC, the networks may make use of different transmission media, but all use the Internet Protocols (IP) with the data in packets or frames.

A frame arriving at a router has to be completely received before it is stored. The contents of the frame are examined to determine the final destination address. The address is compared to the contents of a routing table maintained in the router. A routing table contains data on all of the possible routes to groups of addresses and known destinations. It may also contain the policies or rules about the priorities for connections.

A **static routing table** can be set up by a network administrator so that frames follow a predetermined route from router to router. The routes taken by frames from routers with static routing tables cannot be changed except by the network administrator. If there are problems with routes or the network on the route needed, the frame will either have to wait or will not reach its destination.

Dynamic routing tables are created by the router itself. The router uses the data in frames and data from other routers to create routing tables that can change. The routing table uses the data found in frames to 'learn' where destinations are located so it can send frames directly along the route to the location. Routers also send routing information to each other so that any changes in the network, for example an overload of a router or the loss of a router from the network, can be used to update the dynamic routing table. Routers can use this information to send frames along different routes to avoid the problem and keep the network operating. Both static and dynamic routing tables may be used in the same router. This enables a network administrator to set up preferred routes but also allows for learning from other routers.

The destination address in the frame is compared to the data in the routing table. The frame is regenerated and sent from the router along the connection indicated in the routing table. Routing tables contain a listing of the destination address for frames that do not have a destination shown in the table. This is the default address and any frames with unknown destinations are also sent to this address. The default address is usually another router on the network and can be set by an administrator or by **Dynamic Host Configuration Protocol (DHCP)** from a server. On large networks, for example the internet, the frame could be passed from router to router many times before it reaches a router that knows its final destination. This is the way that frames can make their way across the internet to a destination and back to the source.

Most frames contain a counter, called a 'hop' count or 'time-to-live (TTL)' count, that shows how many times the frame has been passed through a router. The count is reduced by one each time a frame passes through a router, and when it reaches zero the frame is discarded. This prevents frames travelling endlessly from router to router. Sixteen is the typical number of hops before the frame is dropped. Repeaters, hubs and bridges do not use TTL counts.

Frames arriving at a router are placed in queues in a buffer and each frame is processed in turn. Long queues of frames can result in a decreased performance of the network, so routers drop and refuse frames if the queue is too long. Once the buffer is full, any additional frames are discarded. Routers can be configured to prioritise frames in the queues in the buffer. This provides a mechanism for quality-of-service (QoS) that ensures frames for priority services are processed first and not dropped. Services such as VoIP require a consistent and uninterrupted flow of frames so that users are not inconvenienced by gaps, delays or audio loss during conversations.

Routers work at layer 3, the Network layer, of the OSI model because they only examine the IP addresses in the frames in order to create the layer 2 (Data Link) information to send out the frame. Routers do not examine the user data that is being transferred.

Activity 14f

Select, by copying the table and ticking the box, the most appropriate device for connecting the networks shown.

Network	Hub	Switch	Router
Company LAN to internet			
Two sections of a small LAN in a company			
Two separate sections of a company LAN			

14.2.2 How network components work together

To be able to communicate with each other, components on a network must use the same protocols to enable nodes, or hosts, to exchange data. Every device on a network is a node. A host is a node that generates data to be exchanged. Typical hosts are desktop computers, servers, laptops, tablets and smartphones.

Most networks use Ethernet technology, which transmits data between nodes in frames. Ethernet technology works well on wired and wireless transmission media and is used on copper and fibre and adapted for use on Wi-Fi connections. Ethernet works at layers 1 and 2 (the Physical and Data Link layers) of the OSI model to provide a standard structure for the frame of data. The frame is called a 'datagram' and contains its source and destination MAC addresses, error-correcting data and data about the frame itself in a header. There is also some data that is used to check both the integrity of the frame when it arrives at a node and the data from the user. The user data that is being exchanged is carried inside the frame as its payload.

The payload in an Ethernet frame contains the data about the protocols being used to communicate on the network. Usually, these protocols are the Internet Protocols (IP) and the data will include the IP headers with source and destination IP addresses, the TTL, the user data and the other data needed for the packet to successfully be passed through the network from its originating node to its intended destination node.

Using the Internet Protocols, data frames can pass from a NIC in a node on a home LAN, out of the LAN over the internet to a NIC in a server on the internet, and can travel back from the internet into the home LAN. Wi-Fi access points, switches and routers pick up the frames, examine them and pass them along to their destination.

For example, when a user sitting at a laptop looks at a website, the web browser sends a request for a web page from the website on the internet. The request is packaged and sent out on the home LAN by the network interface card (NIC) in the laptop. The NIC constructs a set of data frames, called 'datagrams', containing the data that the user wishes to send and the protocols needed for them to be passed over the network. The frames are transmitted on the transmission medium, which could be a wired connection or a Wi-Fi connection.

The destination address of the web server is probably unknown to the other nodes on the LAN, so the frames are sent to the default gateway address. If the laptop's connection is wireless, a WAP receives the Wi-Fi signals and forwards the frames.

The frames transmitted on the network are examined by the NICs in any switches that are in the network, regenerated and sent onwards. The physical unit connecting the home LAN to the internet will contain a router and a gateway. The gateway has the network default address to which all frames on the network are sent if they have an unknown destination. The gateway and router examine the frames received by their NICs to determine the destination address. The gateway and router will also have a default address of another router to which to send frames if they do not know where they should go. This is usually a default address provided and managed by the user's internet service provider (ISP).

The ISP's routers examine and forward frames to other routers on the internet until the destination address is found. The frame eventually arrives, along with the other frames from the user's laptop, at the NIC of the web server on the internet. The server receives the frames, reconstructs the request and passes the data for the request to the server's web application for processing. The reply is sent back to the user in a series of frames. The frames pass through the routers on the internet and, via the ISP, eventually arrive at the router and gateway to the home LAN. The gateway and router forward the frame on to the LAN from where it is received by the laptop's NIC. The NIC, central processing unit (CPU) and operating system (OS) in the user's laptop process the frames and pass the data to the web browser for display to the user.

Frames are delayed as they pass across networks and the internet. The delays are caused by the processing time taken by the many intermediate nodes (for example the routers) as they store, examine, regenerate, add or remove protocol headers, and forward frames. The overall delay is called the 'latency' and includes the time taken while the frames are queued, processed and transmitted by the NIC and the time taken to travel, or propagate, across the transmission medium.

The latency on IP networks can be as much as several hundred milliseconds, but it is normally only a few milliseconds. Even so, the delays can cause severe problems with VoIP, video conferencing, video chat, and audio- and video-streaming services because they are communicating in real time. Audio- and video-streaming applications can overcome the delays by receiving many frames, with data payload, and storing the data in a buffer. Data is taken from the buffer by the application at a slower rate than it is received. **Buffering** allows the application to play the audio or video without interruption when some frames are slow in arriving. Routing tables and QoS configurations can attempt to ensure that the effects of delays are minimised.

Activity 14g

Why is buffering necessary when watching a movie streamed online?

14.3 Network servers

A **network server** is a node on a network that provides services, data or applications to other nodes (host computers) locally on the network or globally by using the internet. The servers respond to requests from hosts (the clients) in a client–server relationship using the request–response model of networking.

A server is usually a separate, dedicated node, for example a computer system, but may be a software application running on the same host computer as a client of that server. One dedicated computer system may also run many **virtual servers** and make them all available to clients.

Network servers have a high performance with high-speed CPUs, a large memory capacity (RAM) and large, high speed (i.e. fast read and write performance) secondary storage in the form of magnetic hard disks or solid-state disks. Company servers must multi-task and deal with a large number of users and requests. Network servers, web and email servers, and file servers are often rack-mounted, dedicated devices located in data centres.

Data centres are specialised buildings designed to provide all the power, network and telecommunication connections, and the physical storage space needed to hold, and maintain, clusters of servers in **server farms**. Data centres consume large amounts of power running the servers and providing climate control facilities. Large numbers of servers produce considerable amounts of heat that has to be dissipated, but the waste heat can be used to heat the areas used by staff or to generate power to help run the data centre. The design of buildings for data centres has to take into account the requirements of the servers and maintenance staff as well environmental issues.

Where there are few users and the requirements are not demanding, a server may be run from almost any computing device. A file or FTP server might be run from an older, low-performance device, for example an old laptop in a home.

Network servers are used to centralise resources, share resources and services, and centralise security management. Company data, files and documents can be stored on network servers and made available to employees. The security of the materials can be administered and controlled centrally. In a home, central storage of audio and video files can be used to make these available on a home LAN to household members wherever they are in the house.

Activity 14h

Explain why companies have servers on their networks.

14.3.1 Types of network server

Network servers may be designed to provide users with one or more services, such as the storage of company files and databases, websites, the sending and receiving of email and centralised installations of software applications.

Servers work in the client–server model. The server waits for an incoming request for a service from a client, for example a desktop computer or a laptop, and processes it. A response is sent back to the client. The request may be for a computer resource (for example CPU time for a calculation) or for web content by a web browser on a tablet. Usually, the client makes multiple requests from the server. A request for a web page will require requests for all of the page elements so that a browser can display the page to the user. Servers share their resources with clients, but clients do not share their resources with servers.

Media servers send digital audio and video streams to clients for users to listen to and watch. Games servers connect clients so that gamers can take part in multi-player games.

File server

File servers store files and file folders for access by users over the network. Users will use a client device (for example a desktop computer, laptop or tablet) to retrieve files from and store files on the server. They do not do any other tasks for the clients. Database servers store data but also carry out additional tasks, such as searches of data, on behalf of the client computer, so are not considered to be file servers.

File servers are used on LANs in schools, businesses and organisations to provide central storage for data and files. Modern file servers have developed into **network-attached storage (NAS)** systems, which are devices specifically designed to provide access from a LAN to stored files. A NAS system consists of the secondary storage devices and a computing device specifically designed to allow users to connect from the network for access to the files.

File servers, or the NAS systems, used in organisations have a large capacity of secondary storage. The most cost-effective storage medium is still magnetic hard disks. Magnetic hard disks carry huge amounts of storage, with fast access times, at a reasonable cost. While solid state disks can have much faster access times than magnetic hard disks, large-capacity SSDs are still expensive when measured in terms of unit cost per gigabyte. A cache in computer memory (RAM) can significantly increase the speed of access to the files. Because organisations rely on file servers, the servers have to be extremely reliable with continuous operation and be easy to administer.

Access to the file server usually requires authentication so that the files are kept safe and secure. Authentication for the use of a file server in organisations is based on a user's login credentials and access rights to files. A Redundant Array of Inexpensive (or Independent) Disks (RAID) configuration can be used to protect data and files against failures of hard disk, complete hard-disk failure or failure of parts such as damaged sectors. A RAID spreads files across several hard disks that work together.

Most modern operating systems have the protocols used to access file servers already installed. Common protocols include the network file system (NFS) used by Unix® and Linux®, Server Message Block (SMB) used by Microsoft, and Apple's Apple Filing Protocol (AFP). These allow clients to display folders and files to users as network drives or folders, in the same way folders are used as local hard disk storage. The user sees or makes no distinction between the local folders and the networked folders and can move files easily between them.

A file server is not necessarily the same as an FTP server, but FTP access may be included on a file server to allow network administrators to upload, download and manage files.

The use of file server files is not the same as using cloud storage systems. The means of access may be the same, but cloud storage systems are usually accessed using the internet, while company file servers are usually located on a company or home LAN and not accessible over the internet.

Activity 14i

Explain what is meant by NAS.

Web server

A web server stores websites and enables access to the contents of the websites. A website is usually arranged as a set of linked web pages. The content may be static, but it can also be created and altered dynamically to meet the specific requests of users. The contents of websites can be links to other websites, files and scripts to control the appearance of the website, files and scripts for user input and interaction, video and audio files, text, images and animations. A single web server may hold many different websites.

The function of a web server is to respond to client requests by locating the content of the requested website on its secondary storage and delivering it to the client for display. The user makes the requests with a web browser (the client application) using the **Hypertext Transfer Protocol (HTTP)** or the secure version (Hypertext Transfer Protocol Secure, HTTPS), and the server responds to the client, also using HTTP or HTTPS, by sending the content. An error message is sent if it cannot find the requested content or web page. Websites, and their resources, are located by the resource's unique identifying reference, called a 'uniform resource identifier (URL)'.

HTTP and HTPPS work at layer 7 (the Application layer) of the OSI model and are used to create and maintain a connection to a web server using its assigned TCP ports. TCP ports 80 and 8080 are set aside for use by HTTP. HTTPS uses TCP port 483. HTTPS encrypts the exchange of data between client and web server, whereas HTTP does not.

The HTTP request from the client web browser asks for the resource and tells the web server what action to take. The action could be to send the entire resource, to send just a description of any metadata it is using linked to the resource, or to change or delete the resource in rare cases. Most requests are only for the retrieval of resources. Requests to alter a resource, for example a database entry, do not proceed without alerting the user.

A web server responds to inform the client if the request is successful. A code is included to indicate the status of the request, for example code 200 indicates it is successful, code 308 indicates that the resource has been moved and redirects the browser to the new location, and code 404 indicates that the requested web page has not been found.

The response also contains a large amount of information about the web server, the resources and any cookies to be placed on the client device, and how the resources should be exchanged. The main content of the response is the requested resource. When the resource has been received, the client web browser displays the resource to the user.

A client may also request that a web server carry out actions determined by scripts. This is **server-side scripting**, with scripts being written in languages such as JavaScript®, Perl™, Python®, PHP™ and Active Server Pages (ASP). Server-side scripting can dynamically change the appearance of a web page according to the requirements or actions of the user or alter the page content, for example by delivering advertising content. It can also retrieve data from databases for clients.

Server-side scripting enables web pages to be displayed faster because only the server processes the code. No code is downloaded to the client so the client is prevented from having access to the source code generating the page. Code downloaded and processed by the client is called 'client-side scripting'. This requires the code to be compatible with all clients and takes time to download and process so the page is displayed slower. It also allows users to access the source code itself. Clients that cannot process scripts from a web server often fail to show the requested resource or do not display the web page at all.

Printers, webcams, routers and many other computing devices that lack a user interface can have web servers embedded in them. This allows users to remotely connect to the device and configure it using a client web browser, for example on a laptop or smartphone.

Mail server

Email requires a mail client, which is a dedicated software application or a web browser, to communicate with a mail server. Mail servers and their clients work in the client–server model of networking. The mail server is responsible for receiving, storing and forwarding email messages between mail clients. A mail client composes, addresses and sends the message using Simple Mail Transfer Protocol (SMTP) to the sender's mail server, which determines the address to which it is to be sent. The address domain is looked up using SMTP and is resolved from DNS by the server, which sends the message to the recipient's mail server. This server looks up the recipient's local address and places the message in the inbox to await collection by the recipient's mail client. The message is delivered to the recipient's mail client using Post Office Protocol (POP3) or Internet Message Access Protocol (IMAP). It can then be read by the recipient.

Mail servers can use both IMAP and POP3 to deliver messages to mail clients. IMAP has the advantage over POP3 in that messages are only copied to mail clients and not moved to them. Messages are deleted from the server only when a user tells a mail client to do so by explicitly deleting a message in the mail client. When accessed using POP3, a message is moved to the mail client and automatically removed; that is, deleted, from the server.

Accessing messages with IMAP also allows more than one mail client to retrieve messages from a mail server at the same time. A user can simultaneously retrieve emails on a laptop and a smartphone with different mail clients, leaving the messages on the server. POP3 does not support simultaneous multiple connections by different mail clients and always moves messages from the server to the client when they are accessed. This means that once a message is accessed using POP3, it is also deleted from the server.

The use of IMAP means that mail messages can be stored on mail servers for very long periods of time. This can make them accessible to others because most email messages are stored in unencrypted plain text and are easy to read. Further, deleting messages from one server may not remove them from all the intermediate mail servers that have been used to transfer them from sender to recipient.

Applications server

An **applications (or application) server** is software that provides and delivers applications to clients. Clients use a web browser to access the applications server. The required applications are run on the server and displayed within the client browser. Applications servers provide the interface for the client and the resources on the server to run the applications. There is no need for the application to be installed on the client. Applications servers allow for centralised management and maintenance of applications, as well as increased security for the applications and user data.

The applications that are served up by an applications server are usually those that require a high performance from their host devices, for example a database application server would be used in a large company with many clients and users using it simultaneously.

An applications server may be named after the application or applications that it is hosting, for example a web server and a database server are also applications servers. However, while many applications servers can make use of a range of protocols to communicate, web servers mostly use only HTTP and HTTPS to service user requests.

The role of an applications server is to provide access to the services on the host server. For example, a database applications server can show a user a web-based interface to the database running on the server for using when interacting with the database. Similarly, an applications server can provide employees with web-based access to customised, bespoke business applications that are managed centrally on company servers. They can also provide their own bespoke user interface.

Print server

A **print server** collects print jobs from computing devices (clients) on a network and sends them to printers. The client device can be almost any computing device, from a desktop computer to a smartphone, that supports printing. A print server is usually a dedicated computer system with software that uses printing protocols such as Microsoft Network Printing Protocol, Internet Printing Protocol or Line Printer Daemon protocol. These protocols allow client devices to send print jobs across a network to the server, which is then responsible for sending the job to the printer chosen by the user. There is usually no limit to the number of clients that use a print server, so print servers in large organisations make efficient use of printers. A print server can also be configured to add items such as watermarks or source information, for example the name of the user who is printing the document or the company name, to printed documents.

A print server may have more than one printer attached or connected and can direct print jobs to the requested printer. It can also queue print jobs from clients and send print jobs in turn to the printer. The queues can be managed to pause jobs, reorder jobs, enforce printing policies, delete jobs and keep a record of jobs for later analysis or for accounting purposes. However, the use of print servers means that the user who sent the print job may not have control over some of the features or functions of the printer, such as choice of paper tray to use, or may not receive notifications that the print job has, or has not, been successful. Notifications to the user depend on the features of the print server software.

The sharing of printers on a network without a print server, for example in a home where one printer is shared between several laptops and smartphones, means that there is no managing of the printing from clients, so print jobs can be delayed or lost or muddled between users. However, the user does have direct control of the printer features and will receive notifications from the printer.

FTP server

A **File Transfer Protocol (FTP) server** stores files and allows access to them using the File Transfer Protocol. FTP servers can store and allow access to almost any type of computer file. An FTP server may also be referred to as an FTP site.

FTP is part of the **TCP/IP suite** of protocols and works at the Application layer of the OSI model. It provides for a client–server connection. The client is a software application running on a remote device. The server can be a dedicated file storage system with FTP server software or a device running local FTP server software. FTP servers also run on servers that allow access using other protocols, for example a web server using HTTP for user requests may also have FTP server software running.

FTP is not automatically secure since it does not encrypt usernames, passwords, files or data. Additional protocols or methods have to be used if the user wishes to ensure that the data in files being exchanged is kept private. To exchange data securely, either a File Transfer Protocol Secure (FTPS) or a VPN connection is needed.

Large distributions of operating systems and their updates, software applications for installation and documentation can be made available on FTP servers for download. However, FTP has been mostly superseded by using HTTP access from web browsers for downloads or by other download protocols, such as BitTorrent. This is because FTP requires a user to log in to the server, either with their own username and password or anonymously, to establish a connection. The other protocols do not require logins, so the overall process of downloading is less time-consuming. Also, accessing FTP servers and exchanging files through a firewall may be difficult. For example, if the exchange of data by an FTP connection appears to take too long or to be idle, a firewall may decide that it is no longer valid and drop the connection.

Proxy server

Proxy servers are logically located on networks between clients making requests and the responding servers. This means that a proxy server acts as a go-between, or intermediary, between clients and servers on a network. A client's request to a server is intercepted by a proxy server, analysed and processed, and forwarded to the server. The response from the server is returned to the proxy server, analysed and processed, and sent to the client.

Proxy servers change the IP address of the requests from clients to their own IP address. In many cases the change of IP address causes no problems, but some network services, for example FTP and VoIP, which depend on the device IP address to establish and maintain a connection, may not function properly. A firewall may see the changed IP addresses as invalid for the session and the connection may be refused or dropped.

Network administrators can deliberately configure the interception of client requests by a proxy server on a LAN. All requests for particular services are directed to a proxy server that evaluates them and takes any appropriate, also configured, action. Responses are received by the proxy server and sent to the client. This type of proxy server is used in large organisations, schools or colleges to monitor, analyse and control the requests and responses that flow between a LAN and the internet. This is used for content filtering of websites in which requests for inappropriate content are blocked.

A reverse proxy server receives requests from clients and forwards them, for example to web servers, and the responses are sent back to the proxy. The proxy replies to the client with the server's response. The client is not aware that a proxy server is in use. There are several reasons for using reverse proxies. Large numbers of requests can be distributed between the servers to carry out load balancing. Caches of static content, for example photographs, can be stored on the proxy to reduce the load on the servers and speed up response times. A reverse proxy can act alongside firewalls to add security to the system and help to protect the actual servers from malicious attacks.

Virtual server

Virtual servers are software instances of different servers running on the same physical hardware. The physical server hardware is usually a high-performance device with, for example, fast CPUs and large amounts of RAM, with access to high-bandwidth communications via multiple NICs.

Each virtual server has its own operating system and server software, which may or may not be the same. Different operating systems and different server applications can be run on the same physical machine. Usually, administration access to each virtual server is remote via a web interface or with terminal software. This is because the physical server is rack-mounted and located in a server farm or data centre with no peripherals, such as a keyboard or display.

Web hosting is a good example of the use of virtual servers. One physical server, or a logical collection of servers, can host many different websites, each with its own IP address and URL. In businesses and organisations, virtual servers are used to provide applications servers, file servers and print servers.

> **Activity 14j**
> Explain why virtual servers are used for hosting websites.

14.3.2 The role of servers in a network

In a network, the role of servers is to share the workload by enabling clients to share their resources, for example CPU time and RAM, to provide centralised storage for data and to be sources of services to client devices. Network server devices may provide one type of service, such as a file server for storage, or may provide more than one service from the same physical device, for example a web server and an FTP server. Each type of server has a particular role to play and they are used to enable network administrators to have centralised control over the services on their network.

In addition to the servers above, Dynamic Host Configuration Protocol (DHCP) servers supply IP addresses to clients on the network. A DHCP server is used to allocate IP addresses to clients when they connect to a network. A DHCP server can reallocate a client's IP address to a different client if the original client disconnects.

14.3.3 The operations of servers in a network

A server is a device that hosts server software. A client can be any networked device such as a desktop computer, laptop or smartphone. All network servers work by the client–server model of networking.

Request-and-response method

The request-and-response method of communication is the basis of communication between networked computing devices, software applications or both. A device or software application sends out a request for a service or for data, and another device or application is monitoring the network and waiting for requests. When it receives a request, it processes it and returns a reply, which is the response. The devices send and receive a series of messages until all the requested data has been sent and received. A user wishing to view a web page from the internet is a good example of the use of the request-and-response method.

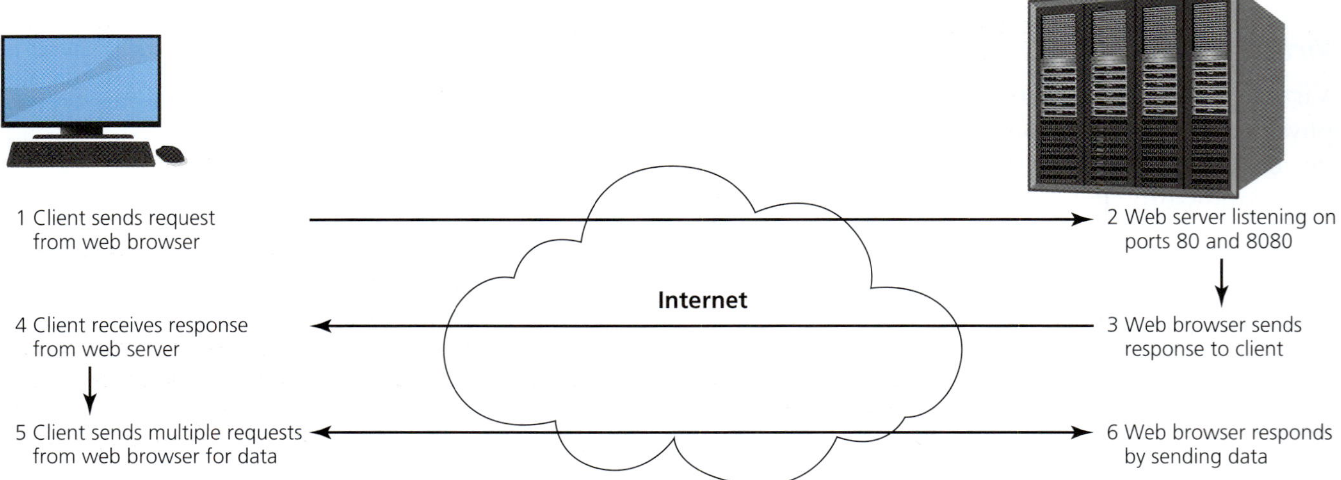

▲ **Figure 14.4** Request and response between a client and a web server

Multiple exchanges of requests and responses using HTTPS between the web browser, the client and the web server determine how and what resources are sent from the server to the client. The web server responds to a request from the client, the web browser, with its own request for information and, in turn, the web browser sends more requests requiring responses from the server. This process continues until the whole of the web page has been sent for display to the user.

Server farms

Server farms are collections of servers connected and networked to appear to users as one server. The servers work and provide services and functions together. Server farms can carry out tasks extremely fast compared to individual servers.

A server farm, or server cluster, can contain thousands of networked servers, each of which is managed and coordinated by a central server unit. The central management server device schedules processes and tasks, assigns them to specific servers, balances the load on individual servers and manages security and updates. Server farms are the basis of data centres and supercomputers and, as with these, the performance of server farms can be limited by the power supply, the network connections and the climate control systems rather than the raw power of the CPUs in the individual servers.

▲ **Figure 14.5** A server farm

14.3.4 The advantages and disadvantages of each type of server for a given scenario

Servers are used by businesses, organisations, schools and individuals to provide services to clients. The advantages and disadvantages of using servers in a network are much the same for every scenario.

▼ **Table 14.8** Advantages and disadvantages of using network servers

Advantages	Disadvantages
• Using servers centralises resources, for example database files, where they can be administered, backed up regularly and safely, and archived as and when required without the administrator having to visit each individual workstation around the organisation. Employees, staff and school students can be reassured that their files on the server are safely backed up without having to remember to do it themselves. • Clients can be added and removed without much reconfiguring of the network. • Server resources can be added or removed without making changes to individual workstations. • Servers can be configured to provide different services to different clients. • Administrators can set up and manage security rules centrally and apply different rules or access rights to different client devices and to different users. • In schools, servers ensure that students' files, student records and other data are managed centrally and kept safe and secure with the appropriate security, backups and archiving carried out. Students, and staff, can use different workstations in different rooms, connect mobile devices and still be able to use the school's network services. • The overall costs in large organisations are less than running hundreds of individual workstations. A few highly skilled IT technicians, while costing a lot of money to employ, can centrally manage the servers. • Installing and maintaining a server for stock databases and the finances, and connecting client workstations, for example point-of-sale terminals, means that the server can be kept safely locked away in a secure back office. • Large organisations, large educational establishments and governments with their own dedicated web servers can make their content available to others over the internet.	• A reduction in network performance occurs if too many requests are made to a server. • If a server fails, the clients on the network needing its services or resources cannot operate. • Server-based networks are expensive to install, maintain and administer compared to standalone workstations. • Dedicated web servers can cost a lot of money, take time to maintain and update, and require highly skilled people to administer. Because of this, most organisations, schools and smaller businesses use web-hosting servers owned and run by third parties. A web-hosting service owns, maintains and administers the servers and rents out their use to others who place their website on the servers. The disadvantages of this approach are that the hosting company has control of the website and the business data, as well as the security and safety of the data. Also, the business website can be stored anywhere that the hosting service places its servers, which may not be in the same country as the business, and if the hosting company is sold or goes out of business, the website may be lost.

Mail servers are almost always owned and run by large organisations, internet service providers and governments. In the past, anyone could set up and run a mail server and use the mail relays on the internet for email, but the continued abuse of email services over the years has made this uneconomic. To successfully run an email server that accesses the internet requires the skills and knowledge to deal with spamming, malware, hijacking emails and other email abuses. While using a private email server keeps emails away from other providers, it does require extensive technical knowledge and the cost of owning a domain name, hosting the mail servers and access to the internet. Small businesses, schools and individuals use the services of larger organisations and internet service providers.

Applications servers can easily be configured to use multiple web servers, database servers and other servers, so are often used in large organisations with server farms and data centres. An advantage of using an applications server in a business network is that it can integrate the services of several servers. A web server and a database server can be used together by an applications server to provide employees with a consistent view of the company data. The use of an applications server ensures that all employees use the same versions of software across the company, so avoiding compatibility problems. An applications server increases the security of software and data because it adds another security layer between clients, employees and the company data in the databases. There is no direct link between the client and the server data and additional security checks are carried out when the client requests resources, meaning that cyberattacks on the data are more difficult to carry out. However, the disadvantage of using applications servers is that, as with other centralised resources, a very large number of simultaneous requests from many clients can lead to network traffic congestion and poor performance.

The main advantage of using print servers is to enable many clients to share a few printers. Print jobs can be sent to the correct printer if there is more than one available. Only a few printers require maintenance, checking for sufficient paper and replacing ink or toner. In schools, students' print jobs are queued until a printer is free to print their work, and their print jobs can be audited and checked to see whether they are allowed to be printed.

The disadvantages of using a print server compared to having individual printers is that not all of the features of specific printers are made available to users by the print server, and the print server may not alert a specific user to a print error, a paper jam, or a lack of paper, ink or toner. The print job then fails without the user knowing until it is too late. In organisations and schools where printers are not placed near workstations, individual print jobs may become lost among the many others coming off the shared printer.

FTP servers are simple to use and administer because they are standardised across all modern operating systems, so almost any device can be used. Separate connections are made for the control of the connection and for the transfer of data, which means that data transfer is fast. In a company, an FTP server can be used to distribute company documents instead of emailing them to each employee. Employees use their login details and download them themselves. For administration purposes, FTP servers allow files to be easily managed but available for general access, for example in schools by students or by employees in company offices.

The main disadvantage of using FTP servers for data exchange compared to other network technologies is that the transfer of usernames, passwords and data is not encrypted, therefore not secure. FTP uses two connections to a server so requires more bandwidth on the network and may fail due to firewalls. There is no recording of date stamps or file attributes so there are few records kept about the files. Also, there may be a limit on the number of simultaneous connections allowed.

The main advantage of using a proxy server is that it can control, analyse and provide usage data for analysis of client requests to other servers and the responses received. In schools, this can be used for filtering content to prevent access to inappropriate websites or other content on the internet, prevent the sending of inappropriate emails in and out of the network, and send alerts to teachers. A proxy server can also hold a cache of frequently accessed content to reduce the time taken to provide a response. Students in a class studying geography may all request the same web pages, so a proxy server will download them once for a student from the internet and then distribute them to other students. The other students will receive their web pages much more quickly than if they were individually accessed from the internet.

A disadvantage of using a proxy server in a school, for example, is that it may create a network bottleneck as it tries to process many different requests from many different clients. For example, students researching many different topics on the internet create a large number of different requests. The time taken to receive, analyse and process each request and its response increases the delays and students may notice a reduced network performance.

The advantages of virtual servers are that there is a significant saving in electrical power compared to running individual devices for each server. The costs of installing, maintaining and upgrading the physical devices is much reduced because there are far fewer devices.

Virtual servers can be quickly moved from one physical device to another if recovery from disaster is needed. If the host server fails or needs to be shut down for maintenance or upgrade, a backup or copy of the virtual servers can quickly be set up to run on another physical device.

However, running different virtual servers on one physical host device means that if the host device fails, all the hosted virtual servers fail. The use of server farms and data centres with many physical devices can overcome this disadvantage, but the costs may be beyond what small businesses and schools can afford. Also, it is possible that, if using rented space on physical devices in server farms or data centres, the virtual servers of one company or school may co-exist with virtual servers from other companies or from sites considered inappropriate for use in schools. While the security systems in use should prevent data from crossing between servers, there is a small possibility that the data from one virtual server may appear on other virtual servers. This is because the virtual servers, which should be completely isolated from each other, use the same RAM and storage systems.

14.4 Cloud computing

Cloud computing is the sharing of data storage, server resources, files, databases, computational power and other computing resources on a LAN, a WAN or the internet.

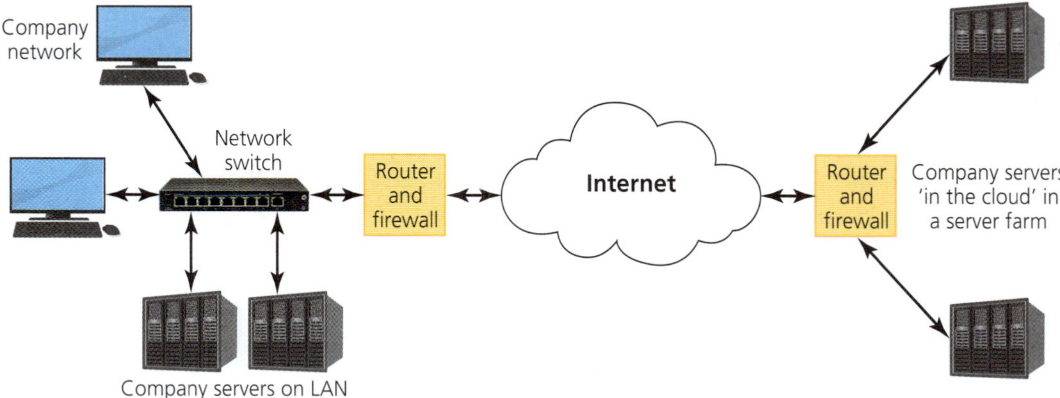

▲ **Figure 14.6** Cloud computing

Cloud computing makes resources available without complex set-up or management tasks. Cloud-based resources are managed by the providers and users are able to use cloud computing with minimal effort. An authenticated login is often all that is needed.

Cloud computing is made possible by the distribution of resources across many servers, located and managed in data centres and accessed using the internet. Large clouds have their resources distributed over many data centres.

14.4.1 Characteristics of cloud computing

Cloud computing is available on demand, over the internet, to different types of client devices as and when the clients require access to the resources. It requires minimal user set-up and can be paid for as it is required or used. Access to cloud computing is device-independent and the physical location of the device can be anywhere with internet access. While clients share the cloud computing resources, such as computing power and storage systems, the resources used by individual clients are kept logically separated from the others. This means that the resources used by one client may be the same as those being used by another, but the clients are not aware of this and are kept securely apart.

The amount of computing resources in use by a client can be increased or decreased on demand. If the client needs more computing power to run a web-based spreadsheet application, cloud computing can supply it without user intervention. When not required by the client, the amount of computational power is decreased. The use of the resources by client devices or users can be monitored and audited. Users can be restricted in the amount or type of resources they use and can be billed for their use.

Cloud computing can reduce the costs of using information technology to organisations and individuals because the installation, maintenance, updating, backups, archiving and other tasks associated with IT provision are carried out by the cloud computing providers. Using several large data centres distributed around the world provides IT resources at lower costs to organisations.

Taking away the responsibility for security of data from the organisation can be seen as a drawback, but cloud computing can provide greater, and more reliable, security of data to organisations because the data is distributed over several data centres. The organisation does not have to spend money providing security for its data.

Key characteristics of cloud computing for sharing computing resources

The key characteristics of cloud computing are its availability over the internet using standard TCP/IP protocols, its availability to many different types of device (for example laptops, smartphones, tablets and desktop computers), the dynamic allocation of pooled resources to clients as and when they request or require them, and the monitoring and auditing of the use of the resources.

14.4.2 Uses of cloud computing

Cloud computing is used to share resources between users, allowing them access over a network or the internet, from a variety of client devices. There are three models of cloud computing:

» Infrastructure as a Service (IaaS) provides virtual host devices to users, where physical servers, in data centres, create virtual machines for users. IaaS is used as a basis for developing, testing and evaluating new operating systems, applications and other software. It is also used to analyse large amounts of data, as in data mining, and for **disaster recovery**. Whole IT systems can be backed up and stored, ready to deploy in case of disaster.

» Platform as a Service (PaaS) provides not only the virtual devices but also the operating system and other software to create a virtual environment in which the user installs and uses their own applications. PaaS is used by, for example, developers of mobile and web applications. The PaaS system provides the environment, such as the operating system, and the programming software and utilities for the developers to code, test and evaluate their products. PaaS can also allow developers to add new capabilities to their applications by providing ready-made, ready-to-use components, for example chatbots, artificial intelligence systems and access to the Internet of Things (IoT). PaaS can also add systems for security, database management and data analysis for the developers to make use of.

» Software as a Service (SaaS) allows users to run operating systems, applications or whatever they wish on their virtual machines. SaaS provides users with virtual devices, operating systems and software applications, for example word processors and spreadsheets. SaaS provides software applications that are accessed using a web browser. Sometimes called 'cloud-based software', applications, for example database management, are rented from providers and run in any compatible web browser. Data storage can be local, for example on the user's own hard disk, or in the user's rented storage area on the cloud provider servers.

Uses of cloud computing by individuals and organisations

A common use of cloud computing by individuals is for file storage. Individuals can rent, or subscribe to a provider of, a file storage area in the cloud. The storage area is distributed across data centres and rented from providers, such as Dropbox, Google Drive™ and Mega, which administer it. Individuals may create an account to use a small amount of cloud storage from these providers free of charge, but larger amounts of storage capacity are paid for by monthly or annual subscriptions. Some providers use end-to-end encryption and some automatically synchronise files between the different devices of a user so their files are always available.

Cloud computing allows users to make use of the processing, storage and files on distributed servers without having the systems on their own devices. Social-media platforms, including email and blogs, are essentially cloud computing services.

Software applications for office tasks, for example word-processing and spreadsheet applications, are available as cloud-based services, for example from Microsoft and Google. These applications can be used to enable users to create, edit and collaborate with other users in real time on documents using web-based interfaces.

Cloud computing can be set up by organisations in several ways. A private cloud is owned and managed by an organisation and access is restricted to its own LAN or WAN. A public cloud is located on data centres owned and managed by third-party providers and is accessible over the internet. A hybrid cloud is a combination of a private and a public cloud system where, for example, applications and shared files exist on the public cloud, but confidential company data is stored on a private cloud.

14

14.4.3 The advantages and disadvantages of cloud computing for a given scenario

▼ **Table 14.9** Advantages and disadvantages of cloud computing

Advantages	Disadvantages
• Individual use of cloud computing has advantages and disadvantages compared to other ways of storing computer data and files. Individuals use cloud storage instead of local, hard disk storage or removable storage because it's managed by others, has a very large capacity, is often free to use and more storage can be paid for if required. Also, it is available from any suitable device that has network or internet access. • Other cloud computing services are used by individuals for access to movies or gaming platforms. Video-streaming services provide individuals with access to cloud-based movies, and gaming platforms allow individuals to play online games with others. As with other cloud-based services, the cost to individuals is less than purchasing DVDs or Blu-ray discs or individual computer games. • Businesses and companies use cloud computing because it allows access to company resources wherever the employee happens to be located. Employees from different geographical locations can collaborate on projects stored in the cloud. The employees can use a suitable internet-enabled device to access the resources. • The resources are automatically updated, backed up and archived and there is very large storage capacity available. Companies can pay for as much or as little storage capacity as required. • Companies can use IaaS, PaaS or SaaS, or a combination of all three types of cloud-computing, to save costs and to ensure access to up-to-date resources. Costs are saved because services are rented for use as required, there is no need to own the infrastructure, software or application, and different combinations can be rented from providers as and when required for each project or task being carried out. The amount of energy used by the company is reduced, so it can save costs of heating, cooling and lighting the space for the servers and the offices for the IT staff and it can save the cost of the electricity to run the servers. The employment costs of specialised IT technicians are also saved as they are not needed.	• To access the files there must be a network or internet connection. • There is a cost for the monthly subscription to the services and the internet access, and the need for a reliable, high-bandwidth internet connection. • The storage of the data and files is not under the control of the individual. Not having direct control of files means that there is always the worry that the provider may deny access or go out of business causing the files to be lost, or that the files can be seen by others because of poor security. • Concerns over security, the requirement for a reliable internet connection with enough bandwidth to carry all of the networked traffic, and the reliability of the cloud provider. • Security issues include the lack of direct control over the company data and files, and the running or storage of company software and files, on physical servers that also host other companies' software and files. Any attack on these physical servers can affect many companies simultaneously. • There is also a greater risk to company data as it is being shared or exchanged between the clients in the company network and the cloud data centres. There are more staff, possibly around the world, accessing the company data, so the risk to the data is further increased as it travels over more networks. The more networks that the data has to travel over, the more it is at risk from being corrupted, lost or stolen. • Companies and organisations running online shopping sites or social-media sites sharing data centre resources with smaller businesses may be attacked and the smaller businesses suffer as a result. For example, a distributed denial-of-service attack (DDoS attack), where the incoming traffic flooding the host servers originates from many different sources, may affect one company's virtual servers and disrupt the virtual servers of other companies at the same time. This is because the virtual servers share the networking resources of the physical servers in a data centre. • Laws and legal regulations referring to data and its storage may vary in different countries. As data is exchanged between various company locations around the world, it moves from one jurisdiction to another and may be subject to different laws. Companies have to abide by the various laws, which may increase their costs and risks of legal action.

14.5 Data transmission across networks 

Data transmission is the physical transfer of data over **communication channels**. Data is sent in the form of signals along a transmission medium, which can be a guided medium such as fibre-optic cables or copper cables, or an unguided medium such as a wireless connection. Data can also be transmitted using lasers to cross a space without any physical or wireless medium.

In networking, data is a digitised representation in the form of a sequence of bits modulated onto electromagnetic waves and carried in communication channels. The channels of communication can be any communication media capable of carrying a signal. Channels include physical media, for example copper wires and fibre-optic cables, and wireless channels, which include radio waves, **microwaves**, visible light, **infrared** and other electromagnetic waves.

The most common method of transmitting data across networks is Ethernet. Ethernet is used on LANs and WANs, can be implemented on wireless networks and forms the basis on which Internet Protocol (IP) networks are built.

14.5.1 Speed of transmission, bandwidth and bit rate

The speed of transmission of data over networks is affected by many factors, for example the method used to transmit the data, the type of communication channel in use, the protocols being used to transmit the data and the number of the nodes through which the data has to be transmitted. The type of nodes, the type of processing to be carried out on the data and the speed at which the nodes carry out the processing of the data are also factors.

The speed of data transfer across a network is referred to as its bandwidth and is shown as a bit rate.

Bandwidth

Bandwidth is the rate of transfer of data over a communication channel. The rate is shown as the number of bits that are transferred in a single unit of time. Bandwidth is usually quoted in terms of the number of bits per second. The communication channel can be the whole, or just a section, of the path used for communication. For example, the bandwidth of the whole network or just the wireless part of a network can be described. A wired network may have a bandwidth of 1 Gbit/s while the wireless connection to it, accessed by a WAP, may have a bandwidth of only 54 Mbit/s.

Bandwidth can also be used to measure the network's capacity to transfer data or the maximum amount of data that a network can transfer. Typical bandwidth tests available for free over the internet attempt to measure these.

Bit rate

Bit rate is the number of bits (binary digits) that are transferred along a communication channel in a single unit of time. The unit of time is usually one (1) second.

Bit rate is, therefore, bits per second or bit/s and is displayed using the SI standards of notation. For example, 1000 bits per second, a kilobit per second, is shown as 1 kbit/s. The lower case 'k' is used here because it is a metric (International System of Units) measurement. Occasionally, bit rate is still shown as bps (bits per second).

The bandwidths made available by different transmission media

Fibre-optic cables are used as media for data transmission because light can travel along them at extremely high speeds and with little signal loss over very long distances compared to copper cables. Fibre cabling can transfer data at around 20 to 40 Gbit/s for most scenarios, but in ideal conditions can achieve bandwidths of up to 100 times this over a single channel.

Copper cables, although many are being replaced with fibre cabling, are still the most common transmission medium for networks and telecommunication lines. Originally, the copper cabling was coaxial with a single core surrounded by a metal braid, insulating and protective coverings. It was used for creating LANs until the need for higher speeds made it redundant. LANs based on early types of **coaxial copper cabling** have a maximum bandwidth of 10 Mbit/s, which is too slow for modern networking. The limitations of maximum lengths of 185 m or 500 m, the cable being stiff and more difficult to install and maintain soon led to the development of **twisted pair cabling**.

Modern copper cabling in networks uses twisted pair cables. The connections are made using pairs of copper cores twisted around, but insulated from, each other. Commonly, the cable has four pairs of twisted cores and bandwidths of between 10 Mbit/s and 100 Mbit/s (designated 10BaseT and 100BaseT) are possible using just two of the pairs. Higher bandwidths, up to 1 Gbit/s, are possible if all of the pairs are used. Sometimes, however, the spare pairs are used for other purposes such as providing power.

Lasers, as well as being used to transmit data along fibre-optic cables, are used to transmit data across free space. 'Free space' means that there is no physical or wireless transmission medium. Free space can be space (as in outside of our atmosphere), a vacuum or just air where there is no actual medium for transmission.

Along fibre-optic cables, phenomenally high data rates of hundreds of terabits per second have been achieved using lasers during experiments and testing, but between 30 and 40 Gbit/s is usual for most commercial scenarios. For use in transferring data in free space, lasers can achieve transfer rates over 1 Gbit/s, but the speed is dependent on a clear line of sight between the transmitter and receiver. Also, there must be very accurate aiming of the laser beams if transfer rates are to remain consistently high. In air, minute particles such as atmospheric dust or water droplets in rain or clouds can disrupt the laser light beams and reduce bit rates. Bandwidths are still less than 200 Mbit/s in tests.

Lasers are being tested between spacecraft or satellites as an alternative to conventional radio or microwave connections. Outside of the atmosphere, in space itself, while there is still dust or particles that can interfere with light beams, the quantity is so much lower that laser beams can reach very long distances without loss of signal. In outer space, bandwidths of up to 10 Gbit/s have been achieved in tests.

The bandwidths made available by different access technologies

The technology used to transmit data affects the bandwidth available for transmission. The following processes all affect the bandwidth available for sending data between computing devices:

» preparing the data
» addressing the data
» adding any extras such as error-correction data
» modulating and demodulating the data on and off any carrier waves
» requesting the re-transmission of lost or damaged data.

Ethernet

Ethernet technology divides data that is to be transmitted into small sections called frames. Frames carry the data, its source and destination addresses and error-correction data. The error-correction data allows for the resending of damaged and discarded frames.

Ethernet was originally designed to use coaxial copper cable but has been developed and improved to be able to use twisted pair and fibre-optic cable as its transmission medium. It can support bandwidths of up to 400 Gbit/s.

Fibre optic

Data is transmitted along fibre-optic cables at high speed because it is carried by light. While the speed of light inside fibre-optic cables is only about 60 per cent of its speed in free space, the high speed, combined with various methods of modulating the data onto the light, still allows for significantly higher data-transmission rates compared to copper cabling, especially over long distances. Also, there are fewer repeaters used in long-distance fibre-optic cabling than in the equivalent copper cabling installations. This means that there are fewer processing overheads when using fibre-optic cables.

Fibre-optic cables can carry many different wavelengths of light at once, with each wavelength providing a separate, independent channel of communication. The total combined bandwidth of all the channels through the fibre can be very high. Data-transfer rates as high as over 100 terabits per second have been achieved under test conditions.

In commercial applications, small diameter fibre that uses one wavelength of light, called single mode, makes bandwidths of 1 Gbit/s available, while larger-diameter fibre using many wavelengths, called multi-mode, can provide from 100 Mbit/s to 40 Gbit/s.

Wireless

Different wireless technologies using radio waves provide differing bandwidths.

Bluetooth has a low bandwidth of about 1 to 3 Mbit/s. It is used for data transfer over short distances and uses much less power than other wireless technologies. Later versions of Bluetooth may appear to have higher bandwidths, some over 20 Mbit/s, but the data is not actually transferred over the Bluetooth link itself. Instead, Bluetooth is used to set up a Wi-Fi link for transferring the data.

Infrared lasers are used in fibre-optic communications to provide the light for data transmission along the cables. It is also used between computing devices but does not pass through obstacles such as walls, so is only suitable for short-range communications. Infrared is commonly used to send commands from remote control devices to appliances, for example televisions or sound systems. It has a low bandwidth and usually carries only a series of pulses of coded data as commands to the appliance.

Wi-Fi is used for wireless networking in homes and businesses and can provide enough bandwidth for video streaming and other bandwidth-consuming activities. Wi-Fi versions and standards are continually developing and each new version provides for greater bandwidths. Bandwidths of 11 Mbit/s and 54 Mbit/s are quoted as possible in homes. In practice, the bandwidth is often lower due to the restrictions imposed by the location of the WAP and the construction materials of the building. The later Wi-Fi standards have bandwidths of up to 1 Gbit/s.

Microwave links are used to transmit data between two devices. Microwave radio transmissions do not easily penetrate obstacles so are restricted to devices located within line of sight of each other. Microwave transmissions are used to carry data between mobile TV cameras, base stations and satellites, and between mobile telephone base stations. Microwave links used to provide an internet backbone in residential areas can have high bandwidths in excess of 10 Gbit/s.

Mobile communications

Mobile data connections allow users of portable devices, for example smartphones, to access internet services using the mobile telephone networks. The first mobile data connections appeared in the 1990s and had low bandwidths of around 200 kbit/s, which was considered fast at the time. However, modern mobile technologies allow users to watch streamed video, access web services and talk with each other over high-bandwidth connections. 3G technology has speeds of around 5 Mbit/s, although some providers offer more (some also provide a lot less); 4G technology can reach from 100 Mbit/s to 1 Gbit/s; and 5G promises bandwidths of 10 Gbit/s or more.

The increased bandwidths available from 4G and 5G connections are possible because they use higher radio frequencies than the earlier generations of mobile data connections and use different methods of modulating data onto the radio waves.

14.5.2 Data streaming

Data streaming allows users to access content immediately without having to wait for it all to be downloaded. For example, video or audio files can be watched or heard on a smartphone as soon as enough data has arrived to display some video frames or to construct some sounds for playback. There is no need to wait for a 1.5 Gigabyte data file of a movie to be completely downloaded before viewing can start. Minor interruptions of the data stream are compensated for by downloading a little bit more data than is immediately required to create a 'buffer' of data.

Continuous streams of data are also generated and used to track events. Financial institutions analyse the continuous streams of data coming from stock exchanges and banks, the websites of online retailers track and analyse the 'clicks' of buyers and potential customers, and large organisations collect, analyse and distribute the vast amount of data that continuously streams from social-media platforms. Other examples of where data streaming is used include computer gaming, where the activities and interactions of players are collected by gaming platforms, and global news companies, which distribute news data around the world. These types of activities make use of cloud computing systems where data is continuously streamed between data centres.

Real-time and on-demand media streams

Real-time streaming of media, for example video and audio content, is the capturing, encoding and distribution of live events over the internet. Examples of live streaming include the streaming of a sports game or a music concert as it is happening. Although an event may be recorded for later use, live streaming does not involve the recording of the event before it is streamed, so the audio and video are compressed and streamed immediately.

On-demand streaming distributes previously recorded content to users as and when they wish to access it. Examples of on-demand streaming are watching a movie or episodes of a old TV series.

Both real-time and on-demand multi-media content can be viewed in a web browser or with dedicated apps on a range of devices, for example smartphones to large smart TVs.

Both real-time and on-demand streaming of media require the content to be compressed so that the data can be transferred to users over the internet. While compressed video and audio content can be streamed over low-bandwidth internet connections, for a good viewer experience internet connections should have enough bandwidth to avoid jerky video, lagging of audio, slow buffering, freezing or complete loss of the data stream.

A bandwidth of about 2 Mbit/s is sufficient for streaming video in standard definition, for example DVD quality; but a bandwidth of about 5 Mbit/s is required for high-definition video, of Blu-ray disc quality, to be streamed properly to the user; and 9 or 10 Mbit/s is required for UHD-quality streaming.

Impact of bit rate and bandwidth on the streaming of audio and video data

The term 'bit rate' is used to describe the bandwidth of a network connection, but is also used when referring to video and audio files. It is used to describe the amount of data per second that is provided to a client, such as a media player, to display a video or audio file.

The bit rate of a video or audio file represents the amount of information that is stored in a single unit of time, usually one second. Just like bandwidth, this is shown as bits per second or bit/s. Video stored at low bit rates produces poor-quality images that lack detail, can be jerky in motion and are pixelated. The higher the quality of video required, the more bits per second must be stored. For example, video with little movement, such as that used for video calls on smartphones, could be stored with as low as 16 kbit/s, but Blu-ray high-definition video would need to be stored at a rate of 10 or 15 Mbit/s. For acceptable-quality internet video, for example from YouTube, a rate of 2.5 Mbit/s is sufficient.

Similarly, audio files are stored at different bit rates. A low bit rate such as 32 kbit/s is enough for just speech, but multi-channel sound on a Blu-ray disc can require up to 18 Mbit/s, as uncompressed audio files are huge.

The bit rate for audio files is calculated by multiplying the sampling frequency by the number of bits per sample and then multiplying by the number of audio channels in use. The result is shown in kilobits per second (kbps).

Audio that is stored or transmitted with insufficient bit rate sounds poor. Humans are more sensitive to problems with audio than with video and can detect sounds with too low a bit rate as having 'warbling' or 'tweeting' bird-like sounds or silences ('drop-outs'). This spoils the listener's experience more than the occasional video problem would spoil a viewer's enjoyment of a movie.

To ensure that a viewer's or listener's experience of the media is good, the transmission medium carrying the streamed content must have sufficient bandwidth to deliver the required number of bits per second. If the multimedia content is not compressed, the required bandwidth would be far too high to be possible. To deliver CD-quality sound as uncompressed audio, a bandwidth of 1.4 Mbit/s would be needed, and for high-definition video, over 1 Gbit/s would be necessary. To reduce the bandwidth required, and to make the streaming of audio and video possible over most internet connections, the content of the video and audio files is compressed. A number of **compression** techniques, using codecs, are used to reduce the bit rate of multimedia files.

If the bandwidth of the connection is not sufficient to transfer the file data fast enough, the client application cannot properly display the video and audio. The viewer will experience jerky video that may stop altogether or poor-quality audio with artefacts such as warbling. If the bandwidth exceeds the bit rate required, the client application can receive and store more of the content in a buffer so that, if a connection problem occurs, the viewer can still watch the video or listen to the audio while there is data in the buffer. If the problem is resolved before the data is used up, the viewer notices nothing, but if the buffer empties before the connection problem is resolved, the video or audio will stop or 'freeze'.

The bandwidth of the internet connection used for real-time or on-demand streaming must be able to deliver enough data for immediate viewing and a buffer of data to ensure a good viewer experience. A minimum bandwidth of 2 Mbit/s is recommended for streaming movies and less for just audio streams, and this can be easily provided by most home broadband providers. Wired networks and Wi-Fi can meet this requirement. 3G mobile connections, which are stated to be able to provide up to 5 Mbit/s, rarely manage to stream multimedia adequately, but 4G and 5G can easily do so. The high bandwidths of 4G and 5G allow the streaming or download of multi-media content with few, if any, issues with buffering.

14.5.3 The properties, features and characteristics of different transmission methods

Different transmission methods are needed for use with the different transmission media used in computer networks. However, all of the methods carry digital data in the form of signals. Where more than one type of medium is used on a network (for example wired and wireless media) or where the signals pass along different media, the signals are converted (for example by routers, gateways or dedicated transceivers) at the physical, hardware level. The communicating user devices, for instance smartphones or laptops, have no knowledge of them. For example, a user checking emails and downloading them onto a smartphone is connected to the internet using Wi-Fi into a WAP. The WAP has a twisted-pair copper wire connection to the router, which is connected to an internet service provider. The ISP uses fibre-optic cables to connect into the wider internet, which is made up of copper, fibre, wireless and satellite communications systems. The emails travel as signals along all these different methods of data transmission and are converted whenever they change medium.

Fibre optic

Fibre-optic cables are extremely thin strands of glass or plastic polymer drawn out to very long continuous lengths. Silica-based glass is used because it can carry a wide range of wavelengths of light. Plastic polymer strands are cheaper to manufacture than glass and are often used for small business and home installations as they are usually restricted to shorter distances.

Fibre-optic strands can be easily broken and, unlike copper wiring, do not survive tight bending. To protect them, they are surrounded by coverings, but even these will not protect them from sharp, right-angled, corner bends. Broken strands are more difficult to repair than breaks in copper cabling. Joins in fibre-optic strands require specialised equipment and skilled technicians, and joins can reduce both the performance and the bandwidth.

The individual strands are surrounded by a cladding, or coating, made of acrylic that has several functions. Its main function is, because it has a different refractive index from the glass core, to trap the light waves along the glass core. It also helps to protect the glass fibre from water and from physical damage. The glass and cladding are wrapped

▲ **Figure 14.7** Fibre-optic cable

in resin, a buffer and a protective plastic outer covering to protect the cores from chemical and physical damage. A protective metal armour may also be added if the cable is to be laid directly into the ground or used in other hazardous locations.

Data is transmitted along fibre-optic cables by modulating the data onto pulses of light. The light is kept inside the fibre core by total internal reflection because the cladding around the core has a higher refractive index than the glass and the light is reflected back into the core with almost no loss at all. Using this phenomenon, light can be made to travel very long distances along fibre-optic cables with little loss of signal. Fibre-optic cables can be bundled together and the different strands used separately or in groups to make the physical connections between locations. As explained when discussing bandwidths, fibre-optic cables can carry many independent channels of communication by using different wavelengths of light at once. The number of wavelengths of light that can be used in a strand is partly determined by the diameter of the strand. Small diameter fibres that use one wavelength of light are called single mode, and larger diameter fibres that use many wavelengths are called multi-mode.

The light used in fibre-optic communications is generated by LEDs or by lasers. Laser light is preferred because it can be created at higher power and is coherent light, while LED light is not. Coherent light consists of a single wavelength while incoherent light contains many different wavelengths.

Lasers can be used to generate coherent light at different wavelengths so that each can be sent independently down fibres. Incoherent light from LEDs with many wavelengths cannot be used like this. LEDs cost less than lasers to manufacture and are used in low-cost installations. LEDs do not produce as much power for the light as lasers so are used for cables over shorter distances.

The data to be transmitted is modulated onto the light waves by the lasers and shone down the fibre. Laser light is more directional than LED light so is used for thinner fibres. Coherent light from lasers allows the fibres to carry large amounts of data and thus have a high bandwidth because it can be modulated in more ways. For example, the wavelength can be altered, the amplitude (size of the wave) can be altered, the phase can be altered, or a combination of all three can be used when modulating signals. Phase-modulation techniques now enable very high bandwidths to be achieved. Laser light is usually modulated by an external device rather than by the laser itself. This allows complex modulation to be achieved on multi-mode fibres so the bandwidth is very high and the light travels further.

The light can travel very long distances without much loss of signal. It is received at the end of the fibre cable by a light sensor that turns the light into electrical signals. Data transmission over great distances can be achieved with fibre optics. Repeaters receive the light signals, which are reduced in power over distance, convert them into electrical signals and then resend them as higher-power light signals to travel over very long distances. However, the cost of repeaters is high when complex modulation is used because of the processing involved. The cost is extremely high when many wavelengths of light are used to increase bandwidth because a repeater is needed for each wavelength.

Optical repeaters that amplify and resend the light itself without turning it into electrical signals are now used. They are less complex, are cheaper to design and manufacture, and need less power and less maintenance because they only receive and amplify the light itself. All the wavelengths of light are repeated at once so there is no need for any separate processing of the channels of communication.

Uses of fibre optics

Fibre-optic cable connections are used in networks to provide links between nodes. Many connections between countries and continents now use fibre-optic cabling as the transmission medium because signals do not degrade, the cables need less maintenance than other media and fewer intermediary devices are needed. Signals on fibre-optic cables are not subject to electrical interference, and it is difficult to tap into the fibres to eavesdrop on communications.

Fibre cables are often used in large company networks but not so often in home networks. Home networks are usually wireless to avoid the use of cables altogether. Also, home routers supplied by ISPs include Wi-Fi and copper cable connections but rarely include provision for fibre optics.

Advantages and disadvantages of fibre optics

▼ **Table 14.10** Advantages and disadvantages of fibre optics

Advantages	Disadvantages
• Its ability to carry very high bandwidth connections and to be able to carry these over long distances without signal loss: Distances of 2 km or more are common, whereas copper cabling, such as twisted pair, is limited to about 100 m. • Use glass as strands: The raw materials for glass are plentiful so fibre can be cheap to manufacture. Often, fibre-optic cable is cheaper to install than copper. • More resistant to corrosion than copper so is cheaper to maintain over long periods of time, especially when used externally. • Take up less space, weigh less and can actually be more flexible (depending on the external protective construction) than the same diameter copper cabling. • Not susceptible to electromagnetic interference nor do they radiate any electromagnetic energy so eavesdropping is not possible. • Difficult to physically tap into without detection: Fibre-optic cables are very secure compared to copper cabling.	• Easily damaged if bent into too sharp a corner as the cores will snap. • Installation and repairs can be expensive because specialised equipment, and considerable skill, are needed to fit connectors and make joins. • Long-distance links require expensive and complex repeaters, especially for high-bandwidth, multi-mode links.

Copper cables

Copper is a metal element that is ductile, which means it can be easily stretched and bent without breaking, is highly conductive of electricity, and is relatively common and easy to extract from copper ore deposits. Its properties make it highly suitable for use as cabling. It can be easily and cheaply made into wires that conduct electrical signals very well.

Other metals are also used as electrical conductors, for example gold or silver is used in contacts for connectors or as tracks between components on circuit boards, and aluminium is used in electrical or telephone cabling. Gold and silver may conduct electricity better than copper, but they are more expensive. Aluminium can be used as wires, but it can corrode more quickly and break more easily when bent.

Copper cables used in networking for data transmission are either coaxial or twisted pair.

Coaxial copper cables

Coaxial cables have a central copper core surrounded by an insulating material (the dielectric), usually made of plastic, and then a shield of

▲ **Figure 14.8** Coaxial copper cable

braided copper. There is an outer protective plastic coating that defends the conducting layers from physical and water damage. The central copper core can be a single strand of copper or made up of thinner strands twisted tightly together. Sometimes the core is made of steel coated with copper. The central core carries the data signals, and the braided copper shield provides a return, or earth, connection to complete the electrical circuit.

Until the late 1990s and early 2000s, coaxial copper cabling was used, to 10Base2 standard, in most LANs for data transmission by Ethernet. This has now been superseded by twisted pair cabling, as 10BaseT.

10Base2 refers to the 10 Mbit/s bandwidth; Base refers to the type of signalling; and 2 refers to the maximum distance of 200 m of the cable length between nodes. In practice, the maximum length was only about 185 m and there was also a restriction of 30 nodes in a network segment. Also, the minimum length of cable between nodes was 50 cm. The connectors used were T-shaped and attached to the NIC with the cable running from node to node. This worked well for the Ethernet LANs at the time, but the low bandwidth prompted the move to 10BaseT, where T refers to twisted, and then to configurations with much higher bandwidths.

Uses of coaxial cable

Although networking has moved to twisted pair cabling, coaxial cables are still used for cable television systems, carrying the TV channels into the home from street cabinets linked by fibre-optic cabling. Also, coaxial cables carry terrestrial TV signals from the antenna to TV sets and carry satellite TV signals from the dish to the satellite TV decoder set-top box. Different copper core and dielectric dimensions, which are precisely set during manufacture, are used for the different purposes.

Coaxial cables work well for some data-transmission uses because they can carry signals over reasonably long distances without signal loss, are resistant to electromagnetic radiation or interference and can still work even when having suffered some damage and corrosion.

Advantages and disadvantages of coaxial cable

▼ **Table 14.11** Advantages and disadvantages of coaxial cable

Advantages	Disadvantages
• Costs less to purchase and install, requires less skill to install and requires less maintenance time compared to twisted pair or fibre-optic cables. • For public telecommunication links, coaxial cables are used because of their better shielding, lower susceptibility to interference and less crosstalk, which allows longer lengths of cable than twisted pair cables. • Coaxial cable is preferred for **downlinks** from antennae or dishes to TV sets and set-top boxes, because it is easy to install and requires almost no maintenance.	• Unsuitable for use in computer networking. • Coaxial cables do not carry the signals required to provide the high bandwidths used in LANs, but twisted pair and fibre-optic cables do. • Coaxial cabling can radiate more electromagnetic energy than twisted pair and this can be used for eavesdropping on the transmissions. • Most devices today do not have the necessary connectors or electronic components to use coaxial cable for LANs but instead are fitted with the connectors for twisted pair or fibre-optic cabling.

Activity 14k

Explain why coaxial copper cable is used for telecommunications links.

Twisted pair copper cabling

Twisted pair copper cabling consists of two cables, each with a copper core encased in an insulating plastic sheath, twisted around each other. The core is a single strand, or several tightly twisted strands, of copper. Twisting cables around each other can reduce interference and noise from electromagnetic sources near the wires because both wires in the twist are affected, which cancels out the interference and noise. Where cabling carries more than one pair of twisted wires, the numbers of twists per metre can be different for each pair to avoid them interfering with each other. Wrapping the pair of wires in a shield can reduce the effects even more. The different number of twists and the shielding also reduce 'cross-talk' between the pairs of wires. Cables with no shielding are called unshielded twisted pair (UTP) and those with shielding are shielded twisted pair (STP) cables. There are different methods of shielding used in the cables (such as by individual cores, in pairs or in groups of pairs) but all STP cabling is more expensive than UTP.

Uses of twisted pair copper cabling

Twisted pair cabling is used in traditional voice telephone connections and in modern networks.

UTP cabling is used in most modern networks. UTP cables are categorised to a number of standards, the most common being Category 5 (Cat 5 and Cat 5e) cables, which can carry data for 100BaseT and 1000BaseT networks with bandwidths of 100 Mbit/s and 1 Gbit/s respectively. Later categories, for example Cat 8, can have a bandwidth of up to 40 Gbit/s.

▲ **Figure 14.9** Twisted pair Cat 5e cable

Cat 5 (and Cat 5e) UTP cables have eight cores arranged in four pairs. 10BaseT or 100BaseT Ethernet networks with Cat 5 (and Cat 5e) cables use only two out of the four of the pairs. The other two pairs are not used. There is a limit of 100 m in length for Cat 5 and Cat 5e cable use in networks. For use with 1 Gbit/s, 1000BaseT, and more bandwidth, all four pairs of cables are used.

Each pair in the cable is assigned a purpose in the data connection and each core has a covering with a different colour from all the others to make connecting the plugs easier. Special tools are available to align and crimp (squeeze tightly) the cables in the plugs so network technicians, and home users, can easily make their own custom-length cables. The common name for the connectors is RJ45, but the correct name is an 8-position, 8-contact (8P8C) modular connector. Single core strands are easier to fit to the connectors, but most cables have stranded cores. This is because stranded cores do not break so easily when the cables are rolled up or are bent and folded around corners when installing.

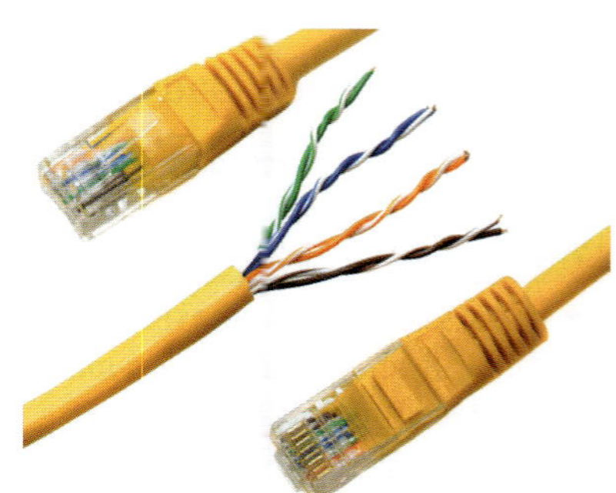

▲ **Figure 14.10** Twisted pair Ethernet cable

Cross-over cables, with the connectors wired so that the cores are wired differently at each end, were once required to connect to nodes such as switches and routers to ensure that each pair was connected correctly at each node. However, because most modern switches and routers can autosense which pairs are in use and adjust themselves automatically, cross-over cables are no longer necessary. Also, NICs can now determine whether 10BaseT or 100BaseT is in use on the network.

Shielded twisted pair (STP) cable is more resistant to electromagnetic interference but is more expensive. It is sometimes used over short distances instead of fibre-optic cabling as there is no need for extra transceivers.

Advantages and disadvantages of twisted pair cable

The main advantages of using twisted pair cables in networks for data transfer are that they provide very high bandwidth connections, the cross-talk between the cores in the cable is reduced and there is less interference from electromagnetic sources compared to coaxial cable. Compared to fibre-optic cabling, UTP cable is now relatively cheap to purchase and easy to install because it can be shaped around corners and through spaces, and most devices have connectors already fitted. Also, UTP cables can be easily assembled by technicians.

Care must be taken during installation and use when bending or moving UTP cables, as deforming the cores and pairs can affect the rate of data transfer. Kinks or sharp bends in the cores can set up electrical reflections that may cause signal loss and data corruption. If cores are damaged in different ways, the losses can be considerable. Coaxial cables are not so susceptible to data loss when damaged, but fibre-optic cables can fail completely when damaged.

Twisted pair cables are more susceptible to electrical interference than fibre-optic cables. They also radiate less electromagnetic energy than coaxial cables so are more difficult for unauthorised eavesdropping.

As with coaxial cabling, twisted pair cables have a plastic covering so are considered fire hazards when used in long installations between rooms and buildings. Precautions have to be taken to include firebreaks so that fires cannot burn along the cables from room to room. Replacing the twisted pair cabling with wireless or fibre-optic links is sometimes a necessity.

Lasers

As well as sending light beams down fibre-optic cabling, lasers can be used for transmitting data where there is no physical or wireless transmission medium to guide the signals. Visible light, infrared and ultraviolet laser light can be used.

Infrared light is used for crossing free space, which is atmospheric air, a vacuum or space far from the Earth's surface. Where it is very difficult, or impossible, to make links with solid cabling (for example using copper or fibre-optic cables in outer space or across hazardous terrain) using lasers to exchange data over free space is an alternative to wireless links.

The data signals are modulated onto coherent IR light and sent out as laser light beams. The transmitter and the receiver of the beams have to be very accurately aligned because the beams are narrow and travel in straight lines. There is little room for error.

In the Earth's atmosphere, the usable distance is limited to about 500 m because particles of water and dust in the air, and the air itself, reflect and absorb the light. Also, any obstacles in the path of the beam will disrupt or block it.

Uses of lasers in data communications

Free-space optical (FSO) communication, using lasers to carry the data signals, is used to provide line-of-sight, wireless, high-bandwidth communication links between remote sites. In cities where laying new fibre or copper cabling is too expensive, FSO can be used to provide high-bandwidth internet links. Improvements in the components used in laser technology have increased the bandwidths possible, reduced the errors caused by obstructions and enabled smaller transceivers to be made. FSO has also been used to connect aircraft, spacecraft and military installations.

Between spacecraft in orbit around the Earth, it is obviously impossible to use wired connections. Wireless connections work well, but laser links are now being developed. With the ability to provide a high bandwidth over relatively long distances in space, even though they require line-of-sight, laser links between constellations of spacecraft or satellites can be used to exchange data such as high-resolution images. The use of lasers means that images can be received much faster by scientists than if they were using radio waves to transmit the images.

▲ **Figure 14.11** Lasers for data communications

Advantages and disadvantages of lasers

Using lasers for data communication provides a high-bandwidth connection. A laser connection across free space is difficult to tap into or eavesdrop on.

The main disadvantages of using lasers for data communications are that the equipment for laser links is expensive and it requires skilled technicians. Laser links must be accurately aligned on setup and must stay accurately aligned. This is relatively easy for stationary installations (for example between buildings) but for moving objects (for example satellites) maintaining a line-of-sight, narrow-beam connection requires complex computer control systems.

Laser beams are narrow and links must be line-of-sight to avoid loss of signal and data loss. The laser beams travel in straight lines and cannot pass around corners. Atmospheric particles, such as dust or water droplets, can reduce the amount of light passing and degrade the signals. Any obstacles in the way of the laser light beams will stop the signals completely.

Laser beams create a safety hazard. Laser light can cause eye damage, so the use of lasers is restricted and care must be taken with installations.

14.5.4 Effect of the method on available bandwidth for data transmission, audio and video streaming

While copper cables carry data modulated onto electrical signals that travel slower than the light waves used in fibre-optic cables, the bandwidth is affected by two factors:

» the number, and range, of different signals that can be sent at once along the medium to carry the data
» the method used to modulate the data.

The techniques that are used to modulate signals with data are beyond the scope of this syllabus but are different for each type of medium.

In networks that have several different transmission media making the links between nodes and between network segments, the available bandwidth on the network is reduced below the maximum possible. This is because there is always some traffic, for example between switches and routers, monitoring the network and there are processing delays whenever the signals cross from one medium to another.

The choice of transmission method affects how fast data can be exchanged. The transmission method determines how fast, and how many, signals can be sent along a medium and this determines the bandwidth that is available.

As previously discussed, twisted pair copper and fibre-optic cables, which are the most common media and methods used in LANs, provide the highest bandwidths for use in computer networking. Audio and video streaming at the highest quality requires high bandwidths.

14.5.5 Effect of the method for ultra-high definition television systems

The development of ultra-high definition (UHD) TV has provided a better experience when viewing movies and television programmes at home, but the transmission of UHD TV requires much greater bandwidth. Even with modern compression techniques, a minimum bandwidth of about 25 Mbit/s is needed to stream UHD video reliably. Only transmission media that can provide very high bandwidths are used for these services.

14.6 Network protocols

14.6.1 Definition of a protocol and why protocols are necessary

An agreed set of rules and conventions that determines how computer devices can communicate with each other in the same network is called a network protocol. Network protocols are necessary to specify the rules and conventions on the following:

» how network devices identify themselves and other devices
» how they create connections between each other
» how they package the data to be exchanged
» how the packages of data travel and find their way between devices.

All devices on a network that need to communicate with each other must use the same protocols to exchange data and make use of the global network that makes up the internet. If a device does not adhere to the protocols used by other devices, it will not be able to communicate with the other devices.

14.6.2 The purpose and use of protocols for exchanging data across networks

The purpose of network protocols is to ensure that data can be exchanged quickly, reliably and securely. A network protocol is said to be reliable if it informs the sender whether or not a data packet has been delivered to the intended destination. Reliable protocols are used when it is important that data is received, for example during a login process. An unreliable protocol does not inform the sender and is used when the loss of some data is not important, for example in video streaming.

Different types of devices can communicate on modern networks if they use the same set of protocols, regardless of what network they are on, the transmission medium in use and their location in the world. The protocols used for communication over modern networks, including the internet, are set out in the Internet Protocol suite, often called the 'TCP/IP' set of protocols. TCP is the Transmission Control Protocol and IP is the Internet Protocol. TCP/IP determines how the data should be addressed, packaged, sent and routed from one device to another.

Transmission Control Protocol

Transmission Control Protocol (TCP) software in the **network stack** is responsible for passing data between applications and the IP software. It controls the flow of the data by setting up and maintaining the logical connection between devices, ensuring that the packets transported between devices are correctly ordered, error checked and reliably delivered.

TCP is connection-oriented and first sets up a connection between devices by a series of 'handshakes', which are short messages detailing how the data is to be exchanged. The client first sends a request, the server or other device sends back an acknowledgement, and the client then acknowledges the server's acknowledgement.

TCP then enables the data to be exchanged, using the details set up by the handshake, in a two-way (duplex) connection between the client and the server. The exchanges are checked for errors and if the packet is not correct, or is missing, a request for re-transmission is sent. This ensures the reliable delivery of data.

After all the data has been exchanged, the TCP connection is ended with another series of handshake messages. These indicate that the data has been successfully transferred or that one of the devices is closing the connection.

An example of TCP in action is when a web browser application requests a web page from a web server and sends the data containing the request to the TCP software in the network stack. TCP sets up the connection by creating a TCP packet with the appropriate header information. In this case, the header will contain details of the HTTP ports to be used, usually 80, the sequence number of the data or packet, and the other required information.

The packet from the TCP software is passed to the Internet Protocol (IP) software for addressing and then passed along down the network stack to be sent to the web server.

When the NIC in the device of the web browser receives data from the server, the packet arrives and is passed up the network stack to the TCP software, where the packet is examined. The checksum in the header is used to check that the header has been received correctly, the sequence number is used to determine the order of the data and an acknowledgement is sent to the web server, if required. If all is correct, the port number in the address field of the header is examined and the data payload is passed to the application, in this case port 80 for the web browser.

Internet Protocol

The Internet Protocol (IP) specifies the following:

» the addressing system for labelling and routing network datagrams, or packets
» how the packets are structured
» how data packets are fragmented and reassembled for transport in networks if, as is usually the case, the amount of data is too large for a single packet.

Every device in a network is assigned a unique IP address to identify it. Packets using IP have headers that hold the source and destination IP addresses, which are used to route the packets through networks to their destinations.

The Internet Protocol system is connection-less and relies on routing information to deliver packets. There is no fixed route set up between devices by the IP software, but enough information is placed in packets for them to be directed (or routed) through any number of networks, from node to node (for example routers) until they reach their destination or are discarded. Many errors, corruption of data, duplications or losses of packets may occur along the way, but the sender is not notified of these under IP.

In networking terms, an IP network is said to be unreliable, which means only that the sender does not get a notification about the delivery of a packet. The responsibility for reliable delivery, and notifying the sender whether or not the packet has been delivered safely, is that of TCP. TCP sets up its connections and deals with errors while using IP to carry its packets back and forth. Each IP packet is transmitted independently to find its own way on the network and this enables many devices to share the bandwidth of the transmission medium.

The address data in the IP header of the packet is used to route it to its destination. On a network segment, a packet is received by all devices and it is processed only if the device is the intended recipient, else it is discarded. If the packet is intended for a device that is not on the local network segment, for example a packet with part of an email message for someone on the other side of the world, the packet is picked up by a gateway and repackaged using the address in the header and sent to a router on the internet. The packet is passed from router to router until reaches its destination. Each router examines the IP headers for the addresses and, if it knows where the destination is to be found, it delivers the packet, but if it does not know then the packet is sent to the next router, and so on. Eventually, the packet is delivered or, if the hop count reaches zero, the packet is discarded and the sender is notified with a message using the **Internet Control Message Protocol (ICMP)**.

TCP/IP can be compared to sending goods across the world using a postal service. TCP is used to make the goods for the parcel with a note of their contents and what they are for. IP constructs the parcel to hold the goods, the TCP goods are placed inside and the parcel is addressed to the recipient. The sender's address is also put on the parcel. The parcel is given to the local depot of the postal service, which passes it from depot to depot, carrying it by road, rail, air or sea, with each depot not knowing the location of the recipient but just passing the parcel on until it reaches another local depot. The local depot delivers the IP parcel, which is opened. The goods inside could have been broken on their travels but, if the parcel is intact, it will be delivered anyway. The TCP goods are examined, and if any are damaged or missing parts, the sender gets a report and is asked to send them all again. If all is well with the goods, the notes are read and the goods passed along to be used for the intended purpose.

▼ **Table 14.12** Purpose and use of network protocols

Protocol	Purpose	Use
Internet Control Message Protocol (ICMP)	• Supports the Internet Protocol by carrying error messages between devices. It does not carry user data. • Sends error messages when a service request by TCP fails or packet is undeliverable.	• ICMP works at the Network layer (layer 3 of the OSI model) as does IP, but error messages from ICMP are often delivered straight to the application that sent the original data. • Used by 'ping' and 'traceroute' or 'tracert' for network diagnostics.

Protocol	Purpose	Use
Address Resolution Protocol (ARP)	• Finds the MAC address of a NIC from the IP address on Ethernet networks. • Replaced by Neighbour Discovery Protocol (NDP, ND) on IPv6, which is used to discover the hosts, routers and other devices on the network.	• MAC address has to be found because nodes are uniquely identified by their MAC address, which unlike their IP address does not change. • ARP lists kept on every device on a network and updated and broadcast across the networks. • APR lists are not shared with other networks.
Inverse Address Resolution Protocol (InARP)	• Finds the IP address from the MAC address.	• Used on frame relay networks for finding the IP addresses when routing packets between networks.
Dynamic Host Configuration Protocol (DHCP)	• Assigns IP addresses to devices. • Informs devices of the network parameters, for example the default gateway IP address.	• DHCP server works as a client–server model, where other devices joining a network send requests using a **User Datagram Protocol (UDP)** packet. • DHCP sends network details to client as a lease that expires, is automatically renewed or is reassigned if the device has left the network.
User Datagram Protocol (UDP)	• Sends packets on networks by connection-less method.	• Used where error-checking of packets is not necessary. • Lack of error-checking reduces the response time. • Used by DHCP clients and servers to exchange network details. • Used by routers exchanging routing data with the Simple Network Management Protocol (SNMP). • Used by using the Domain Name System (DNS) when looking up IP addresses/domain names.
Hypertext Transfer Protocol (HTTP)	• Sends requests from, and media content to, web browsers and web servers. • Makes use of TCP/IP protocols to provide the reliable connection needed for HTTP. TCP provides the reliability and connections on ports 80 or 8080. IP provides for the addressing and routing of the requests and responses.	• Used by world wide web using hyperlinks embedded in markup documents, or pages, to display content to users. The HTTP exchanges are made over a secure SSL/TLS connection that encrypts and decrypts the requests and responses. • Is an application-layer protocol that works as a client–server using request–response. A user, for example a college student searching for information for a project, starts a web browser on a laptop. The web browser, the application, acts as a client and sends a request to a web server and waits for a response. The web server listens for requests and, when it receives them, sends responses back to the client. The first series of requests and responses sets up the client–server connection and the next series contains the requested content. • Uses the Uniform Resource Identifier (URI) and the Uniform Resource Locator (URL) systems, or schemes, to find and retrieve media content for users.
Hypertext Transfer Protocol Secure (HTTPS)	• Creates secure connections using HTTP.	• Encrypts the exchanges by creating an SSL/TLS layer on top of HTTP.
File Transfer Protocol (FTP)	• Transfers files over a network between an application on a device, the client, and an FTP server.	• Used in **active** and **passive** modes to create and maintain connections between clients and FTP servers for transferring files.

Protocol	Purpose	Use
Layer 2 Tunneling Protocol (L2TP)	• Enables private data to be exchanged between different networks across public telecommunication networks.	• Used to create a data link on one network that allows packets from another network to travel across it to another network without being examined or altered in any way. • Used to create WANs where two IP LANs are connected together with a network using different protocols. • Supports VPNs for a data link between devices. • Used to overcome restrictions on networks. For example, if packets from protocols that are forbidden on a network are wrapped inside packets of allowed protocols, they can travel on the network.
Simple Mail Transfer Protocol (SMTP)	• Sends email messages from email clients to mail servers. • Sends email messages between mail servers.	• Used to send email messages. It is not used by email clients for collecting (or retrieving) email messages from servers. • Clients use port 587, servers use port 25. Client use of port 25 is blocked to reduce spam emails. • The client first sets up a reliable TCP/IP connection. TCP/IP is responsible for the addressing, error-checking and routing of the SMTP message. • SMTP connections can be encrypted using TLS even though the actual email is not.
Post Office Protocol (POP3)	• Retrieves email from email servers by web clients.	• Works at the Application layer and is used to access email messages stored on email servers in mailboxes. When an email client connects to a mailbox and downloads messages, the messages are deleted from the mailbox. This is the default setting for POP3 but the client, under instructions from the user, can be configured not to do so. In contrast, IMAP's default setting is to leave messages in server mailboxes. • Uses port 110 and expects email servers to be listening on this port ready to accept a connection. Clients request a connection, the server responds and a client–server connection is set up. This is an example of a request–response scenario. • TCP/IP is used for addressing, routing and error-checking the packets, while the email client deals with the username and login process. • POP3 connections are usually open only for as long as it takes to download the emails.
Internet Message Access Protocol (IMAP)	• Retrieves email from email servers by web clients. • IMAPS (IMAP Security) allows for the encryption of email messages with TLS. StartTLS can also be used when encrypting messages. Port 993 is used when IMAP uses TLS for encryption.	• Application-layer protocol used by email clients to retrieve email messages from server mailboxes. • Allows more than one email client application to access and manage the mailboxes on an email server, whereas POP3 does not. Messages must be deliberately deleted by the user as viewing an email message on one device leaves the message on the server and allows it to be viewed on other devices. • Flagging, or marking, messages on one device flags the message on other devices too. • Clients can carry out server-side searches of email messages so that users do not have to download and look through every email to find items. • IMAP servers listen on ports 443 and 993. • IMAP connections can stay open for longer than it takes to retrieve mail messages.

Protocol	Purpose	Use
Telnet	• Creates a connection between the virtual terminal (the client) and the remote host (the server). • For network diagnosis.	• Application-layer protocol that uses TCP/IP for a client–server, connection-oriented method of communication on port 23. • Used to access devices remotely, such as routers, for configuration. • Grants almost total, unrestricted access to the host server and there is little or no security.
Secure shell (SSH)	• For access to remote hosts. • Secure alternative to telnet for logins to remote servers and to FTP for file transfer.	• Administrators of business and organisation networks use secure shell (SSH) to issue commands to remote servers, manage network devices such as routers and switches, and make secure file transfers between remote hosts. • Used, as Secure File Transfer Protocol (SFTP), for automatic file transfers by FTP but with the exchanges secured with encryption. • Uses cryptography when exchanging 'keys' to establish secure connections. The client starts an SSH connection by making a request to a SSH server, which responds with a series of exchanges of data to set up the keys to be used. SSH uses public key cryptography, and the keys to be used are configured and generated from passphrases entered by administrators. Clients have public keys that can be used with private keys held by host servers. These are used to authenticate the client and host when the connection is being established.
Internet Protocol Security (IPSec)	• Authenticates and encrypts packets of data. • Provides for authentication at the start of a session. • Establishes the encryption keys to be used during a session.	• Protects the data flowing between two hosts, between two networks or between a host and a network. • Can be used to protect just the payload of an IP packet while in transport on a network. • Can be used to encrypt and authenticate an entire IP packet, which is placed within a new packet and sent over a network. This is tunneling and creates a VPN.
Transport Layer Security/Secure Socket Layer (TLS/SSL) *Note: Transport Layer Security (TLS) has superseded Secure Socket Layer (SSL), which is now part of TLS.*	• Provides security for data on networks.	• Works at the Transport and Application layers and is used with other protocols by many software applications to secure data. • Is the basis of the digital certificate system. • Uses public key cryptography at the set-up stage for the authentication of the web browser and server. • Then uses symmetric cryptography with shared, secret keys generated during the establishment of the connection, unique to the session, for exchanging the data, such as text and video, between web browser (client) and server. • Used, for example, when a website is accessed by a web browser and content is sent from a web server. The client (the web browser) and the web server are authenticated using cryptography so the connection is secure and private. All subsequent exchange of the web content is encrypted. This means that other users cannot understand the data being exchanged, and this keeps online financial transactions safe and secure from unauthorised users.

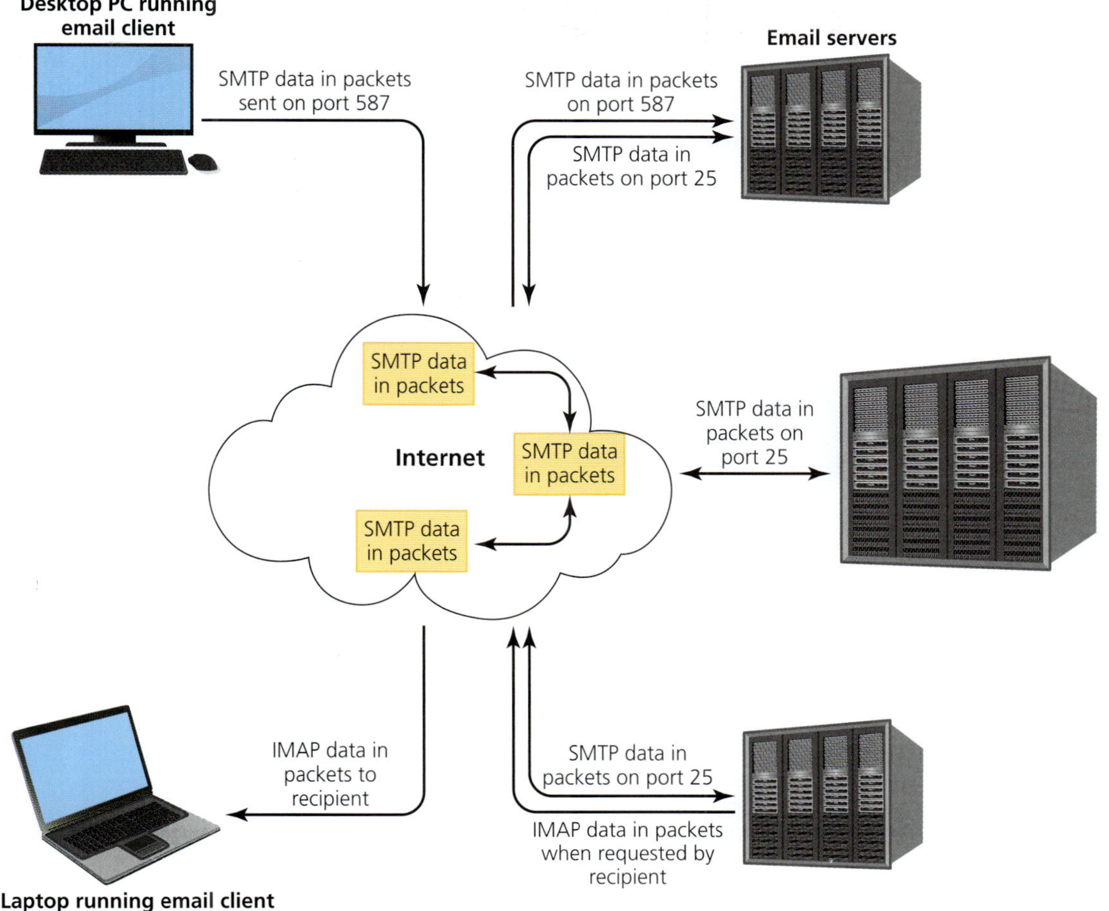

▲ **Figure 14.12** Sending email using SMTP

Activity 14l

1 Why do network administrators prefer not to use telnet anymore?
2 Describe how remote access to configurations without telnet is achieved.

Task 14a

Typing **ping** into a terminal or command line can be used to find out the IP address of a website. **You may need to seek permission to carry out this exercise because you need to open a Windows PowerShell or a terminal window.**

1 Use **ping** to discover the IP address of **google.com**.
2 Use **tracert** to display the route that IP packets take from your computer to **google.com**.
3 Find out and explain why some hops in **tracert** are shown as stars.

Task 14b

Typing **arp** into a terminal or command line can be used to find out MAC addresses. **You may need to seek permission to carry out this exercise because you need to open a Windows PowerShell or a terminal window.**

Type **arp /a** into a terminal or PowerShell window. Explain what each of the columns in the display represents.

Active and passive File Transfer Protocol modes

In active mode File Transfer Protocol (FTP), the server responds to a client by asking the client for the data port to be used for exchanging data. The client sends this information to the server on port 21. The server then sends a message back to the client with a request to make a connection on the client's data port. The most common port used by FTP servers for active mode data connections is port 20.

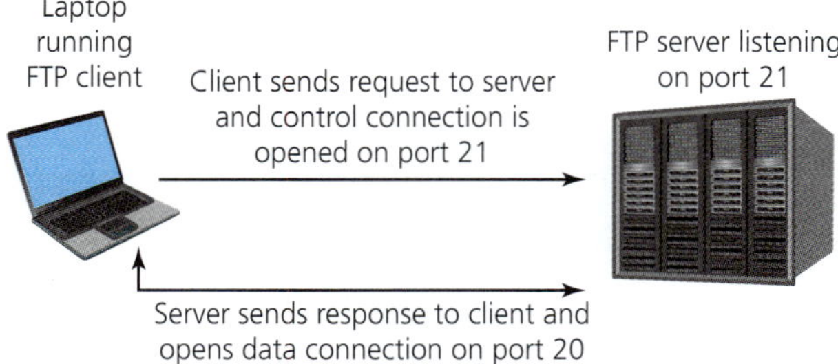

▲ **Figure 14.13** Active FTP connection

In passive mode FTP, the client connects to a data port that the server has already opened and is listening on. In this case, the client has requested the data connection. The client can choose any port available from the FTP server to set up a data connection.

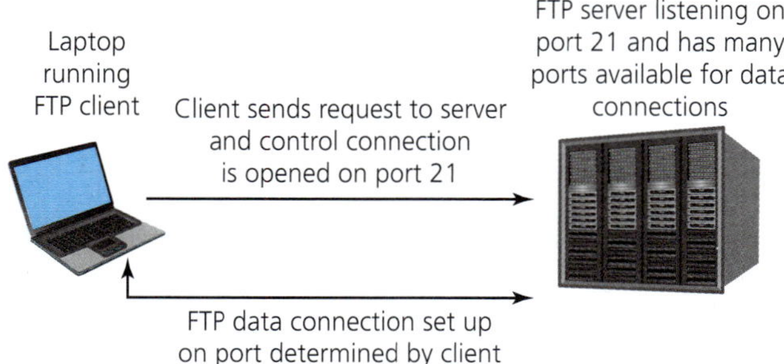

▲ **Figure 14.14** Passive FTP connection

The difference is that in active mode the data connection is started by the server, but in passive mode it is started by the client. This is important because company firewalls are usually configured to block requests from external servers for connections to clients on a network, so they refuse active FTP connections. Passive mode FTP connections are allowed because the data connection originates inside the network and is outgoing.

A username and password are usually required when making a connection to an FTP server. However, many servers allow FTP connections in 'anonymous' mode. Anonymous connections are used on FTP servers that store publicly available files and documents. Most FTP servers allow only limited access to their files with anonymous connections.

FTP connections are not encrypted and the requests and login details are exchanged between client and server in plain text. The data is also not encrypted during transfer. FTP is insecure and all logins, passwords and data can be read by software that can capture them from networks while they are being transferred.

FTPS extends the FTP protocol by adding Transport Layer Security (TLS) support for security. A TLS connection is either made at the very start of the FTP session so it is all encrypted, or the FTP connection is made first and the TLS connection is created and used for exchanging data. Clients can request that a connection be made using FTPS and servers can refuse to make connections that do not do this. Note that Secure File Transfer Protocol (SFTP) is an extension of Secure Shell (SSH) and connections can be used to encrypt commands and data exchanges, but these are not FTP exchanges.

In businesses and organisations, FTP clients can be configured to:

» transfer files automatically to ensure that the most recent are always available
» apply rules that decide which files are automatically selected for transfer
» transfer files automatically based on their name, modification date, file size or file type
» monitor folders for any file changes and transfer files as needed.

Activity 14m

1   Explain why firewalls can cause problems when using FTP.
2   Why is access to some FTP sites called 'anonymous'?

14.6.3 Methods of sending data over a network

At the basic level, data is sent over a network transmission as a series of binary digits carried, or modulated, onto electromagnetic waves. These waves are carried on the transmission media of copper wires, fibre-optic cables or radio, for example Wi-Fi, waves. Network interface cards receive data from the CPU and transmit it along the medium, as well as receiving data from the medium and sending it to the CPU.

Ethernet technology is used to carry data in frames, or datagrams, over the transmission media, providing network services up to the Data Link level (level 2 of the OSI model). Internet Protocol (IP) networks, such as LANs and WANs, are built on Ethernet technology with IP packets carried in Ethernet frames. While IP packets can be carried by technologies other than Ethernet as they travel across the LANs, WANs and the internet, it is the use of packets conforming to the Internet Protocol suite for data transmission that makes the internet work.

There is a limit to the size of a unit of data that can be transmitted on any particular network. It is called the **Maximum Transmission Unit (MTU)** and varies from network to network. The MTU is not the same as the maximum size of an Ethernet frame. While the two are related, each is configured separately from the other. The MTU is set by the network hosts, for example routers, and it is preferable that Ethernet frames do not exceed the MTU. An Internet Protocol packet must fit inside an Ethernet frame, so the sizes of Ethernet frames vary depending on the size of the packets of data that they are carrying. If an IP packet is too large for one frame, it will be divided into smaller frames, in a process called fragmentation, so each fragment can fit into an Ethernet frame. Internet Protocols ensure that IP packets are not made so large that they create frames that exceed the MTU of the network.

In summary, an IP packet fits inside an Ethernet frame that must not be bigger than the MTU of the network or network segment.

Packet switching

In IP networks, for example the internet, the data is divided into small amounts and put into packets addressed with the destination and then sent by NICs. The packets are routed, or switched, across networks to their destination and this is called 'packet switching'.

The packets carry a payload of data along with information about the applications that use it, the address of its destination and the address of its source. The addressing is processed by the NICs and the payload is processed by software applications. For example, a portion of data about a web page would be packaged with the address of the client to which it was sent, the address of the server from where it came and the part of the web page. The addresses would be processed by the server and the client's NICs, and the payload of data would be sent to the client's web browser for eventual display to the user.

In Figure 14.15, a request from the desktop PC for a web page would be broken into packets and sent over the internet. The IP packets can travel from the PC to the web servers and back using any route between Routers 1 and 2. Some packets might travel from Router 1 to Routers B, A, C and E to get to Router 2, while some may use a totally different route.

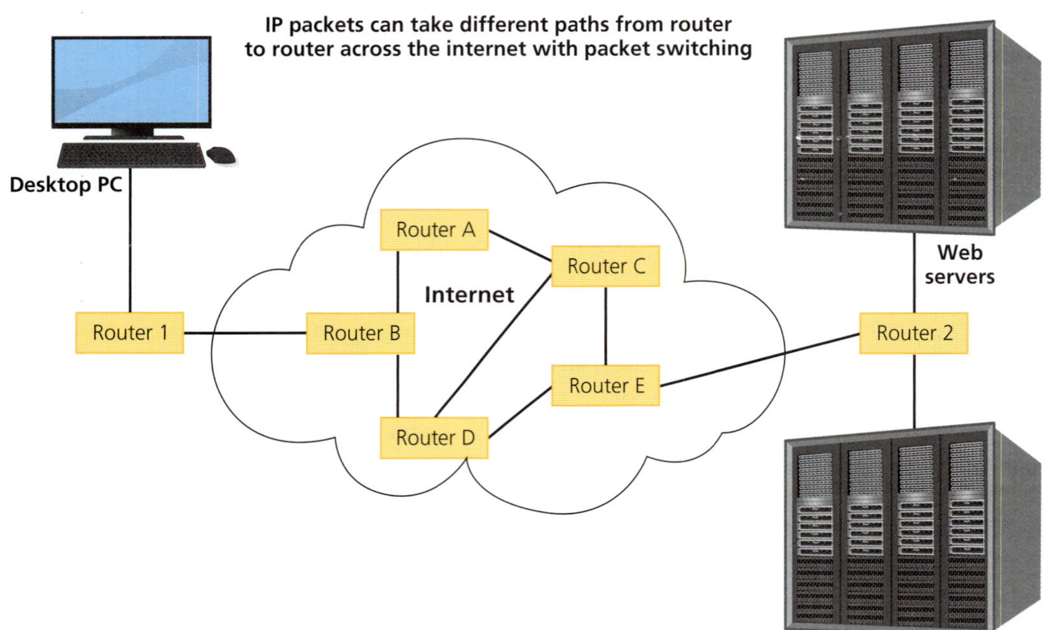

▲ **Figure 14.15** Packet switching

The IP packets are interleaved with packets from other devices travelling over the internet and this makes very efficient use of the communications channels.

Internet Protocols provide the address for use by nodes on networks. Due to the massive increase in the number of devices and the corresponding need for more addresses, the version of the IP protocols is moving from the version used for a long time, Internet Protocol version 4, to Internet Protocol version 6. IPv6 can provide 2^{128}, or well over 3×10^{38}, addresses because it uses 128 bits for the addresses, whereas IPv4 uses only 64 bits. The **packet header** for IPv6 is larger than for IPv4.

While many networks, especially those in homes and small organisations, still use IPv4 protocols, the internet is using IPv6 protocols. However, the two protocols are not interchangeable, which means that IPv4 packets cannot be used on IPv6 networks and vice versa. To overcome this, when IPv4 packets need to travel across IPv6 networks they can be wrapped, or encapsulated, in an IPv6 packet and sent. The converse is also used: IPv6 packets can be sent across IPv4 networks encapsulated in IPv4 packets. This can be managed by the Tunnel Setup Protocol (TSP), which in effect creates tunnels over the host IP networks for the other version to use.

The encapsulation method works quite well but can add large processing overheads and cause delays in network traffic, so other methods are being used. One such method is the use of tunneling with network address translation (NAT) techniques, which translate one version of the IP address into the other.

Modern operating systems can create 'dual stack' networking systems for the devices on which they run. This gives the device both an IPv4 address and an IPv6 address, so it can communicate using either address as needed.

Until the time when there are no more IPv4 addresses in use, data exchange over networks will have to accommodate both versions.

The structure of packets

Internet Protocol (IP) packets are made up of the data payload and a header. The header carries all the information necessary to deliver the payload to its destination, to check its integrity and to pass it to the appropriate software application.

The data carried by IP packets follows directly on from the headers and, as shown in Figure 14.16, the data payload has multiple headers added and removed as it travels across networks.

The structure of IPv4 packets

The header of an IPv4 packet consists of 14 fields and varies in length between a minimum of 160 bits and a maximum of 480 bits.

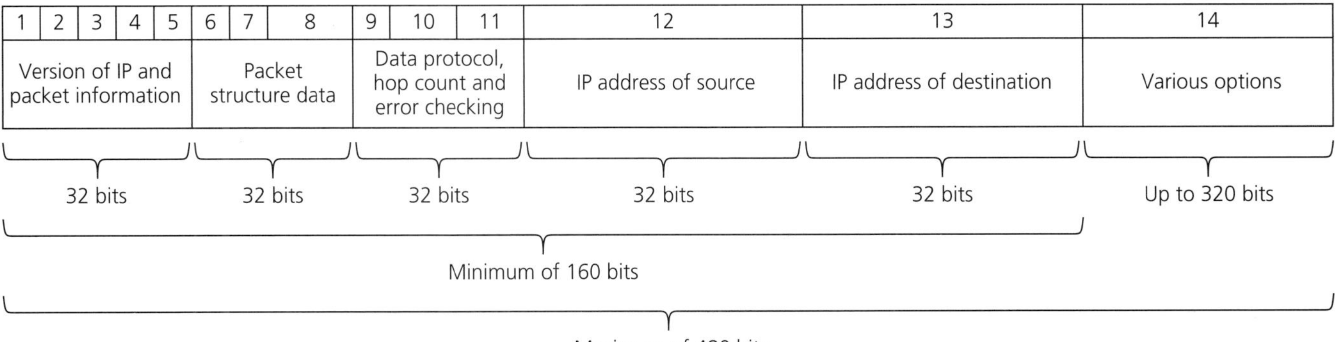

▲ **Figure 14.16** Structure of IPv4 packet header

The first 32 bits contain data that states the version (IPv4) of the protocol in use and the lengths of the header and the whole packet. The second 32 bits have data about whether or not the packet can be fragmented and, if it has been broken up, how it can be reconstructed. The third 32 bits contain the time-to-live (which is used as the 'hop count'), the protocol (for example FTP) in use for the data payload, and the checksum for error checking the header. The next two sets of 32 bits are the source IP address and the destination IP address. Up to 320 bits can be added for various options, such as a time stamp, so the overall length of an IPv4 header can vary. The data payload is added to the end of the header and can vary in length, which is why there is a field in the header that states the total length, header plus payload, of the packet.

The structure of IPv6 packets

IPv6 packets are similar to IPv4 packets, with a header and a data payload. The header is larger for IPv6 packets, consisting of a fixed length of 320 bits, to allow for the two fields of 128-bit addresses. Unlike IPv4 headers, there is no checksum because it is assumed that there is enough error detection provided in the protocols used at the Data Link and Application layers of the OSI model. The positions of the bits used for the various options are referred to as being offset by the number of octets from the start of the header. For example, the source IP address starts at an offset of 8 octets from the start. One octet is eight bits, so the source IP address starts at 8×8, or 64 bits from the start.

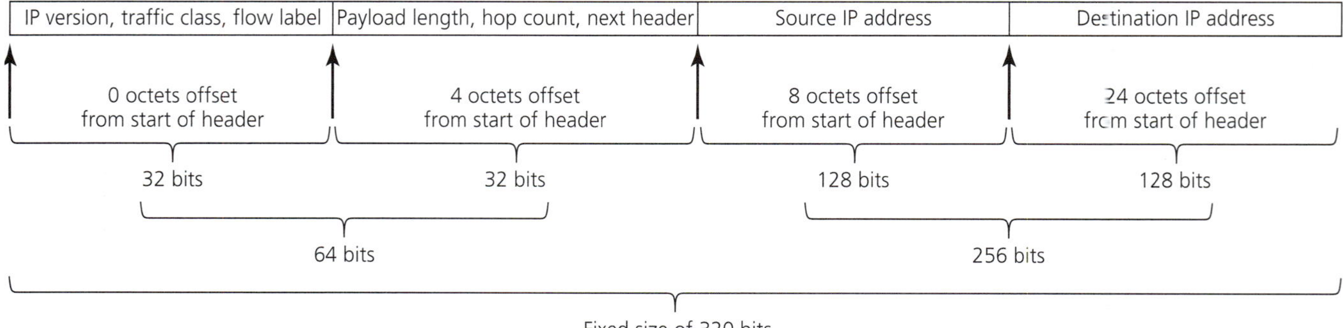

▲ **Figure 14.17** Structure of IPv6 fixed-size packet header

The first 32 bits (four octets) contain the version (IPv6) and some network control data; the second 32 bits (the next four octets) hold the header and payload lengths, the hop count and details of the next header, for example the protocol in use. The next two sets of 128 bits hold the source and destination addresses. When the hop count reaches zero, the packet is discarded and the sender is notified. The final destination node will process the packet even if the hop count is zero.

The two address fields each have 128 bits allocated, which means that the range of addresses available is huge: 3.4×10^{38} addresses are now available for devices to use. In practice, the number is less as many are reserved for various purposes but, even so, there are many more than were available under IPv4.

The size of IPv6 packets is handled differently from IPv4 packets. A minimum packet size is specified by IPv6 and sending nodes are supposed to determine the maximum size of the data frame, the Maximum Transmission Unit (MTU), allowed on the network and to create frames that take advantage of this. The frames should be as close to the MTU size as possible. If very large packets are fragmented by the sending host, the details are stored in the **Next header** field. Routers do not fragment IPv6 packets, so if a large IPv6 packet arrives that is bigger than the MTU, the router drops the packet.

The structure of TCP packets

The **Transport Control Protocol (TCP)** is used to create a connection-oriented link between two nodes, for example a client and servers over an IP network. It is used when transferring data with HTTP (although later HTTP standards such as HTTP/3 now use QUIC, a UDP-based protocol instead) at the **Transport layer** of the network models running over IPv4 and IPv6 protocols.

Transport Control Protocol (TCP) packets have a header that can vary in length. The header is made up of the source and destination addresses, sequence numbers, error-checking fields, and various flags (codes) indicating the status of the packet.

▼ **Table 14.13** Contents of TCP packet header

Octet offset	Number of bits	Field used for
0	16	Source port
	16	Destination port
4	32	Sequence numbers
8	32	Acknowledgement codes (flags)
12	4	Offset specifying the size of the packet
	12	Control flags indicating status of the connection use 9 bits; 3 bits are 'reserved'
	16	Window size indicating the size of the packet that the receiver will accept
16	16	Checksum
	16	'Urgent pointer' indicating the urgent byte in the data – depends on a flag in the **Sequence number options** field
20	32	Various options if required/set – size can vary but 0s are added to make it always end at 32-bit boundary

The address headers refer to the ports used by the sender and receiver for the service required. Port numbers are a way of indicating which service the packet is using, for example ports 21 and 22 are for FTP and port 23 is for SSH.

The sequence number field indicates the sequence of each byte of data carried by the packet. The sequence number can be the sequence of the data within the packet, or within the whole stream of packets, and specifies the order in which the data in the packets is to be used.

Other fields include the offset field, which uses 4 bits and specifies the size of the header followed by a number of flags. Each flag uses 1 out of 9 of the following 12 bits (3 bits are 'reserved' and not used), to indicate the actions to be taken if there is, for example, network congestion, the connection needs to be reset or the packet is the last one.

TCP checks the sequence of the packets and, consequently, the order of the data payload, leaving the delivery and delivery checking to the Internet Protocols (IP) on which it runs. For example, when a file is requested by a client from a file server using FTP, the following stages occur:

» The server uses TCP to break the file into segments called TCP packets.
» Each TCP packet includes the service port addresses (in this case ports 21 and 22 for FTP) and the timer set for an acknowledgement.
» The TCP packet is passed to the IP software in the networking stack.
» The IP software encapsulates the TCP packet into an IP packet that has the destination and source IP addresses.

» The IP packets, each carrying the TCP packet, which in turn carries the FTP payload, are sent from the FTP server to the FTP client.
» If the timer expires before the acknowledgement is received, the packet is sent again.

The address in the IP header gets the packet to its destination. When the client receives the packets, each IP packet is stripped of its IP wrapping and the TCP packet is examined. The TCP software in the network stack checks the integrity of the packet, checks the sequencing of the packets and ensures that they are free from errors and in the correct order. It sends an acknowledgement, a proof of receipt, to the sender. It then sends the data in a stream to the FTP client application. The FTP application checks that the data is error-free and, if so, stores it as a file wherever the user has decided. If errors are found at any stage, or if a packet goes missing, the protocols, TCP, IP and, in this case, the FTP client, can request a re-transmission of the data.

Due to the error checking and facilities for requesting re-transmission of data, TCP is a reliable method of transferring data but it can be slow at times.

The structure of UDP packets

User Datagram Protocol (UDP) packets are used to send messages between nodes (for example when a host requests an IP address from a server) and to be notified of other addresses using the Dynamic Host Configuration Protocol (DHCP). While there is error-checking data and the data assigning the service to be used (in the form of a port number), there is little else in the header. This means that a UDP packet is processed much faster and replies are received quicker.

UDP is a Transport layer protocol and runs 'on top' of the Internet Protocols of IPv4 or IPv6.

▼ **Table 14.14** Contents of UDP packet header

Octet offset	Number of bits	Field used for	Notes
0	16	Source port	If specified, will be used as port for reply
	16	Destination port	Must be present
4	16	Length	States length of packet
	16	Checksum	Optional, used for error checking

The UDP header is 64 bits long and has four fields. The first two fields are the source and destination ports. A networking port identifies the application or service that is using the network. Ports are codes that specify the service for which the data is intended. The port numbers used by TCP and UDP are the same. Client nodes do not always specify these in UDP datagrams but servers must always use well-known port numbers, for example port 123 for Network Time Protocol (NTP) or ports 67 and 68 for DHCP.

The length stores the length of the whole (packet, header and data) in bytes, with a minimum of 8 bytes (64 bits, the length of the header). A maximum length of 65 535 bytes (8 bytes for the header plus 65 527 bytes of data in the payload) is specified. IPv4 reduces this length by its own 20-byte (160 bit) IP header. IPv6 allows UDP packets to be greater than the 65 535 bytes.

The checksum field is optional for UDP and for IPv4, but if the packet is carried on an IPv6 system it must be present and has to be created.

UDP is used for voice, video and audio streaming, as well as real-time applications where packets need to be received quickly and without too much processing. Often, users prefer the loss of packets to the waiting time imposed by error checking and the resending of damaged packets.

The increased network traffic generated by the use of UDP for VoIP or streaming services can reduce the network accessibility for other services, for example database services or POS systems.

Modes of connection

Data can be transferred from one device to another in one of two ways. Imagine two devices that are able to exchange data. The devices can either connect directly to each other with a fixed connection or send the data to another network node that sends it to another, and so on, until it finds the receiving device.

The direct, fixed connection between the two devices is called a connection-oriented, or connection mode, link. Both devices are connected either on the same network or through telecommunications links so that each device knows where the other is located. The connection is kept until all the data has been exchanged. The type of connection that relies on other nodes to pass on the data until the destination is reached is called **connection-less**.

Connection-oriented mode provides a fixed link between devices. This is used in **circuit switching** where communication channels are made and used exclusively by the devices. A traditional telephone call has a switched circuit through the public telecommunications network and only the caller and recipient use it. In connection-oriented data exchange, the devices have to set up a physical (direct) connection or a logical (virtual) link before data can be exchanged. True connection-oriented links are expensive and uncommon for exchanging computer data between devices on a network.

In networks using packet switching, virtual circuits can be set up where a virtual direct connection is established over public telecommunication links by setting up specific routes between devices for data packets to travel. The route appears to be a direct connection to the devices. The actual route is configured when the virtual circuit is started but may not be the same every time.

Permanent virtual circuits are virtual circuit connections that, once created, are not shut down but kept open as dedicated, private links between devices, company LANs or company WANs. Connection-oriented mode is used so that the whole of the connection's bandwidth is available to the devices and no other devices use the connection.

In virtual circuits that use packets of data, the physical transmission media are shared with other network traffic. This means the whole bandwidth is not available, so there are varying delays as traffic is interleaved with other traffic and data packets can be lost. However, the packets arrive in the same order as they were sent so the processing of fragments, and of streams of packets, is reduced. This is useful when a continuous bitstream of data is required, as in voice conversations or media streaming. Connection-oriented mode using virtual circuits is considered to be a reliable method of data exchange because there is provision for notifications to be sent back to the sender of packet delivery, loss of packets and requests for resending of packets.

In a connection-oriented mode, virtual circuits using packet switching send all the data packets over the same route during the exchange of data. This is made possible by giving each virtual circuit its own identifying code or number, so that packets tagged with this number are quickly routed along the virtual circuit. The virtual circuit number is assigned when the connection is first established and is stored by each node until the circuit is ended. This is processed much faster than the connection-less mode, which uses routing tables.

Frame relay, which works at the Data Link layer of the OSI model, is a protocol that uses the connection-oriented mode to transfer data over the public telecommunications systems. Data, voice or computer data that has to travel between LANs, WANs or the internet is carried in frames over public lines using frame relay protocol. This provides a very fast method of data transfer over routes and paths that may change with every new connection. However, the devices and end users are not aware that there are different routes being used. Virtual circuits are established and used for each connection. Frame relay frames carry the virtual circuit identifier numbers in their address fields so that each frame of the data transfer travels along the same route.

The frames used by frame relay protocol vary in size and have fields that determine the start and end of the frame, the address field, the size of the frame and an error-checking field. The address field holds the virtual circuit identifier, in 24 bits, of the connection along which the frame travels or travelled.

Also stored in the address field is the congestion-control data, which determines what happens if the frame encounters congestion on the network. The data held here allows frame relay connections to provide a quality of service (QoS) by allowing devices to monitor the level of use of a connection, discarding frames if the level is excessive or transmitting more frames if the connection is under-used at any one moment.

Transmission Control Protocol (TCP) is also a connection-mode way of transferring data. Before exchanging user data, TCP connections require the devices, a client and a server, to establish a connection by exchanging information about themselves. Servers listen for requests by clients and create virtual connections when data is to be sent between them.

TCP is used by many internet services, including the world wide web (WWW) with HTTP and file transfer (FTP), because it provides the error checking for reliable delivery of packets. Any damaged or missing packets are re-sent. While this may cause long delays in displaying the data to users, for instance a web page may take several seconds to appear in a browser fully, the data is nearly always accurate and complete. The delays caused by repeated requests for packets to be sent again add to the latency.

A connection-less mode of exchange, such as UDP, is used when high-speed delivery of packets is preferred over accuracy, for instance in video streaming.

In the connection-less mode of communication, devices have no direct, fixed link between each other. Each device knows the addresses of the other devices to which data is to be sent but does not know their location, or even how to get to the addresses. Data is sent out on networks and is routed to the other devices using the addresses attached to the data.

In connection-less mode, a message, or data, is sent from one node, a device, to another node without the sender knowing whether or not the receiver is ready to accept the message or even if it actually exists. All that is known by the sender is the address of the recipient to which to send the data. If the data is not delivered, the sender is notified. This is similar to writing a letter, placing it in an envelope and addressing it, then giving the letter to the postal service. The sender relies on the postal service to find the address and deliver the letter or, if requested by the sender, to report back if the letter has not been delivered.

Ethernet is connection-less. Its frames contain source and destination addresses as well as data for error checking. Its frames carry data coded by other protocols.

Internet Protocol (IP) and User Datagram Protocol (UDP) are also connection-less ways of exchanging data over networks. Both protocols transfer data in packets, or datagrams, that have the address of the sender and receiver in headers. This allows the packets to be sent from one node to another, called routers, until the packets arrive at their destination, without the sender having to know where the destination is. It does not even matter what route the packet takes as it travels. If the packet does not arrive at all, or is damaged on its travels, the sender is requested to send it again.

Circuit switching

When a continuous, direct communications link is made between two devices, a circuit is created. Continuous circuits can be made over telecommunication networks by connecting multiple shorter sections together in one very long circuit using switches. This is circuit switching. It was first used to connect traditional analogue telephones together so callers could talk to each other. Switches were used to connect the callers through the telecommunications network.

Circuit switching used in modern data communications does not tend to use physical switches anymore, as in traditional telephony, but makes use of software switches that can be made, altered and removed, as and when required, by configuring routers. When a circuit is created using configured routers over the public telecommunications network, it is a virtual circuit.

For example, in Figure 14.18 the routers B, A, C and E have been configured to form a virtual circuit so all packets from the desktop PC and the web servers travel along the same route.

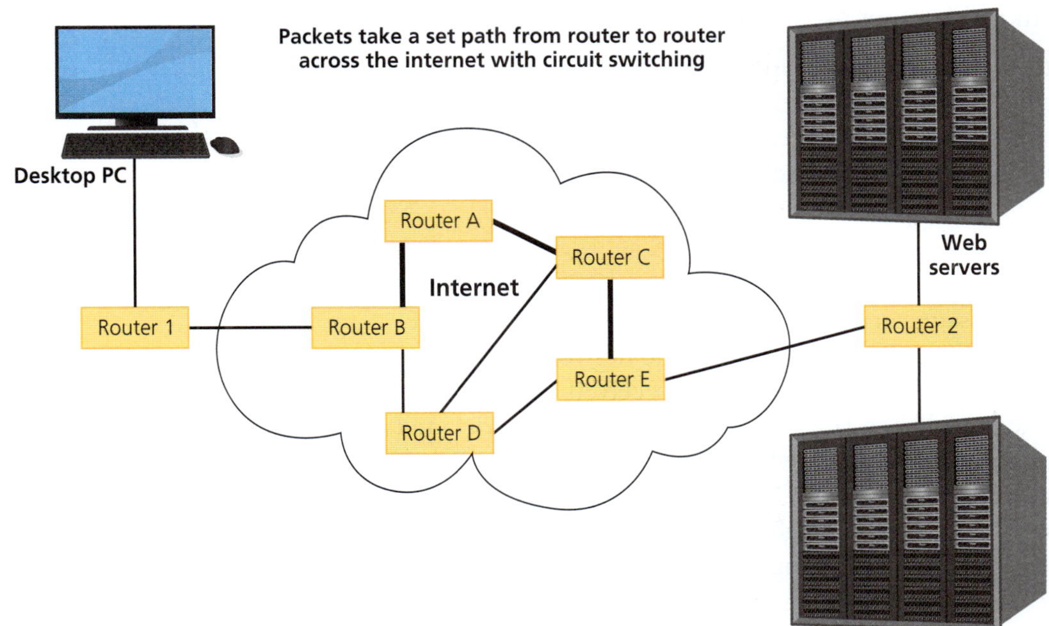

▲ **Figure 14.18** Circuit switching

Virtual circuits can be created and removed as and when users, for example large organisations, businesses or governments, require them for data transfer. Some virtual circuits are set up and exist almost permanently while others are quite short-lived. Virtual circuits can be temporarily created by protocols, such as TCP, on an unreliable, connection-less Internet Protocol (IP) network. While virtual circuits are emulations of real circuits, they do provide the same benefits as real circuits without the drawback of being completely permanent. Circuit switching allows the connected devices to make full use of the bandwidth of the communications channel, but it is not an efficient use of the channel because the channel is reserved whether or not data is actually being transferred.

Use of communication channels

A communication channel is a pathway along a transmission medium over which data flows. A channel can be a real connection, for example a copper wire, a wireless link or a logical connection intermingled with other connections on the same real connection. The intermingled logical connection is called a multiplex and the many logical connections share the same transmission medium. Multiplexing of channels onto a single transmission medium can be achieved in a number of ways. Time-multiplexing switches multiple channels at each end in synchronisation with each other so that each channel has a small amount of time to itself on the single medium. This method is used for voice calls, and if the switching is carried out fast enough, users do not notice it. Time-switching of voice calls, which are digitised, over the public telecommunications system's network allows for up to 30 channels at once.

Wave-division multiplexing creates multiple communication channels in fibre-optic connections where light of different wavelengths is used. Each wavelength of light provides a separate communication channel along a single fibre. Simultaneous two-way, or duplex, communication can be carried out along fibre-optic cables using wavelength multiplexing.

The amount of data that a communication channel can carry is its bandwidth, measured as the bit rate. In electronics and radio, the bandwidth may be expressed in terms of the available frequencies but, although data may be transferred over radio links, this is not usual in data communications.

In packet-switching networks, transmission media are shared between many communication channels. Permanent or semi-permanent virtual circuits to provide private communication channels can be set up using router configurations or by the data streams themselves.

Message switching

Message switching sends the whole, complete message over a communication channel in one go. Each message carries data about the addresses and the error-checking and is sent from node to node across the network. At each node, for example a router or switch, the entire message is received and stored, the data and addresses checked, and then the path to the next node is decided. The message is then sent to the next node, where it is again stored and checked and the next path is decided, before being sent onwards. If there is a series of messages from the same sender, each message may take a different path from the others but each message is sent without being broken up.

Store and forward

Message switching uses the store-and-forward method, so-called because the whole message is stored before it is forwarded. Packet switching is really a form of message switching in that each packet is a 'message' and it is received, stored, examined and sent on its way by every node along its path. However, unlike true message switching, the packets are not the whole of the data that is to be transferred.

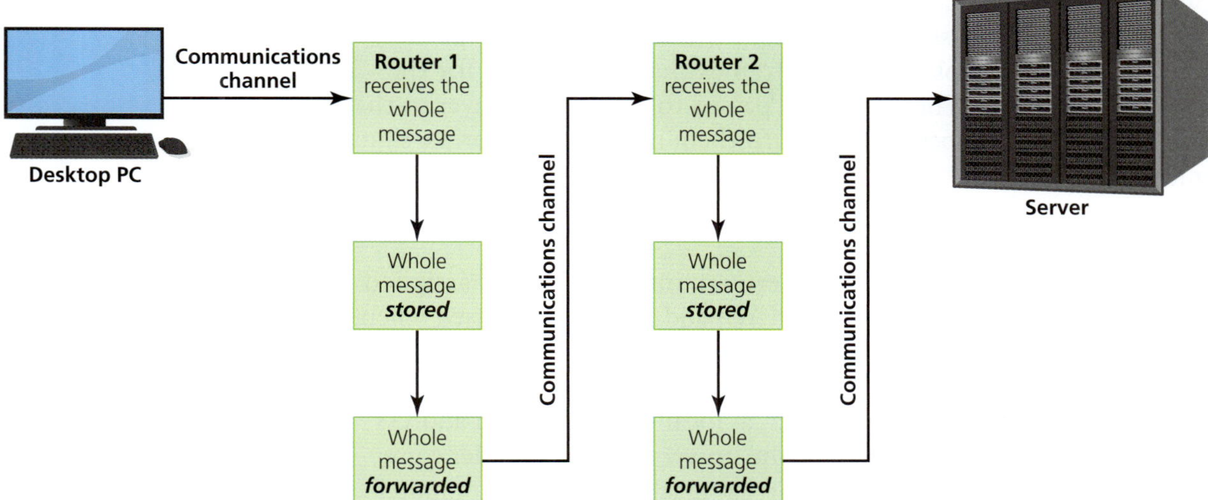

▲ **Figure 14.19** Message switching over a communication channel using store and forward

Store and forward has some advantages compared to dividing messages into small packets. If messages are to be sent to several locations at once, they can be 'broadcast' in their entirety at once, messages can be stored at switches or routers if network congestion is high or the onward path is busy or, conversely, they can be prioritised over other network traffic.

Message switching can introduce considerable delays for the messages. Every node on the path has to receive and store the whole message before it is sent out again. The overall delay can be considerable, with each node operating at different rates and adding its own delays.

Email messages are stored and forwarded by email servers. An email message is received and stored until requested, at which point it is forwarded to the recipient's inbox. The store-and-forward method of sending email allows for email filtering, for example filtering out spam, because messages can be examined for content before being forwarded or not.

14.6.4 The purpose and use of network addressing systems

A systematic, precise address system is an integral part of Internet Protocol (IP) networks. Without its addressing system, an IP network will not work.

Each node on a network has to be identified so that it can receive data from other nodes. Network addressing systems ensure that each node has its own unique, reachable address. While it might be preferable for every node in the world, on every network, to have a completely different address from any other node, it is not really necessary. Just as different roads can have houses with the same numbers, each house can still be identified and reached as a separate place. For example, No. 11 London Road is a different place from No. 11 Cairo Road. The same method is used to identify nodes on networks. Parts of the address of a node might be the same as parts of the address of other nodes, but there is always some part of the address of a node that ensures that its intended messages, or data, go nowhere else.

14

Media access control

Media, or medium, access control (MAC) software is responsible for the interaction between the physical transmission medium (the wire or cable or wireless connection) and the hardware in the network interface card (NIC).

MAC software works at the Data Link layer of the network model and controls how the data from the other software layers, for example TCP/IP packets or Ethernet frames, are encapsulated into frames that are suitable for the transmission medium in use. These are then passed to the Physical layer so that the hardware's electronic components can turn them into the appropriate signals and transmit them.

Media access address

During its manufacture, every computer network interface card is assigned a unique identifying media access control (MAC) address, which is used when sending and receiving frames on a network. These are called 'universally administered addresses' and cannot be changed because they are fixed in the read-only memory of the hardware. However, a 'locally administered address' can be assigned to a NIC that can override the universal address under some circumstances, for instance when running virtual machines on one computer system sharing NICs, each virtual machine can have its own assigned local MAC address. Also, some modern operating systems, for instance Apple's iOS and Android, can generate a different MAC address for a device in order to avoid being tracked when searching for wireless networks.

A MAC address uses Ethernet, Wi-Fi and Bluetooth and is made up of 48 bits, giving 2^{48} (281 trillion) possible addresses. While this is a huge number, the numbers of devices, for instance smartphones, smart TVs, video-streaming devices, computer games systems and many more, that need one or more MAC addresses for their NICs is rapidly increasing. Some wireless networks, for example as used in home automation, use 64 bits to provide for an even greater range of MAC addresses.

A MAC address consists of two sections, the first of which identifies the manufacturer of the NIC and the second the individual NIC itself. MAC addresses are written in various formats, but the most common format groups the bits into six octets, in hexadecimal notation, for example: MM.MM. MM.NN.NN.NN or MM-MM-MM-NN-NN-NN or MMM.MMM.NNN. NNN.

When a frame is created by a NIC for transmission on a network, the MAC address is included in the frame and used to send the frame to its destination. The MAC address can be used to send a frame to a NIC only on the same network. MAC addresses have no use in routing frames to other networks. Frames destined for NICs on other networks must use other addressing systems, such as IP, to find their way.

MAC addresses are used by switches when directing packets, using a stored list of MAC addresses, to their ports. They are also used by WAPs in security to control which devices can connect by using a list of MAC addresses that are allowed to connect. If a device does not have its address on the list, the WAP refuses to allow it to connect.

DHCP servers use MAC addresses to identify NICs in devices and assign them an IP address. Lists of MAC addresses stored on the DHCP server can also be used to assign the same IP address to a device whenever it makes a request.

IP addressing

The Internet Protocol (IP) addressing system enables the exchange of data in packets over and between networks. It provides a unique address for each NIC (in a device) in a network, a way of locating a destination NIC (in a device) and a way of finding a route from the sending device to the destination device, even if the location of the destination device is not known to the sending device. IP addresses can be used in routing packets from one network to destinations on other networks.

The IP software in a device works at the Internet, or Network, layer of the TCP/IP and OSI models and is responsible for the way that packets find their way from one network node to another. The sending and destination addresses are placed into the header of the packet, then error checking and other control data are added and the payload attached. The data payload can be packets of data from other protocols, such as TCP or UDP, which are using the IP container to send data over the internet. The IP packet, carrying the UDP packet as a payload, for instance, is then encapsulated inside an Ethernet frame that has the MAC addresses and sent on the transmission medium.

When the frame reaches a node, the Ethernet frame is examined, the payload of the IP packet is examined, and the addresses used to determine where the packet goes next or if it is to be processed by the node itself.

IP addresses

An Internet Protocol (IP) address has to comply with one of the two current versions: IPv4 or IPv6. Internet Protocol version 4 uses 32 bits for the addresses, while IP version 6 uses 64 bits.

The 32 bits of an IPv4 address are divided into four parts, each of eight bits. When writing an IP address, it is usual, but not compulsory, to write it as four decimal numbers separated by a dot, or period.

For example, a 32-bit address in its full binary form could be 11000000101010000000101000001011, which is 11000000 (which is 192), 10101000 (or 168), 00001010 (or 10), and 00001011 (or 11).

The binary representation is difficult for humans to understand, but when shown as 192.168.10.11 it is much easier. Sometimes IP addresses are shown as octets in hexadecimal, for example 192.168.10.11 is c0.a8.0a.0b. Each of the four octets can be any number from 0 to 255, because the minimum in binary is 00000000 and the maximum is 11111111, giving an overall maximum of $4\,294\,967\,296$ (2^{32}) addresses.

IPv4 address in binary notation:	11000000101010000000101000001011			
Divided into four octets:	11000000	10101000	00001010	00001011
Decimal notation octets:	192 .	168 .	10 .	11
Hexadecimal notation octets:	c0 .	a8 .	0a .	0b

▲ **Figure 14.20** Representation of an IPv4 address

The bits of an IP address are separated into two parts: the network part and the host part. The number of bits of the address that are allocated to the network address depends on the network itself. Networks fall into one of five classes: Classes A, B and C, each of which have different bit lengths for identifying them, and Classes D for multicast addressing and E for future use. The whole range of 2^{32} addresses is distributed to these classes of networks. Not all addresses are available for nodes as some are reserved for experimental purposes or used in routing packets through networks.

The first three octets are the network address. In the IP address 192.168.10.11, the network address is 192.168.10. All hosts on the network must share this part of the address if they are to be able to communicate with each other. The final octet is the host, or device, part and, in this case, it has the unique address of 11 on that network. A host with the address 192.168.11.11 is on a different network, but it can have the same host address of 11 because it is not on the same network. Its network address is 192.168.11. This is similar to having two houses with the number 11 but on different roads.

In a similar way as for IPv4 addresses, an IPv6 address has a network section and a local section. The network section of the address is called the routing prefix in IPv6. Also, IPv6 addresses can be allocated for specific purposes, for example for routing packets, for segmenting networks or for localised addressing of nodes. IPv6 allows for more features, in addition to the large addressing space giving many more addresses, than IPv4. All modern operating systems include support for IPv6 and devices can use the protocols to discover other IPv6 nodes on their local network, generate local IPv6 addresses themselves and gather other information about their local network so that they can instantly and automatically communicate.

IPv6 addresses use 64 bits so there is a much larger number of addresses available. IPv6 also adds new features to the system. The representation of an IPv6 address is more complex than for IPv4 as there are more numbers to deal with. It is usual to show an IPv6 address in eight groups of 16 bits, each group in hexadecimal format, separated with a colon.

An example IPv6 address could be, in binary:

1010111100110011100111111111101111111111111111111111111100001
1111100000000000000000000000000000000001000101000010111
0011100110011010100

Dividing it into eight groups, in decimal, although this is never used to represent the address, it would be:

44851:40955:65535:64575:0000:0000:35374:29492

In hexadecimal, the usual way of representation, it is:

af33:9ffb:ffff:fc3f:0000:0000:8a2e:7334

(It is preferred that lower-case letters be used but upper case is acceptable).

IPv6 address in binary notation:	1010111100110011100111111111101111111111111111111111111100001111100000000000000000000000000000000001000101000010111001110011001100110100						
	1010111100110011	1001111111111011	1111111111111111	1111110000111111	0000000000000000	0000000000000000	1000101000101110	0111001100110100
Decimal notation:	44851	40955	65535	64575	0000	0000	35374	29492
Hexadecimal notation:	af33 :	9ffb :	ffff :	fc3f :	0000 :	0000 :	8a2e :	7334

▲ **Figure 14.21** Representation of an IPv6 address

None of the IPv6 representations are easy for humans to use so attempts have been made to shorten them, for example by removing leading zeros.

Tunneling is an alternative way of carrying the different version of packets over networks. This encapsulates them in a packet of the other version of the IP address. This works for IPv4 on IPv6 networks, and vice versa, but adds to the time taken to process packets.

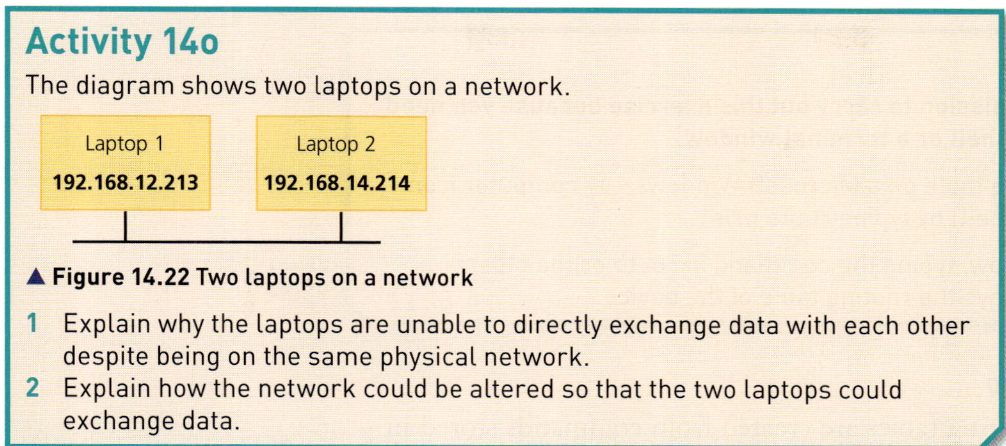

Activity 14o

The diagram shows two laptops on a network.

▲ **Figure 14.22 Two laptops on a network**

1 Explain why the laptops are unable to directly exchange data with each other despite being on the same physical network.
2 Explain how the network could be altered so that the two laptops could exchange data.

14.6.5 Managing network traffic

Network traffic, the datagrams or packets transmitted on networks, has to be managed to ensure that the maximum use is made of the available bandwidth of the transmission medium in order to:

» enable devices to message each other
» enable packets to find and arrive at their destination
» control the movement of packets between devices and networks.

Network managers have a number of tools available for controlling and directing network traffic in their networks, between their networks and other networks, and for configuring the devices (for example gateways and routers) that are used to pass traffic through large networks (such as the internet) and direct the datagrams along their path.

The directing of datagrams (packets) along a path through networks is called routing and packets are said to be 'routed' through networks. The routing of packets through networks is the process of finding a path for the packets to follow so that they can arrive at their destination. The path may or may not already be known and the location of the destination also may or may not already be known. In all circumstances, a path must be found so that packets can be reliably exchanged between devices as and when a user requires. The concept of routing in networks is that routers either 'know', from their databases of paths, the path to the destination or will keep passing the packets along from router to router until a router is found that does know the path to the destination.

Network routing is made up of three main parts:

» Routing protocols that collect and distribute information about networks and network devices.
» Routing algorithms that find paths through networks.
» Routing databases that record and store the paths that have been found by the routing algorithms. These are like a network map and are used to create a routing table. Routers store a database of paths and use it to look up where they should forward data packets.

Routing tables can be manually configured by network administrators or automatically created, and amended, as the router discovers paths. Sometimes there is a mixture of both used by a router. Routers can exchange router data between themselves to keep the tables up to date. Routing tables are also used in network hosts other than routers, for example a gateway or a user device such as a laptop.

All devices on a network have a routing table. A device needs a routing table to be able to determine where to send data packets on its and other networks.

> ## Task 14c
>
> **You may need to seek permission to carry out this exercise because you need to open a Windows PowerShell or a terminal window.**
>
> You can examine the routing table on a Microsoft Windows® 11 computer from the command line (PowerShell) by typing **route print**.
>
> In Linux, in a terminal window, typing the command **ip route** or the older command **netstat -rn** displays the routing table of the device.

Static and dynamic routing

When the router starts, routing tables are created from commands stored in its non-volatile RAM (NVRAM) and are stored in the RAM of the router. **Dynamic routing tables** are not usually stored when a router is powered off, but are recreated by interrogating other routers and put into RAM when the router is started. Routers exchange data about themselves so that each router on a network will have information that describes nearby routers to other routers and describes other routers to those nearby.

A manually configured table is called a static routing table. A table that is created and amended automatically is called a dynamic routing table.

Static routing uses a table of manually configured paths stored by a node, for example a router. Static routing tables are fixed and do not change if the network changes.

A static routing table is used to ensure that a router always uses the same path when sending on a packet, for example to set a default path or route for packets from a router when no other route can be determined by dynamic routing. This ensures that packets will always be passed on, even if that router cannot determine for itself where to send them.

Static routing gives a network administrator full control over where packets are sent, does not use much of the CPU resources of the node and does not add to the network traffic as there is no exchange between routers of dynamic routing data. It does not, however, have any tolerance of faults because any problems with other routers, for example if the next router has a power failure or is busy, cannot be compensated for by finding another route for the packets. Also, any mistakes by the administrator in setting up the table cause the routing to fail. Further, the administrator has to spend time setting up the tables and maintaining them, and the time required can be lengthy if changes are to be made to the network.

Where static and dynamic routing tables are used together, static entries of routes are always used first, which can cause some packets to be sent along unintended paths. The administrator can specifically alter the order that entries in the routing table are used, but this adds to the complexity and can cause more mistakes.

Dynamic routing allows the routing of packets to adapt to any changes or conditions on a network. Routers exchange data about the status of themselves and other routers and about the traffic on the network, and alter their routing tables to compensate for these changes so that the most efficient paths can be found and packets can be routed around problems. If a network path is very busy or congested, or if a router has a power failure, routers use routing protocols to exchange messages about these issues and the tables are updated accordingly.

Selection of paths and the use of routing tables

When a router selects a path for a packet that has just arrived from a node, such as a laptop or another router, the router examines the packet to determine the source and destination addresses in its headers.

The router looks through its routing table for a match to the destination address of the packet and, if it finds it is on a network directly connected to itself, the router forwards the packet to it. If the destination is not directly connected to the router, or the router knows where the destination is but the packet first has to go to other routers, or it does not find the address in its table at all, it must work out the best route, or even the best router, to use when sending the packet on its way. If a static default address is configured then the packet goes there, otherwise a best route is needed. A best route is very difficult to define but is usually one that will get the packet to its destination in the shortest time. It is not necessarily the shortest route in terms of distance.

After examining the destination address and searching the routing table, routers work out the best route or path to the address indicated by the table. Several routes to a destination may be found so algorithms are used to calculate a routing metric for each route. The lower the value of this metric, the better the route, so it is more likely to be used. The routing metric for a route is based upon available network bandwidth, latency, hop count, MTU and many other factors that the router knows about the network and about other routers.

Some of the routes will be known about already, have a metric already calculated and be in the routing table if other packets had previously passed through the router to the destination. Only the best routes are stored. In a Microsoft Windows 11 routing table, the metrics are in the last column for IPv4 addresses but in the second column for the IPv6 addresses. Linux displays metrics at the end of the row.

When the best route to the destination address has been found, the router forwards the packet to the next router on the path to it. The router updates its own routing table and informs other routers about the route and about any alterations.

In summary, the router checks the destination address in the packet header and searches its routing table for a match; if found, it sends the packet along that path. If not found, or if the destination is through other routers, it determines the best route to the next router by working out a routing metric, selects the route and sends the packet. This method allows for network conditions to change or be monitored by routers that update other routers, and the best route for packets to be chosen.

Function of routing protocols

Routing protocols make the rules and conventions for routing Internet Protocol (IP) packets through networks. Routing protocols work at the Network layer of the OSI model or the **Internet layer** of the TCP/IP model, because the packets of data exchanged using these protocols are about network management. Routing protocols determine how routers communicate with each other about networks and how they choose the routes for sending IP packets across networks.

Routing protocols belong to one of two main categories. The protocols that are for routing packets between different LANs over the internet are called Exterior Gateway Protocols (EGP) (or **Border Gateway Protocols** or BGP) and those that are only for routing within LANs are called Interior Gateway

Protocols (IGP). Routing algorithms used by routing protocols work out the 'best' route to use. The routing algorithms used by a particular protocol are fixed so the choice of algorithm to use depends on which protocol is used. Network administrators of large, corporate networks have to carefully choose the protocols they install on their routers to ensure that their network routers make the most efficient use of the available bandwidths and paths.

Activity 14p

The network shown uses the Routing Information Protocol (RIP) to determine the best route for packets across the network from A to B.

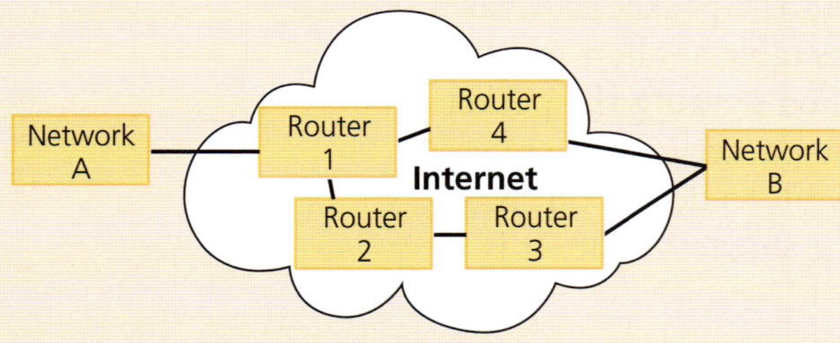

▲ **Figure 14.23** Routing Information Protocol (RIP) network flowchart

1. What would be the best route chosen by RIP and stored in the routing tables for use by packets travelling from network A to network B?
2. What would be the metric for this route?
3. Why might a different protocol choose a different route?

Interior Gateway Protocol

Interior Gateway Protocols (IGPs) work within networks and are used by routers to exchange information and select paths on LANs. IGPs work at the Data Link layer (layer 2 of the OSI model) because they make direct links between devices. Messages are exchanged between routers in UDP packets and are used to update routing tables.

IGPs are divided into two types. One category is based on the 'state' of the first link to the next router, called 'link-state protocols', and the other is based on the 'cost' of a route, called 'distance-vector protocols'. Each category has its own algorithms for selecting routes.

▼ **Table 14.15** Interior Gateway Protocols

Type of IGP and examples	How its algorithm works	Advantages
Link-state protocols **Example: Open Shortest Path First (OSPF)**	• Works out a metric based on the closest, individual links to the nearest routers.	• Routers do not need to collect and store as much data so routing tables are smaller. • Works well for the network links around individual routers. • Increases the speed of searching through the tables and the router works faster.
Distance-vector protocols **Examples:** **Routing Information Protocol (RIP)** **Border Gateway Protocol (BGP)**	• Gathers information about the entire path between networks to a destination using 'costs' such as the delays between routers. • RIP uses the number of 'hops' from router to router or the bandwidth of the link. • BGP uses bandwidth and delays.	• Routing tables can be exchanged between routers when they update each other. • Every router gets more information about more paths through the networks and can use this to determine the best routes.

Exterior Gateway Protocol

Exterior Gateway Protocols (EGPs) are used by routers to exchange their data about paths and links between LANs over the internet. When used in this way, the paths are chosen by sets of rules and policies and not by calculating metrics. The original EGP has been extended and updated and is now properly known as the Exterior Border Gateway Protocol (EBGP).

Border Gateway Protocol

Border Gateway Protocols (BGPs) are the protocols used by routers to communicate with each other and choose the paths for IP packets. When BGP is used by routers within a LAN, it is referred to as an Interior Gateway Protocol (IGP) and when used by routers to select paths between LANs, for example over the internet, it is referred to as an Exterior Gateway Protocol (EGP).

Use of protocol layering

Protocols explained

It is easier to understand how the components of computing systems work together if they are logically divided into functional parts or layers. For example, a typical computing device such as a smartphone has its hardware layer consisting of a touchscreen and NIC, its calculating and logic layer such as its CPU, and its software applications layer such as its email or web browser clients. These do not look like real layers sitting on top of each other but can be thought of as layers to make understanding how they interact easier. In networking, the different functions of the software used to make data exchange possible are also separated into layers. Visualised as 'sitting on top' of each other, the different software components for each networking task make up a network stack. The protocols used by the software in the network stack are layered in a **protocol stack**.

The protocols at the bottom of the stack determine how the interactions with the real, physical hardware are carried out, while the protocols at the top of the stack deal with interactions with the software applications running on the device.

The two models, the IP and OSI, while they have different layers and names for their layers, are broadly equivalent layering models of network currently and have the same idea of stacking, or layering, the different protocols in order of their purpose. Network protocols are arranged in logical layers from the top layer used by software applications (for example an email client) to pack their data (such as an email message) ready for sending down to the bottom layers – the Data Link and Physical layers – which construct the packages and prepare the frames for transmission on the medium using Ethernet technologies.

Visualising the protocols as being in layers, with the data in the message passing up and down, helps to explain how the protocols work with each other. The layers, called a network stack, do not physically exist as real layers as in a cake or a trifle, but are a way of visualising the idea that the network protocol software passes the data from one set of protocols to another.

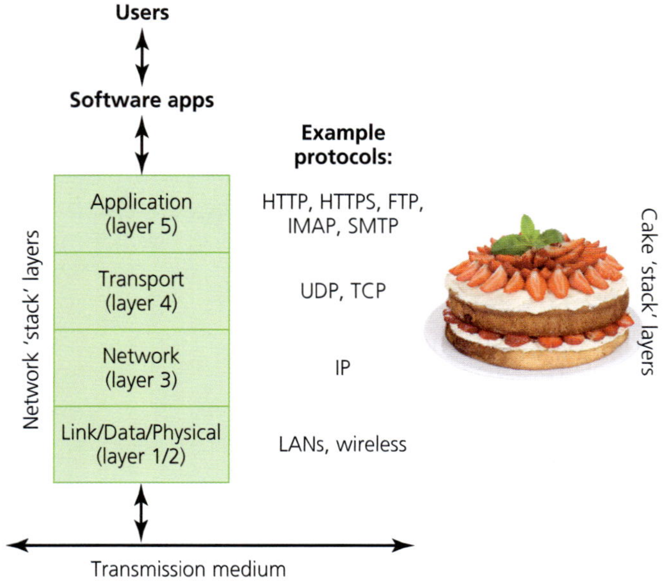

▲ **Figure 14.24** Visualising a network stack

When an Ethernet, or wireless equivalent, frame arrives at a NIC on the transmission medium, it is received by the NIC at the physical software layer, decoded and passed to the next layer up, which processes it. At each layer the protocols are examined to determine where the data ends up.

For example, a frame carrying an email message arrives inside an Ethernet frame and is passed to the Internet Protocol (IP) software to determine whether it is at its destination. It is passed to the Transmission Control Protocol (TCP), for example for checking, and passed up to the IMAP software in the email client application for processing and for display to the user. In replying to the email, the email client application uses SMTP to create the message packet, sends it down the stack of layers to TCP for setting up the error checking and connection, on to IP for addressing purposes and then to the Physical layer for placing inside an Ethernet frame for sending.

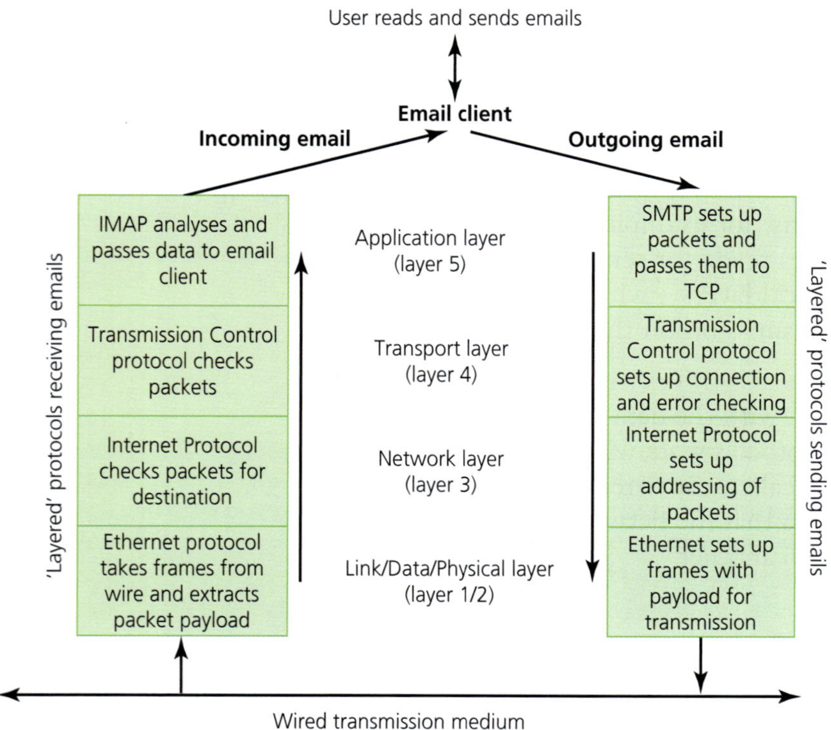

◄ **Figure 14.25** Emails received and sent up and down the protocol stack

Activity 14q

1 Why are the layering models of network protocols called stacks?
2 Why are the protocols in networking arranged in layers?
3 Why are conceptual models used to organise the networking protocols?

Internet (TCP/IP) protocol suite

The TCP/IP suite of protocols, commonly known as the Internet Protocol suite, is a collection of protocols that control the transmission (TCP) and the make-up (IP) of packets. It is one conceptual model of networking protocols and is visualised as being made up of four layers. The layers are the **Link layer** at the bottom, the **Internet layer**, the **Transport layer** and, at the top, the **Application layer**. At each layer, there is a set of protocols that, together, manage the communications on networks by devices, for example smartphones and laptops, and the user's software applications, for example web browsers and email clients.

When a user opens a web browser and uses it to display a web page from a website that is stored on a web server, the web browser uses protocols at the Application layer to construct a request. The Application layer passes the request to the Transport layer. It is passed on to the Internet layer and then to the Link layer for transmission by the NIC. When the NIC in the web server receives the transmission, it is passed up the layers to the website software application, which replies and sends the response back down the layers to its NIC. The transmission is sent to the user's device and back up the layers to the web browser, which displays the web page.

Each layer has a specific function in the journey of the request and response. Software applications use the Application layer protocol rules to construct a request and the data to be sent, and encapsulate, or wrap, it in a packet with extra data about what to do with it. At the Transport layer, the protocol rules are used to encapsulate the packet inside another packet with the headers about the type of connection to be used. At the Internet layer, another wrapping of the packet is carried out to add the addressing and routing information. Lastly, at the Link layer the whole packet is again encapsulated according to the rules in a frame, or datagram, suitable for the transmission medium. The NIC is responsible for turning the whole datagram into signals and transmitting it on the medium. Sometimes this is said to be an additional hardware, or physical, layer, but it is usually included at the Link layer.

Functions of the Application layer in the internet suite

The Application layer contains all the protocols necessary for exchanging data or for providing services for users. Software applications (for example email clients using SMTP), web browsers using HTTP, database management systems, file transfer applications using FTP, and some network services (for example using DHCP) use protocols at the Application layer.

A user wishing to exchange data or use a service on a network interacts with it via the interface of a software application and, usually, knows nothing about what happens when a web page is chosen or an email is sent. Software applications, for example web browsers and email clients, have the responsibility of preparing, sending and receiving the data for the user. They start by using Application layer protocols to create data packets.

Application layer protocols prepare the data and encapsulate it in packets. The TCP port numbers are also included when specific client–server applications are involved. Software applications on a host device are identified to the protocol suite by their port numbers. Specific ports are associated with specific software applications and are known as TCP or UDP ports, for example web browsers and HTTP use ports 80 or 8080, email and SMTP use port 25 and FTP uses port 20.

The Application layer protocols do not refer to, nor do they know anything about, what happens once the packet is passed down the layers. The Application layer protocols are only concerned with getting the data ready and with which software application is using it. When packets arrive, the layer protocols receive them from the Transport layer and check only the data inside the packet for errors.

Packets of data created using the Application layer are passed to the Transport layer.

Functions of the Transport layer in the internet suite

The protocols at this layer are concerned with the type, set-up and control of the connection between the network hosts whose software is exchanging data.

The type of connection is either connection-oriented, set up with TCP, or connection-less, set up with UDP. Error management, flow and congestion control can be used by these protocols. Also, the port numbers added at the Application layer are managed and used to direct packets to specific software applications when packets are passed upwards.

TCP is connection-oriented and it ensures that the host-to-host connection, in effect the software client to software server, is reliable. This means that lost packets are sent again, that any duplicates of packets are discarded and that data is sent to the Application layer in the correct order so that software applications get a stream of data that makes sense to them. All of the error checking, discarding and resending adds overheads to the time taken to exchange data, but it ensures the data that finally arrives at its destination is reliable and ready for the software to use.

The User Datagram Protocol provides for connection-less links between hosts. One example of when this might be needed is for streaming media content where the data arrival time is more important than the reliability of the data. Errors can be compensated for at the Application layer by the software requiring it, but it is important that the data arrives quickly. Another example where UDP packets are used is with DNS lookup, where the IP address of a host or website needs to be found.

Transport layer protocols, such as TCP and UDP, wrap the packet from the Application layer into another packet, add information in headers to the packet (for example about the source and destination ports to determine which software application owns the data) and pass it to the Internet layer for addressing.

Functions of the Internet layer in the internet suite

The Internet layer protocols form the rules that enable the internet to work and set out how packets should be constructed to enable them to be sent across networks and from network to network. The main protocol at this layer is the Internet Protocol (IP) that provides for the address system used on the internet and on most home and business networks such as LANs. Each device, or host or node, on a network is given its own IP address, which indicates the address of the network to which it is connected and its own unique address.

IP provides only the rules for sending packets between hosts and between networks. The management of connections or the delivery of packets to software applications is the responsibility of the Transport layer protocols. IP also lays out the rules for constructing packets with IP addresses. This means that a packet received from the Transport layer is encapsulated in a bigger packet that has the source and destination IP addresses, and some data to error check the header, added to it. When a packet is destined for another network or has to travel across the internet to get to another network, gateways and routers use protocols at the Internet layer to choose a route for the packet. When arriving at a host, a packet is checked at the Internet layer to see if it is at the correct IP address and, if so, the IP headers are stripped away and the packet passed up to the Transport layer. If not, it is discarded. Packets from the Transport layer are wrapped at the Internet layer and passed to the Link layer.

Functions of the Link layer in the internet suite

In computer networking, a link is only between two hosts (nodes). It is the transmission medium joining two nodes and can be a physical cable or a wireless link. A link can also be a virtual link or a tunnel but it is still only between two hosts.

The Link layer defines the rules for sending data between the two hosts over the link between them. It does not extend to other networks through routers but only to the exchange of data between hosts on the same network. The Link layer protocols define only how data packets, in datagrams, get from one host, or node, to the next. The routing of data packets from node to node, for example along a path selected by routers to find a distant node, is the responsibility of the IP layer.

Link layer protocols define the structure of datagrams, carrying the wrapped packets of data, and the rules for sending them on different transmission media. Most LANs make use of the unique MAC address of the NIC in a device, using Ethernet-based protocols to send datagrams between hosts.

Frames on the transmission medium arriving at a NIC are first processed at the Link layer to check whether they are at the destined MAC address and, if not, they are ignored. If they are at the correct MAC address, they are stripped of the Ethernet wrapper and passed to the IP layer, which examines the IP addresses in the packet.

If they are at the intended IP address, the IP headers are removed and the packets are passed up to the Transport layer where they are checked for errors and, if all is well, passed to the Application layer for sending to a software application.

If they are not at their destination IP address, the packets are discarded unless the host node is a gateway or router. In this case, routing protocols find a route, change the IP addresses to the next node and send the packet back down to the Link layer. The Link layer then sends it on its way again in a new datagram.

A packet of wrapped data arriving at the Link layer from the IP layer above is, again, wrapped into a packet, which now has MAC addresses added. This packet is sent to the hardware in the NIC, which turns it into the signals and it becomes a frame to be transmitted on the medium.

Sometimes, an additional fifth layer is shown at the very bottom of the TCP/IP model. This layer is the Physical, or hardware, layer and separately specifies the details of the plugs and sockets for cables as well as how frames are constructed. In other reference texts, the Physical layer is assumed to be part of the Link layer. It does not really matter how this layer is shown.

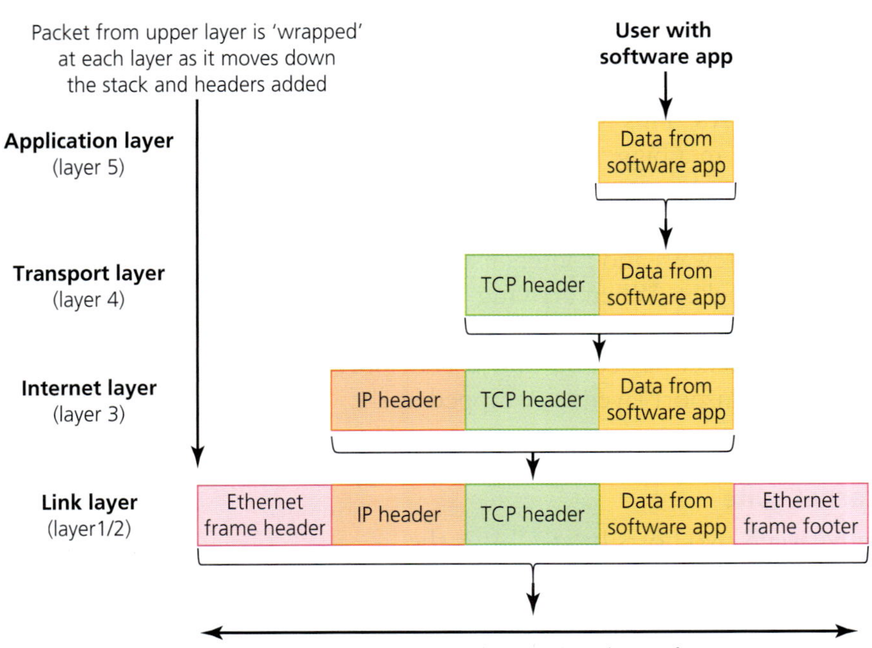

▲ **Figure 14.26** Data has headers added as it moves down the Internet Protocol suite layers

Open Systems Interconnection model

The **International Organization for Standardization (ISO)** sets out the Open Systems Interconnection (OSI) model for communications. The model uses the idea of seven layers that define protocols to be used when devices communicate with each other.

The seven layers of the OSI model are, from the top:

- » layer 7 is the Application layer
- » layer 6 is the Presentation layer
- » layer 5 is the Session layer
- » layer 4 is the Transport layer
- » layer 3 is the Network layer
- » layer 2 is the Data Link layer
- » layer 1 is the Physical layer.

The layer protocols define how each interacts with the other layers to enable communication. For example, the Application layer at the top specifies how software applications communicate with each other and with the other layers and, at the bottom, the Physical layer specifies how the NIC connects and communicates with the transmission medium.

Functions of the Application layer in the OSI model

The Application layer, layer 7, protocols set out the rules for how software applications exchange data and communicate with each other. This layer provides an interface with software that needs to exchange data for the users of the device. For example, if a user wants to send and receive emails using a laptop, an email client uses SMTP to package the email data and IMAP to receive them. Layer 7 protocols are used to pass the packages to, and get them from, the Presentation layer. The Application layer is said to be providing network services to the users because it exchanges data between software applications and the Presentation layer.

Functions of the Presentation layer in the OSI model

The **Presentation layer**, layer 6, is responsible for converting the data from the Application layer into the correct format and applying any encryption that is required. The conversion is required because different software applications, for example email clients and web browsers, use different syntax and formats for their data. Layer 6 ensures that the data to be sent is in the correct format and using the correct syntax for the layers below. Encryption of the data can be carried out at this layer.

When a data packet is received and arrives at this layer from the network, meaning it is passed up from layer 5, the Presentation layer is responsible for decrypting the data and for converting it into the correct format for the software application to receive. For example, a packet of email data is converted to the format that the email client requires, while a packet of web page data is converted to go to a web browser. The Presentation layer protocols ensure data being sent over a network is always in the correct format and is using the correct syntax for that network.

The Presentation layer passes packets to and from the Session layer.

Functions of the Session layer in the OSI model

Layer 5, the **Session layer**, controls the connections between hosts. It adds more ways of checking and recovering data compared to the Transport layer of the TCP/IP suite.

It provides functions for setting up full-duplex, half-duplex or simplex connections and this manages the exchange of data between network hosts. Connection sessions can also be 'checkpointed', which is the copying of a snapshot of the data into memory and using it to check the data. Also, sessions can be suspended, started again and stopped with very few errors.

Functions of the Transport layer in the OSI model

Layer 4, the **Transport layer**, provides more or less the same functions as the Transport layer of the TCP/IP suite. The Transport layer provides the reliability of connections between hosts. It provides for connection-oriented and connection-less links and for the checking and retransmitting of missing packets. It also provides for the segmentation of packets and for their reassembly in the correct order, and the error checking needed for this.

TCP and UDP are classified as working at this layer of the OSI model. TLS works at this layer to provide security. The Transport layer also provides for tunneling and VPNs, for example with IPsec.

Data packets from the Presentation layer are encapsulated at layer 4 inside packets that have headers of data about the type of connection and how to error-check and correct errors in the connection. They are then passed to the Network layer. Packets from the Network layer are error-checked, checked for sequence numbers and, if correct, sent on upwards to layer 5. If necessary, a request for retransmission is sent down the layers to the sending host.

Functions of the Network layer in the OSI model

The Network layer provides protocols for the addressing and routing of packets. Internet Protocol (IP) and the Gateway Protocols work at this layer. ICMP works at this layer for routers to pass messages to each other, as previously described (see Section 14.6.2 'The purpose and use of protocols for exchanging data across networks').

Packets from the Transport layer are again encapsulated into packets with headers containing their source and destination addresses, as explained previously for IP.

Once repackaged, the packets are passed to the Data Link layer, layer 2. Packages arriving from the Data Link layer are unwrapped and their IP addressing information is examined. Packets are, depending on the host, either ignored, routed (using the routing protocols) to another host (back down to the Data Link layer in a new packet) or sent upwards to the Transport layer for processing.

Functions of the Data Link layer in the OSI model

The Data Link layer protocols deal with the mapping of IP addresses (used for data exchange and routing on the internet) to the MAC addresses of the hardware, the NIC, in devices. Protocols at this layer specify how the data packets from the Network layer are, once again, repackaged and made into frames, for example for transmission on an Ethernet connection. When the frames are ready, they are passed to the Physical layer for sending. When receiving a frame from the Physical layer, the Data Link layer unwraps the frame and checks that the frame is at the correct MAC address and, if so, it passes it up to the Network layer. If the frame is not at the destination MAC address, it is ignored and discarded.

Functions of the Physical layer in the OSI model

At the Physical layer, the collections of binary data that make up the frames are converted into the signals that are required for the transmission medium. For example, if the medium is a physical copper wire, the frame is turned into signals using small electrical voltages; if the medium is a fibre-optic connection, the signals are carried by light waves; and if the medium is wireless, the frames are modulated onto radio waves.

The specifications for the Physical layer also detail the electrical voltages to be used, the use of each of the pins on the physical connectors and the techniques to be used for wireless connections.

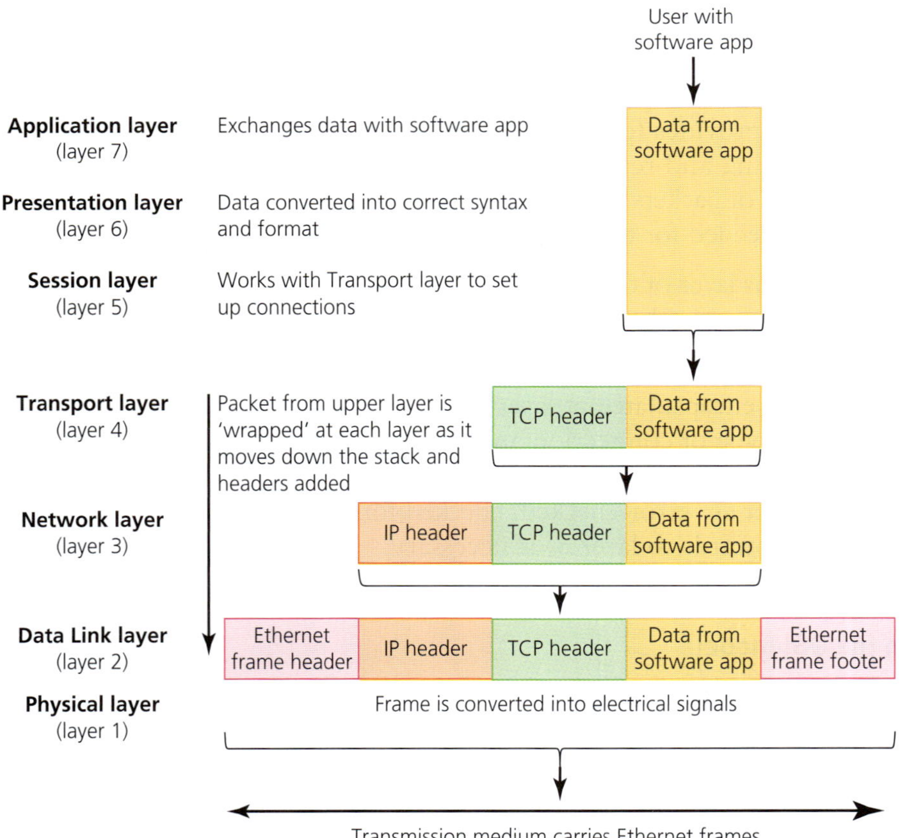

▲ **Figure 14.27** Data moving down the OSI layers

Comparison of schemes

While the OSI model is used for some telecommunications systems, the TCP/IP model is more often used in networking and now provides the standard. Both models are designed to enable software to be developed and used to a set of standards so that all devices can communicate with each other.

The four layers of **internet (TCP/IP) protocol suite** are broadly equivalent to the seven layers of the OSI model, as is shown in Table 14.16.

▼ **Table 14.16** Comparison of the layers in the TCP/IP and OSI models

Internet (TCP/IP) protocol suite	OSI model	Purpose of the protocols in the layer
Application	Application	Interactions with user software.
	Presentation	Conversion of syntax and format.
	Session	Manages connections between hosts.
		The Application layer of the internet (TCP/IP) protocol suite is equivalent to the OSI model's Application, Presentation and Session layers because these deal with how the user's software applications prepare and package data ready for encapsulating and sending. The layer(s) also specify how to receive and unpack the data and pass it to the appropriate software for the user.
Transport	Transport	The Transport layer of both models refers to the reliability, types of connection and checking of packets.
Internet	Network	The Internet layer of the internet (TCP/IP) protocol suite and the Network layer of the OSI model are responsible for the addressing and routing of packets.
Link	Data Link	This includes the Physical layer in the internet (TCP/IP) protocol suite.
		It manages the mapping of IP addresses to MAC addresses, for creating frames for transmission and for receiving frames from the transmission medium.
		It deals with the conversion between the frames and the signals needed for the transmission medium. It also specifies the physical layout of the connectors and wires or details of the radio waves.
(Physical, usually part of Link)	Physical	This layer is a separate layer in the OSI model.
		It is responsible for the conversion between frames and signals.

14.7 Wireless technology

Wireless technology uses radio waves, part of the electromagnetic spectrum, to carry data. There is no physical cable to direct, or guide, the signals between devices.

14.7.1 Methods of wireless transmission of data

Wi-Fi

Wi-Fi is a set of protocols, working at the Link layers of the network protocol models, specifying how data is exchanged by wireless networking. Different versions of Wi-Fi, specified by the versions of the IEEE 802.11 group of protocols, designate the radio bands and the rate of data transfer, the bandwidth, and the maximum power of the radio waves that can be used. They are designed to work alongside the Ethernet system used for wired networking, and the Wi-Fi frame structure is very similar. However, Wi-Fi frames need to carry some additional information.

A Wi-Fi frame carries a data packet by radio in the same way as Ethernet frames carry data packets on wires. The purpose of the frame is to deliver the packet safely from the sender to the receiver. There are several types of Wi-Fi frame. There are frames for carrying the actual user data, control frames and management frames. Control frames are used to set up the channels to be used and to acknowledge that data has been received, and management frames are used when devices join and leave the Wi-Fi network.

Wi-Fi frames carry some control bits, the source and destination MAC addresses and the data payload in the same way as do Ethernet frames. There are additional fields included to indicate, for example, the SSID of the network.

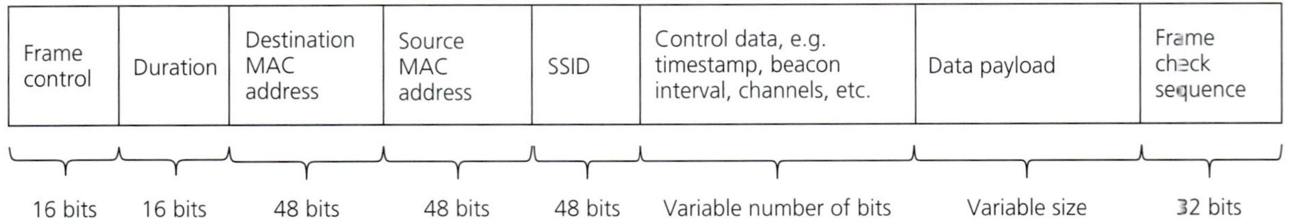

▲ **Figure 14.28** An example of the structure of a Wi-Fi frame

Wi-Fi transmitters are restricted to very low powers so the radio waves do not travel, or propagate, very far or penetrate solid objects, such as walls, very well. This ensures that Wi-Fi can be used in networks that are close together in homes or apartments and not interfere too much nor be accessible to eavesdroppers outside the home. The low power also reduces the possibility of any harmful effects on human health. There is little or no evidence to suggest that Wi-Fi signals are harmful.

Wi-Fi uses the **2.4 gigahertz (2.4 GHz) and 5 GHz bands**, both of which are divided into channels for data communication. A device using Wi-Fi has a wireless network interface controller (WNIC) with a MAC address that can use either, or both, of these bands.

There are fourteen Wi-Fi channels available, but not all are licensed and available in all areas of the world, and channel 14 is rarely used today. There is a range of channels available in the 5 GHz band and, again, the availability of these differs around the world. The differences in the availability of Wi-Fi channel depends on how radio wavebands are used in different countries. Some channels may be used for other purposes, such as military or weather services, and Wi-Fi would interfere with these uses. Using a Wi-Fi channel that is not licensed is illegal in most of the world, so devices are programmed by manufacturers with different software depending where they are to be used.

Wi-Fi network connections using the 2.4 GHz band can work over longer distances than those using the 5 GHz band, but the bandwidth at 2.4 GHz is less than that of 5 GHz. In practice, in a home, this means that 2.4 GHz can cover a larger area but with lower data-transfer rates than 5 GHz. However, because its radio waves can travel further, 2.4 GHz connections are subject to more interference from the Wi-Fi of neighbouring homes.

Wi-Fi nodes can work in infrastructure mode, which means connecting to a central node, or in ad hoc mode, which means connecting directly to each other.

In infrastructure mode, a service set identifier (SSID) is assigned to a network, as explained in Section 14.2 'Components in a network'. All devices on the wireless network must use this SSID but devices need not be on the same band or channel. A WAP allows many devices to connect to a wired network using Wi-Fi. WAPs built into the routers supplied by ISPs allow home users to share a single internet connection on their Wi-Fi-enabled devices, for example tablets and smartphones.

The provision of Wi-Fi hotspots allows users to access networks and the internet while away from home or the office. Wi-Fi hotspots are wireless access points that provide connections to the internet for mobile devices. They are often provided free in the public areas of cafes, hotels, restaurants, libraries, airports and railway terminals, but can be made available by almost any organisation. Wi-Fi hotspots are also available in many cities for use by the public, but mobile networks have taken over this role. Users can connect their smartphones, tablets and laptops to a Wi-Fi hotspot and use the internet without having to pay for mobile network charges. This is convenient but can expose the user's data to security risks.

Ad hoc mode enables the direct transfer of data between devices. It is used for many purposes, for example connecting controllers to games consoles, transferring data from devices directly to printers and downloading photographs directly from digital cameras to other devices. Other devices that capture video and images, for example dashcams in cars, can be accessed directly using Wi-Fi ad hoc mode. The dashcam becomes a Wi-Fi node to which a laptop can connect so a driver can directly download the video and audio.

Bluetooth

Bluetooth is a wireless method of exchanging data over short ranges using radio waves in the 2.4 GHz band. Bluetooth is managed by the Bluetooth Special Interest Group (SIG), which sets the standards for devices using Bluetooth and owns the Bluetooth trademark.

Bluetooth is a separate networking system from the Internet Protocol suite and has its own protocol stack. However, the Bluetooth stack of protocols is more a grouping of protocols than a stack and is designed to be very flexible, allowing applications to use whichever protocols they need.

The Bluetooth protocols are divided into groups. There is a Physical layer made up of the Radio and Data Link layers. The Radio layer is a group of protocols that deals with the physical transmission of radio waves, including the frequencies and modulation to be used and the Baseband protocols that specify the frame format and the addressing. The Data Link layer enables both connection-oriented and connection-less connections to established links between devices. A Middleware layer includes protocols to find information about the devices and some protocols from the IP suite, such as UDP and TCP. At the top of the stack, there is the Applications layer where the software applications that use Bluetooth interact with the protocols in the Middleware layer.

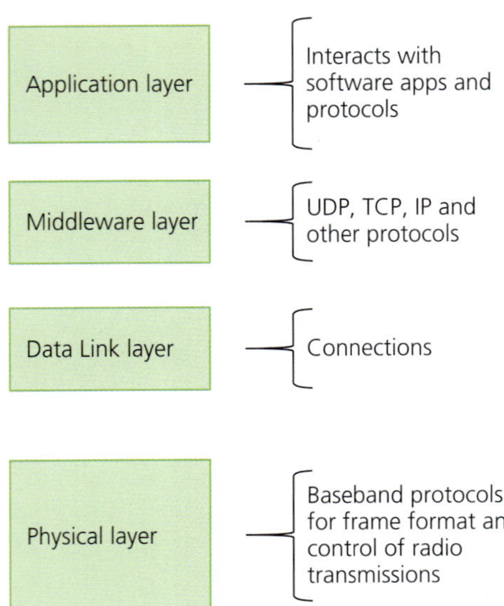

▲ **Figure 14.29** Bluetooth protocol stack

Bluetooth protocols are implemented in most modern operating systems and most mobile devices have the necessary radio hardware. Devices without their own Bluetooth hardware can use adapters or USB dongles.

When devices connect using Bluetooth, they have to be within range, about 10 m, of each other. Up to eight devices can be connected. When Bluetooth-enabled devices wish to connect, they scan the area for Bluetooth signals to 'query' any devices close enough. Bluetooth devices are listening on the radio frequencies assigned to Bluetooth and when they detect a query from another device, a response is sent with the information needed for displaying its details on the querying device. The details usually include the name of the device and the user, if it is a smartphone, for the user to view. The querying and response is handled by the lower layers of protocols and any user messages are sent up the layers for display.

If the user wishes to connect, the receiving device often requires a code to be entered on the connecting device before it will allow the connection. When the code is entered, or permission is granted, the devices connect and are said to be paired. Pairing between devices that have previously been paired is usually automatic, with the users being unaware of it.

Bluetooth connections use spread-spectrum frequency-hopping methods to make use of the 79 channels available in the Bluetooth standard. This means that when the devices establish a connection, they pick any one of the 79 channels and, if it is free at the time, they can connect. If it is not free, they try another and then another until they find a free channel. To avoid interference with, and from, other connections, the devices also change the channel frequently, often thousands of times a second. The later versions of Bluetooth technology can transfer data at up to 24 Mbit/s, but most Bluetooth connections never achieve a fraction of this.

When more than two devices link together with Bluetooth, they are said to form a very small ad hoc network called a piconet. One device is in control of the piconet and other devices can join and leave at any time. Piconets can link together to exchange information between all of the devices in a network called a scatternet.

Bluetooth is typically used to exchange small amounts of data, such as between peripheral components and computing devices. Examples include the Bluetooth connections between keyboards and a mouse to the host computer, or microphones and earphones. Bluetooth headsets, or handsfree systems, connect with smartphones to allow conversations on mobile (cell) phones while people are doing other tasks. Smartphones can use Bluetooth to connect with car sound systems to play music or hold conversations using the system's speakers. In some countries, using handheld mobile (cell) phones in cars while driving is illegal so people use this type of connection to avoid prosecution. In other countries, the use of mobile (cell) phones while driving is not allowed at all.

Users can exchange images or contact details between smartphones connected together using Bluetooth. Remote controls for TV sets or sound systems can use Bluetooth instead of infrared.

Infrared

Infrared (IR) is electromagnetic radiation with wavelengths longer than those of visible light, extending from the red end of the visible spectrum. It is usually invisible to the human eye, but some IR emitted by lasers when used in data communications is just visible. Viewing laser light directly can irreversibly damage the human eye. The use of IR for data transmission by devices must conform to standards published by IrDA, the Infrared Data Association.

IR is used to exchange data over short distances between devices. IR does not penetrate walls or other obstacles and usually requires a line of sight between sender and receiver so it does not interfere with other devices in adjoining rooms. For example, remote controls for TV sets use IR to transmit control signals. The data to be sent is modulated onto IR light by rapidly switching the light on and off so that pulses of IR are sent to the device being controlled. The protocols for this use were set up in the 1980s and are still in use today. TV set manufacturers have been assigned a set of codes to use only with their appliances so each remote control only works with the manufacturer's sets. However, as the codes are publicly known, third party remote controls can be set up to work with any TV set or appliance by either programming the codes into the remote or by copying the IR signals.

The receiver of IR uses silicon diodes that do not respond to ambient infrared light but only to the pulses from senders. This ensures that the data exchange is not subject to interference from sunlight or other IR emitters such as radiant space heaters and that the receiver does not respond to changes in sunlight levels. However, the use of IR remote controllers in very brightly lit rooms or with high levels of sunlight may require the remote to be closer to the receiver.

LEDs are used in remote controls, for example for TV sets, but lasers are used to transmit data across free space. Links can be made between buildings and achieve data transfer rates of about 4 Gbit/s. This can provide links for mobile communications systems or internet connections without the need for more cabling. Again, there is a danger of damage to the human eye, so great care has to be taken with the installations. In fibre-optic cable installations, infrared lasers are used because IR has been found to be the best choice, with the best transmission and the least dispersion and signal loss along the silica strands of the fibres.

Microwave

Microwaves are radio waves, electromagnetic radiation, with short wavelengths from about 1 mm to 10 cm. These are longer wavelengths than infrared radiation and so are longer than the wavelengths of visible light, making them completely invisible to the human eye. While there may be some human cell damage and longer-lasting effects from very high-power microwaves, there is little evidence to suggest that the low-power microwaves used for data communications are in any way harmful.

Microwaves travel in straight lines and, unlike the longer wavelengths used in radio transmissions, are not reflected by layers of the atmosphere and are not conducted as ground waves, so can be used only for line-of-sight links. Microwaves are used for data transmission in wireless networking technologies, for example Wi-Fi, data links and satellite communications.

The microwaves used for data communication have short wavelengths of between about 1.5 and 2.5 cm, which are frequencies of between 12 GHz and 18 GHz. The microwaves used by Wi-Fi are 2.4 GHz, which are also used by Bluetooth, and 5 GHz, with wavelengths of about 12.5 cm and 5.9 cm. Terrestrial UHF television signals are carried on microwaves with frequencies of between 300 MHz and 3 GHz (wavelengths of 10 cm to 100 cm). For comparison, infrared (IR) light has a frequency of around 30 THz with a wavelength of about 10 µm, and visible light is in the 300 THz range, which is about 1 µm in wavelength.

Microwaves are used to provide data links between **ground stations** in a microwave relay network. This provides a sequence of microwave transmitters and receivers that pass data from link to link across long distances. Microwave links can carry high-bandwidth data connections, but many microwave relay networks have been replaced with fibre-optic cabling, which can provide high bandwidth with lower maintenance costs and more security.

Microwave links are used to provide data connections from mobile television studios and cameras to avoid trailing wires and to provide **uplink** and **downlink** connections between remote and main television studios via satellites. News and sports video coverage can be sent back to the main news studio from anywhere that can 'see' the satellite.

Data communication links between two sites require directional antennae because the beams are narrow and the sites must be in line of sight. For terrestrial links between two ground stations, even with the antennae on the top of tall towers or buildings, the effective range is only about 70 km. For links between a ground station and a satellite **transponder**, as long as there is no obstacle in the way, the distance can be tens of thousands of kilometres.

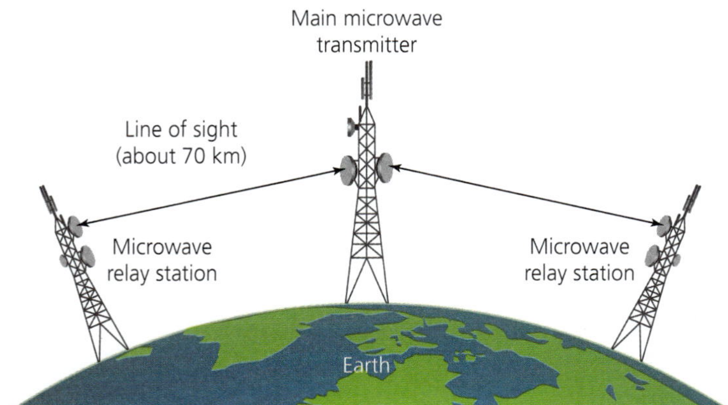

▲ **Figure 14.30** Microwave links

Satellite communications systems use microwaves as carriers of television channels and for data links. Direct-to-home satellite broadcasting can cover a much wider area than broadcasts from terrestrial transmitters and have more TV channels. However, the microwave downlinks are more susceptible to signal loss in poor atmospheric conditions than are terrestrial TV signals.

Radio

All wireless technology uses radio waves. The term 'radio wave' really refers to all wavelengths that are longer (lower in frequency) than infrared (IR), but for convenience microwaves are considered to be those that are shorter in wavelength (higher frequency) than radio waves and are usually referred to separately. In reality, the electromagnetic spectrum is a continuous range from extremely long wavelengths, of 100 000 km, through radio, microwave, infrared, visible light and ultraviolet to the extremely short waves, 1 pm, of gamma rays.

Radio is used for Wi-Fi in wireless LANs, and microwaves for communications links, Bluetooth, mobile networks, satellite links and all the other wireless methods used for data transmission. Packet radio carries data between radio modems in amateur radio systems.

Radio waves with very long wavelengths of about 10 000 km to 100 km (about 30 Hz to 3 kHz in frequency) can pass through sea water and can be used to communicate with submarines. The long wavelengths require very long, tens of kilometres long, antennae that do not fit into submarines so the submarines can only receive the signals but not send them. The bandwidth is extremely small and only allows data such as text messages to be sent.

Near Field Communication

Near Field Communication (NFC) has many uses and applications by individuals and businesses despite the slow rate of data exchange (maximum of 424 kbit/s, compared to Bluetooth at up to 2.1 Mbit/s or to WiFi that can be up to 300 Mbit/s). NFC is limited to about 10 cm so the devices have to be brought close to each other. **RFID** systems can be as much as 200 m apart. Although NFC exchanges can be intercepted, the short range and encryption make it more secure than RFID. Connection times are reduced and there is no need for manual entry or confirmation of the connection details. There is no need for passwords, passcodes or configuration of settings and a connection can be established automatically by NFC, so two smartphones can be touched together and the contacts, images and photographs, text messages and social-network links can be easily transferred. Connections to wireless headphones, media players and speakers can also be made by one tap. NFC can be used to 'bootstrap' and automate Bluetooth or Wi-Fi connections by sharing the IP or MAC addresses, and data is then transferred over Wi-Fi or Bluetooth. NFC can also be used to send images or documents to a printer. When a device is tapped against a panel on the printer, printing starts with minimal user interaction. As a smart card emulator, it allows smartphones to act as payment cards or loyalty cards or as cards used for ticketing on public transport, by being very close to a contactless card reader.

NFC use in health care facilities allows a very secure and easy-to-use transfer of medical records and patient information. NFC tags can be used to track doctors and nurses as they visit and monitor patients so that drugs and prescriptions are more accurately assigned to the correct patients.

14.7.2 Advantages and disadvantages of wireless transmission methods

The main advantage of using wireless networking is that users can access network resources and the internet while mobile because there are no wires to restrict their movements or be a trip hazard. This extends to being able to connect to networks and the internet while away from home, the office or anywhere that has a wireless access point.

Wireless networking uses less physical cabling (there has to be some, for example to connect WAPs) so is less expensive to install and maintain. Wireless networks can be used where physical wiring is impossible or not allowed. For example, wiring the open spaces of a college campus for every student to use would be almost impossible. In some historical buildings it is not permitted to attach any wiring, or anything else, to the structures. Maintenance of the wires is avoided. The cleaning and servicing of the areas around offices with no network wires is made easier.

In supermarkets or stores, checkouts can be moved around when altering the store appearance and remain wirelessly connected to the store network. Wireless payment terminals, such as contactless payment machines, connect wirelessly to the store or provider network, so can be given to the purchaser to use with their bank cards without wires impeding them. Contactless payments use wireless technology and, by definition, there is no physical contact so payment is made faster and more convenient for both the purchaser and seller.

Devices can be added or removed from wireless networks much more easily than with wired networks. Users can move about and their devices can join or leave wireless networks automatically without the need to connect or disconnect wires. In networking terms, this is called scalability. The number of devices in a wireless network can be scaled up or down more easily than with wired networks.

The disadvantages of wireless networking are that connections are short range and less reliable and have lower bandwidths compared to wired networks. While wired networks can reach up to 100 m between nodes, wireless networks using Wi-Fi have a much shorter range. Also, obstacles, and the materials that make up the obstacles, can reduce the range even further. For example, the construction materials used in walls (concrete, brick, steel and plaster) can all affect the wireless range in unpredictable ways and all reduce the range. Wireless network connections allow users to move about, and changes in location alter the environment of the devices and their distance from WAPs so connections become unreliable or even lost altogether. Wireless networks used for individual devices do not have the bandwidth of wired connections.

While adding and removing devices from wireless networks is easier than with wired networks, this can be a disadvantage as there is less control over the devices that come and go. Network administrators can apply password control over who can access the network, but controlling the type and the numbers of devices is more complex.

A further disadvantage is that wireless networks can be less secure than wired networks. Any device can be easily connected and, if strict security measures are not in place, the data being exchanged on the network can be intercepted and viewed by unauthorised users. Managing security on wireless networks is more complex than on wired networks. Also, the range of wireless network signals may extend outside of the company or home buildings, which makes eavesdropping simpler.

The advantages and disadvantages of wireless networking apply to all methods, but there are some that are specific to each.

Benefits and limitations of Wi-Fi

Wi-Fi is used by most mobile devices, for example laptops, tablets and smartphones, because it allows users to connect to networks and use the internet from anywhere that they can find an access point. Connection to a WAP usually involves little more than a Wi-Fi search and the input of a passcode or phrase to connect. Often, the use of public WAPs (hotspots) is free to customers of a business, for example a cafe, so users choose to connect to a free hotspot instead of using their mobile connection, which may be costly. The bandwidth available on a Wi-Fi hotspot may be higher than that of the mobile connection but, even if it is not, it is sufficient for exchanging emails or viewing web pages.

Wi-Fi hotspots are useful when mobile connections are poor or not available in a particular location. With the use of VoIP, and with most telephone calls being carried over IP networks anyway, voice calls can be made using Wi-Fi connections when there is no mobile signal available inside a building. The call is initiated over Wi-Fi and passed to the public telephone system by the provider of the mobile service.

The use of Wi-Fi allows users to access the internet from almost anywhere in the home, add devices with ease and connect appliances so that they can be remotely monitored or controlled. Little technical knowledge is required to connect smart TVs, refrigerators, heating or air conditioning units, security or nature-watch cameras into a home Wi-Fi network.

A feature of Wi-Fi is its limited range and that some construction materials block or attenuate the strength of its signals. This can be a benefit in that the signals do not extend too far so do not interfere with those of other homes or companies, but it is a drawback when there is a larger, or unusually shaped, dwelling or building to cover. If Wi-Fi coverage is restricted then more WAPs have to be used, which adds to the complexity of the network and reduces the ability of users to move freely while maintaining connectivity. Wi-Fi works at its best when the devices have a clear 'sight' of each other and connection performance is reduced when devices are moved further apart.

Wi-Fi radio signals can be susceptible to interference from other radio waves. The radio bands available for use by Wi-Fi are shared by Bluetooth and cordless (traditional, not mobile or cell) telephones, and, while these should not overlap or use the same channels, they can add to the congestion on the bands. Users experience a reduction in the rate of data transfer, with internet access appearing to be slow.

Benefits and limitations of Bluetooth

Bluetooth connections are simple to set up, often requiring minimal input from the user, so are useful, for example when photographs or contact details are being exchanged between smartphones. Bluetooth-enabled devices connect again automatically after being paired once, which is more convenient for users. For example, a smartphone will reconnect to a car entertainment or hands-free system when the user gets into the car without needing to be set up again.

Bluetooth is found in most mobile devices and is free to use. Small networks (piconets) can easily be set up by users. Smartphones, smart watches, headsets, security devices and medical equipment can all use Bluetooth to connect and transfer data. Bluetooth requires little power because its transmissions are restricted to low powers, so battery life is lengthy. The latest standard, Bluetooth 4.0 at the time of writing, specifies not only higher data transfer rates but is more energy efficient in its use of the radio. However, older versions of Bluetooth can use more power and drain batteries quickly.

The main drawbacks of Bluetooth are its short range and low bandwidth. While some Bluetooth connections can reach up to 10 m and some (Bluetooth 4.0) have a specified bandwidth of 24 Mbit/s, this is rarely achieved in practice. Ranges of only a few metres, or even only centimetres, and low bandwidth limit most Bluetooth connections. The low bandwidth means that exchanging photographs is more convenient for users but is slower than other methods of transfer. The restriction to a short range is also useful when transferring data because it makes eavesdropping on the connection more difficult.

Most Bluetooth connections are encrypted and data is secure. However, users must ensure that they control the pairings of their device. Smartphone users can turn Bluetooth on and off and control whether or not the device will allow or refuse connections. Because Bluetooth systems scan for other Bluetooth systems automatically and can connect automatically, it is important that users do control their devices to protect their data.

Benefits and limitations of infrared

Infrared (IR) devices for home use, for instance TV remote controls, are cheap to manufacture and use very little power so battery life can be many months or even years. IR used for data transfer in homes does not interfere with other wireless technologies and is safe to use. When generated by lasers for use in data communications, for example in fibre optics, IR travels long distances without the signal degrading. In free space, lasers using IR can be a cheaper and less disruptive method of creating data links.

A drawback of using IR is that the sender and receiver must be in line of sight because obstacles block the signals. Also, devices cannot usually move about when IR is being used without losing the connection. IR is restricted to short distances and has a low bandwidth, so is unsuitable for transferring large amounts of data between devices.

Benefits and limitations of microwave

Microwave data links can carry large amounts of data across long distances, provided the sender and receiver are in line of sight. Trees, hills, buildings and poor weather, for example rain or snow, can block or attenuate microwave signals. Even clouds or high quantities of pollen or insects can interfere with microwave signals. Microwave signals are also susceptible to disruption by solar activity such as solar flares.

Benefits and limitations of radio

The benefits and limitations of using radio for data communications depend upon which parts of the radio section of the electromagnetic spectrum are being used.

Using radio means that data can be transferred over long distances and at high rates without physical connections. Since there is no physical link, there are no installation costs or disruptions associated with installing wiring or changing building structures to fit wires or cables. Additionally, there are no ongoing costs with maintaining wiring, cables or the structures that hold them, and there is no damage to buildings from installing wires throughout an organisation or home.

At the very highest frequencies (the shorter wavelengths) radio can provide a high bandwidth, but these waves are easily disrupted or blocked by obstacles. At the lowest frequencies (the longer wavelengths) the bandwidth is less, but the signals can more easily penetrate objects. Choosing the most appropriate set of frequencies to use for a particular method of wireless transmission is the job of standards organisations so that manufacturers and users can make the best use of radio. For example, assigning 2.4 GHz and 5 GHz radio bands for use by public users means that Wi-Fi can safely provide good coverage with high bandwidths.

The limitations of radio include:

» the obstruction and absorption of the signals by physical objects
» the need for care when installing to avoid interference between signals
» the need for more security measures to be taken
» the possible risks of injury and damage to human health from the very high frequencies of radio waves.

14.7.3 Security issues associated with wireless transmission

Wireless networking is more susceptible to security issues than wired networks for several reasons. Devices can be added or removed without complex configuration procedures. The radio waves used in wireless networking may spread beyond the company building or the home, so are susceptible to eavesdropping by others. Some wireless technologies do not automatically encrypt the data, so eavesdroppers may be able to read data from intercepted wireless connections. Various protocols exist for the encryption of wireless transmissions and the data being carried.

There are many ways that unauthorised users may attempt to access the data of others by using wireless technology. Bluetooth connections are short-range, use encryption by default and do not carry much user data. However, open and unsecured Bluetooth connections can be used to extract personal information, user IDs and password details from mobile devices. Tapping into, and eavesdropping on, microwave data transmissions is more difficult and these are usually encrypted.

Most of the security issues that concern users and organisations involve Wi-Fi because most computing devices have Wi-Fi technology installed. Also, the internet and LANs use IP networking and Wi-Fi is the most common method of exchanging data between mobile devices.

'Man-in-the-middle' attacks are carried out by setting up a fake wireless access point that appears to be the actual WAP device to which the user's device intended to connect. The fake WAP has two wireless NICs; one connects to the user device and the other to the real WAP. When a device connects through the fake WAP, all its data can be accessed by the owner of the fake WAP. If the data is not encrypted, it can be understood. Public Wi-Fi hotspots are often used in these attacks as hotspots usually do not use encryption. It is the responsibility of the user to encrypt their data or connection, for example by using a VPN that encrypts the connection from end to end.

Some companies do not allow wireless technologies in their networks and do not have WAPs in order to reduce security risks. However, almost all laptops and many desktop computers have wireless technologies already installed and these pose security risks. A laptop connected to a wired network may be configured to allow connections using its wireless technology. The laptop becomes a WAP and permits others to join the network and access the company data. Another method of attack turns a wired laptop into a fake WAP and other company devices then connect to it. Companies and administrators have to set policies and configurations to combat these issues.

Wireless security protocols

Wireless security protocols are intended to provide the security and confidentiality of data to the same standards as those found on wired networks. The IEEE 802.11 standard regulates the security for Wi-Fi. Wired Equivalent Privacy (WEP) was the first protocol designed to do this and is still available on older devices. However, the more secure **Wi-Fi Protected Access (WPA)** is now the preferred choice for securing Wi-Fi links.

Wired Equivalent Privacy

Wired Equivalent Privacy (WEP) uses stream cyphers for encryption. Stream cyphers are symmetric key cyphers, which means that the same keys are used for encryption and decryption of the data frames. The key is used along with a number, generated from the sequence of numbers in the data, to encrypt the data digit by digit in a stream. Originally, WEP used 64-bit keys (40 bits with an additional 24 for initialising the encryption), but later versions used 128 and then up to 256 bits.

When a device connects to a Wi-Fi WAP using WEP, the authentication that requires the key prompts the user to enter a sequence of characters. The characters are usually in hexadecimal notation to represent the key. There are two ways that a device can authenticate with WEP. In Open System Authentication, the client device joins with WAP before the keys are established for authentication. The keys are used to encrypt the data frames after the client is linked to the WAP. In Shared Key Authentication, the same key is used during the set up and subsequently for encrypting all the data frames.

Neither method of authentication is completely secure. The shared keys can be derived from the exchanges at set-up and by analysing the data frames. Stream cyphers should never repeat keys, but the number of bits available for the keys is too low to enforce this. By analysing large numbers of frames for repetitive streams of data, it is possible to determine the keys and render the encryption useless.

Also, WEP is optional and is often not even set up or used by network administrators or home users. This means that Wi-Fi traffic is not automatically encrypted and is open to anyone to read.

Wi-Fi Protected Access

Wi-Fi Protected Access (WPA) has replaced WEP as the standard method of securing Wi-Fi connections and WEP should not be used. It also replaced the original TKIP (Temporal Key Integrity Protocol), which was an attempt to replace WEP on old hardware.

TKIP uses different sets of keys for each data packet to prevent hackers determining the keys from the stream of packets. TKIP is used as part of the WPA protocol.

WPA uses the TKIP methods and has additional checks on the data packets to ensure that they have not been changed during transmission. This is important because it enables the detection of the capture, examination and resending

of data packets by devices between the original sender and receivers. This is
a Message Integrity Check and requires new, more powerful computer hardware
to run, which is why WPA cannot be used on older wireless networking devices.

WPA is implemented in a number of ways. WPA-Personal (also called
WPA-pre-shared key or WPA-PSK) is used for home users and small businesses.
A shared 256-bit key is created for each Wi-Fi device from the wireless network
SSID device and the network passphrase. The pre-set passphrase is entered by
the user as a string of characters and is used with the SSID to calculate a unique
256-bit key for that device to share with the WAP. The 256-bit key is used
to generate a 128-bit key for encrypting the data frames. In large companies,
a separate server may be set up to provide the authentication of WPA. It can
enforce company policies and protect against attacks on the network.

Wi-Fi Protected Setup (WPS) is used to enable simpler connections to WAPs.
The use of a PIN or the press of a button can set up a connection using WPA
between WPS-enabled devices. However, WPS is not secure because most
modern WAPs and routers have WPS enabled by default and WPS PINs can be
determined by **brute force attacks**.

WPA is vulnerable to weak passphrases. The keys used for encryption are
dependent upon the passphrase, the SSID name and the length of the name. If
these are short or weak, then the keys can be discovered by brute force attacks
on the passphrase. The SSID is already visible so, once the passphrase is known,
the keys can be discovered or the network accessed by unauthorised users. This
is why a brute force attack on the PIN used for WPS is a security issue and users
are advised to turn off the feature on home routers.

Another issue with WPA is that if the pre-shared keys are discovered, it is
possible to decrypt all the packets ever transmitted using those keys. Frames
captured by unauthorised users can be decrypted and the data understood. The
keys can be discovered from the passphrase. The passphrase, or Wi-Fi password,
for a free Wi-Fi hotspot is given to anyone who asks for it so, from it, user data
can be captured and understood. This is why users of public Wi-Fi hotspots
are advised to use extra security, such as TLS or VPNs. Better still, do not use
services that require personal information or passwords, such as online banking,
while connected to a public Wi-Fi hotspot.

WPA has been improved over the years and new standards published. The
later versions, WPA3 and WPA4, use more bits for the keys and use different
methods of sharing and generating keys to improve security. WPA3 and WPA4
attempt to secure against the issue of weak passphrases and the use of the
passphrases to discover the keys.

14.8 Mobile communications systems 

Mobile communications systems allow users to move around and change
locations while remaining connected to networks. The connections enable
voice conversations and the exchange of data by users. The link between the
end user and the networks is wireless, using radio waves, without the physical
connections that may be used elsewhere in the system, such as wires or cable.

Mobile communications systems allow users to contact and exchange data with
others regardless of their location. This enables users to work from home, in
their offices and when travelling between locations. Employees and employers
can keep in contact, individuals can message each other and people can respond
to, or be alerted to, emergencies quickly with the use of mobile communications
systems. A medical use of mobile communications devices is for assistance in

contact tracing of people who may have been close to others with infectious diseases, such as Ebola, influenza or other corona viruses.

14.8.1 Cellular networks

Cellular networks enable the use of mobile communications by user devices, for example smartphones and tablets. Most laptops, and many tablets, do not have access to cellular networks and use Wi-Fi. Most smartphones can use either or both.

As described in Section 14.1 'Networks', cellular networks divide a terrain into areas that are approximately hexagonal with a base station in the middle. The hexagonal shape of the coverage is approximate in cellular networks because buildings, hills, trees and other objects affect the propagation of radio signals. Square shapes are sometimes used. The area covered by a base station is called a cell. For this reason, the telephone handsets that use them are called cell phones in many parts of the world. In other parts of the world, the phones are referred to as mobile phones. Cell phones can make voice calls and send SMS messages. Smartphones are cell phones with added features that can use Wi-Fi capability and the cellular network, for example for email and web-browsing apps, social-media apps and many other useful facilities.

Cellular networks enable the apps and services to be used when the user is mobile because the base stations allow devices to move seamlessly between them, provide internet access by connecting into the public communications networks and have sufficient bandwidth. These details have been described in Section 14.1 'Networks'.

Structure of a cellular network

Each generation of mobile networking has meant a new set of standards for cellular networks, but the principles remain the same. Cellular networks use low power and have a shorter range than traditional radio communications systems. In each cell there is a base station, with an antenna, that has a receiver, transmitter and a control unit. The cells with the antennae are arranged so that they are as equidistant from each other as possible. This means that the cells usually appear to be hexagonal in shape.

Each base station has a set of frequencies allocated to it for use in the connections with mobile devices. Between 10 and 50 different frequencies are assigned to each cell base station. The frequencies assigned to a cell are completely different from those of the cells next to it. The transmission power is limited so that the signals do not spread too far into neighbouring cells. This is to minimise the interference between signals from different cells.

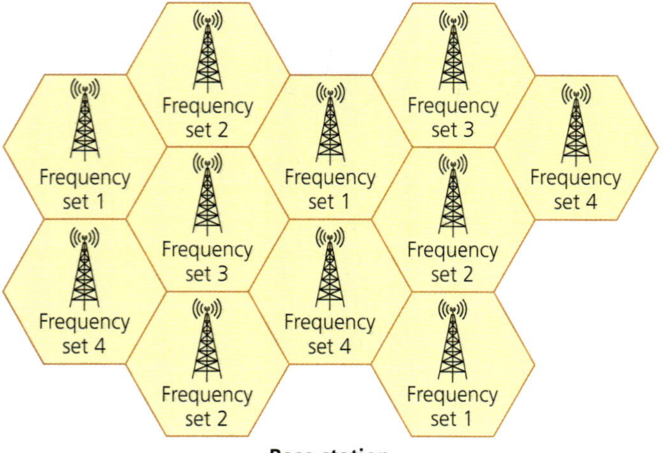

▲ **Figure 14.31** Frequency reuse by mobile (cellular) base stations

The frequencies used by base stations can be reused across cellular networks because some base stations are far enough away from each other to avoid interference. In cities and other areas where the geography does not allow the neat arrangement of cells, other configurations have to be constructed.

However, as the radio waves are not restricted by the theoretical shapes, and since the power of cells that are not next to others can be increased, it is probable that there will be network coverage. In effect, there are larger cells for some areas compared to others. Cellular network providers can adapt their networks and divide their cells into smaller cells by adding more base stations as the number of users in an area grows. In some city areas, cells may be only a few square metres in size. Obviously, this adds to the costs for the provider and many do not aim to cover every square metre of a city. This is because users are moving about and can quickly move into an area that is covered. Each base station is connected to the public telecommunications network via the service provider.

The aim of the structure of the network is to allow users to move from cell to cell and remain connected. A user should not notice any interruption to their calls or data exchanges while moving from one cell area to the next. This is accomplished by the handoff from one base station to another as the user moves. A handoff between base stations means that the user's smartphone automatically changes the frequencies it is using. As the user moves to the edge of the range of one base station, the station requests a handoff to the station with the strongest signal in range. The stations and smartphone automatically set up the call on the new frequencies without the user noticing. Any number of handoffs can happen at any time during the call.

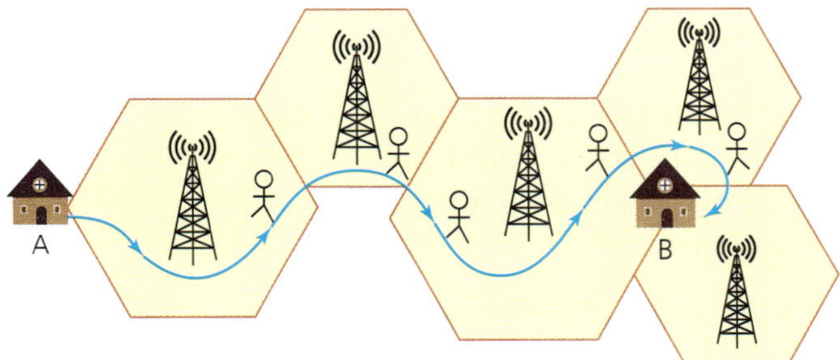

▲ **Figure 14.32** Base stations handing off a mobile (cell) phone as a user goes from A to B

In Figure 14.32, a person walking from house A to house B has handoffs at least four times. There may be more as the person passes through the different areas, as signal strengths depend on many factors. House B is covered by three areas so handoffs will also happen as people move around inside the house.

Use of 3G system for mobile communications

The first global standard for digital cellular networks was GSM (Global System for Mobile Communications). In the early 1990s, GSM set out the protocols and was used for the 2G networks. It included the general packet radio service (GPRS), which is a standard for exchanging packets of data, and which became part of the third generation (3G) standard that replaced 2G. GPRS has a very high latency because voice data has a higher priority than GPRS, but it does provide users with internet access wherever they are in the world. It is often used as a 'last resort' connection when signal strengths are low.

3G networks can revert to 2G connections if the signal strengths are low. This is to ensure that the user can make voice calls even if an internet connection is not possible.

3G networks provide greater bandwidth than the earlier generations and have data transfer rates of over 2 Mbit/s if the devices are fixed installations. However, for mobile devices, the bandwidths are significantly lower. In a fast-moving vehicle, such as a car or a train, a bandwidth of between 128 and 144 kbit/s is usual, but this rises to about 384 kbit/s if the user is walking slowly.

3G networks provide greater security than earlier generations because devices must authenticate themselves to be able to join the network. This means that users can be sure that they are connecting to their intended network and not 'spoof' ones.

3G networks support circuit switching, which can be used for voice calls. However, 3G networks also provide access to IP networks, with packet switching, which allow users to make voice calls and exchange data. With the use of IP, users have access to the range of services found on the internet. While 4G and 5G are replacing 3G as the standard for mobile networks, in many areas of the world 3G is still in use and will probably continue until the 2030s. In some areas of the world, however, for example in China and the USA, government and service providers are announcing that their 3G networks will soon be closed down. Many countries have already closed down and no longer operate 2G networks, and elsewhere, such as India, some providers have already closed down their 3G networks as well.

2G and 3G networks are being shut down because most users are expected to move to 4G and 5G networks. The frequencies used for 2G and 3G can be used for other purposes, and new devices, such as smartphones, do not support older network connections.

Use of 4G system for mobile communications

4G networks are succeeding 3G networks because they have a greater capacity to support more users at the same time, can offer higher bandwidths to users and provide a better quality of voice call. 4G networks are based on packet-switching techniques and can implement IP networking to provide internet access.

4G can provide a bandwidth of up to 1 Gbit/s when users are almost stationary relative to the base station and around 100 Mbit/s if they are moving. This increase in bandwidth allows mobile devices to access all internet services, from VoIP to the streaming of high-definition video. Whereas streaming video over 3G is subject to buffering problems such as jerky or stuttering video and audio, missing video frames or freezing and loss of audio, using 4G removes most of these issues.

Many more simultaneous connections to base stations can be made with 4G compared to 3G. Also, there are more efficient uses of the frequencies, better handoff techniques and seamless 'roaming' between networks. All these factors allow users to move around between areas, countries and continents and still be able to connect to a mobile network. Smartphones should be able to connect to a 4G network anywhere in the world. Roaming allows mobile (cell) phones to connect as a 'guest' to mobile networks other than the mobile network of their service provider. This can happen when the cell coverage of the provider is not available or the user moves to a geographical location outside of the provider's area. For example, a user travelling to another country can connect their smartphone to a cellular network in that country.

> **Activity 14r**
>
> Explain why viewing a streaming video on a smartphone using 4G is preferable to using 3G.

Use of 5G system for mobile communications

5G networks use higher frequencies than 3G and 4G and these provide higher bandwidths. The higher frequencies, however, have a lower range and the coverage of an area can be very small. A movement by a user of only a metre may dramatically reduce the signal strength. A smaller range also means that the cells are much smaller and there has to be many more cells than previously. However, the increased bandwidth and the increase in the number of simultaneous users compared to 4G means that most service providers are moving to 5G networks.

In theory, 5G can currently provide almost 2 Gbit/s but, in practice, this is often between 50 Mbit/s and 1 Gbit/s, with about 500 Mbit/s being the most common bandwidth that users experience. This enables access to all internet services and allows users to stream, or download, high-definition video movies without any of the problems experienced with 3G or 4G connections. 5G networks offer more reliable and higher quality voice connections and the increased capacity for many more users. Users can make high quality voice and video calls and hold multi-user video calls without the interruptions, jerky video and intermittent loss of audio experienced with 3G or 4G.

The very low latency of 5G networks enables the use of devices and tools that require reliable network connections. Remote use is possible over 5G, for example of robotic devices in surgery, vision-enhancement smartglasses and other tools where fast, reliable data exchange is essential. Devices can be interconnected using 5G to create an Internet of Things (IoT) with billions of possible connections. 5G enables smart homes with interconnected appliances, smart cars with 5G connections for navigation and communications, industrial applications in automated manufacturing and many other scenarios that require a very fast and reliable system of data exchange.

> **Activity 14s**
>
> Explain why the low latency of 5G mobile networks is important in robotics.

Mobile hotspot

4G and 5G can be used to create a Wi-Fi hotspot to connect to the internet. Similar in operation to a Wi-Fi hotspot, a device, for example a smartphone, equipped with a connection to a 4G or 5G mobile network, can be set up to allow Wi-Fi-enabled devices to connect and to share the 4G or 5G connection.

Users can set up a **mobile hotspot** with a smartphone and connect a tablet or laptop. The tablet or laptop can have access to the internet via the smartphone's mobile network. The benefits of using a mobile hotspot from a smartphone are that the connection is secure because it uses the private mobile network connection of the user, it can be used anywhere that the smartphone can connect, it needs no wired connections and there is no need to search for a Wi-Fi network connection. However, the drawbacks include a reduced bandwidth because the connection is shared, there may be extra charges made by the service provider or data limits may be reached if a great amount of data is exchanged, and the battery charge of the smartphone will be depleted more quickly.

Internet service providers also supply dedicated devices that connect to the mobile (cellular) networks and provide Wi-Fi for mobile devices to use to connect to the internet.

14.8.2 How satellite communications systems are used for transferring data

Large numbers of satellites orbit the Earth. While some gather data about weather, some are used for navigation, and others are used for exploring inner and outer space, many satellites are used for exchanging data between users on Earth. Satellites used for data exchange can be in a variety of orbits around the Earth. To exchange data, the satellites are linked to ground stations that transmit data to, and receive data from, the satellites and there must be a direct line of sight between the satellite and the ground stations. The connection to the satellite from the ground stations is called the uplink and the connection from the satellite to the ground stations is called the downlink. Permanent links are needed for television and radio transmissions direct to Earth, for environmental sensing and measuring systems, and for connections into the internet by ground-based users. These use geosynchronous satellites.

Many data communications satellites require permanent links to a ground station. This sends up signals used, for example, to ensure that the ground stations and satellites are always visible to each other and the satellite is placed in geosynchronous orbit. The links are used for telecommunications and for remote sensing of activities or environmental measurements on the Earth's surface, for example mapping of terrain or buildings and temperature measurements.

> ### Activity 14t
> Satellites are often described as geosynchronous or geostationary. Explain the meaning of these terms when applied to satellites in orbit around the Earth.

Data from sensors and images taken by cameras onboard satellites are used for observing weather activity, for tracking atmospheric events, for example hurricanes, and for studying the oceans. Data is gathered from sensors that measure factors like the temperature of the Earth's surface, cloud temperatures and the amount of water vapour in the atmosphere. This data can be used along with high-resolution visible light and infrared images in tracking weather patterns as well as weather reporting and forecasting. The data and images are also used to monitor the effects on the environment of climate changes. Data is transmitted along connections that can use satellite links, for example:

» automatic teller machines (ATMs)
» electronic point of sale (EPOS) terminals
» contactless card payment machines and other devices that require connections to credit card providers or banks for authorisation of payment
» e-commerce and online shopping from around the world.

Satellites enable data communications between locations distributed around the world. Alongside fibre-optic cables, satellite links make the connections that enable the internet to span the globe. In summary, internet connections use satellite links to provide global coverage.

Global Positioning System

Satellites are used for the different global navigation systems now in use. The most widely used system globally is the US **Global Positioning System (GPS)** and the following text is based on this system. GPS has a constellation of 31 satellites that orbit at about 22 200 km. They take about 12 hours for one complete orbit, so any location on Earth can 'see' several satellites at any one time. GPS provides location data for military, commercial and individual civilian use.

The data transmitted from GPS satellites can be used by **SatNav** receiver devices to calculate their position. The main purpose of a SatNav receiver is to receive the data transmissions from the satellites and use the data to calculate its position on Earth. This position can be overlaid onto stored maps and a route can be calculated to another location on the map. As a user moves with the SatNav receiver, the device continually recalculates its position and displays it on the map. There is no need for the receiver to have a connection to a mobile or Wi-Fi network because the satellite signals are transmitted direct to the receivers on Earth. The receiving devices do not transmit any data to the satellites nor do they need any other connections to calculate their location. SatNav devices that do have network connections can use additional services (for example map updates, weather and traffic news, alerts to traffic or obstructions on routes), but these are not needed for the basic navigational calculations.

Usually, a SatNav receiver can 'see' between seven and twelve satellites, but a minimum of four GPS satellites must be visible to a SatNav device so it can accurately calculate its location. While a location can be determined with only three usable visible satellites, the location cannot be calculated with high accuracy using fewer than four. The SatNav receiver has to have four pieces of information from the data transmissions: three sets of data about its location and one about the time difference between it and the satellites. Every SatNav receiver has an almanac, which is regularly updated by transmissions from the satellites, that tells it where each satellite is at any one time and calculates its location using trilateration in three dimensions.

Using four satellites, the SatNav receiver can calculate its latitude and longitude and its altitude above the Earth's surface. These can be repeatedly recalculated at a high rate to provide a series of locations if the receiver is moving. It is left to the actual device to display the location in a format that the user requires and can understand. In practice, a SatNav receiver will track as many satellites as it can at any one time. This means that if it loses the signals from one or two, it can still calculate its position from the others.

Errors caused by obstacles such as clouds, buildings, and the leaves on trees, are mostly ignored by the receivers used in cars, for example, as the errors are small or considered not important. However, for use in military, aircraft and shipping scenarios, the errors are compensated for by using more complex receivers that make additional location checks against known locations.

SatNav receivers use the results of the calculations to display the location to users as longitude and latitude coordinates, as a point on a map or in other user-friendly ways. In a moving vehicle, the device can be made to read the details out loud so as not to distract the driver. A receiver can provide other information to users. It can calculate how far it has travelled, how long the journey has taken or will take, its current speed, an average speed over the journey or parts of the journey, and a recording of the route taken.

A SatNav receiver may provide additional features but these require an additional wired (for example USB) or wireless (for instance Bluetooth, Wi-Fi or mobile) connection. Traffic updates, weather news and useful or interesting locations on the route (points of interest), for example a gas (petrol) station or a restaurant, can be overlaid on the map or searched for and the user alerted as appropriate. These are not provided by the satellite transmissions. Directions along a route between two locations (route planning) are the most common use in vehicles and these are displayed on the screen of the device or read out loud to the driver.

The Global Positioning System is used commercially in aviation, road, rail and shipping transportation, agriculture and scientific research. In aviation, GPS provides location and altitude information to supplement other onboard instruments. Airline operators and private pilots can determine routes and file flight plans using GPS coordinates and information. In many areas of the world, aircraft are not allowed to fly unless a flight plan has been filed with air traffic control, so GPS makes this task easier for pilots. Routes can be determined according to destinations, obstacles en route and fuel requirements. Routes can be quickly altered and recalculated using the GPS information and information from weather updates, air traffic congestion and for other reasons. Aircraft can transmit their GPS coordinates to air traffic control and be tracked with accuracy. In the case of loss of contact with an aircraft, its last known position can be known. Its route, using GPS locations, is also recorded in the aircraft data recorders.

In other modes of transportation, GPS is used to plan routes and to locate ships and vehicles. Navigation of ships and boats on the high seas and when entering or leaving ports makes use of GPS to ensure the safety of ships and crew. GPS is used to accurately navigate shipping channels and ports with highly detailed maps that are constantly being updated. Small sailing and fishing boats use GPS to find their way between ports and to fishing areas. With fishing areas being carefully controlled by many governments, accurate GPS location finding is important for the fishers and for the patrol boats.

Heavy machinery used in mining and construction uses GPS for accurately locating and manoeuvring. Positioning markers for mining and for road construction, and programming vehicles to follow a route during mining or construction, enables the vehicle drivers to be more precise in driving. Accurate positioning of roads, levels and gradients is achieved using GPS location devices.

Public transport (for example buses and taxis), delivery vehicles, emergency vehicles and private cars can use GPS locating devices (SatNavs) to plan routes and find locations. Planning the fastest route using GPS, along with traffic information provided by a mobile data connection, can save money for taxi passengers, enable drivers to deliver goods more efficiently, and allow first responders to arrive at emergencies quicker and take patients to hospitals in shorter times.

People use GPS in social and fitness activities. Cycling, cross-country hiking, sky-diving, paragliding and any activity that can make use of location and route finding can use GPS. Tagging photographs and objects with GPS coordinates shows their location.

GPS can be used for tracking objects, vehicles and people. However, a GPS location of an object is only useful if it is known. A GPS tracking system requires both a GPS device and a data connection. A data connection is required to send the GPS coordinates to the person who is tracking the object. This is usually accomplished using a mobile communications network, for example the 3G, 4G or 5G mobile networks. Smartphones already have these and, as most have GPS as well, a smartphone location can be tracked. Objects or people can only be tracked in real time if there is a reliable data connection to send the GPS locations.

For tracking vehicles, for example automobiles in the case of theft, or for checking the locations visited, two methods can be used: active and passive. An active system requires a data connection and sends the GPS information to a central control centre in real time. This is useful for tracking lost or stolen cars or trucks so that they can be found, and for tracking the movements and delivery of expensive or dangerous items, for example currency to banks or nuclear waste to processing facilities. A passive method stores the GPS locations of vehicles at regular intervals so they can be downloaded when the vehicle is back at the depot. This is useless if the vehicle is stolen or lost as the data is lost along with the vehicle. However, it is very useful for monitoring the routes taken by rental cars, buses and delivery vehicles.

GPS is used in robotics to guide autonomous robots in finding their way. Robotic delivery vehicles, drones and robotic farming machinery use GPS to accurately navigate areas. For tracking animals during their migrations or movements, GPS can be fitted into very small devices that also contain transmitters to send data by radio to scientific institutions for study. To save weight in power packs, and power, the devices do not send data in real time but periodically transmit recorded locations back to base. Of course, the GPS receivers have to be able to 'see' the satellites so GPS tracking is not the best choice for tracking animals whose movements are not in open areas.

Global mapping systems

Satellites orbiting the Earth can take photographs, monitor the weather, monitor the seas and provide much information for producing accurate maps. Satellites used for such monitoring are called Earth observation satellites and are usually in low Earth orbits between about 700 m and 2000 m. Orbits lower than 600 m experience drag from the atmosphere and often have to be boosted back into their intended orbit. This requires the use of precious fuel and can shorten the life span of the satellite. The details of the orbits are beyond the scope of the syllabus, but they do have an impact on the type and amount of data that is collected. Different orbits and altitudes of orbits can affect the quality of the images and data collected.

The images from satellites used in mapping vary in resolution. The resolution depends on how many pixels represent an area on the Earth or, conversely, the area on the Earth that is represented by one pixel. The resolutions in use by **global mapping systems** can often distinguish between objects 15 cm apart. Military satellites have higher resolutions.

Satellite imagery is used in creating maps (cartography). Photographs and data from other instruments, for example infrared detectors, can be combined to produce high quality maps for navigation, meteorology, mineral prospecting, fishing, construction and environmental monitoring. The maps can be printed as hard copies or used in computer displays with real-time updates.

Companies that own the satellites license the data to other companies to produce maps, for example Apple and Google use imagery licensed from satellite companies. These maps are made available to the general public to view and use. The maps provided by Google Maps™ combine images from space and street-level photographs. Satellite images are used for aerial viewing of locations while street-level photographs allow close viewing. Combined with the use of computing programming and scripting, users can search for and display places of interest. Businesses and individuals can use these displays in advertising, location finding and reporting.

Satellite imagery produces vast amounts of data, requiring huge data storage capacity. The data can be analysed using artificial intelligence and used, with other sensor information, for climatology when monitoring and researching surface temperatures, moisture levels and other parameters of the Earth's surface.

Maps of vegetation, forests, crops and other plant growths can be monitored and used in scientific research. The effects of climate change on plant coverage and growth can be studied over time. The effects of wildfires on plant growth can be seen.

Land and sea boundaries change over time, glaciers shrink and expand, and volcanoes erupt. The effects of these can be photographed and mapped from satellites and the effects monitored.

Surveillance

Satellites are used in surveillance in many ways. The general public can view images from satellites of their own locations and places of interest, but usually do not have access to the real-time, high-resolution images that can watch individuals. However, some interesting views are available from NASA and other space organisations.

Governments use satellites to watch the activities of other countries at sea and on land. Satellites are too high to be intercepted or interfered with by others so are able to look down on the Earth's surface without much interference. There is an agreement, but no official treaty, that weapons are not allowed in space, so surveillance satellites, unlike aircraft, can currently be deployed to look down on other countries while remaining safe from enemies.

Law enforcement agencies can watch their own citizens using satellite imagery. As resolutions increase, people and vehicle identification details can be viewed and monitored from satellites. While this may appear simple, it requires high-resolution cameras and imaging equipment to be mounted upon satellites, or constellations of satellites, that can be accurately targeted at specific locations while the satellites orbit at high speeds. Law enforcement and military satellites can be used in this way to track criminals or terrorists and their activities. Military operations can be watched using satellite imagery by commanders in real time from very long distances away.

Satellites can be used for surveillance indirectly by using data from systems such as GPS. Tracking vehicles to ensure that they are only being used for their intended purposes is surveillance. GPS provides route instructions and up-to-date traffic information for the drivers, and the tracking of school buses in the USA is used to calculate the most efficient route for transporting children to and from school. However, GPS is also used to ensure that the buses do not deviate from authorised routes and that drivers keep to speed limits, and to check that the drivers do not take too long with their rest breaks. The GPS data can be used to check on drivers and their driving skills and also be used in disciplinary actions. On the other hand, such data may be of assistance to drivers in case of accidents or other events.

Civil rights organisations have raised concerns about the invasion of privacy from satellite surveillance of citizens.

The GPS satellites are also used to survey and monitor the Earth for one important, specific purpose. A frequency is set aside for the GPS satellites to transmit data to US ground stations as a means of detecting, locating and reporting nuclear detonations in the atmosphere or in space near to Earth. It is a way of enforcing the treaties banning nuclear weapons and for detecting nuclear accidents.

14.8.3 How communication data is prepared, sent and received by satellite communications systems

Data is exchanged between satellites and ground stations using microwave radio signals. Satellites can exchange data with other satellites and relay data around the world. A ground station is aimed at a satellite and the signals are transmitted up to the satellite, which receives the data and retransmits it back down to a ground station. As long as there is a direct line of sight between the satellite and the ground stations, the signals can be received by ground stations anywhere. Usually ground stations are long distances apart, so satellite communications are used to exchange data over very long distances. Satellite communications may be used between ground stations that are not far apart to avoid running new cables or when there is data exchange for a short time only, for example for TV coverage of a sporting event or outdoor concert.

Geosynchronous and geostationary satellites provide a permanent link to ground stations that do not have to track the satellite's orbit. Satellites in other orbits appear to move so the antennae of ground stations have to track them across the sky and will lose sight of individual satellites. Constellations of satellites are needed for continual coverage of a location on Earth if medium or low orbits are used.

The data, for example a television or radio broadcast or the internet, is received by a ground station that modulates onto microwave radio signals. The frequencies used are allocated by the region being served and by the type of satellite communication in use to make efficient use of the spectrum and to avoid interference between signals.

A ground station receives data from the source. The source can be the internet, television or radio stations and it is carried to the ground station as signals along telecommunication links. At the ground station, the signals are modulated onto the frequencies assigned for the uplink of the data to the satellite. The antenna, or dish array, is accurately pointed at the satellite and transmits the data to it. At the satellite, the data is demodulated from the uplink signal and modulated onto a different signal for sending back to Earth. The downlink frequency is different from the uplink to avoid interference. The receiver in the satellite is called a transponder because it changes the frequency used for the signals. There may or may not be processing of the data by the satellite, depending on the purpose of the link. Usually, there is only a frequency change because the satellite is acting as a repeater and not carrying out routing or filtering activities.

The antenna in the dish array of a ground station is accurately aimed at the satellite because the satellite is a very small target a long distance away. The strength of the uplink signal is higher than the downlink signal strength because the antenna on the satellite is relatively small. For the downlink signal, a lower signal strength can be used because data link ground stations can have quite large dishes for their antennae. For satellite TV reception, smaller dishes are used to allow fitting to homes, but they must be accurately aligned.

When the microwave signals are received from a satellite, the ground station demodulates the data from the carrier waves and modulates it into a format that is suitable for transmission along the medium used to send the data to the recipient. The transmission medium could be public telecommunications lines or a cable going to a home satellite TV set-top box.

Use of satellites in telecommunications

Satellite links form a major part of the private and public telecommunications links that make up the internet by carrying data over long distances.

Use of satellites for internet access

Geosynchronous satellites provide global coverage for data exchanges over the internet. Satellites can also be used to provide direct internet access to users on the ground. Geosynchronous satellites can enable these connections to reach bandwidths of over 500 Mbit/s for the downlink.

Until recently, the upload connection to the internet for satellite internet used a traditional landline or mobile connection and only the download came via satellite. This was because the uplink to satellites required high precision when aiming the antenna and needed considerable power to transmit. The drawback is that, despite the high download bandwidth, the bandwidth link back to the ISP is much lower and the time between the request for data, for example a web page, and the delivery of the page was much higher, often more than 10 or 20 times that of conventional surface-based internet access. This increased latency is mostly due to the travel times of the signal from the ground station to the satellite and then back to the user's antenna. The delay causes a total latency of 550 ms compared to, for example, a latency of 20 ms for surface connections.

Two-way satellite internet, where both the uplink and downlink are direct from user to satellite, offers greater bandwidths but there are still the delays that cannot be overcome. A request by a user for a web page must travel to the satellite on the uplink, back down to a ground station and then across the terrestrial internet to the web server. The reply travels in the reverse direction back to the user. A signal takes about 120 ms to travel between a ground station and satellite. At least four links are made for each request and response, which introduces a delay of about half a second overall and is impossible to reduce. This is a best-case scenario as the ground stations are rarely directly below the satellite and the atmosphere further slows the signals. A typical latency is around 1 to 1.5 seconds.

The latency or delays may be tolerable to users who otherwise would not be able to access the internet because there is no local infrastructure. The digital divide in regions may mean that local infrastructure is poor or non-existent. Also, the user may be on a ship at sea, in an aircraft or in remote areas, such as high in the mountains or in a desert. In these circumstances, satellite communications are the only way to provide a connection to the internet.

A high latency can affect the initiation of secure connections. Secure request-and-response connections, for example between a web browser and a web server using HTTPS or SSL/TLS connections, include many exchanges of small amounts of data in many data packets. Long delays between the exchanges can result in a failure to authenticate and set up a secure connection. Other TCP/IP protocols may not work properly when there is a high latency in the network. Software in the satellite devices that alters the data about timings or divides the route taken by the packets into apparently smaller sections can make the delays appear to be much shorter so that the gateways and routers do not drop the packets. The delays are still present but the browser and web server only 'see' part of the delay, which is smaller and acceptable. While direct satellite internet access is not very good for gaming due to the high latency, it performs well enough for web browsing, email and video streaming, although most providers apply limits on the amount of data that can be exchanged every month.

However the links are managed, the downlink bandwidth maximum is rarely achieved as the encoding and modulating of the signals means that some users share the links. A typical bandwidth for each user would be around 40 Mbit/s.

Recent developments have placed, or are planning to place, large constellations of satellites in low Earth orbit to provide internet access to areas of the world that either currently have no access, have poor access or are very large, for instance Africa and continental USA. Low Earth orbit means the satellites are closer to the Earth so latency is reduced, but the technology is more complex because connections have to be handed off more often from one satellite to following satellites.

In some cases, a satellite in low orbit, for example the Canadian CASSIOPE satellite, may receive data from a ground station and store it but not retransmit it until it passes over another. This is used where there is no need for a continuous stream of data or a link but when a user wishes to transport large amounts of data from one location to another. The data is uploaded and stored until the satellite passes over the destination ground station. It is similar to posting a USB memory stick carrying the data. With an orbital period of about 90 minutes and passing over each station about fifteen times a day, the system works well for bulk data.

Use of satellites for television services

Satellites are used in both the distribution of television services between studios and centres and for the broadcast of digital television services direct to viewers. Most satellite television services broadcast digital television direct to the user from geostationary satellites. The sources for the television services come from the TV service provider or broadcaster and are encoded into a digital video and audio stream. The high bandwidth available in direct satellite broadcasts allows both standard definition and high-definition television video, along with multichannel soundtracks, to be broadcast. Ultra-high-definition (UHD) TV can also be broadcast. However, the bandwidth is not limitless and broadcasters often prefer to supply more channels rather than higher definitions. More channels result in a greater income from advertisers and some broadcasters reduce the quality of the broadcasts to squeeze in even more channels. In the USA about 100 TV channels, in India about 100 TV channels and in the UK about 200 TV channels are available from satellite.

The broadcaster may supply viewers with its own TV channels, may broadcast channels from other broadcasters alongside its own, or a combination of both. Live TV broadcasts, pre-recorded programmes and other channels are collated and assembled at the broadcaster's control centre into the channels of TV that the viewer can watch. The channels are then sent as feeds to the uplink ground stations. Access to the channels can be controlled by the broadcaster by encrypting some or all of the channels before sending them to the uplink station. Some broadcasters have their studios, control centres and uplink stations close together, but others have them in different locations. Feeds between the different locations can travel over any telecommunications systems or other satellite links.

At the ground station, the feeds from the broadcasters are modulated onto the uplink frequencies using a block upconverter (BUC) and sent to the satellite. A transponder on the satellite receives the signals and sends then back to Earth on a different, downlink, frequency or set of frequencies. Most systems use the DVB-S standard for transmission, which sets out how digital television is broadcast from satellites.

A home viewer has a satellite dish installed so that it points at, and has clear view of, the broadcasting satellite. The dish itself is a reflector designed to capture the signals from the satellite and direct them onto a receiving unit mounted at the focus of the dish. The receiving unit is a low-noise block downconverter (LNB) powered along the coaxial cable by the receiver in the home. The LNB receives the weak signals from the satellite, amplifies them without adding 'noise' and converts them to frequencies that can be sent down a coaxial cable to a satellite TV receiver in the home. Coaxial cable is cheap and easy to install but cannot carry the frequencies transmitted by satellites, so a LNB has to be used to send the appropriate intermediate frequency (IF) signals down the wire. The IF signals are processed at the receiver in the home, which is usually located near or inside the TV set, and the digital television channels are recovered for display to the viewer.

To enable more channels to be carried on a single transmission frequency, satellite TV signals are polarised. Typically, both vertically and horizontally polarised signal are transmitted from the satellite and received by a LNB that is positioned, or configured, to receive one or the other. Modern LNBs can be electrically switched to receive either, but not both at the same time. Each polarised signal can carry many television channels.

Broadcasters can apply encryption to some or all of their TV channels, and some individual programmes carried on a TV channel may be encrypted while others are not. This is to control access by viewers. Unencrypted channels or programmes are said to be 'free-to-air' and can be watched by any viewer with suitable equipment. Encrypted channels and programmes require a decoding system to be viewed. The satellite receiver, or set-top box, is equipped with a slot for inserting a conditional-access card carrying the codes that allow the decryption of encrypted content. The viewer pays the broadcaster for access to its TV channels and content. Each broadcaster has its own encryption method to avoid unauthorised viewing of its TV programming content.

Activity 14u

1 Why do ground stations and satellites have to be in line of sight to be able to communicate?
2 Why do satellite TV pictures in homes sometimes break up or disappear in bad weather?

Use of satellites in radio

Radio stations can be broadcast from satellites in different ways. Stations can be broadcast using the same set of frequencies as Digital Video Broadcasting – Satellite (DVB-S) television. These radio stations require less bandwidth than TV and are often broadcast to duplicate terrestrial radio transmissions or to add extra channels to a national system. This is common in Europe.

Satellite radio in other areas of the world uses a different group of frequencies. In North America, 2.3 GHz is used, which is close to the Wi-Fi frequency allocation, but in other parts of the world 1.4 GHz is used. The system uses geosynchronous satellites over the intended target audience area. This enables listeners anywhere in, for instance, continental USA to access the same radio station at the same time. This system is intended mainly for use in cars, which have receivers fitted into the car stereo system.

Use of satellites for telephones

A satellite phone, or satphone, is similar to a mobile (cell) phone but connects to satellites rather than to terrestrial cellular networks. Both geosynchronous and low-orbit satellite phone systems are available.

Geosynchronous satellites, with three or four satellites, provide continuous coverage over almost all of the Earth's surface, but have difficulties when users are in locations at high latitudes far from the equator, due to the shape of the Earth and to obstacles. There are also delays in the signals, which may intrude into conversations due to the distances between the user and satellites.

Low-orbit satellites only appear in the sky over a satphone for a short time, so satellite radio systems have a constellation of 50 to 100 satellites.

A satphone is similar in appearance and size to a cellular smartphone but does not currently have the same features.

Aircraft and ships use specialised satellite radio receivers and transmitters for satellite radio communications. These can be configured to allow passengers to make voice calls when aboard planes in flight or on ships at sea. The costs to passengers are considerably higher than calls made on the ground via mobile networks. Similar satellite connections can be used to provide internet access to passengers.

14.9 Network security

Network security includes all the procedures and policies that must be carried out and used to protect stored data and files from security threats. Data can be protected against destruction, manipulation, modification and theft if the appropriate preventative methods are used. Security procedures and policies enable individuals and network administrators to track and put in place measures to avoid access to the data and files by unauthorised persons and by malicious software.

14.9.1 Networking security threats

One of the reasons for storing data and files on networked devices is to allow them to be shared and accessed remotely, but this makes them vulnerable. Sharing and remote access must, by definition, allow users to access files from other networked devices. Data and files can easily be damaged or destroyed if care is not taken to protect them. Authorised users can make mistakes, for example accidently deleting a file, and unauthorised users and software can amend, delete or steal data.

Unauthorised access is when files or data are read, viewed or otherwise accessed by persons or software without the required permissions to do so. However, this is not the only threat that requires network security measures. Access to services, files and data may be denied by attacks that prevent authorised users from accessing their files.

The term 'hacking' refers to the action of accessing computer systems, files or data without having the usual credentials to do so. This is not necessarily unauthorised as hackers can be employed to access systems, files or data where the owner has lost the passwords or where owners require and give permission for their security measures to be tested. However, a common use of the term is to describe the act of attempting to gain unauthorised access to systems, files or data.

Address Resolution Protocol (ARP) listings on devices are vulnerable to malicious attacks. The entries in the ARP list can be altered to redirect traffic to an attacker's address by changing the IP address that should be associated with the MAC address to the attacker's address. When packets destined, for example for a company website, are sent, they end up at the attacker's address because the MAC address now has a fraudulent IP address associated with it. The fraudulent IP address could be a look-a-like website that gathers login credentials that the attacker can use for fraud. Controlling access to these is essential.

The generation and distribution of SSH keys is beyond the scope of this syllabus, but SSH clients store the keys for all the hosts that they have ever been connected to. There is no time limit on the use of these keys. Management of SSH keys is a critical job of network administrators because SSH keys, the equivalent of usernames and passwords, are used automatically by devices and many applications. New connections remain vulnerable while the keys are being exchanged because they could be intercepted and then used by others in a 'man-in-the-middle' attack. However, the long-term storage of keys by servers means that a client and server will use the same keys every time they connect. This reduces the possibility of man-in-the-middle attacks for frequent connections. However, a vulnerability of SSH is that the server keys are not discarded but stored for very long periods of time. Usually, the keys are stored in the same logical folder on all servers, so if the server is accessed by unauthorised persons, the keys can easily be located.

Brute force

A brute force attack is the systematic submission of many passwords or pass keys in the hope of correctly finding the actual password. A brute force attack is often automated using software and can work very quickly if the password is short, a common name or a dictionary word because the software will check these first. Using a long password with combinations of upper and lower case characters, unusual characters and numbers can significantly increase the time that would be needed for a successful brute force attack. Other methods of preventing brute force attacks from being successful include having time delays between unsuccessful password attempts, restricting the number of attempts that can be made before the account is locked and adding a multi-factor authentication system or requiring a CAPTCHA answer from the user. Restricting login to specific devices, IP addresses or geographical areas can also be used to protect against brute force attacks.

Denial of Service

A Denial of Service (DoS) attack attempts to make a computing resource unavailable to its legitimate users. It is carried out by disrupting the services of a host on a network, either by forcing the host to stop providing services ('crashing' it) or by flooding and over-loading the host with so many multiple requests that it cannot cope. The latter is the more common type of DoS attack and is usually achieved by multiple attacking systems using up the network bandwidth or host resources. When multiple systems are used to attack a host in this way, it called a distributed DoS attack.

There are several ways to defend against a DoS attack. Rules can be set in firewalls to deny network traffic from attackers, but this is complicated to set up and requires the IP addresses and ports of the attackers to be known in advance or to be rapidly implemented once an attack is underway. Using a firewall to close all unused network ports can help prevent DoS attacks. A 'cleaning' or 'scrubbing' centre, which receives all network traffic destined for the victim of a DoS attack, analyses the traffic and passes on only legitimate requests to the host. All other traffic is denied. Using an intrusion-prevention system to monitor network activity can recognise DoS attackers if they have patterns, signatures or content that is already known.

Botnets

A **botnet** can be used to deny access to services, send spam emails, steal files and data, and allow an attacker to remotely control devices by issuing commands that appear to come from the device itself.

A botnet consists of a number of networked devices that run software, called bots, carrying out repetitive tasks using computer programs written as scripts.

A computer script is a set of commands in a computer language that is stored in a file and can be executed on a computer without being compiled. Scripts are written in languages such as Perl, PHP, JavaScript or Python.

A botnet can be made up of a group of any type of networked devices such as smartphones, laptops, tablets or even home appliances connected in the Internet of Things (IoT). Each device runs a copy of the malicious software, called a bot, that makes use of TCP/IP connections. Botnets typically make use of SMTP, HTTP and IRC to contact each other, send malicious code or disrupt services.

Bots work by creating client–server or peer-to-peer networks between devices. The malicious software is sent over the internet by a cybercriminal, called a bot herder or bot master, by randomly probing IP addresses or by inserting the bot into the hypertext, JavaScript or other code of websites as a Trojan. Sometimes the Trojan carrying the botnet malware is rented or purchased from other cybercriminals. Websites that distribute inappropriate content or illegal copies of software are the most common carriers of malicious software. A device that loads the website or responds to an IP address probe is 'infected' with the bot software. The infected bot also probes random IP addresses and passes the code on to other devices. Each bot regularly updates itself by communicating with other bots.

Botnets based on the client–server model connect to a central command device, called a command and control (C&C) server, and await instructions. The bot herder sends out commands to all the bots, which carry out the instructions and send the results back to the herder via the C&C server.

Peer-to-peer botnets do not have a central bot herder, but bots update themselves with the latest version of the malicious software from other bots. Commands or instructions are carried out and the results sent back to the originator. Peer-to-peer botnets are less vulnerable to disruption if the software is detected or discovered because the loss of a few bots in a botnet can be tolerated. New bots are found to replace them.

Botnets are used by cybercriminals to disrupt services, gather data about a user's activities or personal and confidential information, and commit other cybercrimes.

Denial-of-service attacks can be carried out by botnets simultaneously sending out many thousands of requests to servers. Too many requests can overload a server and prevent it responding to legitimate requests. The server will either stop responding to all requests or take a very long time to respond to legitimate ones. Legitimate users are denied access to its services.

Spam emails can be sent from a user's email clients by botnets. Phishing emails requesting confidential information can be sent, asking users to access spoof websites in order to gather login details that will be used for fraud. Botnets can make a user's devices visit websites without the user knowing. These visits can be used to artificially increase the statistics about visitors to the site.

Botnets can be used to gather data about the keys that a user presses or to directly harvest data about login credentials. Key loggers and spyware can use botnets to send the data back to the originator of the botnet. Botnets have been used to generate Bitcoins on computers without the owner's knowledge. This is stealing the user's computing resources. Some botnets are used to gather social-networking patterns and to collect information about the political views of users. These details can be used to influence the users or others.

Botnets present a threat to network security because they can harvest keystrokes, gather confidential details such as passwords, send spurious or spam emails and carry out denial of service (DoS) attacks.

Malware

Malicious software, or malware, is created to inflict damage on computer systems, applications, files or data. Malware can also be software that intentionally has an adverse effect on the user. Software or software errors that unintentionally cause harm are not malware.

Types of malware do harm in different ways. Adware, **ransomware**, rogue software, scareware, spyware, Trojan horses, computer viruses and worms are all malware. Malware makes use of, and exploits, vulnerabilities in operating systems, software applications, add-ons or plugins (for example to web browsers or other software), poor or incorrectly applied security measures, or inexperience or inattention by users. Devices connected to networks are especially vulnerable to malware. Chapter 5 of *Cambridge International AS Level Information Technology* discusses malware in more detail.

Malware spreads across networks in different ways. Worms transmit their own code to other computers on a network; viruses can be carried by being embedded in other, often innocent-looking, software that is exchanged; and emails can carry viruses in attachments or have links to websites with malicious code that is transmitted when a user clicks on the link and opens the website. In whatever way the malware is transmitted, it is a threat to the user's data, files and computer systems.

Malware presents various threats to computer files and data, particularly those stored centrally on networks. Malware that gathers login details for the storage areas, such as cloud storage, and targets these systems, for example a data centre, can very quickly collect or damage vast quantities of data. The most obvious threats to files and data stored on networks are deletion or destruction, manipulation, modification and theft. Malware spreading across networks can very quickly delete large quantities of data. Data can be manipulated and changed so that any information based on it is false. Data can be stolen, often without the user being aware, by sending it to others. The thief can use the data, such as login details, to access bank or credit card accounts to actually steal funds or use them to pay for goods delivered to an altered address, or even use them to create a new, or copy, identity and apply for government documents, as in **identity theft**.

Ransomware blocks access to, or encrypts, user data and then requires users to pay a ransom for the encryption keys. It works 'best' when targeting network, or cloud, storage systems. Because many different organisations may share a server farm or resources on data centres, for instance data storage, a ransomware attack may simultaneously affect many organisations. Data may be stolen for industrial espionage purposes. Networks make these actions possible because devices are interconnected and can exchange data over the internet, and the cybercriminals receiving the data can be based anywhere in the world.

Malicious actors

Malicious actors, or threat actors, are people that perform acts intended to harm the computer networks, systems and devices of individuals or organisations. They also, as a consequence of their actions, cause harm to the individuals and organisations themselves. Malicious actors can be cybercriminals, state-run agencies, company employees, competitors or individuals seeking attention or trying to make an impression on others. Each type of malicious actor has its own motivation. Cybercriminals creating scams or carrying out hacking may have a financial aim, state-run agencies may be spying for information, rogue company employees working from inside a company or for competitors may be seeking a business advantage, and individuals seeking attention may be attacking computer systems just for the so-called fun of it, by spreading misinformation or 'trolling' others. Other malicious actors may have an ideological motive, seeking to spread their own ideas or to provoke terror. Identifying the motive or goal is one of the ways that malicious actors can be identified.

Structured query language

Structured query language (SQL) is used for programming and for managing data in relational databases. SQL is used to access, return and present the results of the access to a user, for example when a user searches for a payment, item or any other piece of data on an online database. However, if the database is not properly secured or there are vulnerabilities in a web application, SQL can be used to attack the database and discover login details and sensitive and personal information, such as usernames and passwords or credit card and bank account details. By entering malicious SQL commands into the search or login fields of an online form, a process called SQL injection, malicious actors can gain unauthorised access to the sensitive data. The data can then be used for other malicious acts including identity theft, stealing money from bank accounts or fraudulently using credit card details. Social-media sites, online retailers and university websites are the systems that are most frequently targeted by SQL injection attacks.

One way of protecting against SQL injection attacks is to ensure that each account that has access to a system or database has only enough access rights to allow it to carry out its task, for example an account that only needs to read and view data should not be allowed to edit or write to the database. This is the principle of least privilege. Another way is to use SQL commands that are stored in the system and fixed, and to avoid dynamic SQL. This prevents cybercriminals from adding their own SQL commands into the system's own SQL code. Also, using reputable software developers to create web and database systems and keeping these properly updated to correct any software flaws or vulnerabilities helps to prevent SQL injection attacks from being successful.

Network policies

Organisations have **network policies** to set the rules on how their network is used. A network policy can include the following:

» the types of technologies allowed and not allowed (for instance whether or not wireless devices or USB memory sticks can be used)

» what users can and cannot use the network for (an acceptable use policy)
» the responsibilities of every user when using the network (including how to keep data safe and secure and what to do if a breach or virus attack occurs)
» the security protocols to be used (such as a password policy).

It also serves as a basis for enforcing legal action if the rules are broken.

14.9.2 Impact of network security threats on individuals and organisations

Impact of data destruction

Data destruction is the process of making sure that it is impossible for data to be recovered. Data may be destroyed as a result of a malware attack. This could have a serious effect on a company's business, which may suffer as a result of the loss of orders or customer data, or on individuals, who may lose personal or sentimental images and photographs.

Ransomware is used to maliciously encrypt data; it is decrypted only when the owner pays for the keys. The ransom can be millions of dollars and there is no guarantee that the keys will be sent by the cybercriminal or that they will decrypt the data. Companies can go out of business if the price of the ransom bankrupts them, if the data is never retrieved or if customers do not trust them again. Individuals may also find the cost of the ransom too high or not receive the keys to decrypt their data.

Data may have to be destroyed for many reasons. Data-protection laws and regulations may require data to be destroyed, for example when it is no longer needed or is out of date. A storage device may have to be replaced and its data removed before it is sold, disposed of or recycled. Intentionally destroying data can be quite difficult. Removing data completely from hard disks is sometimes almost impossible, because reformatting, reusing or overwriting data may not completely remove it. It is often necessary to resort to physical destruction of the device. However, while this may seem a good method, data can still be retrieved from the magnetic platters of hard disks after physical deconstruction and from the chips used in SSDs even after they have been shredded.

There are methods of data destruction using software. Overwriting the entire hard disk sector by sector with ones and zeroes several times is used to securely destroy its contents. Another method is to use encryption provided by the disk manufacturer to store the data. Encrypting all the data with an encryption key of at least 128 bits renders the data impossible to retrieve or be understood without the key. To destroy the data, the key is discarded by the encrypting software. This is called cryptographic erasure and is deemed highly effective in ensuring that data is lost forever. While the data remains on the disk, it is irretrievable.

However, ensuring that every copy of the data has been destroyed may not be possible because the data may have been copied, stored in the cloud or in backups and archives, or remain in caches on many other networked or users' computing devices. It is also possible that developments in computing power could discover encryption keys used to cryptographically erase a disk.

Impact of data manipulation

Data manipulation is the changing of the format of data so it is displayed in different ways. This can be visual presentation (for example presenting data in a table or a chart) or by using the data in calculations or spreadsheet formulas to create new data.

For example, the data you have is 10. It can be displayed as 10 or ten. It can be manipulated by dividing it by 2 to get 5 or by adding 2 to get 12. The original data remains as 10 but is manipulated in various ways.

Data can be manipulated in many ways to extract information. Manipulating data can be intentional and for a good purpose. In businesses and organisations, intentional manipulation can extract information from databases using, for example, SQL. SQL is a text-based Structured Query Language and is used to create reports and manipulate the data in a database, for example run queries, edit data, insert data, edit tables and edit records. Individual users can set up filters or queries on their spreadsheets and databases to find information, and online retail websites can manipulate data about the goods for sale when users look at or search their websites.

When data is manipulated by attackers, it causes victims and others to doubt that their own version of events is correct. Some attackers do not concern themselves with the victims that they are attacking but do it to create chaos and misunderstanding. Other attackers may be paid by rival organisations to undermine a business. Governments or politicians may attack others for political gain. By causing others to doubt the real data, a process called 'gaslighting' in some countries, governments' and politicians' credibility and reputations can be seriously damaged, public opinion altered and elections influenced.

Impact of data modification

Data modification changes the actual data. Data modification attacks are often carried out by employees or insiders who have an argument or disagreement with their employers. These attackers do not need to bypass or break through security measures because, as employees, they already have login credentials and access to the data. Other attackers can exploit loopholes in the security or acquire login credentials through social-engineering techniques.

For example, the data you have is 10. It is stored as 10 in a spreadsheet cell. The contents of the cell are changed by a user from 10 to 12. The original data has been modified to be 12.

Data modification attacks are made because they undermine the confidence that customers and others have in a company. Changing the stock market values for a company or even altering the display screen, either decreasing or increasing the values, can have serious effects on the company's business. Attacks on companies that supply goods can disrupt supply chains for manufacturing or food processing. Altering the health records of patients could result in the wrong medication or dosage being given. In extreme cases, patients could die as a result.

Maliciously modifying data, in addition to manipulating data, can undermine customer confidence in businesses, alter public opinion and influence others in decision-making. Often, it is very difficult to detect a modification of data or how much data has been modified. This creates even more distrust and loss of confidence in the data and in the organisations that own it.

After a data modification attack has been detected, it can be very time consuming and costly to restore the data to its original form. All data, backups, archives and archive data must be checked and compared with source materials, if available, to attempt to correct the data. Even after this has been done, victims cannot be absolutely sure that all the data has been corrected. This can have long-term consequences, such as distrust and uncertainty.

Impact of theft of data

The theft of data from a company, organisation or government can lead to fraud, bankruptcy and embarrassment. Stolen credit card or bank details and other financial data can be used by thieves to commit fraud by using the details to buy goods or steal money from accounts to carry out other fraudulent and criminal activities.

The financial impact of these activities is not only immediate losses if money is stolen or goods bought by the thieves and charged to the credit cards, but also the fines that may be imposed. Many governments impose fines on organisations and businesses if they lose or have their data stolen due to security breaches. Under data-protection laws and regulations, those that store data about people must take every precaution to keep it safe and secure. Very high fines, for example over $1 000 000 for breaches of security with health data in the USA, can cause financial hardship for companies and may cause them to go out of business.

Also, when a data breach has occurred, organisations must notify those affected. This can add to the already high costs of the loss or theft of data. Employees can become stressed by the extra work and concerned about their employment and prospects in an affected company. The cost in terms of time taken by employees to process and send out notifications can further reduce employees' productivity on other tasks. Customers lose confidence in businesses that lose data or have data stolen. They may take their business away, so the company that lost the data becomes less profitable. Potential new customers go to competitors, which will gain customers and increase their market share.

Ransomware can deprive organisations of their data. Paying the ransom may retrieve the data or it may still be lost, and not paying the ransom loses the data. This can cause organisations to go out of business or close down. Repairing or recovering data may involve computer downtime and this is costly to businesses. Data breaches that involve large quantities of data may involve extra costs such as the need to open new call centres and set up new websites to answer queries from those affected. Those customers or individuals affected by the data loss may be offered compensation or may seek compensation through lawsuits, which also is costly.

Organisations that have data stolen may face further long-term impacts in that they may have to upgrade, improve and alter their security procedures and policies. New, more secure software versions may have to be installed if existing ones are shown to be vulnerable or out of date. The impact of data theft on individuals can be financial loss, loss of reputation or embarrassment if the data is shared or made public.

Impact of identity theft

Identity theft is a deliberate use of another person's identity. Identity theft may be for financial gain or other benefits. Identity theft involves the theft of personal information, for example names, dates of birth, driving licence numbers, passport numbers, credit card and bank account details, PINs, social security or national identity numbers and electronic information such as passwords, electronic signatures or biometric data. Details used for identity theft can be gathered from social-media postings, web browser histories, brute force attacks on passwords for logins, spyware, botnets and social engineering.

The person whose identity has been stolen may or may not be aware of the theft. However, usually there is a financial or social loss to the individual. Financial identity theft uses the person's details to buy goods or obtain financial services that the victim is eventually expected to pay for, or it affects the victim's credit rating so that the victim cannot access financial services. The financial gain for the thief may be used to finance other crimes, for example terrorism, drug dealing or illegal immigration.

Identity cloning is when a person assumes another's identity in real life or social situations and pretends to be that person using their stolen data. This could be to avoid being caught for other crimes or to use that person's lifestyle or benefits, for example to claim medical insurance or tax benefits. If an identity thief uses another person's details to be employed, the employer will report the tax details to the government, which may then investigate the real person. The impact on the real person can be devastating and take years to put right.

14.9.3 Prevention of network security issues

Effective security measures and the enforcement of company security policies can help to prevent unauthorised access to files and data stored on networks. Measures to ensure that data is kept safe and secure only work if every employee uses them and sticks to the policies made by the company. Similarly, the data of individuals can only be kept safe and secure if security measures are used.

Physical methods of security are designed to keep devices, terminals and other network equipment out of reach of any unauthorised user. Software security measures are designed to prevent unauthorised users from electronically accessing data and files using networks.

Physical methods of preventing unauthorised access to data

Some of the simplest methods of physical security are the most effective. Physical methods involve an interaction with hardware components, for example a lock and key, inserting a USB device into a port or plugging a cable into a router. Some physical interaction may also involve software, for example biometric readers for identifying staff when they access doors, but these are really input devices and not physical preventive methods.

Locking the doors to computer rooms, to the rooms that house servers and routing devices, to the buildings that house data centres; and securing windows, loading bays, skylights and other openings, are the most obvious methods that are used. Having a company policy listing who is allowed to have keys and which areas the keys will unlock, and keeping track of the keys, is fundamental to the security of any building, but is most important where confidential data is stored. Placing security guards at the entrances to the building and to secure areas is also important. Entrances to buildings can have physical barriers through which staff must pass to gain entry to the building where they work. The barriers can be manned by security guards who check the credentials of employees or there may be an electronic system where passes and IDs are checked by software. Sometimes both are used.

Security cameras that survey entrances and access doors to computer resources are useful and may be watched remotely. Security guards can monitor many different areas from one room that may not even be in the same building. Any suspicious activity can be seen, and security near to the camera location can be alerted. Automatic alerts can be set up to raise the alarm if movement is detected in areas that should be empty of people. Video from the cameras can be analysed and used with facial recognition software to identify persons who have been seen by the cameras.

The use of electronic means of controlling access to areas is now common. RFID tags and biometric data read by readers near the doors can control where employees or staff are allowed to go and where they are not. Keeping logs of who has accessed which doors and areas, and monitoring these with security cameras, can ensure that the level of security is high. However, these logs and cameras are only useful if they are accurate, up to date and continually monitored. If not monitored in real time, an employee may use another's ID tag, steal data and get away before the theft is noticed. Analysing details from logs after a theft does not prevent the theft from happening because it is too late. People, or AI, must monitor the entry by employees and create alerts if unusual behaviour or activity is noticed. Employees must be trained to log out of their computers when away from their desks, to lock doors when leaving rooms and, most obviously, not to write their passwords down, especially onto papers left near their computer. All these aspects must be included in a company policy and enforced.

The physical method of requiring an access PIN or smartcard also relies on software to work. There is a physical barrier, for example a turnstile, and a reader close by. There may be an entry reader and an exit reader so that control can be exercised over both entry and exit. The barrier is fitted with locks controlled by the software and with sensors that monitor its position. The barrier should also be fitted with a mechanical override as a safety feature so no one gets locked out, injured by the barrier or locked in if there is a fault or failure.

When a person requests entry, or is sensed as approaching the barrier, the person presents the entry credentials, for example they tap a smartcard on the reader, enter a PIN or allow a scan of the biometric parameter such as a fingerprint. The data of the credentials is read and sent to the software for comparison with an

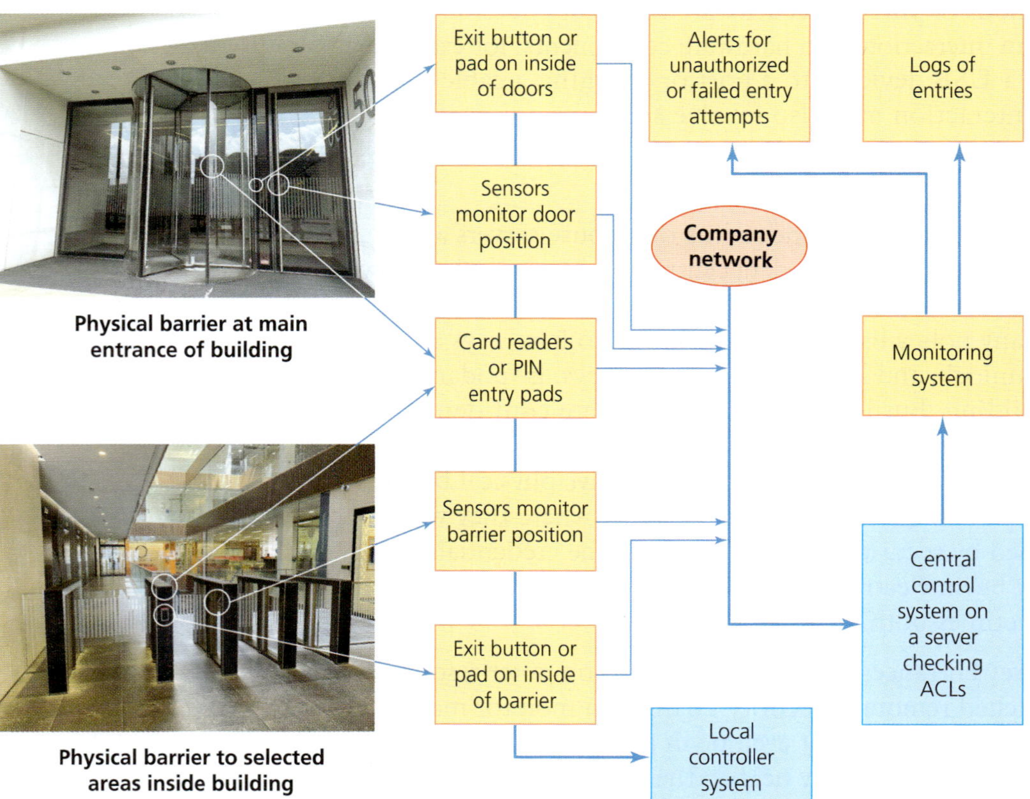

▲ Figure 14.33 Schematic diagram of barrier control at entrance to a company building

access control list (ACL). The software can be running on a dedicated server and is connected using a TCP/IP network, but other dedicated links are sometimes used to increase the security of the system. The software may also be running within the controllers at the barrier and checking ACLs stored centrally. If the credentials match with an entry in the ACL, the barrier locks are released, the person is alerted of this by a sound or a change in colour (or both) and the barrier alarm is temporarily disabled for the duration of the entry. The duration of the unlocking and disabling of the alarm is long enough to allow the person to pass but usually not long enough for two or more people to pass. The use of the credentials is recorded in a log file for later analysis, if required.

Physically securing devices in their locations with restraining cables or locks, or preventing external devices from being connected to the devices, for example removing serial or USB ports, can prevent unauthorised access. Many networking devices do not require any authorisation of USB memory sticks or extra network cables when they are inserted, so access to the physical ports or plugs should be made difficult. Desktop computers and laptops are also vulnerable to such connections and many companies forbid the use of USB devices in their computers.

An extreme physical method of securing a network from unauthorised access is by creating an 'air gap'. Air gapping is used to secure information that must remain totally secure, secret or confidential. An air-gapped computer device has no NICs at all and an air-gapped network has no connections to any other network. It is completely isolated from other networks, from the internet and from the public telecommunications network. No network interfaces, wired or wireless, have any connections other than to other, authorised, devices on the network. This method is used by government agencies, military organisations, agencies that control critical systems (for example nuclear power plants), public transport systems, air traffic control and medical computer installations. However, this method requires a very strict adherence by employees to policies. Employees who try to use USB memory sticks or connect their mobile devices, for example a laptop, to the network can compromise its security. Some malware is specifically coded to exploit such breaches of policy.

Other physical methods include storing backups on devices, disks or tapes that are not easily accessible to others, ensuring that hardcopy confidential data is shredded or burned when no longer needed and not dropped into waste, or trash, bins for collection by cleaning staff. Confidential shredding policies and services protect data, such as access logs or changes in login details, from being used to access company networks.

Software methods of preventing unauthorised access to data

Software methods of preventing, or attempting to prevent, unauthorised access are called 'logical access controls'. Logical access controls apply to systems, programs and applications, processes, data and information.

The logical access methods, used for attempting to prevent unauthorised access, can be found within the following:

» operating systems
» software applications or apps
» specialised security applications
» the management software of network devices or telecommunications systems.

Biometrics

The **biometric** systems that control access to buildings, areas or rooms are logical access controls. The biometric details of an employee are recorded and stored in a database, and the database is searched when a set of biometric data is read by an input device. If a match is found then access is granted; if no match is found then access is denied. Biometric data may be used to control access to various parts of a building or the physical barrier at the main and staff entrances.

Biometrics are measurements of human body characteristics and the calculations that can be carried out using the measurements. To be useful in security, biometric measurements must be sufficiently unique to the individual to identify the person among many others. The measurement must not change over short periods of time, for example a person's weight or height changes often. Suitable characteristics for use in biometrics include fingerprints, retinal colour and patterns, palm and handprints, facial characteristics and DNA. Some biometric data can be attributed to behavioural characteristics such as typing speeds and patterns, walking movements or voice patterns. However, behavioural characteristics are not sufficiently unique to an individual to be used to identify a person to a computer system. Nor would it be practical, for example, to measure walking movements and use them to access a storage device.

The factors that are considered when choosing characteristics for use in biometrics include:

» how easy the characteristic is to measure
» how permanent the characteristic is
» how unique the characteristic is to an individual (and whether every individual shows that characteristic)
» how acceptable it is to collect the data from an individual
» how easy it is to forge or copy the biometric data from the characteristic.

Different characteristics meet these criteria to different degrees; some will be more acceptable than others and some will be easier to measure than others. For example, collecting DNA samples at the entrance to a building is slow, not really acceptable to staff and the readings take time to process, while retinal or handprint scans are quick, acceptable to staff and the results of the readings are available very quickly.

Activity 14v

Research each of the following methods of biometric authentication: fingerprint recognition, facial recognition, iris patterns, retinal patterns, voice recognition, keystroke patterns and handwritten signature recognition, and on a three-point scale of High, Medium or Low, rate each for: costs to implement, user friendliness, permanence, accuracy and security. Present your results in a summary table.

To successfully use biometrics as a security method, it is first necessary to collect and store the data about the characteristics from every person who will be using the system. This is called enrolment. Collection and storage of the data has to be acceptable to the persons affected and comply with data-protection laws and regulations. The collected data for each person is stored in a database and used to calculate reference data that is stored in a template for each person. Margins of error are also calculated for each person dependent on the characteristic used. For example, while most fingerprints are unique to a person, skin damage, wear and cleanliness of skin can affect the reading taken at a door. The biometric system will allow some errors or differences before requesting a retry or denying access. These are stored in the biometric systems as risk scores representing the possible likelihood that it is the genuine person or the likelihood of it being an imposter or someone pretending to be the real person. The storage, transmission and use of the biometric data must be secure.

When a person tries to gain access, an input device with sensors reads the biometric characteristic and sends the data to the software of the system. Biometric systems are often used in addition to PINs, smartcards or RFID cards, or passwords. The person is required to input a PIN or password, or use the RFID card, before the biometric data is read. This is to prevent more than one person using the same PIN, smartcard or RFID card to gain entry. Only the person who should be using it will have the correct biometric data.

The biometric data that is collected as the person tries to gain entry is processed to compensate for artefacts, for example smudges or marks on the fingerprint reader, and sent to the system so that a match in the template database can be found. If a match within the acceptable margins of error using the scores is found, entry is granted. If not, several retries may be requested and, if these fail to find a match, entry is denied and an alert generated. The person may be directed to a real person to be checked, then refused or allowed entry.

As well as being used for security in buildings, access to many computing devices can be controlled by biometrics. Voice recognition and fingerprint reading can be used for access to smartphones, tablets and laptops. On set-up, a user enrols themselves into the logical control software on the device by speaking or having a fingerprint taken. To access the smartphone, the user then only has to speak the command word or touch their finger against the sensor in the reader.

▲ **Figure 14.34** Biometric system using fingerprints for authentication

Biometric methods, for example the use of fingerprints or facial recognition, allow the use of **zero-login** authentication methods. A zero-login system requires users to provide only biometrics for identification. Users do not have to remember complex passwords. Systems using zero-login authentication can recognise individuals by their characteristics, for example by voice recognition, facial features or other unique biometric measurements. Zero-login authentication can use typing speed, how hard a screen is tapped or even measure a heartbeat. This may appear still to be science fiction, but Apple's iPhones use fingerprint biometrics for unlocking and the iPhone X uses facial recognition, measuring over 30 000 points on the face.

Zero-login methods can be extended to remote logins. Logging in to a server means sending login credentials over the network or internet between clients and servers. Using zero-login methods can reduce the vulnerability of credentials during transmission. With zero-login, the authentication software can run locally, for example on the user's smartphone, and only the probability, or risk score, of the identification being acceptable is sent from the client to the server. No credentials or details of the user's location, behaviour or biometrics are sent. The decision by the server whether or not to authenticate is based on the risk score. Zero-login methods are still under development but offer the possibility that thefts of confidential and personal data, such as user IDs and passwords, from servers and the consequential financial and personal harm to the victims will become far less frequent.

Anti-malware software

Protection against malware falls into three phases: before an attack on a network, during an attack and after an attack.

Anti-malware software scans every file as it is accessed or when a user instructs it to scan. Anti-malware software works by analysing and checking software and files against databases of known malware of all types. Files of particular types are always scanned, for example .exe or .zip files, because these are known to be used as carriers of malware. It also acts in a heuristic mode by detecting patterns or types of activity by software that are known to be malware actions. Heuristics allow anti-malware, especially anti-virus, software to detect new viruses by their activity. If a new piece of software appears to be accessing every file, for example every email message, it is likely to be a virus or other malware. There is always a possibility that the anti-virus software is incorrect and that the software is safe, but false positives are better than losing data.

Before a malware attack occurs, protecting the files and data on networks requires that the malware be blocked at the point of entry to the network. Entry points to a network are the gateways and routers that connect to other networks and to the internet. There are entry points at WAPs where smartphones, tablets and laptops connect using Wi-Fi. Also, users may attach external devices, for example USB memory sticks, to their laptops and create entry points for malware.

Monitoring entry points for malware uses software that captures and analyses the traffic moving into, and out of, the network. Datagrams can be analysed and whole files examined for malware. Firewalls examine each datagram as it passes into, and out of, networks and can use predetermined rules to decide what to do with the datagram.

The activity of gateways and routers can be recorded by taking snapshots of their memory at intervals. If malware is detected, the snapshots can be used to determine when and where it entered the network and used to track its progress. Network administrators can use the information to locate and remove the malware.

Malware that avoids detection on entry can be detected by advanced anti-malware software that tracks every file that moves between and around networks. The software, for example Cisco's Advanced Malware Protection software, tracks and monitors the activity of files. Any malicious behaviour is detected and reported to the network administrators for analysis and action. The alert includes the origin of the file as well as its activity.

Other anti-malware software can compare files on the network with databases of known malware, or malware characteristics, and send alerts to administrators. Files detected as suspected malware by this, the router-monitoring software or the file-tracking software, can also automatically quarantine the suspected malware for analysis or deletion in a **sandbox**.

Sandboxing is used when analysing malware so that its activities and the data it may have harvested can be investigated. A sandbox is a computing environment that isolates the malware and restricts the resources it can access. The activities of the malware can be studied without it doing any harm. Sandboxes can be isolated devices, virtual machines or operating systems running in restricted modes.

Anti-virus software

Viruses are a specific type of malware. Protection of files and data from viruses on a network involves using **anti-virus software** at the points of entry, on devices connected to the network and on storage devices for live, backup and archive data. Also, files and data stored on cloud storage systems, for example in data centres, need to be protected. The anti-virus software must be kept up-to-date and in use. Any detections, or suspected detections, must be reported to administrators and acted upon very quickly.

Strict company policies are required that all networked devices have anti-virus software installed, that it is updated regularly and always in use, that all files are scanned when accessed, and that all anti-virus software runs in a mode that can detect unusual patterns of activity as well as actual viruses. Also, the policy should include a ban on downloading files from the internet or that a safe and secure method be used, for example downloading to a sandbox area and scanning all files before access. Employees should be trained in the use of anti-virus software, how to recognise a virus attack, how to respond to it, who to report it to and how to recover from an attack.

Anti-spyware software

Spyware is a specific type of malware. Spyware differs from a computer virus in that spyware is designed to remain undetected on devices, while a virus is designed to copy itself and spread across networks. The term 'spyware' covers adware, keyloggers, rootkits and cookies. Spyware is used to record and track activity on computer devices, for example:

» logging the keys that are pressed
» recording screens
» recording the user's audio and video using the microphones and cameras of the device
» taking remote control of the device
» capturing the user's browser history.

The information is used to target advertising, harvest login credentials, scan for social-media contacts and instant messages, and gather information that can be used against the user.

Scanning files and data as they enter a network for spyware and regularly scanning devices can help to prevent spyware from gathering information for unauthorised persons. **Anti-spyware software** must, like all anti-malware software, be regularly updated and used. Also, companies should establish policies that control the use of the microphones and cameras installed in smartphones or other mobile devices. Use of microphones and cameras by spyware can record conversations that can later be sent to others.

Anti-spyware software should, as with other anti-malware software, monitor the entry points of networks and regularly scan individual devices for spyware. Some spyware can disable firewalls or anti-virus software, so extra measures must be taken to monitor these.

Encryption

Using encryption to protect files and data stored and exchanged on networks is an effective method of preventing unauthorised access. Connections to networks by Wi-Fi and via VPNs use encryption to establish the connections and secure data while it is being transported over the network connection. Encryption of data uses keys that transform the data into unintelligible text and back again into its plain text form. Both symmetric and asymmetric cryptography is used in protecting computer data. Data can be encrypted when stored on local storage systems, such as hard disks or USB memory sticks, network storage systems or in the cloud. However, data is especially vulnerable to attack when it is being exchanged between devices over networks or the internet because it can be intercepted. It is important to secure the connections and the data so that it is unintelligible to unauthorised users. Encrypted connections are difficult to intercept and encrypted data is impossible to understand without the required keys.

Symmetric encryption uses the same keys for encrypting and decrypting data. The keys are a shared secret between sender and recipient. The keys have to be exchanged between sender and recipient so they are known to both. The main drawback of this type of encryption is that keys can be intercepted and used to eavesdrop on the data exchange.

Asymmetric cryptography uses a recipient's key that is known to everyone to encrypt the data before it is sent. The recipient decrypts the data with a private key known only to the recipient. Only the corresponding private key can decrypt data encrypted with the public key. The security relies on the recipient keeping the private key secret.

Symmetric cryptography requires less processing power to encrypt and decrypt data than asymmetric cryptography, so when a connection is established between devices to exchange data, asymmetric cryptography is used to share symmetric keys known only to the sender and recipient. The extremely secure asymmetric cryptography keeps the exchange of the symmetric keys secure from eavesdroppers. Data is exchanged using the symmetric keys, which are discarded when the connection is closed. This use of symmetric and asymmetric cryptography is used by TCP/IP software, for example Transport Layer Security (TLS) and its predecessor Secure Sockets Layer (SSL), which work at the Transport layer for secure data exchange over networks.

Digital signatures that are used to authenticate messages and documents make use of asymmetric encryption. Asymmetric cryptography used in secure data exchange is vulnerable to 'man-in-the-middle' attacks. To prevent such attacks, digital signatures are used, managed, distributed and revoked by Certificate Authorities, which are trusted to verify the holder of the public key. Certificate Authorities issue a digital certificate that shows that the digital signatures and keys are all valid and correct.

Digital certificates

Digital certificates are verified by digital Certificate Authorities, for example DigiCert. Security using TLS relies on these digital certificates being accurate, up to date and correct during the set-up stage to create secure connections. In summary, a TLS connection is started with a handshake to authenticate the client and server, and to share the keys for the encryption. All subsequent data is encrypted with the keys.

The use of TLS to encrypt exchanges between web browsers and web servers is the basis of Hypertext Transfer Protocol Secure (HTTPS). HTTPS uses TLS to authenticate web servers by exchanging digital certificates when web browsers connect. This tries to make sure that the web browser is connecting to the intended website on the server, that the exchanges are protected against 'man-in-the-middle' attacks and that they are kept private and safe from eavesdropping.

When a web browser requests a response from a web server, the authentication is in two steps. The browser has a list with a large number of digital certificates that it uses first to check the issuing authority is valid and then to check the certificates of the server. If a website has no digital certificate, has an outdated certificate or is not using HTTPS, modern web browsers (for example Firefox and Chrome) display warnings to their users. These warnings are designed to protect users and are deliberately made intrusive by appearing across the whole web page. The website's security information can be displayed in the address bar using colour changes and textual information, for example green address bars for full certification or a text warning that it is 'not secure' if not.

Access rights or permissions

Access rights or permissions are used to control who can view, change and use a filing system. Users' access to applications, folders, files and menus can be controlled by setting the permissions.

Access rights or permissions on files are:

» being able only to read the file
» being able to read and write to the file
» being able to run or execute the file.

For folders (called directories in some operating systems), 'read only' means that the folder contents can only be viewed and the file information is hidden, while 'read and write' means that the files in the folder can be altered, written to and deleted. The 'execute' permission on a folder is seen as the ability to search the folder to find a file when the filename is already known. This means that a user can find a file if the user knows what the file is called, but cannot see any other files. To be able to use files in a folder, the access permission for the user on the folder must be set to 'read'. Using access permissions on network folders is a useful method of ensuring that only those employees who are allowed to access files can actually find them; other employees cannot see them, view them or use them.

Administrators can set the permissions for folders and files so that users have varying degrees of access. This protects files and data by allowing, for example, an assistant writing a company letter to access a document folder and the letter template to copy it but not to alter or delete it. A data-entry clerk would have access rights to enter data into a database but not to extract or export it. Some files in a folder can be hidden altogether from those without the right to view them.

With careful setting of access rights, administrators can ensure that users can view, use and modify files only if they are allowed and required to do so. In this way, files and data can be secured against access by unauthorised employees.

Firewalls

Firewalls present a barrier between one network and other networks. Firewalls can also be set up on individual host computer devices. The firewall barrier is used to monitor and control incoming and outgoing network traffic. Firewalls on host devices may be part of the operating system or specialised software applications. Network firewalls are located at gateways between LANs, WANs and the public telecommunications systems that make up the internet.

Hardware firewalls are dedicated devices connected between the networks, and software firewalls are applications running on the same hardware as other network services. Both hardware and software firewalls capture, analyse and direct network traffic according to pre-set rules. Firewalls intercept data packets and examine the contents. The data packets are either allowed to pass through the firewall and between networks or they are blocked and discarded. This is called 'packet filtering'.

The firewall may operate at the Network layer or the Application layer of the TCP/IP suite.

Network layer firewalls compare the data packet's IP addresses and port numbers to a set of rules configured by the network administrator. The rules are Access Control Lists and contain IP addresses and port numbers. The rules can include what to allow or what not to allow, or a combination of both. A common configuration is to have an allow list and include a configuration line to 'deny all' other traffic. This means that only addresses on the allow list will be passed through the firewall. TCP/IP packet headers can contain an option, in the 320-bit optional part of the header, to record the route, so many firewalls and routers immediately discard packets with these options because they can be used to probe networks to reveal details that could possibly be used maliciously.

Firewalls can be set up between network 'zones', for example servers may be in a separate zone from user devices. A network configuration may have servers in one zone, users in another and private databases in another zone, with firewalls in between the zones keeping the networks secure. Firewalls can ensure that access from the internet to servers (for example web servers, email servers and VPN servers) is allowed but access to private servers is not. Company employees can access both. The firewall has several NICs installed, for example one on the internal zone and one on the external networks, and ACLs are applied to each NIC to control where the packets can go.

A **demilitarized zone (DMZ)** is a logical section of a local area network that can be accessed from outside the LAN while the remainder of the LAN is secure behind a firewall. Web servers, mail servers and FTP servers that are available to external users can be placed in a DMZ. VoIP servers are also placed in DMZs. However, any server that stores confidential data, for example a company database server, should not be placed in a DMZ. A DMZ is created between two firewalls. One is the internal firewall placed between the company network and the DMZ, and the other is the external firewall that faces the internet from the DMZ. Figure 14.35 shows an example of this.

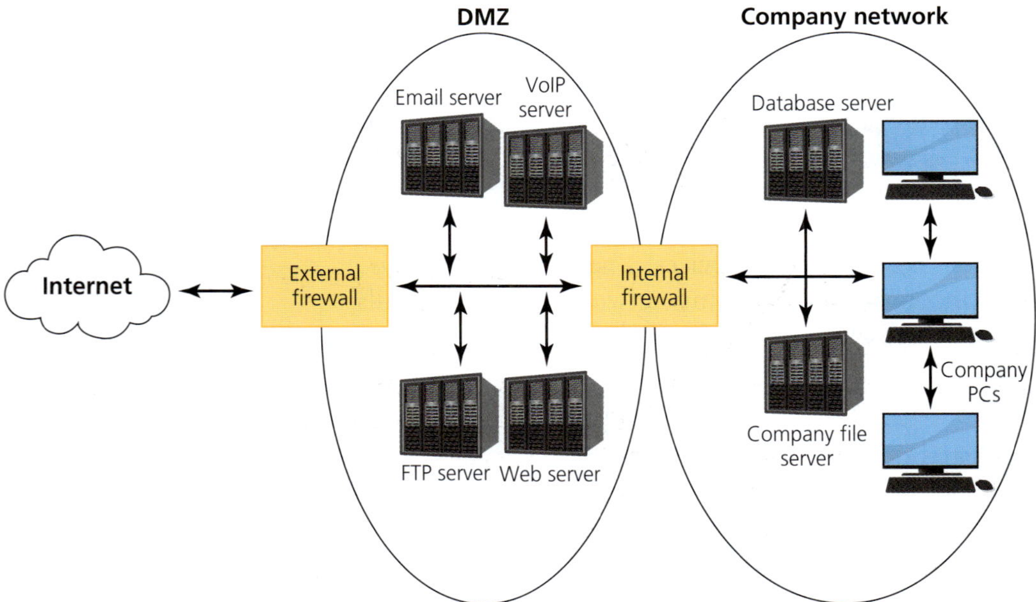

▲ **Figure 14.35** Firewalls creating a DMZ

14.9.4 Advantages and disadvantages of the various methods

Ideally, data should be protected by a range of methods as no one method is completely able to prevent unauthorised access.

Physical methods

Using locks and keys, physical barriers and security guards at doors and entrances to buildings and areas within buildings allows only authorised personnel to get near to the computer systems. This means that easy access to networks is denied. If physical access is denied, then unauthorised access has to be from a remote location, which is more difficult than simply inserting a USB memory stick. However, remote unauthorised access is more difficult to detect and prevent.

Keys to doors can be lost, copied or stolen and used by people who are not authorised to use them. There are no access records for the use of ordinary keys so additional security is needed, for example posting a guard near a door to check that the user who has the key is allowed to open the door. Security guards must be trusted to be loyal to the company and to carry out their duties properly. If guards cannot be trusted or are lazy, then the security they provide is poor and probably worse than not having them at all. Security guards may be bribed to allow unauthorised people access to areas, or they may fall asleep or read newspapers when they are supposed to be watching camera screens, and this can lead to the theft or loss of company files and data.

Employees and visitors to companies may complain that using security cameras and facial recognition software violates their privacy. Accusations of violations of privacy can be overcome by displaying notices that cameras are in use, but these notices also serve to warn unauthorised persons of their presence.

Software methods

Biometrics

The use of biometrics has the advantage over other methods, such as logins and passwords or PINs, in that the details cannot be forgotten and are always carried by the person. Login credentials can be lost, changed or stolen and used by others, but this is very unlikely with biometric data. Biometric data is very difficult to forge and not transferable to others. Other users cannot gain access by borrowing, for example, a co-worker's fingerprints or retina. Biometric systems are usually fast and convenient for the user. Once the initial data has been recorded, access systems using biometrics usually work quickly and reliably.

The use of biometrics has disadvantages. Data must be collected from everyone involved, processed and stored, and made available for the system to use only in accordance with data-protection laws. The actual biometric data itself must also be kept secure from unauthorised users. Biometric hardware can be costly to purchase, install and maintain in order that it works reliably and accurately. For example, retinal scanners and fingerprint readers must be kept clean and free from scratches that might introduce errors. False positives, for example allowing access to those not authorised because the retinal scan is within error tolerance, and false negatives, for example refusing access to a legitimate worker, can disrupt working environments. Eyelashes, contact lenses and changes in light levels can affect the accuracy of retinal scanners.

Biometric devices may be unhygienic and spread diseases if not cleaned between users or regularly. Fingerprint scanners require close contact with the skin and retinal scanners require faces to be brought very close to the device. This can increase the risk of infection with diseases carried by users.

The use of biometrics can invade users' privacy or be biased against, for example, people with disabilities who cannot supply the data required from physical parameters, or those who have been injured. Cuts, burns or wear may change the fingerprints from time to time meaning they may not meet the stored criteria in the system.

Anti-malware, anti-virus and anti-spyware software

The benefits of anti-malware software are that it provides real-time and on-demand scanning of incoming and stored files and data. Malicious software can be isolated, quarantined or deleted and users alerted to the presence and risks of the malware.

A drawback is that anti-malware software can take up system resources, for example storage space or CPU time, and noticeably reduce performance. Another drawback is that anti-malware software developers are always 'playing catch-up' because the software works by identifying patterns or characteristics that are already known about malicious software. New malicious software may cause damage or steal data before the anti-malware software recognises it. This is why anti-malware software must be kept up to date and used regularly.

A 'drive-by download' is a software application, app or program that, without the user's consent and often without their knowledge, is automatically downloaded onto the computer. The program code may run without any user input, but when it does, it infects the computer with Trojans, botnets or other malware that can steal or gather personal information. 'Drive-by downloads' exploit vulnerabilities in web browsers, outdated software, or operating systems, so it is essential that users keep their software and systems up-to-date.

Encryption

Encrypting data as a way of protecting it has advantages and disadvantages compared to, for example, preventing unauthorised users from gaining access to devices or storage systems.

The main advantage is the encryption is done on the data itself, so there is no need to use secure, encrypted connections. The data can be sent over public telecommunications systems between networks by, for example, email. Even though email itself is not encrypted, sending encrypted attachments is secure. Data is kept confidential and private because only those with the encryption keys can gain access.

However, using encryption has some disadvantages. The management of encryption keys is more difficult than managing user login credentials. If a user loses or forgets a password or PIN, these can be reset or changed and access granted to the user, but if data encryption keys are lost or stolen, then the data itself is at risk. If the keys are exposed or made known to others, all the data that uses the now-compromised keys will have to be decrypted, if possible, and encrypted again with new keys. If the keys are lost and cannot be remembered, then the data is also lost. It cannot be decrypted without the keys. While it may be possible to discover encryption keys if they are forgotten, the computing power, time and expertise to do so is very costly. Administrators must ensure that encryption keys are kept safe and secure.

The encryption of data makes its use by software applications and systems more complex. Adding an encryption and decryption requirement for every time an application uses, amends, saves or sends data increases the need for computing power and storage resources. These add to the complexity of systems, to their installation and regular upgrade costs and to their maintenance requirements.

Encryption systems must all be compatible with each other so that data can be encrypted and decrypted with no loss of integrity. If one application encrypts data in a manner that other applications cannot decrypt, then the data cannot be exchanged. For example, if a user working with a database sends encrypted data to a database server, it has to decrypt it for storage in its own encrypted format. If another user accesses the data, the data has to be decrypted again. All such systems must be compatible. Further, these processes require considerable computational power in the computing devices. The increased costs to the company can be significant.

Encrypted connections over TCP/IP networks are effective but complicated to initiate and maintain. The use of secure protocols requires many request-and-response messages to be sent, for example between clients and servers. This adds to network traffic and needs the careful configuration of firewalls to allow the messages to pass freely. Networks need to have low latency so that requests or responses do not 'time out' and get dropped. Encrypted connections pose problems for satellite internet systems, where the long distances that the packets travel add to the latency. Complicated amendments to the timing flags in datagrams have to be made, which makes equipment more complex and expensive and needs a high level of expertise to configure it.

Access rights or permissions

Access rights, or permissions, for files and data must be carefully administered. Using these to control access and protect data can be complicated. Some operating systems, for example Microsoft Windows, allow for the access rights on parent folders to be passed down to the folders within them, while other operating systems may not, and the Encrypting File System used in Microsoft Windows encrypts the files but not folders. Also, applying different access rights to different software applications or processes may prevent automated services, for example routine backups, from accessing every file when it needs to. Backup services may have to be granted additional access rights and this can add to the complexity of configurations.

However, properly configured access rights to files and access to the folders can help prevent unauthorised users from finding and accessing data. It can also help to protect data from mistakes and carelessness by employees.

Firewalls

Firewalls provide protection against malware that can steal, damage or delete data by analysing the contents of data packets for threats. Packets that do not meet the criteria set out in the firewall ACLs are discarded and network administrators are alerted. This provides real-time monitoring of network traffic. The monitoring can also alert administrators to company employees who are sending data out from the network when they should not be doing so or who are contacting external services that may be harmful to the company.

Spyware and other malware can be spotted by analysing data packets for their destination and source addresses. This is in addition to any anti-malware software, which would also be looking for patterns in activity. The advantage of a firewall is that it can spot the spyware in the data packets passing into the network before they can cause any harm.

Firewalls can block services provided by Domain Name System (DNS) websites that provide information about who owns a domain or where it is located. From DNS-lookup services, some websites can gather information that might be useful to unauthorised users and attackers. A firewall can block these services and prevent the information being available.

However, firewalls can be costly to set up and maintain. The configuration of firewalls for a large network requires expertise, so expensive IT technician support is required. Hardware firewalls are more complex than expensive to set up and maintain compared to software firewalls. Firewalls can restrict users and employees in their activities. Individual users are less affected than are employees of large organisations. Strict company policies can place tight restrictions on what employees can do, so some may attempt to find other ways to carry out tasks. For example, blocking FTP makes simple file transfer more complicated, so an employee may use a USB memory stick to transfer a file between the office and home to continue working. This action completely defeats the purpose of the firewall and may introduce malware.

Firewalls add to the delays in the travel of data packets so decrease the overall network performance. Software firewalls running on user devices in the background use CPU time, RAM and data storage resources. Hardware firewalls running on dedicated systems do not do this but do add to the latency of the network.

Using a firewall without additional anti-malware software is not recommended. Malware may not be detected by firewalls if it is carried in software updates, downloaded software or utilities that do not come from authorised or verified websites.

14.10 Disaster-recovery management

Disasters in information technology are a result of threats and hazards that have caused damage or destruction to computing resources and/or to IT services. IT disasters are caused by technical faults or failures of the services or equipment, accidental or intentional activities of the humans using the systems, or by acts of nature, for example floods, hurricanes, earthquakes or fires.

Technical faults and failures of equipment include:

» power failures or outages
» faults in components of equipment (for example routers or servers)
» cable faults
» interruptions and interference (for example to Wi-Fi signals).

Human activities also cause IT disasters, for example:

» the accidental deletion of user credentials from a server
» the accidental mis-configuration of a backup service so that it is not carried out on schedule
» the intentional sending or downloading of malware onto systems.

Natural disasters can also cause data loss as well as put humans in danger, for example a flood destroying a data centre or an earthquake or fire destroying a company building where data is stored.

Disaster-recovery management prepares for disasters, reduces the effects of a disaster and plans for the restarting of the IT services in the shortest time possible so that the company business can start again and carry on its activities as before. A disaster recovery for IT is part of a larger disaster-recovery plan for the business. An IT disaster-recovery plan should include three aspects. Firstly, how to prevent disasters from happening in the first place; secondly, how to detect possible threats and hazards that could cause IT disasters; and finally, how to recover from the disaster by correcting and restoring IT services.

Prevention of IT disasters includes identifying threats, determining the risks from the threats and taking measures to control the threats. For example, the protection from unauthorised access, the protection from malware, and the monitoring and maintenance of the hardware that runs the systems and networks, all require knowledge of the possible threats, how likely the threats are to pose a risk and how to prevent them from causing damage to files and data.

14.10.1 Identification of threats and risks

A threat to IT services is a factor that could have a negative impact on the services. It is not under the control of the organisation, administrator or users and is usually, but not necessarily, from outside the organisation. A risk is the possibility that something might affect IT services in a 'bad' way. It is the uncertainty about anything, such as a threat from malware or the installation of a new component, that might affect the performance of systems or the integrity of stored data, for example.

Identifying threats to ourselves is a process we do every day and often depends on knowledge that we already have. For example, when crossing a road, we already know about the dangers of traffic, or when boiling water, we already know that it can scald us. Similarly, identifying threats to our IT services relies on prior knowledge of what can damage data and files.

The threats to IT services are many but can be grouped into several categories that are known about. There are technical problems, which may arise from power outages or the possibility of not being able to cut off power in an emergency, temperature increases or decreases beyond the equipment tolerances, or electrical fluctuations such as increased or decreased voltages in power supplies.

There are the threats caused by users accidentally or deliberately changing, amending or otherwise interfering with the proper functioning of hardware or software. For example, threats could arise from the unauthorised connection or disconnection of devices, the unauthorised access to data to manipulate, modify or damage it, downloading malware to amend, steal or delete data, introducing spyware to discover login credentials, or accidental file deletion.

Physical threats could be from vandalism, theft of equipment, sabotage or extortion, as well as trip and fire hazards, and these should be considered and measures taken to reduce the risks. In data centres, for example, there are threats to the infrastructure from improper maintenance, poor fire-suppression methods, failure to keep entrances and exits clear for emergencies, and other threats that apply to most buildings where employees work.

Environmental threats to buildings and personnel include volcanic eruptions, floods and earthquakes. The risks from these are dependent on location, but other threats can affect buildings anywhere, for example fire, severe weather and infestation with rodents such as rats. Dust and insects can also pose a threat. In addition, new construction in the area can create vibrations, noise and electrical interference that may threaten the exchange and integrity of data.

Security threats to files and data on networks are identified by analysing data about authentication attempts, access to resources and other logs. Traffic patterns are analysed and detailed information about malware attacks or intrusions are gathered for analysis and used to detect and assess potential threats. When threats are discovered and identified, the risks and vulnerability to the threats are examined. Vulnerability refers to a known weakness that can be used, or exploited, by an attacker. For example, it is well known that a user should log out from a workstation when moving to another or going on a break, but often the user forgets. The workstation, the files, the data and the network resources are now vulnerable to the threat of unauthorised access by anyone who sits down at the workstation.

Checking and testing for vulnerabilities is essential for the security of files and data. Testing for vulnerabilities can include checking that the company policies and training are up to date and enforced, that anti-malware software is updated and running, that files and data stored in the cloud are accessed and stored securely, that there is an appropriate disaster-recovery plan in place, and that all employees and users are aware of what to do in the case of a disaster occurring.

Analysing the risks of threats and the vulnerability to them is critical in ensuring that data and files are secure and that, if a disaster occurs, the files, data and IT services can be restored as quickly as possible to a state before the disaster happened.

> ### Activity 14w
> 1 Describe what an IT administrator understands by a 'threat' to an IT system.
> 2 Describe two ways in which physical threats could damage an IT system.

Risk analysis

Assessing or analysing risks considers both the threats and the vulnerability to the threats. Risk analysis looks for potential threats and considers how vulnerable the files, data and IT services are to those threats. Vulnerability can include how often the threat is likely to occur. Risk can also include how expensive it is to protect against the threat. If the threat is that of an employee leaving a workstation without logging out, but there are only two employees who take a break at the same time every four hours and all confidential data is encrypted, then the vulnerability is still there but the risk may be low. Employing a security guard to watch over the employees and check that they log out every time might be considered too expensive and intrusive.

To carry out risk analysis, the threats must be identified, the vulnerability and the likelihood that the threat will happen must be assessed, who or what will carry it out must be determined, and the cost of preventing the threat from happening must be considered. All these contribute to risk management and disaster recovery.

Perpetrator analysis

A perpetrator is a person who does wrong or carries out activities that cause harm to others. Cybercrime perpetrators carry out illegal activities that put computer systems, IT services, files and data at risk. Analysing what cybercrime perpetrators do and how they carry out their activities helps to assess the threats from cybercrime. There are many types of cybercrime perpetrators but all have similar goals and motivations, for example to cause damage to IT services, to steal data, to try and impress others and to make financial gains.

Perpetrators are the attackers and those who carry out cybercrimes. Perpetrators can include, for example, script kiddies, crackers, hackers, terrorists, business competitors and (foreign) governments that carry out the crimes or intrusions into networks or file-storage systems.

A **script kiddie** is an unskilled individual, often a young person, who writes scripts and program code, or very often uses other people's scripts or codes, to attack computer networks and systems. Lacking the skills to create their own programs, the aim of a script kiddie is to impress others by damaging systems, files and data.

A **cracker** is a person who gains access to networks, or software applications, by taking advantage of security flaws to, for example, use the resources or software without paying for the use. Crackers are usually not as skilled as hackers, do not have the skills to create their own software applications and are aware that their activities are illegal. They often act with malicious intent.

Each type of perpetrator of cybercrime has a different level of skills that can be identified by an analysis of the attack. The aims of the attack also vary with the type of perpetrator.

The companies that have the potential for being attacked, or for being the victim of an attack, employ skilled IT technicians and programmers to analyse potential attackers and their aims, in order to create the plan for disaster recovery. The allocation of company resources to analysing and creating a disaster-recovery plan that enables recovery from cyber threats depends on the likelihood of perpetrators succeeding in an attack on the company. An analysis will determine what resources need to be allocated, for example if anti-virus or anti-spyware is required for each workstation, or what **intrusion-detection** systems can be deployed to combat the type of perpetrator identified by the analysis.

Industrial spies and insiders aim to gather company secrets by accessing confidential data, for example designs, and selling it to competitors. An industrial spy might try to gain unauthorised access remotely by avoiding firewall controls or using phishing techniques to learn login credentials, while an insider, an employee of the company, would use their own credentials to access and steal the data. Hackers gain access to confidential data by various methods, often by exploiting programs written by others or by exploiting security loopholes. Crackers use fake or copycat software applications to steal login credentials. Cybercriminals target confidential data to use it for financial gain. They can make persistent threats over long periods, often for long-term financial gain, or can act for immediate gain. Cyberterrorists can cause permanent harm to IT services or use them to cause harm to others.

Cybercrime perpetrators often do not need a high level of IT skills to carry out their activities. The software and tools that enable cybercrime are available on the internet or from other cybercriminals. Tracking and analysing the tools used by cybercriminals can assist in preparing to defend against the threats. By collecting and studying the details of cybercrimes, investigators can create profiles and reports about the technology, tools and methods used to carry out attacks.

Identifying the transmission protocols, TCP/IP addresses and ports, and analysing the logs of network traffic and of the packets that pass through routers and firewalls can discover information that can help when assessing the risks and threats from the type of cybercriminal that might target a company.

The type of organisation carrying out the risk and **perpetrator analysis** is considered when assessing what type of perpetrator might attack. A financial institution, or an organisation with a large amount of funds, for example, is more likely to be attacked for financial gain by international cybercriminals than a supermarket chain that sells grocery items.

Monitoring and analysing the skills, activities and records of employees can identify potential perpetrators of cybercrimes that could lead to IT disasters. It is essential that employers are aware of what IT skills their employees possess, what their employees are allowed to do and what activities they carry out with company data. While it may not be necessary to know all the precise details about individual employees, being aware of the access rights of the data-entry clerks, for example, can help assess the potential for these clerks to become perpetrators of activities that may put company data at risk.

Risk testing

Risk testing is carried out to establish the extent to which threats pose a risk to IT services, files and data. Risk testing can ensure that, in the event of a disaster, the risks and the potential damage can be mitigated. By analysing the potential threats and risks, a comprehensive disaster-recovery plan to cover all eventualities can be created.

Quantifying the risk

Threats, vulnerabilities and risks can be quantified so that they can be prioritised for investigation, for monitoring and in preparation for any events. Procedures and policies can be created to help defend against them and for recovery if they happen. The amount of time and effort that companies put into their defences against threats can depend on the importance of the threat to the company.

A simple but effective method of quantifying threats, vulnerabilities and risk is to use a **risk matrix**. This displays the likelihood and impact of the potential threats in a highly visible and easily understood way. For example, the impact of the activities could be described as having a catastrophic, critical, marginal or negligible effect on the company. The risk of these activities happening could be described as rare, unlikely, possible, likely or certain to occur.

Some of the accidental or intentional activities that modify or destroy data can include accidental deletion, intentional deletion, software error, cyberattacks, hardware failure, power outage and malware. The activities are placed in the matrix according to the assessment of the impact they could have and their likelihood of occurring. A possible matrix is shown in Table 14.17.

▼ **Table 14.17** Risk matrix of impact of events

Likelihood of event	Impact of event			
	Negligible	Marginal	Critical	Catastrophic
Certain				
Likely				
Possible		Power outage	Software error	Cyberattack
Unlikely		Hardware failure	Malware	
Rare		Accidental deletion		Intentional deletion

Risk matrices are useful but rely on the creator to assign threats and make decisions that may not be reliable. For example, in the table, accidental deletion of data may be more likely in some areas of the company than in others, but this is not shown in the table. Other ways of quantifying risk use different categories, for example risks can be shown as low, medium, high, very high and extremely high, and these are used to determine the actions needed.

Risks can be assigned a monetary cost, for example for mitigating them or recovering from them, and this can be used to quantify risks.

Rating	Description
Low	The risks are acceptable to the company but measures to further reduce risk should be carried out.
Medium	The risks may be acceptable in the short term but future plans must be made to reduce risk and mitigate them.
High	The risks are unacceptable and measures must be taken to reduce risk. Mitigation must be carried out as soon as possible.
Very high	The risks are unacceptable and measures must be taken quickly to reduce these risks and mitigate them.
Extremely high	The risks are completely unacceptable and measures must be taken immediately to reduce these risks and mitigate them.

Another way to quantify risks is to gather the views of experts. A group of experts examines and assesses the risks and each gives a judgement, for example in terms of a value between 0 and 1, of the severity of the risk or the probability of the threat happening. The values are aggregated and used to quantify the risks. This method is subject to personal opinion, but it can be useful in assessing risks because experts rely on their knowledge and experience of previous scenarios.

14.10.2 Control of threats

Threats to IT services are reduced by taking measures to control them. Appropriate and proper use of security measures and procedures can reduce, or mitigate, the effect of threats. The threats are still there, but the possible effects of them are reduced. Mitigation of threats is an attempt to reduce the negative impact, or the damage done, by a threat if it happens. Where prevention is not possible, for example the unauthorised copying of a file, encrypting the file before it is copied will mitigate the damage done by ensuring that the unauthorised copy cannot be understood.

The hardware and software applications used in IT systems are exposed to threats, for example power outages and malware, but taking precautions and security measures can mitigate the damage caused. However, if the measures are not appropriate or are too restrictive, users may be inconvenienced and their productivity will be reduced. For example, email is not encrypted and may expose company data to others, but configuring a company firewall to block email from entering or leaving the company network would make the jobs of employees working in, for example, the sales department much more difficult.

Threats from fire can be controlled by the use of automatic fire detectors and suppression systems, for example using fire-extinguishing gases, foam or the correct type of fire extinguishers. Damage from unstable power supplies can be mitigated by voltage controllers or 'smoothing' systems that regulate power supplies. The climate in the rooms where devices are operating can be controlled by air-conditioning, cooling or air-flow systems to control threats from over-heating. Threats from electrical storms can be mitigated by the use of adequate grounding and lightning-conduction systems.

There are threats from accidental manipulation, modification or destruction of data by users. This can make data unavailable or inaccurate, which will have a negative effect on the business. Data also can be accidently disclosed to unauthorised viewers. These threats can be prevented, or guarded against, by proper company policies and procedures, backups and employee training.

Threats can be controlled by gathering data by monitoring and analysing activity on networks, monitoring employee activity, ensuring that employees and other users are aware of the threats and how to detect them, and ensuring that users know how to react to threats so as to reduce the damage that could be done.

How to detect threats and prevent disaster

Detecting threats to IT systems, files and data can mitigate the effects of disasters, or prevent them happening.

Threat detection can occur in a number of ways. Alerts can be generated by, for example, firewalls or anti-malware software, or by unusual activity on accounts, such as repeated failed attempts to log in to an account. This is the detection of anomalies or unusual activity. The process of 'threat hunting' can be used to detect threats that can avoid the traditional methods of detection and prevention searches by analysing the large amounts of data collected by monitoring activities and user behaviour on networks. They can be detected by specialised detection systems that alert when, for example, activity is compared to previous activity records or snapshots, and anomalies, or unusual differences, are found.

Signature-based threat detection looks for known patterns of suspicious activity, for example in the exchange of certain types of data packets. It has the drawback that it cannot detect new threats because no pattern exists for them.

Suspicious activity on, for example, a wireless access point (WAP) could include unusual beacon frames being detected, or multiple, unusual and new SSIDs being seen, or changes in the MAC addresses within data packets. These could alert administrators to the presence of an unauthorised WAP being set up to gain access to company data.

The behaviour and activities of users and employees is monitored and analysed to provide a baseline of the 'normal' activity and behaviour that is expected on the network. For example, if a login from San Francisco is detected and files are accessed at 4 a.m. from an employee who only works between 9 a.m. and 5 p.m. in London, then it is likely to be a threat to security.

Traps can be set for intruders into networks. Areas of the network or files and data can be disguised as critical information that intruders might look for. When an intruder tries to access these, an alert is sent to the network administrator warning of the intrusion, the threat and a log of the details. These can be used to find the intruder.

Intrusion detection can be used at the point of entry to networks, such as at gateways or routers, for example using firewalls, or at the host devices. Firewalls, or other software, analysing data packets for malicious activity, can be combined with systems for analysing very large quantities of data, such as artificial neural networks, to detect or predict intrusions. Network traffic patterns can be analysed and used to detect intrusions. Intrusion detection by individual host devices works by monitoring changes in files or data and raising alerts if these appear to be suspicious. This is most useful when monitoring operating systems or configuration files that are not usually altered. Changes to configuration files or system files could indicate that an unauthorised user is attempting to add or amend the files to gain access.

Detecting threats from insiders uses the analysis of patterns of employee behaviour to create alerts when the patterns change. Establishing a baseline for employee behaviour by, for example, using machine learning from large amounts of data, can allow alerts to be triggered if a deviation is detected. This system requires very large amounts of data to be analysed to be effective, but it is used to protect cloud storage systems, where there always is a large amount of activity. There are several general principles for user behaviour analysis that can be applied to detecting insider threats. These principles include putting employees or users into groups based on their activities, for example data-entry clerks, gathering data about their activities over a period of time, such as what files and data the clerks access regularly, and linking behaviour to precise user activities, for instance when and from which device the files are accessed. Also, using machine learning without human input can refine and improve the baseline data so that more accurate detection and fewer false alerts occur.

Typical indicators of a possible insider threat are unusually large file or data downloads, accessing data that is not relevant to the employee's tasks, unusual remote accesses and unusual contact with external users.

How to restore after a disaster

Recovery from a disaster and data restoration by organisations and businesses rely upon implementing a previously prepared disaster-recovery plan. A disaster-recovery plan should include measures designed to prevent threats from becoming a reality and causing a disaster, measures designed to detect threats, and measures to be taken after a disaster to correct, restore or recover from the disaster.

The most important measure that can be taken to protect data and ensure it can be restored is to regularly create, and safely store, backups at frequent intervals. User data, company databases, spreadsheets and other documents must all be safely backed up. Also, whole systems (for example operating systems) gateway and router software and configurations, firewall configurations and all other business-critical systems must be stored safely, and securely, in a backup.

If a disaster happens to company or user data, a recent backup can be used to restore the data to the state it was at the time of the backup. Any data entered, created or amended after the time of the backup will, however, be lost and will have to entered or created again.

After a disaster, the cause must be discovered and corrected. Physical disasters may require the safety of employees to be considered, because a fire, for example, would damage more than the data. Hardware may have to be replaced in the event of failure, malware purged and logs analysed before the data can be restored.

How a backup is restored will depend on the technology used for the backup. Backups stored on network-attached storage (NAS) systems can be quickly copied to the 'live' systems, but external HDDs will have to be connected and the restoration may be slower.

Backups can also restore whole systems. For example, a tape backup of a server can be used to restore the server in the event of an attack or a hardware failure. Critical servers will be mirrored or running as virtual servers and can be replaced almost instantly if there is a disaster to one. A mirror system is a duplicate that can share the load when demand is high or can replace a failed system to keep the services available while the failed system is repaired.

Recovery from cloud computing disasters can be faster as the providers will automatically have created backups. These can be made available almost immediately provided the whole data centre is not affected. Cloud computing can mean that company data is stored in different locations, so physical threats are confined to one location. Cyberattacks may also be restricted to one location or, conversely, may affect all locations at once, but the risks are the responsibility of the cloud provider. The provider should have state-of-the-art infrastructure, anti-malware and cybercrime defences as well as automatic backups in place. Restoration after a disaster can be automatic and very fast. However, it is important to ensure that the attack is properly understood and that a fast, automatic restoration does not further compromise data or compromise the backups.

A malware attack requires the scanning and removal of malware from all systems and devices, as well as any attached storage systems. All backups and archives should also be scanned and any malware removed. This must be carried out before data is restored, because otherwise the restored data and the backups may also become infected.

An attack by ransomware, a specific type of malware, is very difficult to recover from. If paying a ransom does not produce keys that decrypt the affected data, the data is lost and can only be restored from a backup. Using a recent backup to restore data is an essential means to recover from a ransomware attack. A very recent backup may mean that no ransom will need to be paid, but that will depend upon the Recovery Point Objective (RPO) and the Recovery Time Objective (RTO), as explained later in this section. After a ransomware attack, all passwords and login credentials must be changed and the malware purged from the system. It is also recommended that any affected files are sent to law enforcement agencies for analysis so that they can try to trace the perpetrators. Using perpetrator-analysis techniques can help identify the attackers.

The restoration of files and data should be tested regularly to ensure that it can be done. There is little point in creating backups that cannot be used. It is well known that tape systems for backups can be difficult to use if the drive mechanism is not carefully maintained or is changed. Tapes recorded on one drive may not be properly read by another. If a drive is changed after a backup but before a disaster, the data may not be restored properly. After any hardware change or software upgrade, it is important to carry out tests to ensure that systems for restoring data still work properly.

14.10.3 Strategies to minimise risks

Minimising the risks of threats to data means taking steps to keep data safe and secure. Part of a strategy to minimise risk is to regularly examine the existing policies and plans. Existing documentation about threats, risks and disaster planning should be reviewed and amended if necessary. This will find and remove uncertainties, ambiguities and errors in the documents to try to reduce confusion among employees. High-risk areas and procedures can be identified and the likely impact of disaster occurring in these areas assessed and documented. Highly critical and medium risks can be addressed and the need for mitigation can be planned and implemented quickly. Any low-risk threats can be addressed later. Also, the required system monitoring can be put in place.

Although a good strategy for minimising risks is not to carry out the risky activity, this is usually not productive or convenient. The risk of a threat from an external attack on a network can be removed altogether by complete isolation from the internet, a strategy used by the military, but this is not practical for businesses that rely on emails and file transfers for their income. Most organisations minimise risk by careful planning and implementing measures to protect their files and data.

Developing and implementing a disaster-recovery plan that includes strategies for identifying threats and risk, controlling threats (**threat control**) and minimising risks, and recovering from a disaster is a method of protecting data from manipulation, modification and destruction by accident or by unauthorised users.

Protection of power supplies

A sudden disruption, fluctuation or interruption in the supply of electrical power can damage electrical circuits and the components in computing devices, and cause loss of data. Also, computing devices do not work at all without electrical power.

Data loss can occur because when electrical power is suddenly lost, not only are the contents of RAM lost, but the sudden loss of power can damage storage devices. A magnetic hard disk could be writing data to sectors when power is lost, so not only is that data lost but the HDD sector could be damaged. The read/write heads travelling across the disk platters may literally crash onto the disks and physically damage them. A modern HDD should retain enough power to safely 'park' the heads in the event of a power outage but, although modern HDDs are more protected from such damage than older HDDs, a sudden loss of power may possibly leave the heads in a vulnerable location.

SSDs do not have moving parts, but sudden electrical interruptions will still lose data that has not been written to the disk. Even worse, SSDs often require that whole blocks of data, for example a 4 Kb file of data, may have to be written in an 8 Kb, or even a 16 Kb, write operation, equivalent to many sectors on a magnetic HDD. A power outage during the write could lose more data than if it were a magnetic HDD. However, modern SSDs include a type of 'emergency power supply', for instance tantalum capacitors in Samsung SSDs, that store enough power to allow data-writing operations to be completed in the event of a power outage.

Modern operating systems, for example Microsoft Windows and Linux, include file systems that attempt to protect data that is being written to disks against sudden power loss. For example, Microsoft's NTFS and Linux's EXT3 use a system, called a 'journaling file system', that keeps a record of the changes it has to do to a disk when storing data. This journal is updated every time a disk-write operation is requested and can be used to recover from a sudden power loss. When power is restored, the journal is read and as many of the write operations as possible are carried out. Unfortunately, despite this, while damage to the disks is minimised, data is lost more often than not.

More common than complete power outages in many parts of the world are problems with the condition or status of the electrical supply. Noise on the power lines and distortion of the waveforms of mains electricity can cause devices to malfunction. A momentary or continuous reduction or increase in the supply voltage, or AC frequency, can cause devices to stop working (freeze) or become unreliable until shut down and restarted. Occasionally, a device or one of its components, or even its own external power unit, is permanently damaged so it will not restart.

Computer devices must be protected from power outages and deficiencies in the power supplied to them. Internal or external power units are used to charge internal batteries in devices to be used when mobile, for example smartphones, laptops or tablets. Many devices do not have batteries installed but rely on a reliable power supply from the utility company. The devices stop working if the power goes out. Organisations and companies, and some home users, cannot rely on utility companies to supply electricity without interruptions. Power may be suddenly lost due to an accident, or a power line fault, or other reason. To protect against this and the loss of data that may result, emergency generators can be used. The generators run on oil or gas and provide electrical power while the main supply is being restored. Monitoring of the incoming power supply allows the generators to be started quickly.

Emergency generators are expensive and mostly beyond the means of home users. Also, while emergency generators may start quickly, the power is still momentarily interrupted and this may cause damage and data loss. To protect against even momentary fluctuations or interruptions in power, many companies, and some home users, invest in uninterruptible power supplies (UPS).

There are several types of UPS technologies. A UPS can simply be a collection of batteries that provide the power to computer systems, network devices and data-storage systems. The batteries are kept charged with mains, utility company, electricity. If the utility company supply is cut off, there is sufficient power in the batteries to keep the systems running for a short time and for the UPS to send instructions or signals to the attached devices instructing them to automatically shut down properly. Other types of UPS are kept charged but are only activated for supplying power if there is a drop or loss of voltage from the external supply.

A benefit of using UPS for power is that short drops or spikes in power are almost eliminated from the power supply to the computer systems. However, this only really applies to the UPS systems used in large, commercial installations, for example data centres. A UPS designed for a home or small business user has drawbacks. A small UPS, for example one designed for use in a home installation or small company, can actually output poor quality AC power with noise and harmonic waves, and these can cause interference with communications and reduce performance by devices.

Password and access controls for data and file protection

Access control consists of three parts: authentication, authorisation and audit. Authentication is proving that a person is who they say they are; authorisation is specifying their access rights, or what they can and cannot access; and an audit is the record-keeping documentation or log of the person's activities with the resources.

Authentication can use physical and electronic keys held by a user, physical barriers that a user has to pass while monitored by other people, biometric data unique to the user, or passwords or a passcode memorised by a user. Only the person who is who they say they are will have the correct authentication. For example, usernames or user IDs identify the user to a computer system, but only the possession of the correct password authenticates the user. There are three types of information that can be used for authentication: something that only the user knows (for example a password or PIN), something only the user has (for example a smartcard), or something that only the user can have or be (for example a fingerprint, a retinal scan or other unique biometric measurement). These are called credentials. In addition to these, if a user is well known to a security guard, the user may be allowed access to a building even if they do not have, at the time of entry, the required smartcard or have forgotten the PIN. This is similar to having forgotten to carry the key to an office and being let in by someone who knows the user. A human element to authentication is usually possible despite it being a weakness in the security system.

Authorisation depends on pre-set rules that grant the authenticated user access to computer resources. Some systems authorise users to have access to all resources just by logging in with an ID and a correct password, but this does not provide the level of control over access to data that most organisations would like. Using login credentials can successfully control access to buildings, areas and computer devices, and to whole systems. However, it is sometimes necessary to control access to specific files or file folders. Administrators have the ability to determine who can and who cannot view certain data while, at the same time, allowing access to devices and systems.

Controlling access to folders or directories on networked systems is managed by using access control instructions (ACIs) that detail the access to individual files for individual users. Administrators configure the access rights to the folders on a server using ACIs that link to the authentication information about a user. This can grant access to a whole folder, folders within the folder (sub-folders), files within the folder and even to the configuration files that may be stored in that folder. Very specific configurations for each user can be made. Further, file permissions (access rights) can be set for a specific user, a group of users (for example all data-entry clerks) or all users who might need to store or read files in that folder. In addition, ACIs can be used to control access by computer devices to the files. For example, a specific device may be identified and granted or denied access depending on its IP address or hostname.

ACIs are stored in the folder itself and consist of the identity of the file that is being controlled, the permissions applied to it and the identity of the user identified by their authentication. Software running on the server uses the ACIs every time an attempt is made to access a file. By using these access control methods, files and data can be protected and made available only to those who are authorised to view and use them.

Using passwords to protect folders is not usually secure unless the folders are encrypted and the password used as an encryption key. For home users, encrypting a folder and keeping the key secret keeps the folder and its contents secure. It is important not to lose the encryption key because, if lost, the data cannot be recovered. For this reason, organisations and businesses usually do not allow their employees to encrypt data using their own encryption systems. If the employee lost, forgot or refused to disclose the key, the data would be lost to the company.

Passwords on individual files are a good method of keeping the contents secure, especially when sending them between networks over public telecommunications systems, but, as with all passwords and encryption, the password or key must be kept secure and not forgotten. It is almost impossible to recover Microsoft Office® files that are password protected if the password is forgotten. Organisations that use passwords to secure Microsoft Office files are advised to use a decrypt tool on the files before they are password protected. This tool can discover the private key needed to unlock the files if the password is forgotten. In a way, it defeats the point of using a password, but it does defend against forgetfulness.

Protection of data and software from malware

Data stored on network storage devices, or in the cloud, is vulnerable to malware because connections are made to other devices that could allow malware to be spread. Protection from malware has to be at the points of entry to a network and at the point of connection between the storage devices and the network.

At the point of entry, firewalls are essential to analyse every data packet for malware and for unauthorised access. Malware can be identified and discarded, and any unauthorised attempts to introduce software that might be or contain malware can be blocked. Blocking insecure protocols, such as FTP, can prevent malware from being downloaded from websites. Restricting access to inappropriate websites is also a way of preventing malware from entering a network.

Email messages that carry attachments can be blocked at the point of entry in case they carry malware. However, this may inconvenience users on the network, for example sales representatives of a retail company, so procedures for scanning incoming and outgoing emails are automated at the sending or receiving devices or by the email servers.

Anti-malware software, for example anti-virus and anti-spyware, is used to detect, remove or isolate, and alert against, malware. For small organisations or home users, installing, using and regularly updating anti-malware software on every computing device is usually sufficient, but larger organisations should, in addition, use a centralised set of anti-malware software applications called a security suite. This allows central configuration and administration, which is easier and more effective than managing a large collection of computers. Company policies for the exchange of data, for example, always use encryption for email attachments to other employees, scan emails before they are sent, scan incoming files for viruses and never use USB memory sticks. This will help to protect data from malware.

Ransomware is a specific type of malware that is designed to deprive users of their data until a ransom is paid to the attacker. Ransomware is often 'delivered' by phishing emails or by 'drive-by downloads'. Ransomware identifies the storage devices attached or available to the computer and encrypts the files on the storage devices. If a ransom is paid, the attacker may provide the keys to decrypt the data. Recovering the data without the encryption keys can be difficult or impossible. Protection against ransomware involves training employees, and educating users, to be aware of possible attacks and of suspicious emails or other activities. Regular training sessions and awareness reminders about cybersecurity threats can help employees to combat ransomware attacks.

As with all malware, protection can be improved by performing regular updates and applying security patches, being cautious with and not following suspicious links to websites in emails, opening email attachments with caution, checking that emails come from reliable sources and using anti-malware software.

Regular, and frequent, backups of data will not prevent ransomware attacks, but they do provide a way to avoid paying a ransom. If ransomware encrypts data, the encrypted data can be deleted, the ransomware removed and the original data restored from the backup. It is necessary to keep the backups on storage devices that are not connected to the computer so that they are not also affected by the ransomware. Networks can be protected from ransomware by taking similar precautions to those for any other type of malware.

Protection of data and software from unauthorised access

Unauthorised access to data and software can result in modification or loss of data. Networks are vulnerable at any point of entry. Organisations and businesses, as well as individual users, need strategies, procedures and policies to protect data. Having strict policies and enforcing them helps to guard against unauthorised access to data.

Company policies relating to security do not refer to the actual technology or software to be used, but make directives about acceptable and unacceptable activities that employees, and all users, must follow. Network security policies specify how employees should access and use the resources safely and securely and also how they should react and behave if a security breach occurs.

Policies specify how and where computer devices should be used by employees, as well as how and when updates and security patches should be applied to device systems, for example operating systems, software applications and other network systems. The policies should state that all network services not being used must be disabled on the device. For example, a data-entry clerk using a desktop computer has no need to use **ping** or **traceroute** to discover network services and devices. Disabling features such as these will make attempts at unauthorised access to other devices much more difficult.

Employees should also be required to sign a non-disclosure agreement that covers the details of the computing devices and other hardware in use by a company. This helps to prevent others from using knowledge about the devices to breach security measures. For example, knowing that laptops with a Linux operating system are being used gives a potential unauthorised user important information about how data is accessed and stored.

The use of VPNs should be strictly controlled. Access by employees to a company network from home must always be via a VPN and direct access over the internet should not be allowed. Policies should dictate that VPNs and appropriate security methods are used, for example end-to-end encryption for connections.

Policies governing the TCP/IP ports that administrators can allow to be open are important. Having a policy assists the administrator in the configuration process. Closing all ports other than those required for the business to operate will help prevent access by unauthorised users. For example, a company firewall does not need the FTP ports to be open if the company only uses email and HTTP or HTTPS for web access.

The configuration of wireless LANs should be strictly controlled. WAPs allow connections by mobile devices, for example tablets, by employees and visitors, and policies should specify how mobile connections are to be allowed or refused. Employees need mobile access to company resources but visitors do not. The configuration of separate Wi-Fi networks in organisations, one or more for employees and a 'guest' network for visitors, should be part of a security policy. Guest networks are commonly used in hotels, cafes and restaurants to provide internet access for visitors and customers. Home users can set up a guest network by configuring the home router supplied by the ISP to allow internet access but restrict access to other networked devices, such as printers or storage areas.

A strict policy that specifies how a company's firewalls operate can ensure that administrators configure the firewalls correctly. This is useful where there are multiple entry points to the networks and multiple firewalls. Details that could be included are, for instance, the requirement that application firewalls be used if transmission speed is important, or that only packet filtering is to be used for remote access to database servers. If transmission speed is not important, then a stateful configuration should be used.

Backup strategies

A business has a **backup strategy** so it can recover from a disaster, for example an accidental or intentional deletion of data, and return to normal operations as quickly as possible. The strategy includes the cost of the systems, how often backups are to be carried out, who is to be responsible for carrying out the backup procedures and where the backups are to be stored. It also considers what data is to be protected and for how long backups should be stored. Individual users should also have a backup strategy to recover important data, for example family photographs or financial details, in case these are damaged or lost.

Backups should include not only company files and databases, for example of customers, but also operating systems, software applications and configuration files, so that a disaster recovery is accomplished quickly. Backups can be stored on local, networked or removable storage devices, and a backup strategy should include decisions of which to use. Local or USB storage is not suitable for networked resources. Individual files can easily be backed up to USB memory sticks, but these are easily lost, do not have enough storage capacity and introduce malware if improperly used.

Most organisations use network-attached storage (NAS) devices or storage area networks (SAN), which are dedicated storage systems for backups. Cloud storage is also used. These all have the advantage of being separated from the user devices, so any misfortune that happens to the device does not affect the data or backups. Tape storage systems are also used and the tapes stored at a different location for safety. Tape systems can be expensive, may become unreliable if the drives are not maintained properly, and the tapes will take time to arrive from the remote location if needed. Tape systems are more suitable for restoring complete systems rather than for individual files.

It is also important to devise a strategy for backups to include the files and data stored on the local drives of the laptops and tablets of employees. Employees must copy the files and data to storage areas that are part of any automated backup process. Files and data that are not included may be lost if the device is damaged or stolen.

Strategies for backups consider the minimum time between backups that is acceptable to the business operations. This is the **Recovery Point Objective (RPO)** and considers how much data loss can be acceptable. For example, can the business afford to lose a day's worth of data, an hour's worth of data, only a few minutes or even less? The time depends on the type of business or organisation. A bank will have a much shorter RPO than, for example, a book seller. The RPO is the interval of time that can pass before the amount of data that is lost becomes unacceptable to the business. For example, consider if a backup is taken every hour on the hour and a data loss occurs at 55 minutes past the hour. That means 55 minutes of data is lost because the latest backup, taken 55 minutes previously, would not include the data generated during the 55 minutes. However, if the business's RPO was 1 hour, then that would be within the limit allowed. The backup strategy would include the most appropriate RPO for the business. For an individual user, an RPO of days or much longer is often more than sufficient.

The **Recovery Time Objective (RTO)** is the maximum time allowed for data recovery from the disaster before the work of the business is seriously harmed. It sets a target time limit for recovery. The actual recovery time (called **Recovery Time Actual** or **RTA**) may be shorter or longer than the RTO and depends on the nature of the disaster.

The details of the backup procedures are also found in the backup strategy documentation. This includes:

- how many copies of the backup are to be kept
- what type of storage medium is to be used
- where it is to be located or kept
- if it is to be networked or cloud-based
- how often the storage medium can be reused, if at all
- who is responsible for the procedures or for checking that automated procedures are working correctly.

Backups can be made of whole systems, folders or individual files. A 'system image' is a snapshot of the system and is a file that contains an exact copy of all the files, data and information needed to recreate the system. It is possible to create an image of a complete hard disk (or SSD) that can be used to recreate the HDD (or SSD) in case of hardware failure, malware attack or operating system file corruption or failure. Disk images are especially useful for restoring user devices. A corrupt file system on an HDD may lose all the data and the operating system of, for example, a laptop. A disk image can be used to restore it to a working state from before the problem arose. As with other backup methods, the restore is to the time when the image was created, so any later data is lost. Disk images are most useful when restoring operating systems. User data can then be restored from other, more recently used, backup media.

Several strategies for backup schedules have to be created. A 'first in, first out' requires a number of backup media sets, for example tapes, to be used, each set being stored while the next is used. When the last media set has been used for a backup, the first media set is reused by overwriting the stored backup data. The more media sets that are used, the longer the period of time that can be covered. Backups at set intervals can be removed from the cycle and stored as archives. For example, using a new media set every day for a month and removing a set every seven days to archive means that a restoration can be made to the day before or to a previous time if necessary. Retaining older backups has the benefit of enabling restoration to a time further back than one day in case there are errors or problems in the previous backups.

A strategy with three or more backup cycles working together is common. A daily, weekly and monthly rotation of the backup media sets, based on the 'first in, first out' system, ensures that backups are available for restoration purposes at appropriate intervals. This is often called the 'grandfather-father-son' (GFS) strategy. Again, some backups are removed from the cycle for safekeeping and archiving.

More complex strategies, such as those based on the Tower of Hanoi (or Brahma or Lucas Tower) puzzle, use a recursive method of cycling backup media sets. The mathematics behind the Tower puzzle are relatively easy to follow but are not in the scope of this syllabus. However, this method tries to ensure that different media sets are rotated in a way that ensures that backups are not overwritten too frequently. The drawback of this method is that it gets more complex to schedule as the number of backup media sets increases and that, as time passes, the backups available from further back in time become fewer and fewer. It is important to remove backup media sets at intervals, as in other strategies, but this disrupts the scheduled reuse of media.

Whatever strategy is used to rotate the media sets, it is important the backups are safe and reliable. Backups should be kept away from the original storage systems, stored safely and securely, and there should be regular testing of the recovery to ensure that the backups actually work.

Practice questions

1 Discuss the advantages and disadvantages of networking the computers in a financial institution such as a bank. [8]

2 Describe the contents of a network policy. [6]

3 a Explain why all devices on a network use the same communications protocols. [2]

  b What is the difference between a reliable and an unreliable protocol? [2]

4 Describe how Near Field Communication (NFC) can make the job of health care professionals easier. [6]

5 Describe how the request-and-response method of network communication is used when viewing web pages on a laptop. [6]

6 a Explain how a WAN is distinguished from a LAN. [6]

  b Explain why data travelling on WANs may be less secure than data on LANs. [4]

7 Describe three examples of client–server networking. [3]

8 Routers can make use of both static and dynamic routing tables.

  a Describe two drawbacks of using only static routing tables. [2]

  b Describe two drawbacks of using only dynamic routing tables. [2]

9 192.168.20.1 is the IP address of a laptop on a company network.

  a Explain why no other device on the network should have this address at the same time as the laptop. [2]

  b Describe two ways that an IP address could be assigned to the laptop. [2]

10 Perpetrators of cybercrime can include script kiddies and crackers.

  a Describe what is meant by:

    i a script kiddie [2]

    ii a cracker. [2]

  b Explain how perpetrator analysis can help in preventing and recovering from cyberattacks. [2]

15 Project management

In this chapter you will learn:
- ★ the stages in the project life cycle and how they relate to each other
- ★ how projects are initiated, planned, executed, monitored and controlled, and closed
- ★ the different types of project management software
- ★ how project management software is used to support a project through its life cycle
- ★ the tools and techniques, such as Gantt charts, PERT charts and the critical path method, that are used to manage projects.

Before starting this chapter, you should:
- ★ be familiar with the terms: project, collaborative working, scheduling and review.

A project is a collection of tasks that are carried out to achieve a goal. An example of a project is the development of an idea, for instance for a new smartphone app, from the moment it is thought of by a programmer or software development company or from when a client asks a company to develop an idea, to when it is in use on people's smartphones. Project management is the use of skills, knowledge, technological tools and techniques to carry out and complete a project. Project management aims to ensure that the initiation and planning stages, the executing and monitoring stages, and the project closure stage are completed successfully.

15.1 Project life cycle

The main stages in a **project life cycle** are the initiation, planning, executing, monitoring and control, and project closure stages. There may be sub-stages within these stages or more stages depending on the project. Stages are sometimes called process groups because within each stage there are many processes, **tasks** and **activities** that are carried out.

15.1.1 Project initiation

A new project idea, or conception, begins with a brainstorming session of potential developers, clients and end users to exchange thoughts, suggestions, views and ideas. A **mind map** may be produced to represent these ideas and thoughts. The mind map is then developed into a document that can form the basis of a new project.

The objectives of the project, what is to be delivered at the end (that is, the product, such as the app) and how long it will take to complete (the duration) are decided and documented in this stage. Responsibilities for the different aspects of the project, for example the tasks and activities, are also decided at

15

the **project initiation** stage. A SWOT analysis (of the strengths, weaknesses, opportunities and threats to the company) may be carried out to determine and clarify whether or not the company's business is able to complete a project. Documentation is created that is often known as the project charter. This documentation states the objectives, scope and stakeholders in the project.

Activity 15a

1   Describe what is meant by 'project objectives'.
2   When are project objectives decided?

Identifying objectives

The objectives of a project should include a brief description of what is to be achieved, for example a new smartphone app to measure the number of steps taken by the user, and must be specific, measurable, achievable, realistic and time sensitive, or SMART.

Specific objectives describe what the end product should be, why this is to be produced (what its purpose is), who will work on the project and what resources are required; also, they should specify any time limits for the work. Measurable objectives would set a budget for the project, for example how much money is allocated for the development and how long each stage should take. The objectives should be achievable or should be reconsidered. The objectives need to be realistic and relevant, for instance is it necessary to create a new app in just one month? Time-sensitive objectives are necessary, for example deadlines, for each stage and for the end product to be completed and available. These are important because, without deadlines, a project becomes unfocused and may never be completed. Unrealistic and unachievable objectives will cause a project to fail. The objectives are documented as part of the project in the initiation documentation and in the project's charter.

Activity 15b

1   Describe what is meant by 'SMART objectives'.
2   Describe what is meant by a 'project deadline'.

Project scoping

In project management, the scope describes what has to be done to complete the project and enables a project timeline to be set and resources to be allocated. When scoping a project, information is gathered about:

» the objectives and goals of the project
» the types of work and the methods used to complete the project
» the satisfaction of the potential end users
» any constraints or restrictions to the project's progress
» the need to allow for any changes, for example in user requirements, that may occur during the project.

Project scope can also include what is to be excluded from a project, such as stating that certain features will be excluded, for example that no personal details will be recorded.

A **product breakdown structure (PBS) analysis** is often carried out to determine what has to be created and delivered at the end of the project. For example, if a new software app is to be created, the components of the new app will be listed and documented.

A well-researched and documented project scope guards against project scope 'creep'. Scope creep is where changes are made after the start of the project, but the changes have not been anticipated or accounted for in the initial planning. This can extend the timeline. **Feature creep** is where more features are added, for example a new smartphone designed to record the location of the smartphone is being developed but more features are added during the project. While scope creep may include feature creep, scope creep is where the project keeps expanding to cover other aspects, which affects the whole project. This can cause projects to fail. Scope creep usually occurs because of vague objectives, poor management or poor communication between those in the project. Documentation about the project scope is included in the project's charter.

> ## Activity 15c
> 1 Explain what is meant by 'project creep'.
> 2 Explain why project creep happens.

Stakeholders and their needs

The **stakeholders** in a project are everyone who is involved, especially those who are likely to gain from a successful outcome. The project managers, the team working for the managers, the clients who asked for the product to be developed and initiated the project, and the potential end users are stakeholders in a project. Internal stakeholders are those who are within the business, for example the owners and employees of a software development company. External stakeholders can include investors who buy shares in or put money into the company but take no part in its running, such as customers and end users. Identifying who the stakeholders are is important because, at any stage in a project, managers need to know who is responsible for tasks, who is going to make decisions, and who is likely to be asking about the progress of the project. When the stakeholders have been identified, their interests and responsibilities are documented and the documentation included in the project's initiation plans. Documentation about stakeholders is included in the project's charter.

The different stakeholders all have interests in a successful outcome to a project. The directors and managers need to be involved in the decisions that affect the project outcomes and need the project to be financially successful. The project team need to be made to feel that their contribution is valuable and appreciated to increase their motivation, which makes the project more likely to be successful. External stakeholders such as suppliers need to be paid for their goods and contributions so have an interest in helping to make the project successful. Financial investors need the project to be successful so they make a profit on their investment. Clients need the project to be successful so that the product meets their requirements.

Resources required and management

During the initialisation of a project, a financial analysis is carried out to determine the costs to the company of developing a new product and the benefits of its development (for example, will the product increase sales and make more profit), and to suggest a budget for the project. The resource requirements are analysed to find out if the company has the necessary skills, expertise and staff to be able to develop the product and complete the project. Other resources are investigated, for example the software required to develop the product and manage the project, such as the **project management software** to be used. Documentation about the resource requirements is also included in the project's charter.

Success criteria

Success criteria are the measurable outcomes required from a project by the client, stakeholders and end users. Success criteria must be concise and specific so that they can be measured, and they usually cover the scope, costs and timings of the project, the benefits of carrying out the project, and the requirements of the stakeholders. Examples of well-defined success criteria are 'The product must be completed in 10 days' and 'The user interface must have a blue background'. Success criteria that are vague, for instance 'The product must be efficient' or 'The user interface must be easy to use', cannot be measured properly and often result in stakeholders being dissatisfied with the product.

Success criteria are defined at the start of the project. Documentation about the success criteria is produced at the same time and includes the success criteria, how and when these will be measured, who is responsible for measuring them and who they will report to. The whole project team should be aware of the success criteria and should refer to them throughout the project.

High-level task schedules

In a **high-level task schedule**, detailed planning of the scheduling of tasks is left to the **planning stage**, but during the initiation of a project it is useful to gain an idea of how the project will run. In the first stages of a project, during the initiation stage, a high-level schedule is created to be a basis for the detailed planning that comes later.

High-level schedules can be based on the product breakdown structure (PBS) analysis and the **work breakdown structure** (WBS), which are used to determine what needs to be done and when. A high-level schedule may be sketched out at initial planning meetings and developed as the details become clearer.

	High-level project schedule				
Task	September 2020	October 2020	November 2020	December 2020	January 2021
Initial planning	⟷				
Meet the stakeholders	⟷				
Work out the budget	⟷				
Appoint the development team	⟷				
Plan the project	⟷	⟶			
Execute the project, design and develop prototypes			⟷		
Evaluate the prototypes			⟷	⟶	
Production				⟷	⟶
Evaluate project					⟷

▲ **Figure 15.1** A high-level schedule sketched during planning

A high-level schedule lists the **milestones**, identifying the important points in time during the project. These milestones may be changed later in the planning but are useful as a starting point.

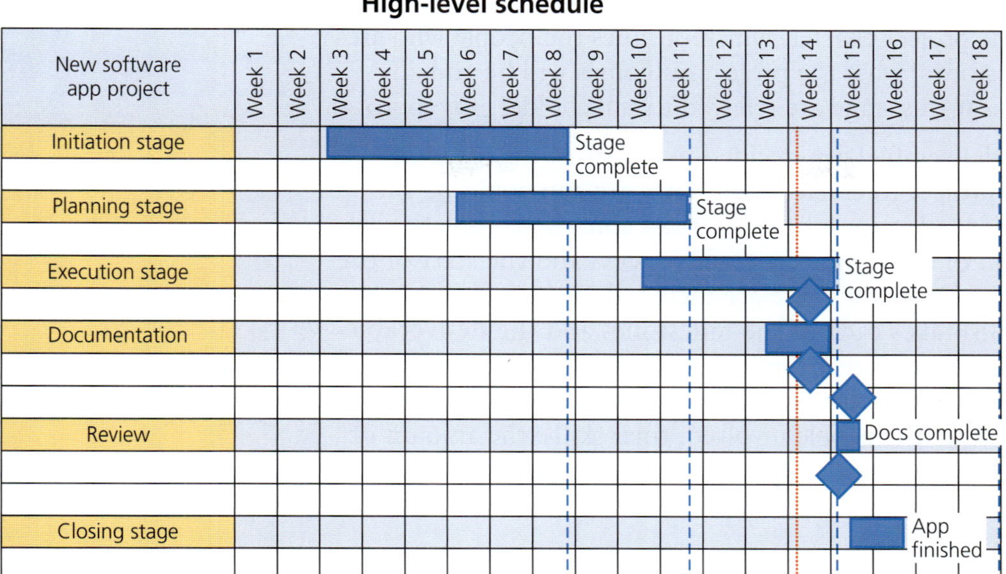

▲ **Figure 15.2** A high-level schedule

Timescales and deadlines included in a high-level schedule are shown to stakeholders so that they know when to expect the product to be delivered. Listing and analysing the tasks to be carried out shows which tasks can be grouped together, which tasks can be carried out simultaneously and which ones have to be completed before others can start. This forms the basis for the critical path analysis that comes later.

A high-level schedule is an overview of the whole project and can act as a project master schedule, showing the major milestones and the most important **deliverables** (the parts of the product that are completed along the way). It can be used as a means of checking that the project is progressing as planned.

High-level schedules are used to create other, more detailed, schedules that break down the major tasks into sub-tasks and activities, and show their timelines and **dependencies**. Dependencies show how tasks follow on from other tasks and show which tasks, or sub-tasks, need to be completed before other tasks can begin. It can show managers who is responsible for the main parts of the project.

The higher-level schedules are sub-divided and expanded to become detailed with the tasks, activities, dependencies and timings. **Working level schedules** are created to be used in the **critical path method (CPM)** for determining dependencies and the longest time that they will take to complete.

When initial plans are completed, they are reviewed to ensure that there are no errors or mistakes and the outcomes can be achieved. After any necessary changes or errors are corrected, another review occurs. This is called an **iterative review** and the project continues to the next stage only when the initialisation stage is properly finished.

15.1.2 Project planning

The purpose of the planning stage is to describe in detail the tasks and their costs, and the time and resources needed to carry out the project. With careful and detailed planning, everyone involved knows what to do and when, or by what time, to ensure that the project reaches successful outcomes. The outcomes depend on the project, but in the development of a new software app, for instance, successful outcomes would be a working app that customers like and buy.

Project planning should result in a detailed plan for the project from start to finish. It should include the project charter, which outlines the people who are involved, the budget, milestones, what the project outcomes will be and any risks. Changes in the scope at this stage are referred to stakeholders for approval.

Milestones and the main deliverables are decided in the planning stage of a project. Milestones in a project to create a new software app could be the completion of the designs, production of a **prototype** app, the completion of the testing, the completion of the user acceptance process and the start of the distribution of the new app. Dates for these are included in the plan. A work breakdown structure (WBS) takes each of the milestones and the **deliverable** products and divides them into tasks that can be assigned by the project manager to a person or a group of people to carry out. WBS is a useful tool for project managers, who have to consider the people involved, their skills, the amount of work to be done and the deadlines for tasks to be completed.

Scheduling tasks

A schedule of tasks lists every task, and the activities within the tasks, that have to be done to complete the project. Project managers and the project team attend meetings to consider the project scope and the deliverables, and to document the tasks. The tasks are put into sequence, and the tasks that can be done simultaneously and those that depend on other tasks are identified. It is important to ensure that these listings are accurate because trying to start a task before essential components are available is not possible. An example would be trying to test a module that interfaces with specific features of a user interface before the user interface is ready.

The links between tasks are identified and documented. A **Gantt chart** can be used for this. Specialised project management software, or a spreadsheet, can be used to create Gantt charts. Project management software can enable Gantt charts to be automatically created and amended with ease, while using a spreadsheet application requires expertise and skill with spreadsheets and is more difficult. After the tasks and links are decided, the milestones can be set. A milestone must be an event or time that is recognisable and achievable. Milestones are placed at regular intervals throughout a project so that progress can be measured. When the milestones are in place, timescales are decided and set.

There are a number of scheduling tools that are used in planning and managing projects. A **Performance Evaluation and Review Technique (PERT) chart** is used to show, and analyse, the tasks, activities and timescales involved in a project. A PERT chart shows which tasks are dependent on others, which tasks run concurrently (can be carried out simultaneously) and the links between them. From a PERT chart, the timings of the project can be analysed and measured. The longest time that a project will take from start to finish can be determined using the critical path method (CPM). Additional allowances in the timings (floats) can be shown and monitored. For example, if a set of tasks could take anywhere from two to three weeks, the variation can be shown as two weeks plus an extra week if required and included in the calculations as a **float** time. By including all possible timings, the critical path, which is the longest time the project could take with all the tasks completed, can be calculated.

PERT charts and CPM can be applied to individual tasks within a project as well as to the whole project. With a very large project, Gantt charts, PERT charts and CPM can become difficult to follow unless the project is broken into smaller parts. A high-level schedule may have many Gantt charts, PERT charts and CPMs described within it.

Activity 15d

The table shows a list of eight tasks for creating a computer game. Complete the table by listing the tasks in the order in which they would be carried out. One task has been done for you.

Task	Order to be carried out
Testing the game	
Writing the test report	
Designing the game	
Writing the user guide	
Coding the game	
Writing the test report	
Releasing the game	
Planning the project	First

Resource scheduling

Resource scheduling lists the tasks and activities that need to be carried out to complete a project and then determines what is needed, and when it is needed, to complete the tasks. Examples of resources include the skills of the developers, the time they will need, any components that need to be sourced and the IT systems and applications software that might be required. The constraints for each task have to be considered. The developers may not be able to devote all their time to a new project, so additional, external developers may be required. Deadlines and a lack of skills in the company may also be constraints.

A review of the resource scheduling is carried out to check that the resources will become available when needed. If the check reveals errors, these are corrected and a further review is carried out. Iterative reviews continue until the schedule is fit for purpose.

Other inclusions in the planning stage

When staffing, milestones and deliverables, and analysis and documentation of the tasks have been considered, the project budget can be allocated. The money available for the project has to be shared out so that each part of the project is properly funded. Planning includes how and when to keep external stakeholders informed and provision for changes to be made to the project. It also includes an assessment of the company's ability to carry out the project and how to assess and manage any risks to the project.

When a project plan is completed, it is iteratively reviewed to ensure that there are no errors or mistakes and that the outcomes can be achieved. When this has been done, the plan is deemed to be fully acceptable.

15.1.3 Project execution, monitoring and control stage

The **project execution** stage starts with the completion and acceptance of the planning tasks, for example final acceptance of the project charter. The project manager gathers the human resources (for example designers and developers) and the necessary resources (for example the IT resources) needed for the project and begins the actual allocation to tasks.

At this stage, the project deliverables are produced. These are the end products that have been requested by the client and can be, for example, sold to customers. This stage is usually the longest stage in the project. It usually takes up most of the time and the resources. During this stage, managers monitor and control the team members and resources to ensure that the project progresses according to the plan.

Project monitoring and control includes keeping track of where in the project the team has got to (that is, what tasks are completed and what are still to be done), and identifying the work done, problems and ensuring that steps are taken to solve them. The tasks are monitored and compared to the project plans to ensure that the plan is being carried out. Project managers check the work on specific tasks and monitor how long work is taking for specific tasks. Developers may be required to complete and submit timesheets or records of their progress to the manager for analysis.

Cost management involves monitoring the hours worked, for example by developers who are working remotely. Their hours and expenses have to be included in the project resource management and monitored during this stage. Risks are managed by identifying them at each task, documenting them and taking action to avoid or mitigate them. Risk management continues throughout the execution stage and regular reviews of risks, for example with developers, is essential. During this stage, managers have to ensure that the required goods and services, for example cables, plastics or specialised software, are sourced and available when needed as detailed in the project plans.

The deliverables are tested in this stage. Repeated testing and review, for example by the client and potential end users, ensures an acceptable end product is created. Iterative testing of prototypes corrects any errors, for example in software code, and improves the product.

Near the end of the stage, the deliverables are ready and available to the client for review. This is Acceptance Testing and involves the client checking that the products meet the **design specifications** and are 'fit for purpose' (that is, are what was planned and do what is required of them). The completion of an Acceptance Test to the satisfaction of the client means that the project is almost ready to move to the closing stage.

The last task in the execution stage of a project is the review stage. During the review, the deliverables that have been produced are documented to show that the project is proceeding according to the planned schedule, that it is within the budget, that any risks have been avoided or mitigated, that any problems have been resolved and any changes were properly managed. It also states whether or not the project can proceed to the closing stage.

The stakeholders in the project must be kept informed with progress regarding timelines, the tasks and the expected milestones. It is important to keep stakeholders well informed to avoid disagreements and disappointments if there are problems during the execution stage. Project management software (PMS) can generate reports with summaries of any analyses and automatically send these to selected stakeholders.

15.1.4 Project close

Reviews are conducted after each stage to ensure that any problems have been dealt with and that the stage is complete and stakeholders are satisfied that the next stage can go ahead. During a review, project managers can ensure that all company policies, procedures and risk analysis requirements have been followed. Also, they can make sure that all goods and services have been accounted for and suppliers paid. Reviews ensure that the project continues to meet the requirements of the clients and is within the project scope. Any new methods that are discovered can be carried over to other stages by reporting them in a review, and lessons can be learned from tasks that did not go well so mistakes are not repeated.

A formal **project closure** stage is important to guarantee that all the work has been carried out, all the deliverables have been produced, all the tasks have been completed to the satisfaction of the stakeholders and every stakeholder agrees the project has been completed.

For example, if the development of a new software app was the aim of the project, the project closure process will show that the development team has actually developed the new app, it was fully tested, and has been accepted by the software development company that employs the developers and by the client who requested it. The project manager and the stakeholders check that all the tasks are complete and that nothing has been overlooked. Lastly, the project manager and the stakeholders agree that the contract for the development has been fulfilled. Without a proper closure review and report, some companies may not consider a project to be over and may keep returning to the development team for changes or corrections to a new software application when, in reality, this is the responsibility of after-sales departments, service technicians or IT support. The **acceptance criteria** is a list of requirements set out in the project plan that lists the conditions that must be met before a project is considered complete.

15.2 Project management software

Project management software (PMS) is a software application, or a collection of software applications, that provides the tools to carry out the different stages in a project:

» the planning, organisation and scheduling of tasks
» the allocating of resources and their costings
» decision-making
» communicating between managers, stakeholders and other interested parties
» creating the required documentation.

It also enables collaboration between the project team members.

15.2.1 Types of project management software

There are various types of PMS available, with each type designed for a particular type of user or categorised according to how it is used or what it can do. For example, project managers are often, but not always, working in an office at a desktop computer so have different requirements from a home-based software developer working as part of a team. PMS may be web-based, cloud-based or suitable for use on mobile devices. Some types of PMS are designed for individual use while others feature collaborative tools. A project team member will choose, or be allocated, a type that is most suitable for their role in the project.

In smaller companies, or when there are a small number of people involved in a project, a personal or individual type of project management software is used. This type is often a single-user PMS because often only one person uses it. A personal PMS may not contain all the tools found in more elaborate types and is not usually suitable for managing large, complex projects. A personal PMS would be used at home to manage lifestyle or home projects.

A desktop-based PMS is usually a single-user application that runs on a desktop or laptop computer with no ability for more than one person to use it at the same time as others. A project manager would make use of this type because the manager would not wish others to use the software or the desktop PC running it.

Web-based project management software runs on a networked server and is accessed using a web browser. A smartphone, tablet or laptop computer can be used to access the PMS software and use its tools. The web browser acts as a client while the PMS application is the server. Web-based project management software may not have all the features of a desktop PMS or a version run from an applications server. Client–server networking is described in Chapter 14 'Communications technology'.

Mobile project management software may run as an app on a smartphone or tablet. It may access data stored centrally or in the cloud using mobile connections. Project managers or team members can access the PMS and update details when away from the office. For example, a project manager visiting a construction site may update the progress while on site.

Single-user project management software is designed to allow only one person to edit the data at any one time. Collaborative PMS is designed to allow multiple users access to the data and to be able to edit it. Users share the data and edit other's data to, for example, allocate resources, amend deadlines, update reports and carry out tasks in the project. Changes in the data are visible to others almost immediately. The data, or a part of the project, where one team member is working may become read-only and unavailable to others until the member has finished.

Many PMS applications have features or tools that allow data to be easily presented in visual formats. This is to try and avoid **information overload**, where there is too much data and information to understand easily and quickly, for example where data may be changing frequently or arriving in large quantities. An example would be a project manager receiving hundreds of emails from team members about the status of a project. The information in the emails and any attachments is difficult to read, understand and monitor. Another example of information overload is the frequent updating of social-media posts that generate so many alerts or notifications that the receiver cannot attend to them all and so misses important information.

Activity 15e

1 Explain why information overload can hinder a project.
2 Describe ways in which the effects of information overload can be reduced.

15.2.2 Use for supporting projects

Planning

With the ability to share online, PMS can create interactive Gantt charts that planners can collaboratively create, task schedules can be altered by dragging (or moving) the task details and allowing the software to update the views in real time, tasks can be assigned to teams considering their workload and skills, and plans can be shared online with teams and stakeholders. Shared plans can be viewed and commented upon by team members and other stakeholders for all to view and consider.

PMS can allow more efficient allocation of personnel and resources by interacting with other company projects. Spreadsheets from company departments, such as finance, can be imported and used to support the planning. Stakeholder details, for example their email addresses and telephone numbers for contact, can be included in PMS software, task lists can be added and the personnel assigned to the tasks can be linked. Budgets and schedules for tasks and milestones can be planned.

Scheduling tasks

PMS contains tools that put project activities and tasks into a sequence, show the dependencies (which may be complex), and calculate the longest times for the project and the possible paths through the project. A project schedule lists the start and end dates of tasks and activities, milestones, dependencies and the project deliverables. The project's work breakdown structure and scope are considered when developing a schedule. The scheduling of the allocated resources and costings can be calculated and automatically updated if changes are made. Purchasing and using, and training employees to use, project management software for scheduling can be expensive for small projects. However, the use of PMS to present the data in a readily understood manner so that managers can react to problems and risks or make changes as the project progresses can help keep costs down.

Scheduling tools include the generation of Gantt charts from lists of tasks and activities, the creation of PERT charts that show a project as a network diagram of parallel, or concurrent, and consecutive tasks with their timings and dependencies, and resource histograms for planning the deployment of different staff and critical path analysis. The critical path method can show the shortest path through the tasks of a project with the longest time needed for the project.

Costings

Project cost management estimates and manages the project budget, documenting all transactions and payments before and during the project. PMS can assist with costings by automatically calculating, displaying, alerting and reporting on expenditure for accounting purposes.

A project has direct costs and indirect costs. Direct costs are:

» the salaries or money paid to the people working on the project, including for the cost of their time if they are not already employees
» the costs of IT equipment, services and software
» the cost of materials
» possibly the costs of transporting materials and people to and from where the project is being undertaken.

Indirect costs are:

» the overheads, for example rent or upkeep of buildings and office space
» office equipment such as pens, paper and other office supplies
» utility bills
» provision of facilities for the staff
» costs of, for example, cleaning and security staff.

Additional costs such as insurance against unforeseen problems or natural disasters and the costs of project management tasks have to be included in the project budget. Management tasks can add about 10 to 15 per cent to the cost of a project. Careful planning reduces these additional costs.

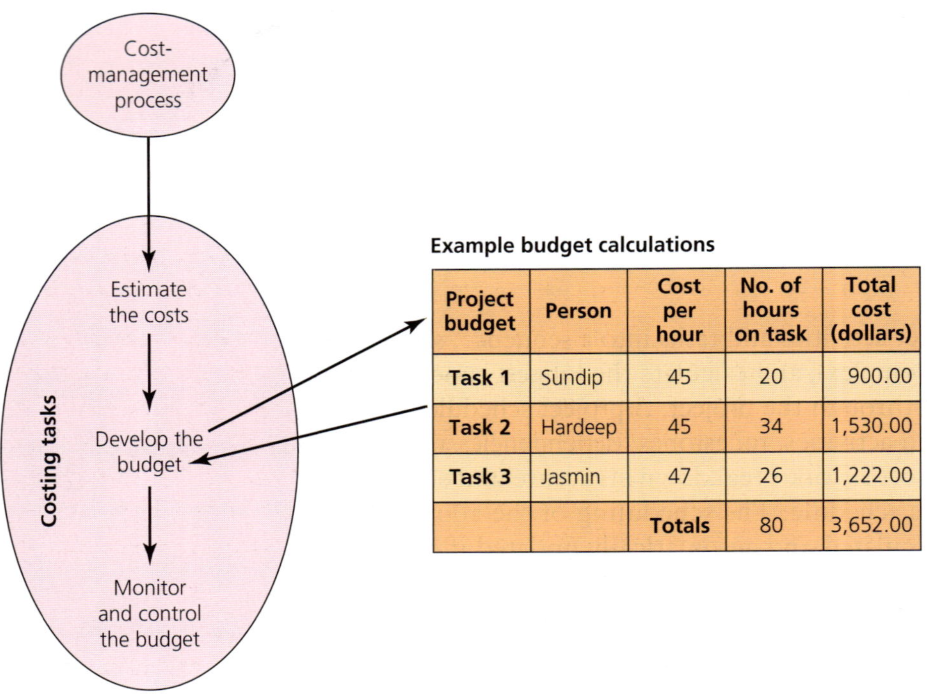

Example budget calculations

Project budget	Person	Cost per hour	No. of hours on task	Total cost (dollars)
Task 1	Sundip	45	20	900.00
Task 2	Hardeep	45	34	1,530.00
Task 3	Jasmin	47	26	1,222.00
		Totals	80	3,652.00

▲ **Figure 15.3** Managing the costing of a project

PMS can help in managing costs by importing data (for example costs and calculations about profit and loss) from spreadsheets, projections and estimates of costs, and can account for inflation or price changes during the project. Importantly, using PMS to manage and track costs allows managers to monitor the expenditure and track it in the budget during the project. As the project progresses, the costs of each task and activity are entered and updated, and the PMS can display the cost details for the tasks in charts or graphs to monitor expenditure. Complete tasks, overdue tasks and those spending more than their original budget are flagged up, alerts are generated and reports are created in a format that project managers and other team members can easily understand.

Archived and historical data can be used to help estimate the costs of the new project. Every activity and task can be examined in detail and their costs used in calculations, using pre-determined formulas, to produce an overall estimate of the project cost. A worst possible (most expensive) cost and a best-case scenario (least expensive) cost can be produced for discussion during the initialisation and planning stages.

Communication

Managers need to be aware of the progress and performance of team members and PMS can provide the means to gather and communicate information during a project. Project managers can access the PMS and extract the information, for example what tasks are due for completion or starting in the next week. Automatic reports and alerts, for example on how long tasks are taking to finish, can be generated and, if required, sent to managers, team members or other stakeholders. The reports can be emailed from a distribution list. Different lists can be configured for different alerts or reports. Reports can be automatically printed and prepared for mailing to stakeholders or customers.

Collaborative working

Collaborative project management involves all project members in the planning, monitoring and execution of the project from start to finish. Those who are carrying out the tasks are often also responsible for the monitoring and control of their own tasks. Each team member has the same importance as all the others and there is no clear hierarchy between team members.

Traditional project organisation has separate tasks that come together at the end to complete the project. In **collaborative working**, tasks are linked together and each one provides a contribution to the final end product. Any changes, amendments or additions to the end product are made and tested and communicated to others by team members as the project progresses.

PMS is used in collaborative working to access a central database and storage system of all the activities, tasks and documentation needed and generated by the project. The tools available allow everyone to have access to the project information and to be able to modify or comment upon it. While email, spreadsheets and other office applications can be used, for collaborative project management specialised software is used. Specialised PMS includes tools for brainstorming so that ideas, views and comments can be shared and displayed to others, recorded and analysed.

Calendars that can be shared are important. Individual calendars and group calendars that are shared and visible to others allow meetings and events to be organised at times that are suitable for all. **Diary-management software** allows the scheduling and re-scheduling of meetings, and the automatic sending of meeting reminders and alerts of potential clashes in appointments. It also reduces the number of errors and mistakes made when trying to arrange meetings between large numbers of busy project members. Diary-management software can be linked or integrated into PMS. Online diary-management software allows access from any web-enabled device.

Online meetings and sharing documentation allow progress to be discussed and team members to be updated even if they are distributed around the world. Tele-conferencing, video conferencing and online chats can contribute to collaborative project management. Tools are included for document sharing, creating content for the project that can be shared between team members and communication between team members. Additionally, search tools are provided for retrieving information, policies, procedural documents, contacts or archived materials that may be relevant and marking or flagging these for others.

Tools for creating **Kanban boards** allow work to be managed at an individual or group level. A Kanban board uses cards or boxes to represent tasks or activities. The cards or boxes are placed in columns that represent stages, and as the work progresses the cards or boxes are moved from left to right into the next column. Sometimes there are horizontal rows representing specific workflows across the stages.

Activity 15f

1   Describe the benefits to a project manager of using diary-management software.
2   Describe the drawbacks of using diary-management software during a project.

Decision-making

Different types of decisions are made during a project, from major decisions that affect the whole project made by project managers or company directors, for example whether to employ more software developers, to routine decisions that are made from day to day, for example deciding whether or not to allow employees to take a leave of absence from work, which are usually made by team leaders. PMS can link the necessary data and information and make it available to decision-makers in formats that can be readily understood and exported to other documents. It can assist in making decisions by ranking, prioritising and selecting from different options, as well as providing information on the possible outcomes of decisions.

The ability of PMS to gather, store, retrieve and analyse data and information progress allows project managers to use 'agile decision-making' that includes all team members working together to make decisions that affect the project. It requires access to all the information about the project, frequent reviews on progress and frequent consultations with stakeholders, and allows small, but important, changes to be made without lengthy delays as the project progresses.

15.2.3 Advantages and disadvantages of project management software for supporting projects

PMS provides software tools for collaborating with, and informing, team members in real time by sharing progress information. This enables decisions to be made more quickly, increases the motivation of team members and alerts them to further issues or problems as soon as they occur.

Documentation can be shared between members and can be quickly edited, updated and corrected for errors. Project status reports can be automatically created and shared, allowing for transparency in the progress of the project and the quick identification of problems. PMS can access data and create and distribute reports in many different formats suitable for managers, accountants and team members, for example, who can use these to monitor the project's progress. This helps keep projects on schedule.

Overruns and overspends can be discovered more quickly and project costs managed and controlled by using tools to allocate and monitor budgets. Risks can be assessed, documented and managed using PMS.

Many PMS tools do not need extensive training to use, and some can be intuitive and simple to use by project managers when accessing and displaying important data. This allows managers to concentrate on managing the project. However, using project management software on small projects may make the process more complicated. Very small projects may not need PMS at all and the time and effort taken to set up the software adds to the costs.

PMS can be expensive to purchase, install and maintain. While some features can be easy to use, training and expertise is required for complex projects and this can take up valuable time and money. However, failure by a manager to be adequately trained in the software may lead to a project being more costly, not being completed on time or failing altogether.

Too many automatic alerts can be time-consuming to process by team members and can impede the work of software developers, for example, who spend time dealing with routine alerts rather than on their tasks. Sometimes more time is spent configuring alerts to warn members about upcoming events or issues than dealing with the issues and events – this can slow the progress of the project.

Each type of PMS has its own advantages and disadvantages compared to the other types. Desktop-based PMS contains all of the features and tools required but may not allow easy sharing of files or data. Often, this type is single-user only so collaborative working is difficult. However, single-user PMS is useful for a small project with very few team members. A project manager would use desktop-based PMS if the project manager is based in an office.

Mobile PMS systems would be useful for those project managers who visit sites or different locations but may not have all the features of office-based systems. Web-based PMS has its data stored centrally so it can be accessed by all the members of the team and updates or amendments are shown to everyone almost immediately. However, storing data centrally, or in the cloud, makes it more vulnerable to unauthorised use or loss, especially when being transmitted between devices. Cloud-based data can be accessed from almost anywhere, so the project team and the stakeholders can access it from home or from offices around the world. Most web-based project management software systems are compatible with each other or can run in a web browser so data can be easily shared between team members, but the file format is not always compatible with desktop-based PMS. This can make working on the project offline difficult as the files may not open in local PMS applications. Web-based PMS systems are easy to learn and use, with online help available, and the data is automatically backed up by the service provider.

15.3 Tools and techniques for project management tasks

PMS integrates the tools needed for managing and carrying out a successful project. Instead of keeping track of separate documents and spreadsheets, trying to move data between them and to manage backups and archives of these, managers and project workers can access all the required data, and display and share it in useful formats easily and quickly. PMS tools can increase productivity and monitor large projects, for example creating new software applications, far better than separate software applications. For very small projects, for example creating a short pamphlet or programme for use with a college theatre production, the use of complex PMS may not be the best solution. Spreadsheets and other software applications can carry out most of the tasks associated with PMS for small projects.

Project management tools and techniques include Gantt charts, **Performance Evaluation and Review Technique (PERT)**, critical path method (CPM) and analysis, calendars, timeline tools, work breakdown structures, status tables and others to plan, execute, monitor and control projects.

Work breakdown structures

A work breakdown structure (WBS) is created at the very start of a project, before the detailed planning of the tasks. It includes every job, activity and task needed to produce the deliverables from a project and can be used to create Gantt and PERT charts. A WBS is also a deliverable and can be developed early in a project by mind mapping. The mind mapping technique is useful when starting to plan a project. Mind maps can be created by hand by individuals, or collaboratively in groups where each person adds their ideas and comments in specialised software for inclusion in PMS. Mind maps focus the ideas around a central idea or topic, with sub-topics radiating out from the centre, eventually ending in 'twigs' with an indivisible activity.

WBS can be based on the project deliverables or on its stages. All the documentation about the project deliverables, scope, charter and plans is gathered and used to identify the job and work that needs to be done. Tasks are repeatedly broken down into activities and jobs until a job cannot be broken down further. Then team members, costs and other resources can be assigned.

Each part of a WBS is called an element and these are arranged in 'legs', with the tasks listed under each heading, for example colours under 'Create the user interface' would be further broken down into activities and jobs that might be even further subdivided. The final WBS would consist of several levels of tasks each with further sub-divisions. Additional information can be added or extracted from data entered into the lists and tables in the PMS.

The complete set of tasks, and all the activities and sub-activities, can be documented into a WBS dictionary that becomes part of the planning documents of the project. It is not possible to include every single detail in the diagrams, so the WBS dictionary is the reference document and it must contain every single detail of the project's tasks.

15.3.1 Gantt charts

Gantt charts are used to show a project schedule visually as a linear time chart. They work well with the **waterfall method** of development where each project stage must be completed before the next starts. The project tasks are placed on the vertical axis and the start times and end times of the tasks on the horizontal axis. The duration of a task is set and so, by inference, the end time is shown. Gantt charts can show the dependencies between the tasks that run sequentially, in parallel or overlap.

Sequential tasks run after each other because one has to be finished before the next can start, so one is dependent on another. Parallel tasks are carried out simultaneously and are independent of each other, meaning that the number of people available may be insufficient for the tasks to be done at the same time, for example. Overlapping tasks may or may not be dependent on each other, but one task starts before another ends, for example some document writing can start while a task review is still going on.

A Gantt chart will display alerts or warnings if inconsistent data is entered. For example, making two tasks dependent on each other creates a loop. An example of this could be the start of a task that depends on the coding of an interface being completed in order to begin testing it. If the start of the coding is made dependent on its testing being completed, then an impossible loop is created. When the Gantt chart is created, the PMS will show this as an error and 'demand' that it be fixed before allowing the user to continue.

Activity 15g

1 A project to develop a new piece of software has six tasks, as shown in the following table.

Task number	Task name	Time allocated (weeks)	Depends on
1	Research the market	2	
2	Plan the project	2	1
3	Design	3	2
4	Develop software	4	3
5	Test	2	4
6	User documentation	5	2

Draw a Gantt chart to show the project schedule.

2 Explain why a Gantt chart is used to display the project schedule.

3 Which tasks are concurrent in this example?

Using and interpreting Gantt charts

Gantt charts are created by first listing the tasks, which tasks they depend on, the start times and the expected durations. Other data about the tasks can be added and displayed to show as little or as much information as the user wishes. A Gantt chart has a timeline across the top of the chart that shows the times and dates, and underneath the tasks are illustrated by horizontal bars. The bars show the duration of the tasks and are often shown in different colours to represent their status. Often, the critical path is shown in red. If the critical path changes with amendments to the task, the PMS automatically alters the colours. The length of the bar indicates the duration, but other data can be added, for example a percentage value to show how far the tasks have progressed.

For example, a list has been made of seven major tasks in the development of a new software app. For convenience, and to make the chart fit in a window, the duration of each task is much shorter than in reality. Entering the duration and the start time automatically fills in the Finish column. The dependencies are added and the list looks like Figure 15.4.

	Name	Duration	Start	Finish	Predecessors
1	Design the app	5 days	26/08/20 08:00	01/09/20 17:00	
2	Code the app	15 days	16/09/20 08:00	06/10/20 17:00	1;4
3	Test the app code	5 days	07/10/20 08:00	13/10/20 17:00	2
4	Create the user interface for the app	10 days	02/09/20 08:00	15/09/20 17:00	1
5	Release the app to the public	5 days	14/10/20 08:00	20/10/20 17:00	3;7
6	Create the advertising for the app	5 days	02/09/20 08:00	08/09/20 17:00	1
7	Advertise the app	20 days	09/09/20 08:00	06/10/20 17:00	6

▲ **Figure 15.4** A list of tasks in the development of a new software app

The Gantt chart calendar can be customised. Saturday and Sunday are shown here as non-working days. The PMS can change these or not show them at all. The number of working hours in a day is set as eight hours by default but, again, this can be changed. The Gantt chart created from this list is shown in Figure 15.5.

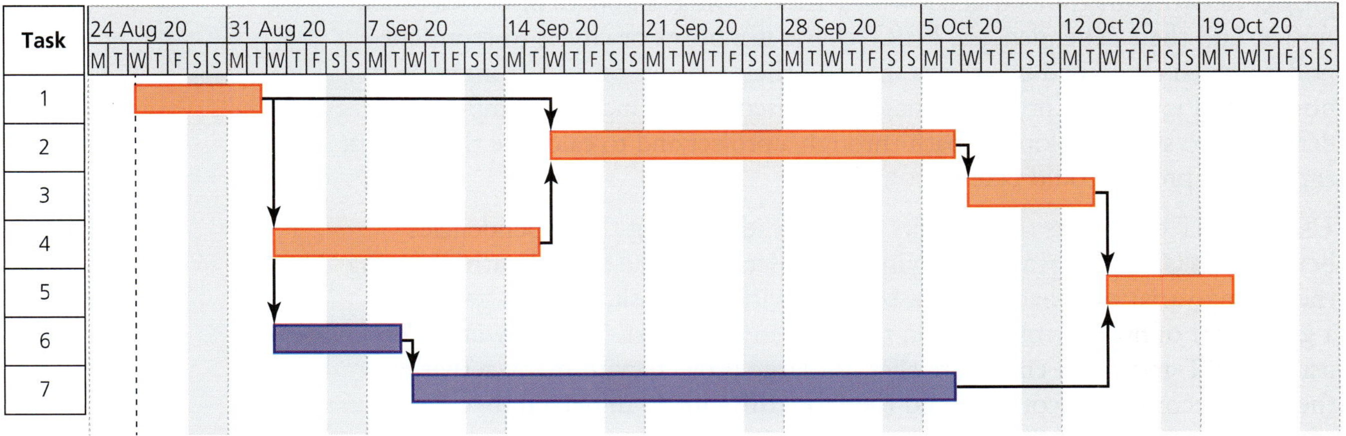

▲ **Figure 15.5** A Gantt chart

Tasks or events can be set as milestones. In this example, altering the entry in the list for task 5, the app release to the public, into a milestone changes the Gantt chart. The milestones appear as black diamonds on the timeline (see Figure 15.6, over the page), and the path through the project is shown by arrows with the critical path shown in red. Other tasks, 6 and 7 here, are shown in blue.

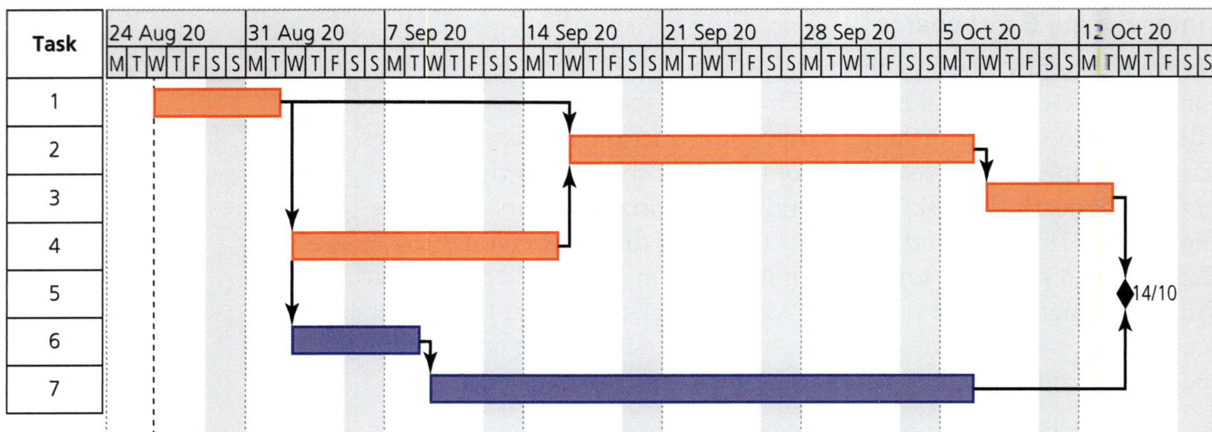

▲ **Figure 15.6** A Gantt chart showing a milestone

Project managers can make and easily visualise changes in Gantt charts by dragging the bars or by amending the data in the task list.

15.3.2 Performance Evaluation and Review Technique and critical path method

The Performance Evaluation and Review Technique (PERT) is a tool that represents the tasks and events needed in a project. Tasks and events are unique to a particular project, but PERT represents them all in a similar way so that they can be easily understood. PERT provides a visual representation of the links and dependencies within a project. By using PERT, a project manager can analyse the tasks and activities needed to complete the project successfully. A project manager can work out how much time is required for each task to be completed and, from this, find out the minimum time needed to complete the whole project successfully. PERT focuses on the events in a project and can be created without the project manager having to know every detail. The data from a PERT chart can be exported and used for detailed scheduling and reports.

PERT and the critical path method (CPM) are both used to plan, schedule and control a project and both use precedence diagrams. Precedence diagrams are project network diagrams that show activities, timings in boxes or circles called **nodes**, and their dependencies. PERT charts use boxes or circles to represent nodes, which are joined by arrows showing the sequence or path from node to node. CPM is used by project managers to discover and highlight important deadlines, to show the longest path through a project and to calculate the shortest time that a project could take to complete.

Using PERT charts has the benefits of clearly identifying the dependencies between tasks, discovering, displaying and calculating the critical path, showing the timing variations that can be associated with each task, and can gather a great deal of project information together in one chart. The drawbacks of using PERT and CPM charts are that, for large projects with many activities, they can become very complex and therefore difficult to understand.

Creating PERT charts

The creation of a PERT chart starts with identifying the tasks and the milestones. The terms 'tasks' and 'activities' in project management are often used interchangeably, but a task is a major job that needs to be done in a project. Tasks are divided into smaller activities, or assignments, that are required to complete the task. A milestone in a project is an important point in time or an event that marks progress through a project. Milestones are shown as circles or boxes in PERT charts and in Gantt charts they are indicated as diamond shapes.

Activity-on-node diagrams and **activity-on-arrow diagrams** are used in PERT charts. An activity-on-node diagram puts the tasks on the nodes, in boxes, and the timings, along with other task information, inside the boxes, with arrows used to represent the flow through the project. For example, a PERT chart using activity-on-node to represent a project to create a new software app with a number of milestones and tasks could use the data as shown in the table.

▼ **Table 15.1** Using activity-on-node to represent a project to create a new software app

Event number	Event at node	Duration (days)	Dependent on event number
1	Design the app	5	
2	Code the app	15	1,4
3	Test the app code	5	2
4	Create the user interface for the app	10	1
5	Release the app to the public	5	3,7
6	Create the advertising of the app	5	1
7	Advertise the app	20	6

A PERT chart produced manually from the table is shown in Figure 15.7. The nodes (circles) show the events and the duration (in days in boxes on the circles), and the path is shown by the arrows.

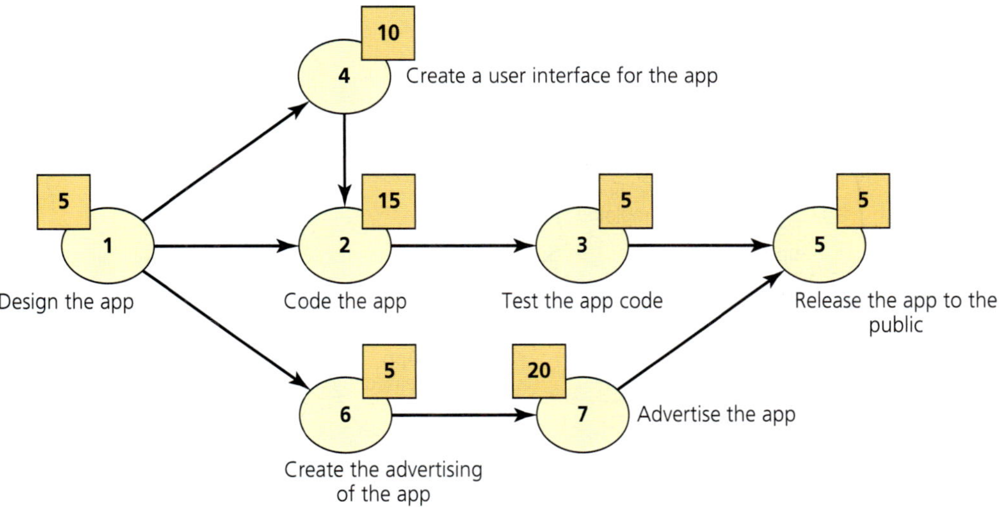

▲ **Figure 15.7** A manually created activity-on-node PERT chart

In the manual chart, all the information and the paths have to be input carefully and checked thoroughly. If the critical path is required, it must be manually calculated and drawn. However, using PMS, the PERT chart automatically shows the data and the critical path is shown by red boxes. Figure 15.8 shows a PERT chart that was created in PMS.

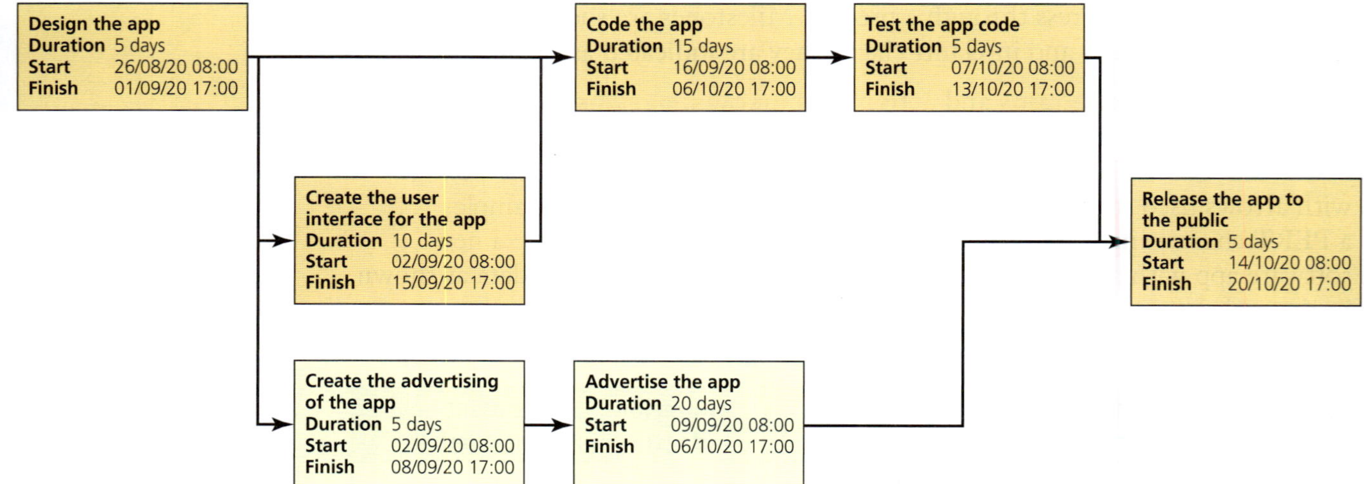

▲ **Figure 15.8** PERT chart created using PMS

'Activity-on-arrow' PERT charts use arrows to represent the tasks to be carried out to move from one milestone, shown as a circle, to the next. This type of PERT chart was more commonly used when the charts were done manually, before PMS became available. Activity-on-arrow PERT charts are less flexible in use than activity-on-node charts. The quantity of data that may appear on the arrows in complex projects can make them cluttered with information and more difficult to interpret. The PERT chart for the creation of the new software app is shown as an activity-on-arrow chart in Figure 15.9.

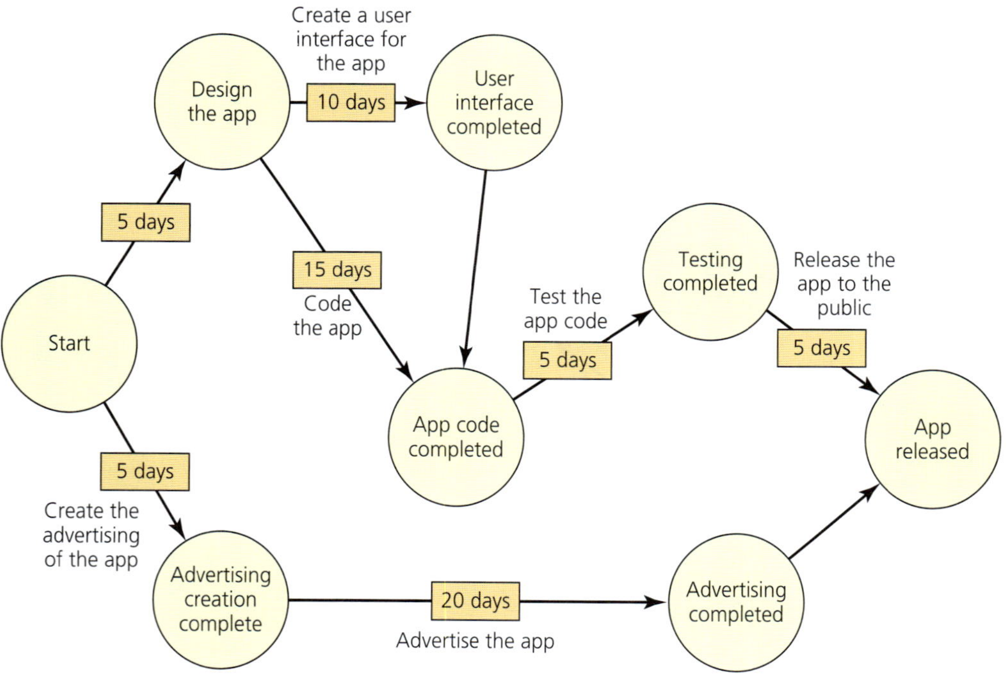

▲ **Figure 15.9** Activity-on-arrow PERT chart

Most PMS produces activity-on-node PERT charts because the tasks, and all their associated data, are entered as a list with precedencies or dependencies. Entering the data as a list into the PMS is much easier and less prone to error than manually drawing it onto a chart. Also, changes are easier to make on activity-on-node charts than activity-on-arrow PERT charts.

Activity-on-node PERT charts emphasise the tasks, whereas activity-on-arrow charts focus on the events or milestones. Focusing on the tasks means that there is, or should be, only one PERT chart, or network diagram, that works for the project. This makes calculating the critical path clear. However, a focus on the milestones may mean that more than one network diagram can be drawn, making the calculations more difficult.

Using and interpreting PERT charts

Each of the components of a PERT chart has a purpose in the use and interpretation of the chart. Nodes are symbols used to hold and show information about the tasks, activities, events and milestones that make up a project. A node can be shown visually as almost any shape, but a box, circle or triangle are the most common. A diamond shape is also commonly used when representing a milestone.

Tasks

A task in a PERT chart is a set of activities that are necessary for the completion of a project and that uses resources, for example time, raw materials, skills and, often, electrical power. The terms 'task' and 'activity' are often taken to mean the same thing, but that is not always accurate. The task, for example of testing the app code, has a number of activities that can make up the test task, for instance creating the test plan, testing the code and reporting the results of testing.

Milestones

Nodes can represent milestones, which are significant and important events that mark the progress through a project. Milestones are used to determine whether or not a project is on schedule. Project management software can be configured to set up alerts or reminders if milestones are approaching, have arrived or have been missed. By linking information about timings to milestones, the critical path can be calculated and displayed to project managers to monitor the project's progress.

A project milestone is a measure of the progress through a project, while a deliverable, which might occur at a milestone and be one of them, is an output from the project. The most important deliverable from a project is the intended final product, for example a new app for a smartphone, but there are usually deliverables at intervals during a project. The final designs, the user interface or user guide, and the test reports, can be considered to be deliverables.

Dependencies

In a project, some tasks depend on other tasks and cannot be started, or finished, before those tasks are started. The way that the tasks relate to each other is called the dependencies. Tasks with finish-to-start dependencies can only start when a preceding task has finished. For example, coding a software application can only start after the design is complete. A finish-to-finish dependency means that a task cannot end until another task also ends. For example, the design of a user interface cannot end until the navigation buttons are all in place. Tasks with start-to-start dependencies cannot start until another starts. For example, testing cannot start until coding has started. A start-to-finish dependency means that one task cannot finish until another task has started. For example, creating and amending the advertising material may not be fully completed before the advertising starts.

Whatever the type of dependency, each task has a predecessor upon which it depends and a successor that depends on it. The dependencies are entered into a list or table, often called 'predecessors', for each task in the PMS and used to generate Gantt and PERT charts and to discover the critical path. A typical list of dependencies, the predecessors, can be seen in the table in Activity 15h.

Activity 15h

A project is made up of twelve tasks of differing durations and dependencies, as shown in the table.

Task number	Duration (days)	Dependencies
0	Start point	
1	1	0
2	2	1
3	3	1
4	3	1
5	4	4,6
6	4	4,7
7	4	4
8	5	2
9	6	5,6,7
10	8	8
11	10	3,8
12	5	9,10,11
13	End point	

1 Draw or sketch an activity-on-node PERT chart as a network diagram to show how the tasks depend upon each other.
2 Add the durations of each task to your sketched diagram and use these to determine the critical path.
3 What is the shortest time in which this project could be completed?

End points

There are start points and **end points** for tasks, activities and sets of tasks. These are usually logical points and show when a particular set of activities has begun or ended.

Timings

The definition and setting of **timings** for tasks and for the overall project are included in a project plan for use in PMS to create the Gantt and PERT charts and a work breakdown structure. Each task, and its activities and sub-activities, is analysed and assigned its timings. Three important timings are entered for use in PERT calculations and from these a fourth timing is calculated. These are shown in Table 15.2.

▼ **Table 15.2** Timings used in PERT calculations

Name of timing	Description	Represented in calculations as
Optimistic time	The minimum possible time Assumes all goes to plan	t_o
Pessimistic time	The longest time for a task or activity to be completed Assumes that everything that could go wrong, will go wrong	t_p
Most likely time	Realistic estimate of how long the task will actually take	t_m
Expected time	Calculated from these timings using probabilities and assumptions	t_e

PERT uses probabilities, which are beyond the scope of this syllabus, to assume that an average time, t_e, of t_o, t_p and t_m, is a reliable estimate of the time that a task is expected to take. PERT also assumes that, for each task, if the timings are estimated six times, one estimate will be t_o, four estimates will be t_m and one will be t_p. The assumptions are used in the formula to calculate the expected timing, or duration, of a task. The formula multiplies the realistic time (t_m) by 4, adds it and the other two timings, the optimistic time (t_o) and the maximum possible time (t_p), together, and divides the result by 6, to give an average of the times, which is used as an estimate of the time. The formula is usually written as:

$$t_e = \frac{t_o + (4 \times t_m) + t_p}{6}$$

For example, if the optimistic time (t_o) for testing the code of a new software app is 6 days, the most likely or realistic time (t_m) is 12 days and the maximum possible time (t_p) is 15 days, then the calculation for t_e, observing the usual order of mathematical operators, gives an estimated duration of 11.5 days:

$$t_e = \frac{(6 + (4 \times 12) + 15)}{6} = 11.5$$

These calculations can be automatically performed for every task, or activity, in the project by the PMS as the timings are entered and are used for the duration of the tasks when creating a project Gantt chart.

▼ **Table 15.3** Table of timings for a new software app project

Task no.	Name	Predecessor	Time estimates (days)			Expected time (days) (t_e)
			Optimistic (t_o)	Most likely (t_m)	Pessimistic (t_p)	$t_e = \dfrac{t_o + (4 \times t_m) + t_p}{6}$
1	Design the app		2	5	9	5.166666667
2	Code the app	1;4	10	15	30	16.66666667
3	Test the app code	2;4	6	12	15	11.5
4	Create the user interface	1	7	10	16	10.5
5	Release the app	3;7	2	5	8	5
6	Create the advertising	1	2	5	9	5.166666667
7	Advertise the app	6	15	20	25	20

Float

Several additional timings have to be considered when planning a project. As well as the expected duration (t_e), the earliest possible start time (E_S), the latest possible start time (L_S), the earliest possible finish time (E_F) and the latest possible finish time (L_F) are decided for each task and the overall project. **Slack** and float can be calculated from these timings.

Slack and float are often used interchangeably, but they are not quite the same thing. Slack time allows a task to *begin later* than was originally planned, while float time allows a task to *end later* than originally planned. Slack time has no work being done on the task because it is *starting late*, while float involves *work being done* on the task as it *overruns*. However, while the differences are important during the actual project because they affect what people can do, in most project management software, slack and float are not shown as separate entities and the term 'float' is used, as here, to refer to both. Another timing, the lag, is sometimes included as a waiting time between tasks.

There are two types of float: there is **free float**, which is the amount of time that a project can be delayed without affecting the next task that is dependent on it, and there is **total float**, which is the amount of time that a project can be delayed without affecting the finish time of the overall project.

Free float is often not considered as important as total float, because free float can usually be absorbed by the slack time of other tasks. Unless the free float of all tasks exceeds the total float, the project will still finish on time. Free float is sometimes referred to as slack.

Total float can be calculated as the *difference* between the *latest start* time (L_S) and the *earliest start* time (E_S):

$$\text{Total float} = L_S - E_S$$

Total float can also be calculated as the *difference* between *latest finish* time (L_F) and *earliest finish* time (E_F):

$$\text{Total float} = L_F - E_F$$

Free float is calculated as the *lowest early start* time of its following, dependent tasks *minus* its own *early start minus* its *duration*. This gives the amount of time it can be delayed without affecting the next task. Usually, free float cannot exceed the total float for the task.

$$\text{Free float of task} = (\text{Lowest } E_S \text{ from following tasks}) - (\text{Task } E_S) - (\text{Task duration})$$

Determining the float times can be a lengthy process when carried out manually, but most PMS will do the calculations for a project manager.

Critical path calculations

The critical path is used to show the longest path through a project and to calculate the shortest time that a project will take. Adding together (summing) the durations of each task on the critical path gives the shortest possible time for the project. However, critical path analysis can be used for much more than just finding the shortest time for the whole project. Using Table 15.3 and adding the expected durations (rounded to the nearest integer), a PERT chart can be drawn manually. In this PERT chart the tasks are shown on the nodes as boxes with spaces for the earliest start (E_S), earliest finish (E_F), latest start (L_S) and latest finish (L_F).

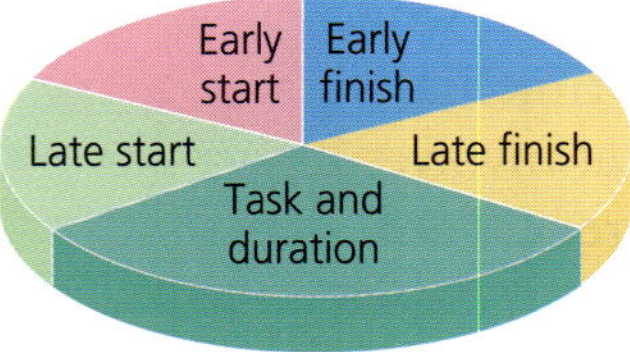

▲ **Figure 15.10** Alternative ways of showing timings and durations on nodes in a PERT chart

Figure 15.11 shows the complete PERT chart and its dependencies with the tasks and durations entered.

Task number	Task	Predecessor	Expected duration (days)
1	Design the app		5
2	Code the app	1,4	17
3	Test the app code	2	12
4	Create the user interface for the app	1	11
5	Release the app to the public	3,7	5
6	Create the advertising for the app	1	5
7	Advertise the app	6	20

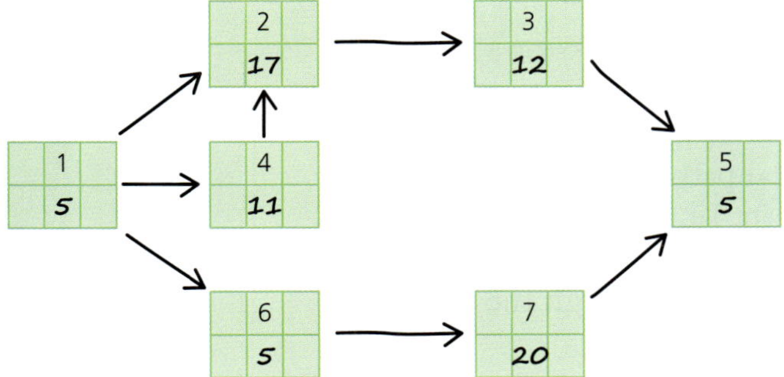

▲ **Figure 15.11** A PERT chart with dependencies and durations from the table

Two passes, forwards and backwards, through the PERT chart are carried out to complete the values for E_S, E_F, L_S and L_F. The forwards pass in this chart goes from the start at Task 1 to the end task, which is Task 5.

» In Task 1, E_S is zero (0) and E_F is five (5). E_S is zero because it is the first task, E_F is 5 because that is the duration so it cannot finish before 5. The values of L_S and L_F are not entered yet. Task 2 was left to last here because it is dependent on two tasks, 1 and 4.
» Task 6 is done next. Task 6 has an E_S of 5 because it cannot start until task 1 is finished with its E_F of 5. Task 6's E_F is 10, the E_S of 5 plus its duration of 5.
» Task 4 also has an E_S of 5, the E_F of task 1, and its E_F is 5 + 11 or 16.
» The E_S of Task 2 has to be 16 from the E_F of task 4 because it cannot start before 4 has ended. When a task is dependent on two or more tasks, the highest E_F of the preceding tasks is always used as its E_S. The E_F of Task 2 is therefore 16 (from Task 4) plus its duration of 17, which is 33.
» The E_S and E_F values for remaining tasks are completed in the same way. Task 5 has an E_S of 45 because it is the higher of the E_F values from its two predecessor tasks.

The resulting PERT chart is shown in Figure 15.12. From the completion of the forwards pass, the overall duration of the project is now known. It is 50 days shown as the E_F for Task 5.

A backwards pass though the chart from Task 1 to Task 5 completes the values for the latest start, L_S, and latest finish, L_F.

» The values for Task 5 are entered first. The L_F is the same as the E_F so is 50. The L_S is calculated by subtracting the duration, here it is 5, from the L_F, so it is 45.

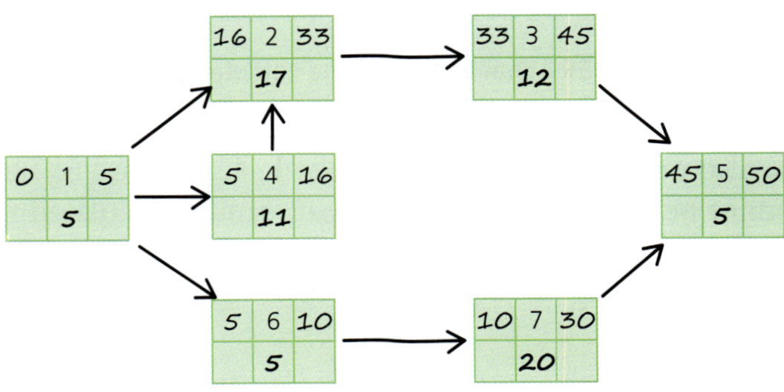

▲ **Figure 15.12** PERT chart showing E_S and E_F values after the forwards pass is completed

» The L_S of 45 is transferred backwards from Task 5 to both Task 3 and Task 7 to become their L_F. For Task 7, the duration of 20 is subtracted to give an L_S of 25. When this is transferred back to be the L_F of Task 6 and the duration of 5 is subtracted, it gives Task 6 an L_S of 20.

» For Task 3, the L_S is 45 minus 12 or 33. Transferring this back to Task 2 and subtracting the duration of 17 of Task 2 gives Task 2 an L_F of 33 and an L_S of 16.

» Similarly, Task 4 now has an L_F of 16 and an L_S of 5.

» For Task 1, there are three values that could be chosen: 16 from Task 2, 5 from Task 4 and 20 from Task 6. Choosing the *smallest*, as this is a backwards pass, gives Task 1 an L_F of 5 (from Task 2) and when the duration of 5 is subtracted, it gets an L_S of 0. An L_S of zero is logical and correct.

From this, the critical path can be determined. If any one of the tasks on the critical path is delayed then the whole project is delayed. Delays in tasks not on the critical path may not affect the whole project. All nodes with equal values for early finish (E_F) and late finish (L_F) lie on the critical path. In other words, if $E_F = L_F$ for a node then the node is on the critical path. In the example, nodes 1, 4, 2, 3 and 5 have $E_F = L_F$ so are on the critical path. The critical path, shown in red on the PERT chart, is therefore this route through the example project:

Task 1 → Task 4 → Task 2 → Task 3 → Task 5

Total float, the amount of time that a project can be delayed without affecting the finish time of the overall project, and free float, the amount of time that a project can be delayed without affecting the next dependent task, can be calculated from the chart. As described previously, float is the difference between the latest start time (L_S) and the earliest start time (E_S) or the difference between the latest finish time (L_F) and the earliest finish time (E_F). For example, Figure 15.13 shows the calculations of total float for the software app PERT chart.

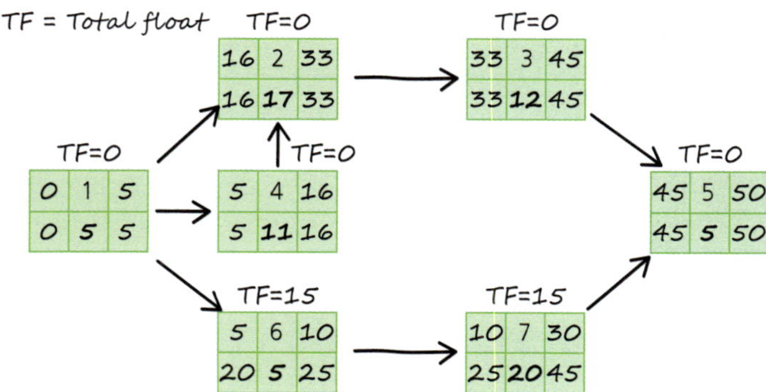

▲ **Figure 15.13** PERT chart showing the calculations of total float for each node

In the diagram, it can be seen that the tasks that have a total float of zero (0) are also on the critical path of tasks 1, 4, 2, 3 and 5. The free float, which cannot be more than the total float, can be calculated from the values in the diagram. Free float is calculated by taking the lowest early start (E_S) of all the following tasks minus the task's own early start (E_S) minus the task's own duration.

Free float = Lowest E_S from following task(s) – Task own E_S – Duration of task

The calculation of the free float for Task 1 is shown in Figure 15.14.

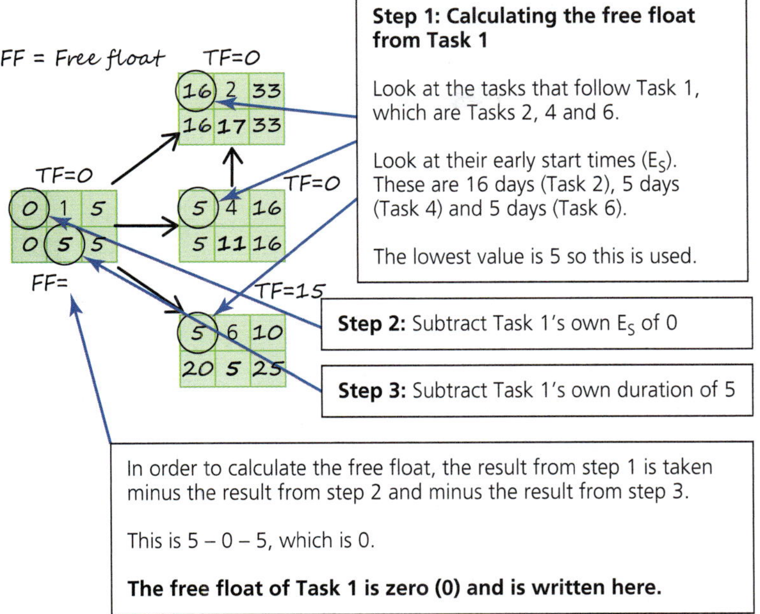

▲ **Figure 15.14** Calculating the free float

The calculations for each of the tasks in the example are shown in this table.

▼ **Table 15.4** Calculations for each of the tasks

	E_S from following task	E_S	Duration	Free float
Task 1	5	0	5	0
Task 2	33	16	17	0
Task 3	45	33	12	0
Task 4	16	5	11	0
Task 5				0
Task 6	10	5	5	0
Task 7	45	10	20	15

Task 5 does not have any following tasks, but it is on the critical path and its free float cannot be more than the total float, so its free float must be zero (0). Project management software automatically does the calculations and shows the critical path, but it is important to understand how the calculations are carried out. Figure 15.15 shows the PERT chart with all the float calculations.

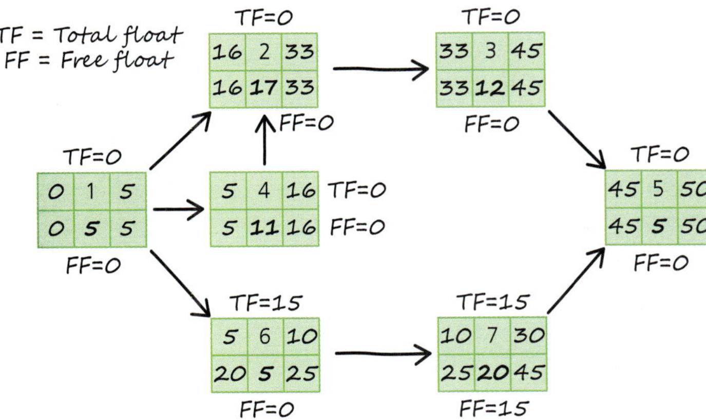

▲ **Figure 15.15** PERT chart with floats

Activity 15i

1 Using a labelled box for each of the nodes and using the table in Activity 15h, draw a PERT chart for the project. Enter the durations and use the chart to calculate the total float and free float for the tasks in the project.
2 From the chart, which tasks are on the critical path?

Use of critical path analysis

Project managers use PERT charts and project network diagrams to identify the critical path through the project. This shows the longest path for task to task and the shortest time that the project will take to complete.

Analysis of the PERT charts and network diagrams of a project using the critical path method shows which tasks can be carried out simultaneously because they do not have to wait for other tasks to be finished and shows which tasks are dependent on other tasks. A project manager can allocate resources, for example a team of programmers, to the tasks according to the dependencies. If several modules of a new software app are needed before the user interface can be coded, the team of programmers can work on the modules and finish them before starting work on the user interface. However, if work on the user interface does not depend on the modules, work on it can start at the same time, but the project manager will need to employ more programmers and allocate them and other resources to the user interface work. This may add to the costs of the project but may also reduce the time for completion. Analysing the charts and using CPM can allow a project manager to investigate the different project scenarios. For example, is the extra cost worth it, or can the project be allowed to take longer to reduce costs?

The project manager uses the early start (E_S), early finish (E_F), late start (L_S) and late finish (L_F) timings for each task to calculate float times and uses these to monitor and manage the project and to understand the effects and implications of any changes or delays to the tasks in the project. The values of the free float and total float are used to determine whether or not delays to tasks will affect the following task or the whole project. Project managers can allocate or move resources accordingly to ensure that the project keeps to its planned timelines and schedules.

Sometimes, a project manager decides or is required to complete a project before the critical path analysis indicates that it is possible. There are many reasons why this could happen, for example because a client demands the project be ready sooner than planned, or an important team member has to move to another, more important, project. The project manager has to take 'crash action' and try to re-plan the project by altering the durations, costings and other resources for each task to reduce the overall time for the project. This may cause the critical path to change to include different tasks. The changes and their effects can be seen almost immediately by the project manager.

Workflow control

Workflow is a structured and repeatable set, or pattern, of tasks or activities. A workflow consists of the steps required to complete an activity, the resources needed to complete the activity, the team members responsible for the activities and how the activities interact with each other. Figure 15.16 shows an example workflow for creating the user guide for a new software application.

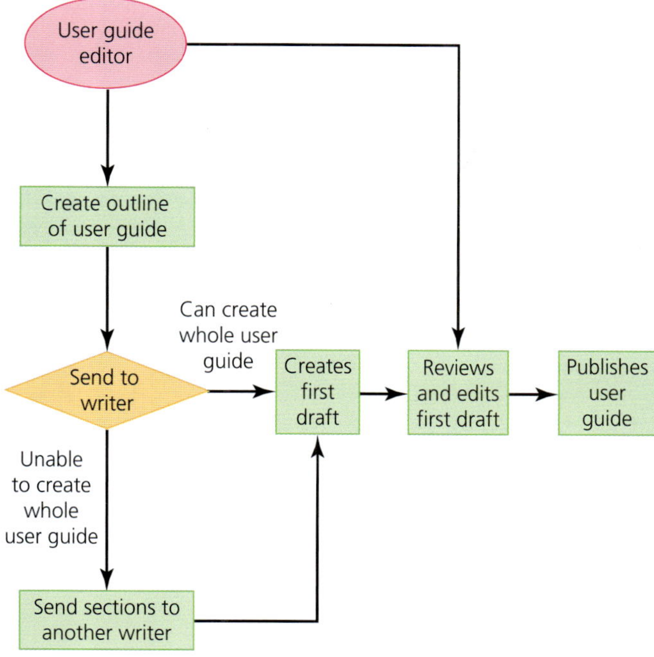

▲ **Figure 15.16** Example workflow diagram

Creating a workflow for the tasks in a project, and for the activities within tasks, shows the critical path through each of the individual tasks. Analysis of the path shows where and when resources should be allocated, and can be used to calculate how much the activities will cost. Adding automatic alerts and reminders into the details in the PMS, linking these to diary-management applications to alert managers and team members, for example to forthcoming milestones, and automatically sending orders for resources can help to ensure that projects progress on schedule and end on time.

15.3.3 Advantages and disadvantages of the tools and techniques

Advantages and disadvantages of Gantt charts

The use of Gantt charts to represent tasks, sub-tasks, activities and milestones in a graphical layout provides a visual overview of a project's schedule so that timeframes and dates are clearly visible. Project managers can see the schedule by day, week, month and year, which enables them to manage their teams more effectively and to allocate resources as required. The status of a project can be easily checked as the chart can show the percentage of the tasks that have been completed, tasks still in progress and those yet to be started. Gantt charts are also useful when making presentations to teams or other stakeholders to show how a project will progress or is progressing. However, Gantt charts require skill and effort to create and manage, and it can be very time-consuming to keep updating the chart. Gantt charts do not show all the details of each task and may be difficult and time-consuming to amend or recalculate if tasks need to be changed.

Advantages and disadvantages of PERT charts

Using a PERT chart provides an overview of a project so that the project team can identify their roles and any bottlenecks or restrictions can be identified before work starts. PERT charts enable the study and analysis of individual tasks to determine what activities are to be carried out in that task and what resources might be needed for each activity within the task. Critical paths can be determined by using PERT charts to show the minimum time required for a project, which enables deadlines to be set and monitored. This, in turn, means that clients will get their products on time. PERT charts can be changed to investigate alternative pathways through projects, to discover if time, costs or resources could be utilised better.

PERT charts focus on time management and may not be appropriate for all projects. If a project is very large and complex, PERT charts can become too large and difficult to manage and take up managers' valuable time. If new tasks need to be added or changes made to the product during development, a PERT chart can be time-consuming and difficult to amend. PERT charts are not suitable for managing projects that share resources with other projects because it is difficult to combine the details of all the projects into one chart.

Advantages and disadvantages of CPM

CPM determines the timings and duration of a project, which enables managers to schedule the use of resources and personnel, monitor tasks and so control costs during the project. Project managers can calculate the time needed to complete tasks and this helps them to predict the completion dates of the tasks and of the whole project. Any delays or problems can be anticipated and managed by monitoring the critical path and re-allocating resources or personnel to ensure that the project is finished on time.

Using CPM can be time-consuming and confusing if a project changes during its progress because PERT charts and pathways may need to be redrawn and times recalculated. Determining a single critical path may not be possible if a project contains tasks that are very similar in duration, so project managers may not find the use of CPM helpful. CPM does not allocate resources or personnel, so a project manager has to be skilled in the interpretation of the results of CPM.

Practice questions

1 Describe the activities that take place when scoping a project. (8)

2 Explain why acceptance criteria are important. (6)

3 a Explain the difference between total float and free float. (6)

   b Explain why tasks on the critical path should have a total float of zero (0). (3)

4 Discuss the benefits and drawbacks of using project management software for managing a project. (8)

5 What are the drawbacks of using the critical path method when managing a project? (6)

6 Complete the diagram by copying and filling in the project stages and using arrows to show the flow of the project. (4)

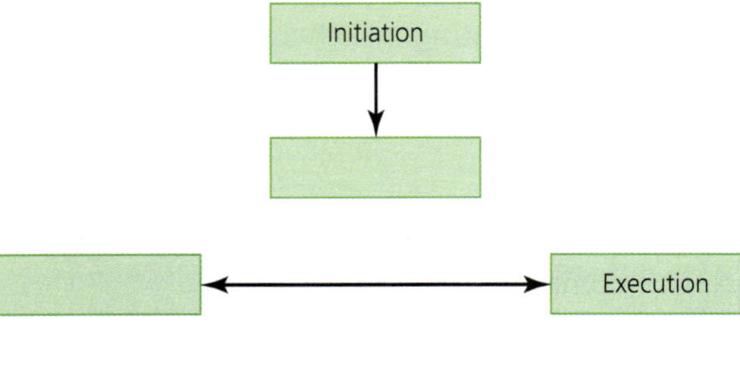

16 System life cycle

In this chapter you will learn:

- ★ the stages in the system life cycle
- ★ how a systems analyst researches and gathers information
- ★ the specifications needed to create a new system or software application
- ★ how designs for new systems or software applications are created
- ★ how a new system or software application is developed and tested
- ★ how a new system or software application is implemented
- ★ the types of documentation that accompany a new system or software application
- ★ how a new system or software application is evaluated
- ★ how a new system or software application is maintained
- ★ about prototyping and the methods of software development.

Before starting this chapter, you should:

- ★ be familiar with the terms: client, developer, programmer, end user, specification, verification and validation.

A **system** is a dependent collection, or group, of functional elements, items or processes that interact together to form a complex unit or structure. In information technology, an information system is one that is able to gather (or collect) information, and process, store and distribute the information.

A computer system can be the collection of hardware that forms a desktop computer or a laptop, or it can be the software that runs on a computer, or it can be a combination of both hardware and software. For example, a web server system is the physical hardware and the web server software. In the wider world, a railway system could be the entire rail network including the physical tracks, the rolling stock, the computer systems that control it and the people who work on it. Because of the wide definition of the term, when a new system is being designed and created, the system itself has to be carefully defined.

16.1 The stages in system life cycle

In a **system life cycle**, planning is followed by the analysis, design, development and testing, implementation, evaluation and **maintenance stages**. Careful planning before the actual development work is started, thorough testing and evaluation with accurate documentation, followed by ongoing support and maintenance will make the new system a technical and commercial success. Each stage depends upon the results or product of the preceding stage. For example, the design stage cannot start until the results of the analysis and the **specifications** are produced. System documentation is produced during each stage to record and support the creation of the new system. Finally, the results and findings of the maintenance of the new system are used to add to the knowledge and understanding of the planning stage so that designs and development of other new systems can be improved or amended. The relationship between the stages is shown in Figure 16.1.

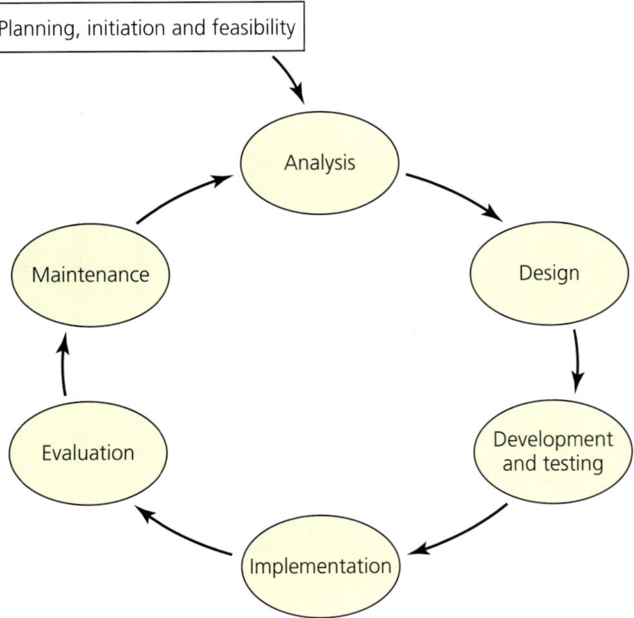

▲ **Figure 16.1** The stages of the system life cycle

Activity 16a

1 Explain why the stages in the development of a new system are shown as cyclical.
2 Explain why a smartphone is considered to be a computing system.
3 List as many of the components of an air traffic control system at an airport as you can.

16.1.1 Preliminary work

Planning and feasibility

Planning occurs before the actual development of a new system is started. The planning stage is similar to the initiation and planning stages of a new project. The resources, costs, timings and other requirements are considered to find out whether it is possible, or feasible, to create the new system or whether it would be better to replace an existing system; whether the new system will be an improvement; and whether it is even worth creating.

A **feasibility study** is carried out to assess whether the system can actually be created. Several factors are considered during the feasibility study. A feasibility study analyses the technical ability of the company or business that intends to develop the new system, checks the legality, for example whether it abides by data-protection laws, and checks whether the company can afford it and will make a profit.

A feasibility study is documented in a report describing an existing system, if it exists, what it is and what it does, outlining any problems with the existing system, outlining what the new system must be able to do and suggesting any alternative solutions to the problems. Only when management is satisfied that the feasibility study has completely evaluated the potential for a successful development will the development of a new system be allowed to proceed and be fully planned. The planning stage initiates the project for the new system, sets out the schedules for the development, and the 'life cycle' begins.

16.2 Analysis

The **analysis stage** gathers information by carrying out fact-finding investigations during the design stage. The information is gathered by systems analysts from individuals or groups using questionnaires, interviews, observation and document analysis. If a system already exists, it is investigated to find out exactly what problems it has. If the system does not yet exist, the analysis stage finds out exactly what is required of a new system. Both problems and requirements are researched when a system is to be replaced with a new one. The gathered information and the planning documentation are used to create user requirements and the system and design specifications.

The systems analyst is responsible for:

» gathering the facts and information about the existing system and its problems
» collecting the views and opinions of the users
» defining and understanding the requirements of the users
» suggesting alternative solutions to problems and choosing the best solution
» calculating the benefits and the costs of the solution
» producing the specifications.

The analyst also has to ensure that the specifications are precise and detailed but can be understood by the developers and user, and has to plan the evaluations and maintenance of the new system. To be able to carry out these responsibilities, a systems analyst must have interpersonal skills, analytical and technical skills, and good management skills. These skills help to ensure that each stage of the system life cycle successfully delivers its data and products so that the following stages can go ahead.

Activity 16b

1 Explain why companies carry out a feasibility study before starting a project to create a new system.
2 Explain why the analysis stage in the system life cycle is so important.

16.2.1 Methods of research

Several methods of gathering information are available to systems analysts.

Questionnaires

Questionnaires enable the gathering of information from many people in a shorter time than, for example, interviewing people. Questionnaires can be printed and handed to people or made available online. Printed questionnaires can be completed almost anywhere but are not always returned to the questioner. Unlike during interviews, printed questionnaires can be made anonymous and people are often more honest or open in their answers when they are not identified. Online questionnaires require the respondent to have access to a computer device, for example a laptop, and this may not be convenient or possible in some working environments. Some workers may have access to computers at work and do not want to use their own time and devices for answering questions about their jobs.

A questionnaire can use either closed or open questions. Closed questions, for example multiple choice or selections, are easier to analyse, but they must include the choices that are most relevant and important. Online questionnaires can have customised questions for particular groups of users, and questions can be set to appear only when certain questions have been answered. Also, the route through the questions can be customised and changed for each user depending on their answers or responses. No two questionnaires gathering information from a group of workers need be exactly the same and this can avoid collusion between respondents. The results from online questionnaires can be analysed automatically and made available very quickly to the analyst. The numbers of questionnaires available can be increased or decreased if the size of the audience changes.

Using questionnaires is less expensive than other methods of gathering information because the costs associated with the time, and possibly the travel, of the interviewers, document readers or observers are much reduced. After the questionnaire has been prepared and is ready for use, few people are required to hand out and collect the copies. The results of a questionnaire can be collected more quickly and entered sooner than those of interviews or observations and, for online questionnaires, there is no need to enter them into a computer system. Questionnaires can be targeted at specific groups of people, employees or users, who can fill in and return them in their own time and when it is convenient.

Questionnaires, however, have disadvantages compared to other information-gathering methods. If the questionnaires are anonymous, follow-up research on individual responses is not possible and some respondents may not provide honest answers. Anonymous comments may be deliberately dishonest and may have a bias or hidden agenda that is difficult to discover. Some questionnaires may not be fully completed or may not be returned, so the data collected will be incomplete.

It is not possible to precisely customise specific questions to individuals, as in an interview, nor can explanations be requested or additional questions asked to clarify answers. There is a complete lack of any facial expression or body language information for the analyst to use that might help interpret the answers.

Questionnaires may be difficult to answer if the respondent has disabilities. Apart from the physical difficulty of completing a paper or online questionnaire, cognitive impairment may make it difficult to understand the questions and formulate the answers. If the person receives help and suggestions in making their answers, then the validity of those answers may be in doubt.

Interviews

Interviews can be either structured or unstructured, formal or informal. To gather the information that is required and be successful, the interview depends upon the skill of the interviewer. A structured interview has a set of standard questions that the person being interviewed has to respond to in either an open and descriptive or a closed and objective form. An open, descriptive form allows the person to talk about the question and elaborate their answers as they wish, but a closed, objective form of questioning restricts the answers. In an unstructured interview, the interviewer acquires the information by asking questions without necessarily having a set order. The questions depend on the interviewer being able to decide the questions as the interview proceeds.

Interviews have advantages over other methods of gathering information. Interviews are able to collect information from people who are not able to communicate effectively in writing or from those who may not have the time to complete questionnaires. They can collect qualitative information, can deal with complex topics or subjects and allow follow-up questions immediately, so that any problems can be identified or topics explored quickly and in depth. Interviewing can minimise misunderstandings in the answers because the interviewer can ask additional questions or the user can elaborate if required.

However, interviews have disadvantages too. They are time-consuming for both the interviewer and the person being interviewed. Interviews can be very long because the interviewer and interviewee may take time to discuss the questions and answers. The interviewer has control over the interview and needs the skills to determine and direct when it is time to move to the next question or terminate the interview. If a large workforce has to be interviewed, this can take a long time.

Face-to-face interviews can be expensive to set up and conduct as travel and timings may make the meetings difficult to arrange. Also, the information gathered may be biased and untruthful but, on the other hand, this can be checked and questioned at the time of the interview. Some people may not wish to be interviewed at all so their knowledge and opinions are omitted from the collected information.

The quality of the data collected by interviewing depends upon the skills of the interviewer to extract the information from the interviewee. The interviewer might be biased and this may affect the information collected, and the sample size may not be large enough if all the users or workforce have not been interviewed.

To overcome some of the disadvantages, telephone interviews can be conducted. This is cheaper than face-to-face if the interviewer and interviewee are located some distance apart, is often quicker to arrange and conduct, or the interviewer can simply make the call when it is convenient.

Group interviews can also shorten the overall time taken to interview large numbers of people. However, in group interviews some people may not respond, may not get the opportunity to respond, or may influence others and introduce bias into their views and answers.

In all interviews, especially where there are groups of people being interviewed together, the interviewer has to carefully record the information. This can introduce inaccuracy and bias if the answers are not recorded accurately. Using voice recording or conducting telephone interviews that can be recorded is a method of reducing inaccuracies in the recording of the information from interviews. However, the major disadvantage of interviewing is that the information gathered has to be transcribed into a format that can be entered into a computer system for analysis. Often, information obtained from interviews is qualitative and not in a format that can be readily analysed by computer software.

Observation

Systems analysts use **observation** to watch the workings and requirements of the current system. Information is gathered by observing the employees, for example workers and management, and the events and objects that make up the system to find out:

- how tasks and jobs are carried out
- how many employees are involved in the tasks
- how long the jobs or tasks take to be completed
- whether there are any problems with the tasks
- how data is entered into the system
- how the results, for example reports, are output from the system.

Observations of a system can be carried out either directly by watching and following employees as they carry out their tasks or indirectly by using video cameras to watch remotely. Sometimes, both methods are used.

Using observation to gather information about a current system has a number of advantages:

» Some tasks may not be fully documented, so observations can reveal problems that may not be obvious from analysing documents.
» The observer can see first-hand exactly how the task or process is carried out.
» The observer can quickly understand how the task or process works and how the employees complete it, or whether they have difficulties with it.
» The observer can make accurate measurements of the times taken by employees.

The disadvantages of observation, however, include the effects on the performance of the employee or employees being watched. Often, employees find ways to carry out tasks that work better for them but are not quite how the company wants the task done. When being watched, the employees may not carry out the tasks in the way that they normally do but may stick rigidly to the set procedures. This makes the observations unreliable as a record of the employees at work. Also, observations take up much of the observer's time, and the observer may even hinder the employees by being in the way or asking questions. The use of indirect observation, with remote video cameras, may involve many hours, or days, of watching multiple recordings of tasks and processes and can take a great deal of time to do.

Document analysis

Document analysis is used by systems analysts to discover how data moves through a system. The system may be a database management system (DBMS) or it may be the whole organisation that is analysed. In a DBMS, data is entered, processed and output. For a whole organisation, there is documentary evidence of how data arrives and gets into the organisation and how it leaves, for example letters, money and resources come in, and reports, profits and products leave. It is not usually possible to gather all the details of this type of information by questionnaires or interviews, but document analysis shows how data is collected, entered, processed and reported as it passes through. In other words, the systems analysts follow the 'paper trail', although many of the documents are electronic records.

The documents studied by an analyst include input screens and reports, calendars, purchase orders and receipts, records or transactions, queries on data to create reports, spreadsheets, employee work records, project scopes and schedules, and all other relevant documents that relate to the system being analysed, for example error reports and maintenance reports. For a whole organisation, all the documents listed may be analysed, but for a specific system, for example a database management system, only those that relate to the operation of the DBMS will be analysed. In the context of information technology, it is the documentation relating to a new IT system that is studied so that the flow of data through the system can be discovered.

The main advantage of analysing documents over other methods of information gathering is that it reveals the flow of data through a system. Also, the type of data and its formats as it moves around the system can be deduced.

However, the analyst may not be made aware of any data that does not appear in the documents, for example some data may not be entered separately because a data-entry clerk is required to enter the overall total from an invoice instead of the cost of each separate item. For this reason, document analysis may not reveal small details about the data.

Activity 16c

Select, by copying the table and ticking the correct box, which research method would be the most suitable for gathering information from each group of people.

Scenario	Questionnaire	Interview	Observation	Document analysis
Customers who eat at a restaurant				
Workers in a factory assembling computer monitors				
Managers in a bank				
End users of an accounting system				
Delivery drivers of a parcel-delivery service				
Designers of a new set of system reports				
Borrowers of books from a library				

16.2.2 Specifications

After the information about a current system has been gathered and analysed, it is used to create the user requirements specification, the system specification and the design specification. These documents are created at the beginning of the design stage.

User requirements specification

A **user requirements specification** sets out the details of an agreement between the user and the development team about what the new system will do. The user is not necessarily the same as the user who actually ends up working with the system; in this context the user is the client who wants the new system to be developed. However, in some cases the two are the same, as the people ordering a new system may be those using it for work. In the system life cycle, it is important to be able to distinguish between the two: a client is who the system is being developed for and the user is who works with it. For example, the client for a new database management system could be the company that runs a car rental service, and the user would be the service agent at the depot who takes customer details and assigns the car.

The development team is usually represented by the managers of the company developing the system rather than the actual programmers. The client and developers negotiate and agree on what will be developed. A user requirements specification can be the basis of the planning, scheduling, costing and procurement of resources for the development process.

The user requirements specification becomes part of the final contract between the client and the developers and is made available to other stakeholders. It ensures that the client cannot demand features that are not in the specification and the developers cannot claim that the new system is complete unless all the features mentioned in the specification are included. This, obviously, does not prevent the client and developers negotiating changes or additions at a later date.

A user requirements specification is a definition of what will be produced and it is used at the testing stages to check whether the new system is doing what it is meant to do and what the client wants it to do. The specification will state what is to be produced and why it is needed. A user requirements specification must be as detailed as possible to ensure that the client's requirements are properly met, and the details must be measurable so they can be checked at testing.

The contents of a user requirements specification include an overview of what the client wants, specifying what the new system should do and how it should do it. A description of the current system and its problems is included in the overview along with the defined purpose of the new system. The purpose of the new system is set out in detail and would be to replace and solve the problems with the old system. The purpose would include how the users would be affected and how they would benefit from the new system. For example, the new database management system for a car rental service would provide the service agents with more detailed and up-to-date information on the whereabouts and availability of the range of cars, allowing them to assist potential customers more quickly.

The location of the new system would be specified if it were a large, whole new system for a company that had a number of offices. If it were a software application, the location could be specified as local or stored in the cloud and the cloud service provider may also have to be specified.

In the user requirements specifications, the details of what exactly the system has to do are stated (the functional objectives). This is the goal statement. For example, if the new system was to be a website for online sales on a global scale, it might have to provide customers who visit the site with up-to-date and accurate product information, provide customised web pages to different customers depending on what they are looking for or their preferences, provide an artificial intelligence-based assistant, provide a secure system for creating accounts for ordering and paying for goods and, possibly, an order history service.

Where a new system is to be developed for a company, for example a new DBMS for a car rental company, the user requirements specification describes the new system in the context of the company and often details how each part of the company relates to the new system and how they must be able to use and benefit from it. The relationship between the people and company departments (called the system externals because they are not part of the actual system but use it) and the new system can be shown as a context diagram.

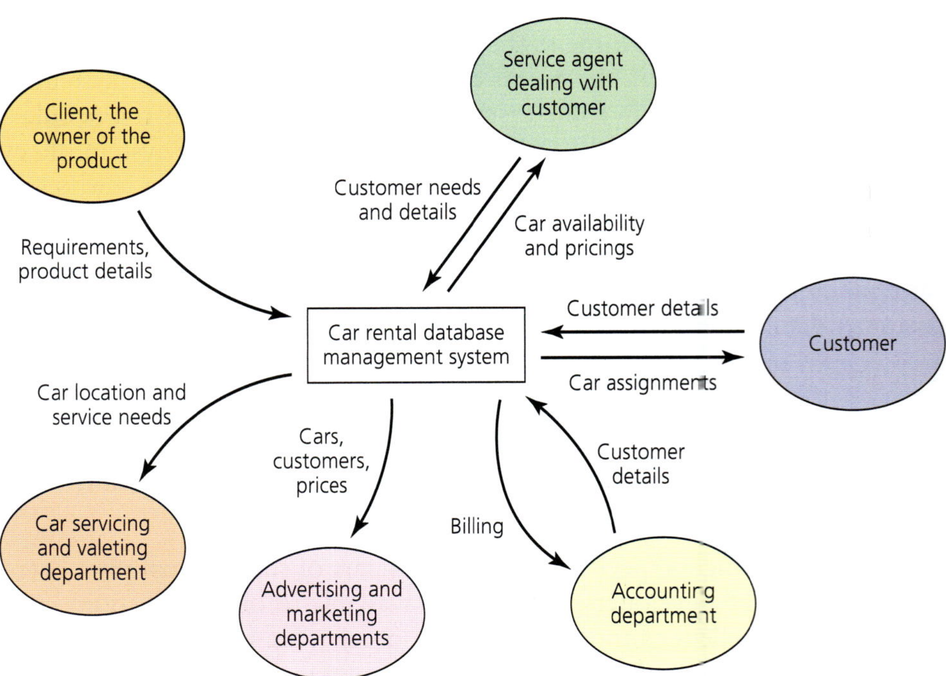

▲ **Figure 16.2** A context diagram for a new car rental database management system

The client is the owner of the system and is the person, or company, that ordered or commissioned the new system from the developers. The client may wish to amend the requirements during the development stages or after the deployment, and this is negotiated with the developers. The service agent is an actual user of the system and is an employee of the client or owner. In the car rental database example, the service agents access the system to input customer details and search for suitable cars for rent. The customer is any user of the system who is not an employee of the company but may be able to, for example through websites, search the database for cars to rent. The accounting department uses the database to bill the customer for the car rental and the marketing department may use customer details, car availabilities and prices for advertising or targeted marketing purposes. The car service department uses the database to determine when and where cars need maintenance or cleaning. These details, customised for each new system, form part of the context in which a new system is to be developed and later used.

> **Activity 16d**
>
> Why is a user requirements specification created?

System specification

A system specification details and describes the functionality and behaviour of a system. It describes everything that the new system should do and how it should behave when doing it. It also describes the data that will be used by the system. A system specification is intended for use by the developers, who base their designs and production on the descriptions in the document, and for the client and users of the new system, who review the document to ensure that it accurately describes what is to be developed.

The technical terms used in the system specification should be clearly defined so that the client and developers know exactly what is being described. Not all clients are skilled technically so may need help in defining and describing what, for example, a new software app must do.

An overview of the system, using the defined technical terms, is included to describe what the new system will be and how the new system will meet the user requirements to solve the problems found in the current system. In this section, the functionalities of the new system are described so that developers know what to produce.

The types of end users should also be included. In general, the end users will be the system maintenance technicians, the company employees and public users, each of which will have different access rights to the data. These will be specified.

The operating environment is specified. This states what hardware and operating systems a new software application requires and will be able to run on. Hardware and software requirements are specified that enable the new application to function correctly and interface with other software and allow the new application to be maintained.

A system specification also includes details of the features and functions that are to be included. This would specify the performance and security levels required, detail the functions and features required by each type of end user, and describe in detail exactly what tasks, for example, the new database must be able to do.

Also, the interfaces with the new system must be detailed. A user interface that is easily understood and logical to use by end users would be described in precise detail along with any interfaces with other software applications. A new application has to be easily accessible by end users but also has to run alongside, and interface with, other applications (such as if it is accessible over networks, web browsers using HTTPS, query languages such as SQL, FTP services) and other communication protocols.

The system specification also details the non-functional requirements, for example security requirements including authentication methods for access by end users, data integrity, confidentiality of data and system availability. Performance requirements, such as the target time taken to respond to end user requests for data and reports, and any error messages that alert end users to delays, problems or other issues when accessing data, are also specified. A detailed description of how the new system will be kept safe is often included, for example how protection from power loss is to be arranged.

The system specification will also contain details of how maintenance and upgrades are to be carried out.

Design specification

The **design specification** is usually produced at the same time as the system specification, towards the end of the analysis stage. It also covers how the system will work and look and how end users interact with it, but it is intended for the developer, or developers, to follow in order to create the new system exactly as required. A design specification enables the developers to actually start work on creating the new system.

The hardware and software to be used is specified. The computer hardware needed to develop the system is specified so that the developers all use the same computer environment, and the software to be used is specified to ensure that they use the required programming language. This ensures that the final product will run in the computer environment for which it is required by the client.

For a new database system, a design specification would contain all the details about the inputs and how they are used, for example radio buttons, dropdown lists, or how to upload images of customers or their ID documents. It would also contain the details and arrangements of screen layouts, such as the colours, font styles, information text, margins and shadings that are to be used. The layout and design of tables, and the relationships between tables, are specified. Also included are details of the processing that is required. Validation rules and any required error messages are detailed. Outputs are detailed and designed. The layout of screen displays, reports and printed materials are specified so that the developers can ensure they are exactly as the client requires.

At this stage, diagrams are created that model the system. These include system flow diagrams and data flowcharts. A **system flowchart** shows the flow of control through a computer system and where the decisions are made that affect the flow. Data flowcharts show how data enters, flows around and exits a system.

The design specification describes what has to be produced but does not necessarily explain how it has to be created. It is usually left to the developers to decide how to create the product using the design specification to guide them. For example, the design specification may state that a new database has to accept the accurate input of customer details by a service agent, but it is the developers who decide how they will write the code for this feature.

16.3 Design

The **design stage** follows the analysis stage using information gathered and analysed by the systems analysts.

16.3.1 System processing

During the design of a new system, the flow of work through the system and the processing of the data as it flows through the system from input to output are decided. Flowcharts are used to display, analyse and document a system. A system flowchart shows the workflow and a data flowchart shows how data enters, flows around and exits a system. A flowchart can be used to represent any workflow process. Flowcharts are drawn using a set of agreed symbols. For this syllabus, the symbols that must be used when creating system flowcharts and data flow diagrams are listed in the current syllabus.

Name	Description of how symbol is used	Symbol(s)
Input or output	This rhomboid shape represents where data is input, for example entered by typing or being scanned, or output, for instance displayed on screen or printed.	
Process	This shape represents where a process, action or a function is carried out.	
Single document type	This shape represents a single document that is printed or appears as a report.	
Multiple document type	This shape represents more than one document.	
Magnetic disk file	This cylindrical shape represents stored files that are accessed by direct access, for example on a magnetic disk or solid-state disk. In some charts, the symbol is drawn on its side.	
Magnetic tape file	This represents a file stored and accessed by sequential access, for example on magnetic tape, such as a backup file or files.	
Display	This shape represents a step that displays information.	

System flowchart

A system flowchart illustrates how an entire system operates by showing how the work flows between all the operations or processes in the system. It also shows the relationships between all the inputs, processing and outputs, and where documentation has to be created. System flowcharts have additional symbols.

A system flowchart is meant to inform developers about the design of a system so it should use the standard symbols in order to be easily understood, should not make excessive use of colour as this can distract from the information, the sizes and the spacings of the symbols should be consistent so as to appear neat and easily readable, and branch directions should be consistent. Decision symbols can be larger as they often need more space for labels, the connectors can be used to avoid cluttering the chart with lines and it can be spread over several pages rather than reducing the scale of the chart to fit on a single page.

A system flowchart for a new system for a car rental business is shown in Figure 16.3. It shows how the different operational processes are related and interact.

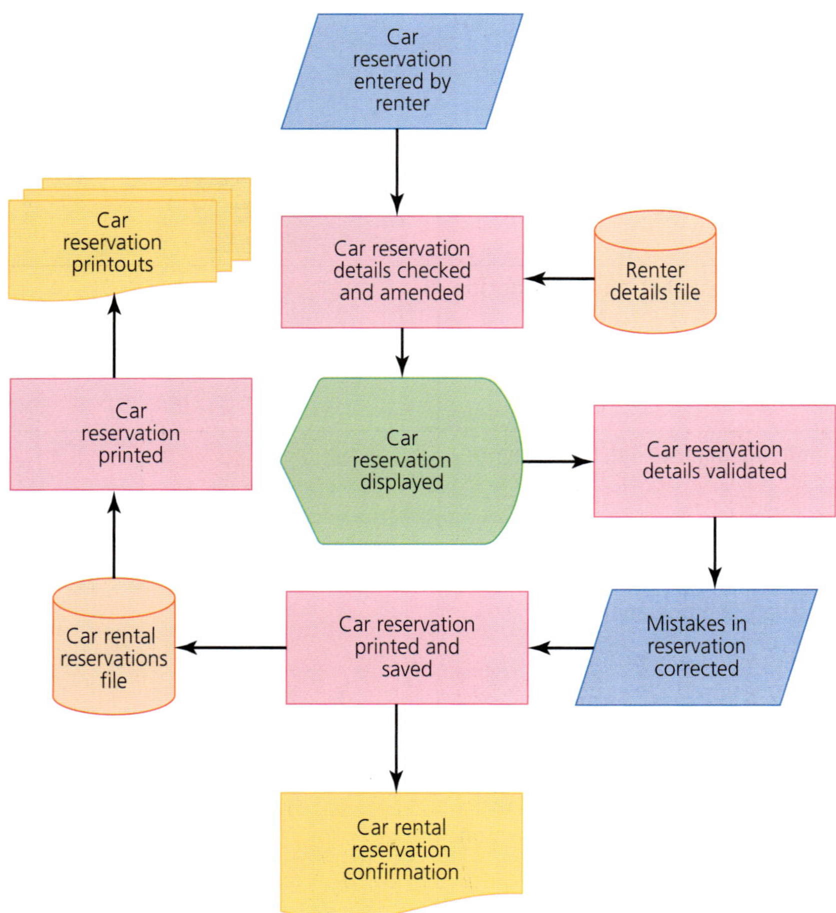

▲ **Figure 16.3** System flowchart for an online car rental system

It starts with the interactions between an agent and a customer and ends with confirmation, or not, of a successful rental of a car. The chart shows an overview of the system and each component would be broken down into a more detailed flowchart before the developers started work.

Activity 16e

1 Using Figure 16.3, which shows the system flowchart for an online car rental system, copy and add to the diagram to indicate where output in the form of hard-copy documents would be produced.
2 Draw a system flowchart, using the symbols shown in this textbook, for when a new user, who does not have an account, tries to log in to an online account.

Data flow diagram

A **data flow diagram (DFD)** shows how the data flows through a system. The inputs, processing and outputs of each entity within the system are represented, but the control of the flow of the data is not shown. DFDs are used by developers when designing a new system to ensure that the data flows between processes so that the new system performs as the client requires.

DFDs can be very detailed and difficult to understand so there are different levels. The levels start at zero, which shows an overview. Subsequent levels show more and more detail until the full data flow is shown. The number of levels depends on the system being shown. For example, a level-zero DFD for a car rental system shows the main areas as blocks.

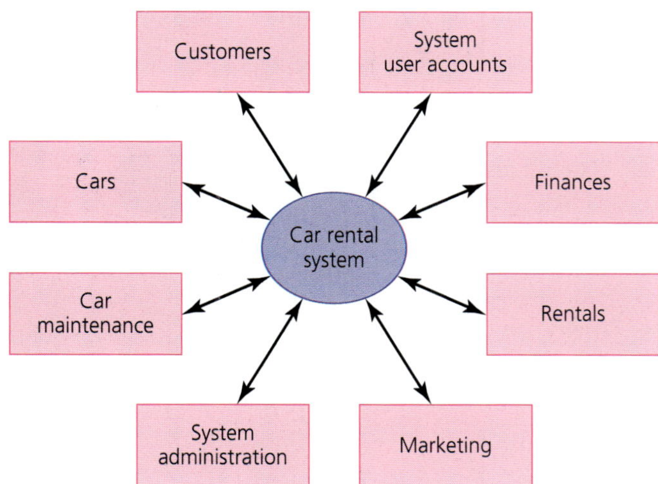

▲ **Figure 16.4** Level-zero data flow diagram for a new car rental system

As the levels increase and show more detail, more symbols are required, for example to show data stores and processes. The symbols used in DFDs for this syllabus are shown in the current syllabus arrangements.

Name	Description of how symbol is used	Symbol
Process	This is where data is changed, or manipulated, from an input to an output. It is usual to shown the name of the process in as few words as possible to express what it does so that readers immediately understand the process.	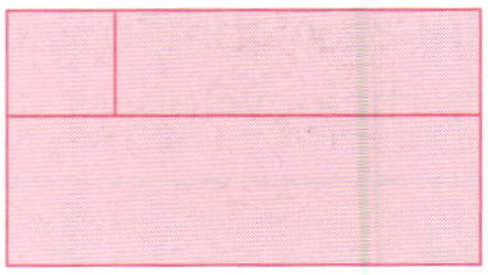
Data store	This is where data is stored for later use. The store can be a datafile, a folder with documents or any other means of storing data. Data stores can be given a reference to indicate whether the store is made up of a computer file, a temporary file that is deleted after processing, or a set of documents. The same data store can be shown several times on a DFD and this is indicated by a double vertical bar on the left-hand edge.	
Data source or destination	This shows where data enters or leaves the system. It could be an entity that is external to the system being developed.	
Duplicate data source or destination	This allows the same source or destination to be shown more than once on a complex DFD.	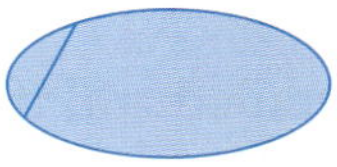

Figure 16.5 shows a level-2 DFD for the car rental system using the symbols in the table above and, as more detail is added and the diagram is completed, the level increases until every aspect is covered.

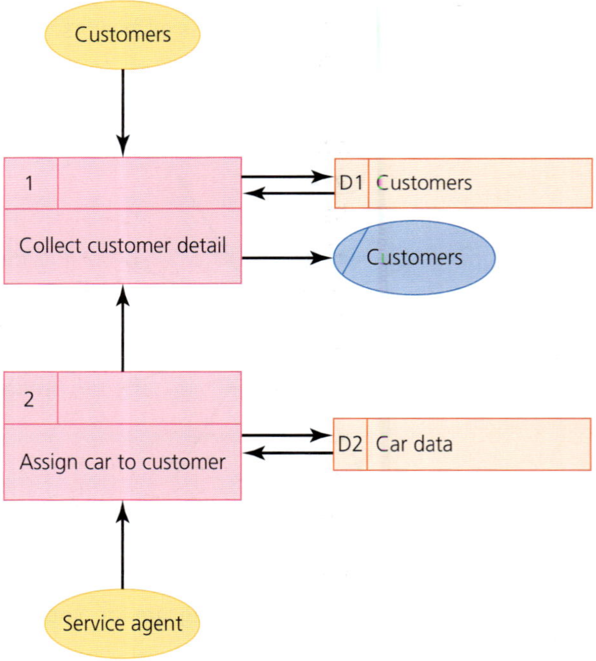

It is important to remember that a DFD is not a flowchart. Flowcharts show how the flow of data is controlled, while DFDs show where data goes. Decisions and loops are not shown in DFDs; only locations where data is transformed. Most designers start the creation of a DFD by identifying the main inputs and outputs and working from these inputs to the outputs. Some designers work from outputs to inputs; it does not really matter as long as the transformations are shown along the way. They then refine it to show more detail. Once an arrow has been drawn to show where the data goes, it can be labelled with information about what data is being moved. When the DFD is completed, many designers draw alternatives to ensure that they have included and shown all the details they require.

System flowcharts and DFDs are the logical design charts. If a physical device is being designed, for instance a new smartphone, there would also be the physical design charts and drawings to consider. In the car rental example, only the logical designs would be required. When the flowcharts and diagrams are complete, they are reviewed, and when found to be satisfactory, are made available to the developers for use during the work.

▲ Figure 16.5 Level-2 data flow diagram for a new car rental system

Activity 16f

Draw a level-zero (0) data flow diagram for a system in a coffee shop where customers can order drinks. The drinks are prepared in a kitchen area by the staff and reports of the orders are sent to a manager. The manager orders the ingredients from a supplier.

16.3.2 Data storage

In a computer system, data storage is required for data that is used for the processes or data produced by the processes in a system. In the context of an information system, data can be stored in any form that is retrievable for use at a later date. Computer files, for example a database, or documents stored as files or on paper as hardcopy, are data stores. Some data stores are 'persistent', which means that they are stored for some time, such as a database, and others are only temporary, for example a temporary file created, used and deleted by the system when printing a document. Systems can create temporary files for many reasons, but common examples include printing, editing documents and the use of 'swap files' to make more efficient use of memory in a computer device.

The use of temporary and persistent data storage has to be considered when a new system is designed. Designers consider what temporary files are to be created by which processes, where they are to be temporarily stored and what medium is to be used. When considering persistent storage and temporary storage, the size of the files, the type of access (serial or direct) and the characteristics of the storage devices to be used are considered. It is also important to consider the safety and security of the data during storage. While the safety and security of persistent storage is important, temporary files must also be kept safe and secure and properly deleted when no longer required.

Databases

Database concepts are covered in detail in *Cambridge International AS Level Information Technology*, but the structure and the data entry and output from a database in a new system are considered and designed in every detail during the design stage of the system life cycle. Failure to properly plan and design a database in a new system could make the system fail to work correctly or even at all.

As DFDs are created, entities are decided and entity relationship diagrams are created. Every detail of the tables, fields, relationships between fields and how the data is entered, stored, processed and output is carefully planned and designed at this stage. Software developers and programmers work to plans and designs included in the design specifications, which are used with the detailed user requirements and system specification when developing and programming the new system. The developer and programmer follow the designs and specifications and create what the client requires.

Input and output

When designing the input to a database, or any new system, the systems analyst uses the specifications to decide:

» what data has to be input
» how the data must be arranged
» how the data must be coded or represented

» the instructions and dialogue that have to be shown to users when data is entered

» data that requires verifying as it is entered

» data that must be validated as it is entered

» how the verification and validation will be carried out

» how users will input the commands to direct the new system to make it carry out its intended functions.

A very important consideration when designing a new system is how the data that is required enters the system. Data has to be input so that it can be processed and produce the required output. Data can be input in a variety of ways. Methods of data entry include manual typing at a keyboard or keypad, automatic scanning, scanning of printed documents, image capture or text-to-speech.

Output of processed data from a system can be in a variety of formats, including on-screen display of the results of processing, hardcopy printout, audible text-to-speech output for a vision-impaired service agent to use when interacting with a customer, or a computer file that is exported for analysis or accounting purposes.

Input forms and reports are used in most systems to collect data and to output it in a format that can be easily understood by users.

16.3.3 Input forms

The method of data entry is determined at the design stage. It is important that the data that is entered is accurate and reasonable otherwise the results of the processing will be of no value to the end users. Designers provide input forms to enable the accurate and easy entry of data by data-entry clerks and the collection of data from other end users, for example when registering for online shopping, and to allow the data that is being entered to be verified and validated as it is entered.

Verification is checking to ensure that the data that is entered is the same as the data in the original source material, and validation checks are carried out to ensure that the data is reasonable. These checks must occur when the data is entered, so forms are designed with this in mind. Appropriate checks and suitable messages to help the user are features of well-designed input forms.

Features, elements and layout

The features of input forms are important characteristics, for example that the form uses left-aligned labels for ease of reading or has validation of data entry. An element of a form is the functional unit that carries out the feature, for example the text is formatted as left-aligned or the input box uses a presence check to ensure that it is completed by the user. The elements of an input form are the items or objects that make up the form. This is almost the same meaning as when the term 'elements' is used when writing HTML code. However, the term **HTML element** refers specifically to a component that adds content and formatting and is defined by a start and end tag. For the purpose of discussing input forms, an element is a component part of the form, whether it be online or on screen.

The components, or elements, that make up input forms are:

» the labels, which can be images or text to give instructions or ask questions

» the input fields where data is collected from the user

» action buttons that take commands from the user and carry out a task, for example moving to the next page or submitting the data

» feedback in the form of text or images that tell the user that the data has been entered successfully or that it needs to be altered before it can be accepted
» the structure of the form, which is the order and connection between the input fields.

A component associated with the input fields is the validation that is used if the data needs to be checked for reasonableness.

An input form is customised to suit the scenario in which it is to be used and is designed to match the context of the system. A well-designed input form has features that enable ease of use and try to ensure that the end user enters the data accurately and does not abandon the form because it is too complicated or difficult to use. The latter is especially important when a form is used to collect data from a potential customer. If the form is too complicated or too long, a potential customer will give up and abandon the form and the company has then lost a customer.

If the amount of data to be collected is large, users should be shown a good reason for completing the form. For long forms that collect much data, splitting them into more than one form (multistep forms) reduces the number of fields that customers have to complete in one go. This can be achieved by having forms split over more than one web page or screen and using a 'next' navigation button at the end of each page. Including a progress bar reassures users that they are successfully working their way through the form and that any disruption will not, or should not, lose all the data already entered.

Questions or data-entry fields on a form should be in a logical order and grouped together in appropriate sections. For example, all fields relating to personal details are usually grouped together. Also, for example, those fields holding the customer requirements about the type of car to be rented are grouped. When collecting data about a customer's credit card for payment, the logical order is the name of the cardholder, credit card number, expiry data and lastly the security code. This is because, apart from the name of the cardholder, the data is shown on the card in that order.

If a field is optional, this should be made very clear to the user. Many users do not like wasting time on questions that they do not need to answer. Others are quite willing to supply additional data.

The positioning and alignment of the text for questions or instructions and the arrangement of the data-collection fields should be considered during the design stage. In many languages, left-aligned text for labels has been shown to be best for readability and reducing the time taken by users to complete forms. Right-aligned labels provide a strong visual connection between the field and the label because components near to each other appear connected to a viewer, but the right-hand edge of the label may look untidy if not fully justified. Most questions and fields are completed more quickly if they are placed just under each other and not side by side on input forms or screens. In-line labels, for example when the question is next to the input area, are sometimes used to overcome this issue but this can reduce the number of questions and fields that can fit on a page. Putting the label in the field can save space on the form but, because these labels disappear when data is entered, the user cannot check that what is written matches what was required. Also, the presence of a label in an input field might cause a user to assume that the field is already populated.

In whatever way the labels are positioned, they must be locked to the input area so that they can be used in web browsers. Resizing a web page in a browser or displaying it on different screen sizes on devices such as laptops or smartphones can move elements around, so the labels and input areas must stay together.

Spacing the input fields on the screen to make proper and appropriate use of 'white space' also makes forms easier and more comfortable to complete. A form that is not cluttered and does not have all the fields squeezed into the top or bottom of the screen is easier to read and more likely to be completed correctly. The text family used (for example styles, font, size) should be consistent throughout the form for easier reading.

Colours should be used carefully so that there is a suitable contrast between the labels and input areas and the background. Colours should also be chosen so that they are easily distinguishable by those with normal colour vision and by those with difficulties distinguishing between colours. While colours are often avoided, they are useful for highlighting important fields. Changing the colour or contrast of the field in use clearly shows that the user has to focus on the input to that field. This visual notification along with a flashing cursor helps the user to easily find their place again after a pause.

A feature that is useful in reducing the length of forms and is often used in questionnaires is conditional logic. This is where the next question displayed is dependent on the answer to the previous one, so the user only has to answer questions that are relevant. Unnecessary questions do not appear to the user. The choices can be made more obvious by the use of images to make the form more accessible, easier to follow and more enjoyable to complete. For entering data into a database management system by a data-entry clerk, conditional logic buttons can be used to provide navigation around the database.

Accessibility features enable people with disabilities to use the forms and enter data more easily. Appropriate use of colour, or its avoidance, is important. Many people (more men than women) have some colour-vision deficiency, so the use of colour as a warning, for example red as an alert to validation errors, may result in the error being difficult to see and to correct for many users. Icons or text should be used for error messages. However, designers should include additional measures, for example sounds or text-to-speech, for those who have visual impairments. In addition, clear explanations of the questions, or of the data required, are a feature of well-designed input forms. These help to reduce mistakes and incorrect data being entered.

Designers should ensure that online forms work properly on all major web browsers. With new systems being designed for online access on different devices, it is important that the forms appear similar and work in the same way in different browsers. It is also important that the form is easy to see and read in different lighting conditions. Online access and remote working mean that the conditions that employees work under are not predictable.

Input forms designed for use on a desktop screen are often designed so that there is no need to use a mouse to navigate between fields or pages. The **Tab** key can be used to move around, and skilled users, for example data-entry clerks, can enter data without having to move their hands from the keyboards and so can enter data faster. If the form is designed for use on different devices, then it is important that the layout and navigation be customised to the device and screen. Apps can determine the type and resolution of screens so forms can adjust automatically. For example, on-screen keyboards can be made to show only numbers when numerical data is required.

There are many more features of input forms that have to be considered by designers, for example:

» the speed of the flashing of cursors, progress bars or other elements: fast flashing may trigger seizures in end users
» the speed of moving from one page or screen of the form to the next after a button is pressed: too fast a speed may mean that the user does not notice and clicks another **Next** button too soon, or too slow a speed might encourage repeated clicking of a **Next** button, meaning the user moves on by several pages or screens and misses questions
» automatically moving to the next question: this is usually avoided because it confuses end users
» the possible use of visual prompts instead of written text: human brains process visual clues or images much faster than textual messages or instructions.

Buttons can be used to make actions happen. When a user clicks or selects an action button, the form can, for example, move to the next page, back to a previous page or submit the data. Action buttons fall into two categories:

» Primary action buttons are those that the user is expected to select, for example a **Next** button or the **Submit** button.
» Secondary action buttons are the alternative options to expected action, for example a **Back** or **Previous** button or a **Cancel** button.

It is usual for designers to specify that there is a visual distinction and a separation in position on the form between primary and secondary buttons to avoid mistakes by the users. For example, placing the **Submit** and **Cancel** buttons close together in the same size and same colour makes it difficult for users to easily tell which is which. Designers put customised labels on action buttons, for example **Create your account** rather than **Submit**, because it makes the form appear more user-friendly and gives more information to the user. Designers often omit a **Clear** or **Reset** button from input forms collecting customer data because if it is accidentally selected or used to clear data that needs to be amended and all the data is cleared from the form, the customer may give up and go elsewhere for a product or service. Instead, validation with appropriate error messages is used to check each field as the data is entered so that the user does not have to go back through the whole form to correct the data if the **Submit** button refuses to work.

Checking the data collected

The data collected by input forms must be accurate and reasonable if the output from a system is to be of use to the users. For use in a computer system, data has to be of the type and in the format that the system is designed to process and output for the user. To ensure that the data is accurate and reasonable, and in the correct format, the type of input fields and validation of the data can be used to control the data that can be input by a user. For online forms that run in a web browser, the validation can be programmed in JavaScript so that the entered data does not have to be transferred to the server to be checked and the user is given immediate feedback if there is a need to amend the data. For forms that collect data that is being entered directly into a database system, the checking and validation can be carried out by the database management system as the data is entered into the form or as it enters the actual database table. The creation of data-entry forms for databases is explained in detail in Chapter 10 of *Cambridge International AS Level Information Technology.* Checking data as it is entered into an input form can be by the type of field or by validation checks on the data.

Text fields are the most common type of input fields. These, as the name suggests, allow the input of text (for example names) in single lines or in multiple lines or paragraphs, for example as a comments box. For a single-line text box, the type of data that can be entered can be controlled by using input masks, for example when a telephone number or a serial number of a product is required in a specific format.

Input masks check which characters are allowed in a specific position in the data that is input. Any characters, such as letters, punctuations or numbers, can be specified or disallowed. When a user types a character, the control automatically checks the entry and allows or rejects it depending on the criteria set. If a user types a letter where a number is required, the entry is rejected. Input masks are not validation routines as they do not show an error message, but they do help to reduce data-entry errors.

If a serial number is in the format TX4356S, an input mask that allows data to be entered only as two letters followed by four numbers and a single letter is used. The check is, in this example, that only letters are allowed in positions one and two, only numbers in positions three, four, five and six, and only a letter in position seven. Setting an input mask for a telephone number is more complex as their format varies around the world, but the input can be forced into specific country formats, for example through the requirement for a country code at the beginning of the number.

It is not usual to control the input to multi-line or paragraph text boxes as these are often asking for longer sentences, for example comments, from users. However, the number of characters that is allowed to be entered is usually controlled to restrict the amount of data that has to be stored.

Validation routines can be applied to text entry fields to ensure that required fields are not left blank. A presence check generates a message or alert if the field has been omitted by the user. More complex formats can be checked by validation routines. The format of an email address, or a post/zip code, can be checked to ensure that the user has entered it in an acceptable format. For example, an email address must consist of a username, an @ symbol, and a domain name, which must have the required component parts.

Limit or range checks can be carried out on numerical data. A limit check sets an upper or lower limit for the data, for example the number of people in a rented car cannot be fewer than one, whereas a range check sets both an upper and a lower limit on the data, for example the number of people using the rented car must be greater than zero but fewer than five. Length checks can be set to control the number of characters entered into a field, for example when entering a UK credit card security code into an online shopping form there must be exactly three numbers. The use of input masks and validation routines for controlling data entry into databases is explained in detail in Chapter 10 of *Cambridge International AS Level Information Technology*.

Form controls are used to force users to input data in a specific manner. Form controls include dropdown boxes, radio buttons and check boxes. Dropdown boxes use a pre-set list of choices that the user must choose from. While this forces the user to choose data that meets the required format for entry, the list may not include all the choices that the user may wish to have and may be too restrictive. Using dropdown lists for the numbers in dates, with separate lists for day, month and year, ensures that only numbers, and the correct amount of numbers, are used for the date. This method also ensures that there is no confusion between the date formats used in different parts of the world.

Dropdown boxes can have almost any number of choices, but extremely long lists are not helpful to the user. For example, choosing your country of residence involves a long list of countries and scrolling down can be tedious if you live in a country that starts with a letter at the end of the alphabet. A check of the geolocation of the user, if the form is online, for example from the IP address, can enable the country of residence to be extracted from the list and placed at the top for ease of selection. This can avoid any errors when selecting from a long list and increase the accuracy of the data. This is referred to as 'smart default' because the input field defaults to a value that depends on other factors. Defaults are usually not used by designers because sometimes end users leave the defaults in the fields and do not input their own data because it allows them to complete the form more quickly. It is not possible to check for this as the field will have a valid entry and can be submitted. However, the data that is collected is not reliable.

Radio buttons are used when only one choice is required from a small range of choices. The user cannot choose more than one option because any choice automatically excludes the others. It is usual to use dropdown boxes if the number of choices exceeds five or six because the user can become confused or the form becomes cluttered with the buttons.

Checkboxes are used where more than one choice from a fixed selection is required from the user, for example selecting the features required in a rental car, such as SatNav, entertainment system and air conditioning. Checkboxes have an attribute that is set to 'on' when clicked and 'off' when clicked again. When the form is submitted, only the checkboxes that are in the 'on' state are submitted.

None of the checks actually determines whether or not the data that is entered by the user is correct in the sense that it is 'right'. For example, a question asked of customers who are renting a car could be 'Have you ever had to pay a fine for speeding?' The answer to the question is 'yes' or 'no' and the input form might collect the data using a radio button that only allowed one of the two choices to be entered by the customer. The form would collect the data in the correct format, as one of two choices, but has no way of knowing if the customer has given a correct (true or honest) answer. A customer with several fines for speeding could still answer 'no', so the data would be in the correct format but the answer would be wrong. Similarly, a text entry of a name can be checked for the correct format of the characters being entered, but it cannot check for mis-spellings. It is important to understand this when designing and creating input forms.

As long as the data meets the requirements of the form by being of the required type and format, it will be accepted. This means that important data must be verified to check it is the same as the original source. Requiring that some data be entered twice (double-entry verification) means that the form or system can compare the two sets of entries and alert the user if there is an error because they do not match. This is often used for email addresses. Designing in a step where the customer is asked to read the data that has been entered before submitting the form or presenting the form with the data for a final, visual check, is a way of trying the verify that the customer has entered the 'right' details.

Other form controls are available for different purposes. For example, online forms for collecting user data may allow the user to send files, such as image files for photographs. HTML has constructs to collect the file information when the user selects a file and send the file when the form is submitted.

16.3.4 Output reports

The output from a system is when processed data, as information, leaves the system. It is usually the reason for the development of the system. It is used to evaluate how well the system carries out its intended purpose. The output from a system is also the information delivered to a user to carry out their tasks, or as a result of their tasks, and can take many forms, for example on screen, printed hardcopy, video or audio. It is the job of the systems analyst to ensure that data is processed and output in the ways that the client and end users require. During the design stage, prototype output reports are sometimes created and shown to clients and end users for feedback on their suitability.

The systems analyst works with the client and users during the design process to determine what information is to be presented and what the purpose is of its presentation to the user, how much information is to be presented, when it is to be presented, and in what way it is to be presented after the processing of the data that has been input. A data dictionary is used as a basis for each output form. A data dictionary is a file (often a document) containing descriptions of, and information about, the structure of data in a database.

It is important to determine what information is to be output to which users so that only relevant and useful information is shown. Some end users will not be allowed access to confidential information so this should not be output to them. The purpose of the output is important because it determines the range of information that can be output to a user. Not having all the relevant information available, for example to a project manager about costings or the skills of software programmers, can impede project planning. The output of processed data information is useful only if it is available at the appropriate point in its processing. For example, the results of the analysis of costings are useful when a project is being planned but not after it is completed.

The method of output and distribution is important because it determines how much and what type of information can be represented and how it can be sent to users. The amount and type of information that can be displayed on computer screens is different from that on printed hardcopy. Most computer displays are landscape-oriented whereas most printed reports are portrait-oriented, so information shown in a web page on screen may not appear correctly when printed on paper. The systems analyst considers the method of output that is appropriate for the type of end user. For example, a printed hardcopy of a progress report may be required for sending to stakeholders, but an on-screen version would suffice for a project manager.

Output on computer displays and on hardcopy should be simple to read and understand by the intended users. Output reports should include a suitable title, sections as appropriate, any tables or columns should be clearly labelled, and for on-screen reports, any fields that are displayed should be clearly labelled and users should be able to easily find and use navigation aids. Navigation aids such as **Next**, **Back** or **Previous** and an exit from the report should be present.

The method of output can depend on who the end user is and what they will be using the output for. On-screen lists are suitable for lists of customers or stock levels. Electronic output reports in the form of documents that can be printed or emailed are appropriate for some output but not others. For example, a list of cars currently out on loan to customers can be generated at any moment by a manager for display on screen, but a monthly report on the numbers of rentals should be in a form that can be emailed or printed.

Output reports used by managers and end users within a company can contain summaries or be very detailed or a combination of both, and are used for decision-making and reporting to supervisors or stakeholders. These reports may be electronic and not intended to be printed so can contain animated graphics, slideshows and video. The positioning of the different elements, such as graphics, text or charts, on screen is determined by the systems analyst to present the information as appropriate for the type of end user. For hardcopy printed reports, animations and video cannot be presented.

The cultural conventions for reading should also be considered. In many areas of the world, reports are read from top to bottom and left to right, so the flow of information should follow this. Where the system is designed for use in languages that read from right to left, for example Arabic or Hebrew, output reports should follow this.

16.4 Development and testing

The **development** stage of the system life cycle is when the work begins on creating the product using the design documents from the previous stages. For a new software application or system, software developers and engineers will start to write the code according to the details set out in the designs. This is the start of the production of a new system and includes the testing of the components and system.

Testing that checks a software component or module does what it is supposed to do is called unit testing. Testing must also check that units work properly together in a new system and this is called sanity testing. Regression testing, or stress testing, ensures that the new system works properly over a period of time and can reveal problems that occur only when the system has been running for a long time. Functional testing, carried out by testers who did not develop the system, is used to ensure that the new system works as intended.

Software developers and engineers test the new code and software components or modules as these are created. A **test plan** is used to ensure that the new system is properly tested according to the procedures and processes specified by the systems analyst. During testing, errors, defects or faults are usually placed into categories that note the importance of the problem. Some errors can be ignored for the moment, but some will need to be dealt with urgently. This is similar to the categorising of problems found during the maintenance stage (see Section 16.9 'Maintenance' later in this chapter).

An example of the type of categorising that might be found in the test plan is shown in Table 16.1.

▼ Table 16.1 Categories of importance of errors

Level of importance of error	Possible impact on the system
Level 5: Critical	This error will cause one or all of: • the system to fail • file corruption • data loss.
Level 4: High	This error will cause the system to lose some functionality or features.
Level 3: Medium	This error will cause the system to lose some functionality or features but can be worked around if necessary.
Level 2: Low	This error will have little effect on the system's functionality or features.
Level 1: Trivial	This error has no impact on the system's functionality or features but may, for example, alter the appearance or positioning of elements on screen.

16.4.1 Test plans

A test plan is written early in the development cycle. It is written by the designers so that the software developers and engineers know what they should be testing. Exactly when and by whom the test plan is written depends on the end product, but it is not good practice for the people writing the code or modules to write their own test plans for the whole system. Software engineers and developers can write test plans for the code or modules that they are creating so as to test that their code works as it should, but test plans for the system as it develops or is finished should be written by analysts or designers.

Test plans are needed because, while code and modules can be tested by programmers as they are created, test plans ensure that the new system has been developed to the specifications and works exactly how the client, analysts and designers expected. Test plans are important because they guide the developers, programmers and testers when carrying out tests, assist those people who are not part of the testing team to understand the testing, and enable managers and clients to be kept informed of the progress of the development.

A test plan acts as a guide, or a set of instructions, on how the testing of the system is to be carried out to ensure that the requirements set out in the designs are met. It also defines what tests are to be carried out and how the results are to be recorded, so that they are used by the developers to ensure that corrections or amendments can be made. The test plan also defines how retests are to be carried out to ensure that any corrections resolve issues shown up in testing.

For software development, there is an IEEE standard that details what should be included in a test plan. After a title, there is an introduction that identifies and describes the test plan in outline form so that a person reading it knows what the plan covers. The plan then includes the items to be tested and those that are not to be tested. A full list of the functions that must be tested is included so that everything is checked. The features, if any, that will not be tested are included so that the testing does not stray from the new system.

The methods to be used are detailed in the test plan. How many tests, or repeats of tests, are specified and the tools or software to be used for the tests is included. The method of tracking the results of the tests is specified so that the testers know how the tests will be reported. The people who will carry out the testing are also specified.

The pass or fail criteria for each item to be tested are defined so that the testers know exactly whether or not the component or system matches the specified requirements. There must be no room for doubt or guesswork about the results of the testing. The test plan also contains the criteria that determine whether or not the testing needs to be paused so that errors or defects can be corrected before testing resumes.

The deliverables from testing are specified in a test plan. They include a final report summarising the test results, as well as periodic (for example daily) reports on the testing of modules or other components. Specifying these in the test plan ensures that they are produced.

The tasks that are to be carried out during testing are shown in a test plan. These include the tests to be carried out and the production of the test reports. The plan also shows who is to carry out these tasks, who is responsible for ensuring that they are carried out properly and who is responsible for informing others of the test results. It also shows who has to be informed by listing their details. A schedule of the tasks in the testing stage is often included.

The resources required for testing are also detailed. The hardware and software requirements should include the device specifications and the software versions that are to be used. Any training that is needed to enable the testing of a new system is included in the resource details.

When creating a test plan, it is essential to customise it to the scenario so that the tests and documents are appropriate and relate to the requirements specification. The first step in writing a test plan is to analyse the end product. The next step is to use the analysis to define the test strategy, which includes the full scope of what is to be tested and the types of tests to be carried out, as described previously. Who is to carry them out, when the tests will be done and the criteria for pass/fail and suspension of testing are added to the plan. The list of the resources and the schedule is written and the deliverables specified.

A test plan can be started by noting down what is to be tested and why and when, who is to do the testing and where, and to whom the test results are to be sent.

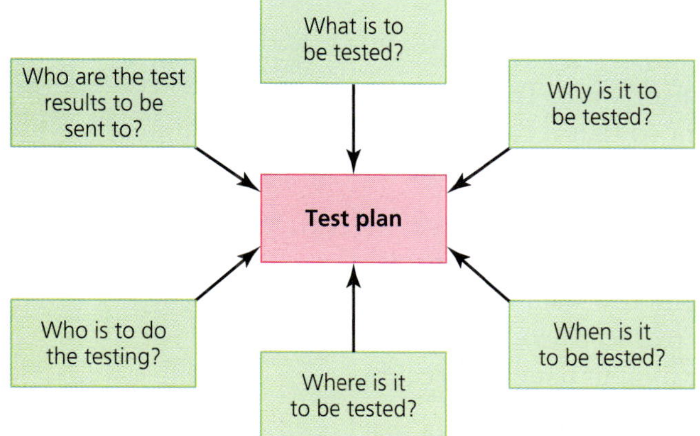

▲ **Figure 16.6** Initial creation of a test plan for testing a new software app

This is developed into a plan that covers the entire scope of the development and is included in the testing policies and strategies that become part of the specifications.

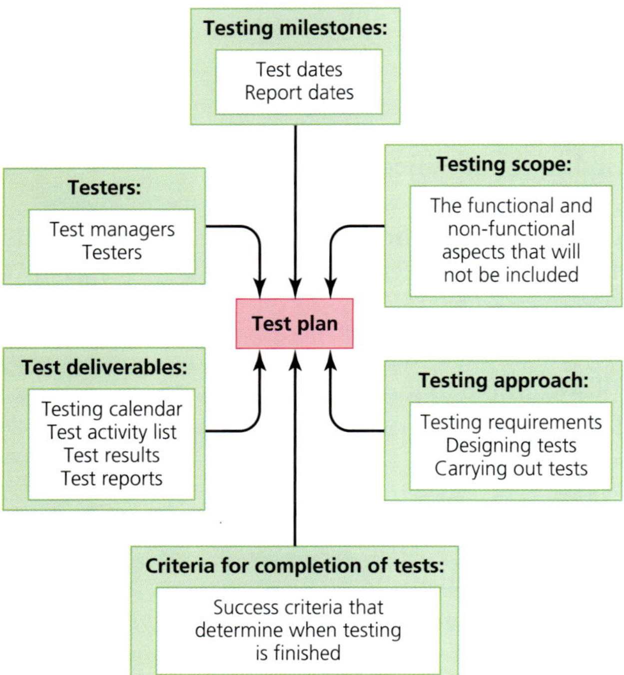

▲ **Figure 16.7** Developing a test plan for testing a new software app

When all the aspects of the test requirements have been determined and documented, the final test plan consists of:

» its purpose
» the scope and objectives of the testing
» who will be the testers
» the methods of testing
» the timings of the tests and the milestones in the testing stage
» the test success or acceptance or testing exit criteria
» the test deliverables for the developers.

User acceptance testing is also documented, but this is carried out by end users after all the other testing is completed.

The test plan also includes the method of recording the tests and the results. Usually, a table is used, as in Table 16.2 opposite.

16.4.2 Test data

Test data is data that is used when carrying out tests on software. Test data must be carefully chosen and prepared. This is important so that it can uncover errors, defects and problems. Also, test data must be able to confirm that the system works as expected and produces reliable results.

Test data can be created manually by the testers, generated by software tools or taken from existing systems that use similar data. The test data is usually representative of the data that is expected to be used in a real scenario. Test data is used to determine whether or not the system can deal with input data as expected.

The data that is input into a system falls into two main categories: normal data that is accepted and can be processed, and abnormal data that should be rejected by the system. Other categories are sometimes described, for example extreme data, which is data at the edge, or on the boundaries, of normal data. A system under development is tested with a range of data, or data sets, that test all, or as many of the possible inputs as possible, to determine whether or not the system can deal with unusual or abnormal input as well as expected, normal or extreme input.

▼ Table 16.2 Table in a test plan for recording tests and test results

Test no.	Type of test	Element, screen or file to be tested	Name of test	Date of test	Purpose of test	Test data or test conditions	Expected outcome or result	Actual outcome or result	Actions required	Report outcome to:
1	Web page display in web browser	Collect user details for new account page	Rendering of page	22 October 2020	Page renders correctly in common browsers	Display in these browsers: • Microsoft Edge Ver. 85.0.564.70 • Mozilla Firefox Ver. 80.0.1 • Google Chrome Ver. 86.0.4240.75 • Safari 14	Three rows, one each for name, email address, telephone number. All coloured white. Rows aligned vertically. Blue background. Company name and logo at top of page, in black.	• Edge: as expected • Firefox: rows aligned • Chrome: as expected • Safari: rows not aligned	Code to be amended/ tested for Firefox and Safari. Code to be tested again in all browsers.	Head of testing team. Head of developer team. Project manager.
2										
3										
4										

16

A system is tested using valid data, or data sets, to check whether or not a new DBMS, for example, accepts the data, processes it correctly and stores it as specified, for instance in the correct fields in the correct database tables. Normal and extreme data is used in these tests. This tests that the functionality of the new system is operating correctly.

Invalid data is used to test how the system deals with data that is not acceptable. Abnormal data could include numbers where letters are expected, out-of-range values for dates or ages, or negative values where positive values are expected. This tests how the system deals with these errors and if, and how, the required error messages are generated and displayed.

Tests with blank or no data are useful to determine how the system deals with this situation. A test scenario for a new DBMS can be set up to import data from external files where there are gaps or blanks. While this is an example of abnormal data, explicit testing for blank data is often carried out to ensure that appropriate error messages are still generated.

Data can also be prepared and used to test scenarios where large quantities of data are to be input and processed by the system. This tests the performance of the new system 'under load' and gives an indication of how it might perform when deployed in a real situation. For example, a new DBMS might correctly handle valid and invalid input from a few users under test conditions but not when many users are working. This must be discovered at the testing stage and not when the system has been installed by a customer.

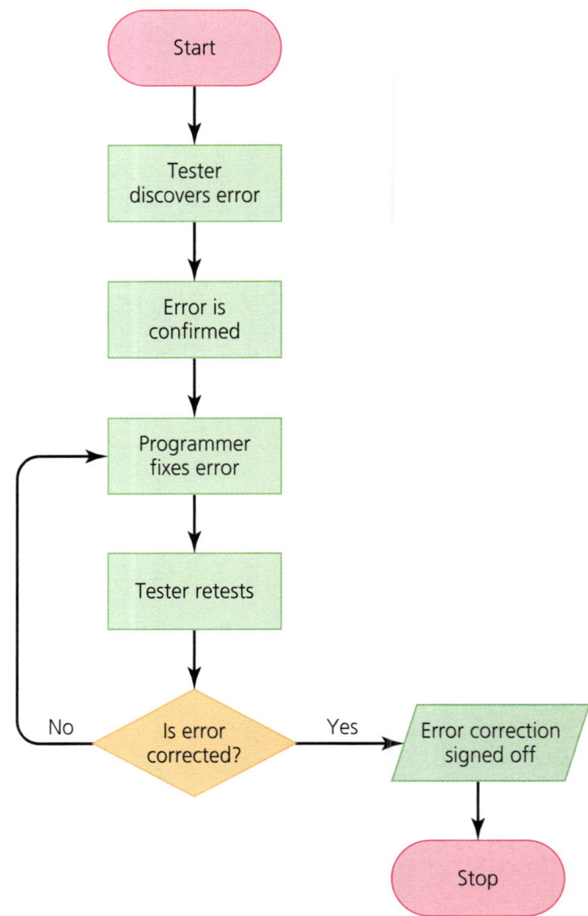

▲ **Figure 16.8** Flowchart for carrying out testing of a piece of code

Figure 16.8 shows a typical testing procedure as a flowchart. Each part, component, or section is tested. The final product is also tested.

Activity 16h

A new online calculator for the quantity of paint needed for spraying doors is being developed to work in web pages displayed in browsers. The area of the door and the thickness of the paint needed is used to calculate the quantity needed. The height and width of the door and the paint thickness are input for the calculation. The maximum height allowed is 2 m, the maximum width allowed is 1 m and the thickness cannot be more than 2 mm.

1 Briefly describe how these inputs could be collected, and checked, in an online form.
2 Give one example of each of the following types of data that might be used in testing:

   a Valid data
   b Extreme data
   c Abnormal data.

16.4.3 Alpha and beta testing

Alpha testing is carried out by testers who are employed by the company, for example the software developers, and is carried out at the developers' place of work. Alpha testing is carried out on the new software, or system, as it nears completion but before it is released to end users. Sometimes, there is a pre-alpha stage where prototypes are tested in-house by the developers.

For alpha testing to be carried out, the requirement and specification documents are usually available to check what is expected, test data is prepared and ready for use, and a test environment, for example a location with the required hardware, is made available. Any software testing tools and a test tracking system for reporting the results are also required.

Alpha testing is finished when all the tests have been carried out, all defects and errors have been corrected and the test summary report has been produced and delivered. At this point, the developers make sure that no additional features are required by the clients, and the testing is ended.

Beta testing follows from alpha testing and starts only when the alpha testing is completed. A version of the new software or system must be available as a 'beta' version for testing and this must be ready for release to end users external to the developing company. Beta testing is carried out by the client or by end users who are not employees of the company developing the system. Beta testing is not carried out at the company's site but at the client's or end user's location because it does not require a special testing environment.

Beta testing is over when all problems with the new system have been resolved or those remaining are considered unimportant by the developers or client. A feedback report is usually prepared from the client and end users, and a summary of the beta testing results is made into a report.

The differences between alpha and beta testing are summarised in Table 16.3.

▼ **Table 16.3** The differences between alpha and beta testing

Alpha testing	Beta testing
Carried out by employees of the company developing the system	Carried out by the client or end users of the system
Occurs at the developers' location	Occurs at the client's or end user's location
Reliability and security are not checked	Reliability, security and robustness are checked
May take a long time during the development	Should not take a long time
Problems or faults can be corrected quickly or immediately by the developers	Problems or faults are corrected and new versions made available so may take time to reach testers
Tests product before moving to beta testing	Tests product in real-life situations

Alpha testing is important because it detects problems before end users have access to the new system, but testing of the full functionality is not possible as the system is still being developed. Beta testing is important because it allows end users to provide feedback under real-world conditions to improve the system. However, the recruitment of suitable beta-testers can be difficult because the testers need to be competent in the use of the new system and must be able to keep their findings confidential.

16.4.4 White box and black box testing

White box testing checks the internal functioning of a system. It is considered to be low-level testing because it tests every line of code, every branching of the program flow, every path and every condition in the system. To carry out white box testing, a tester must be technically competent and know the paths and logic used in the system. Expert testers with considerable experience are usually used to white box test new systems. White box testing can begin at an early stage in the development, for example there is no need to wait for a full user interface to be completed before testing of the logic can begin.

Black box testing is considered high-level testing because it tests how the new system functions and behaves without the need to know how it works. Testers input data and try out the functions to receive output but do not need to know how the system produces the output. Black box testing can detect errors in the user interface, such as problems with data-entry forms, performance problems such as the system appearing too slow, and missing functionality such as not creating a specific type of printed report. Black box testing may have the drawback that not all conditions, logic or paths through the system are tested because the tester may not be able to test all of the conditions that will be met in the real world.

The differences between white box and black box testing are summarised in Table 16.4.

▼ **Table 16.4** The differences between white box and black box testing

White box testing	Black box testing
The internal workings of the new system are known and understood by the tester	The internal workings of the system are not known to the tester
Low-level testing of the code statements, logic branch, paths and conditions	High-level testing of functionality
Carried out by the developers	Carried out by testers other than the developers
Programming knowledge and skills required by testers	Programming knowledge and skills not required by testers
Access to detailed designs needed	Access to requirement specifications needed

Advantages and disadvantages of white box and black box testing

The advantages of white box testing include the thorough and complete testing of all the code, which ensures that the code is as error-free as possible before it is released to users. White box testing can be automated and tests can be run quickly to check for mistakes or errors. This is useful for testing code that was working but now produces errors. White box testing allows developers to fix an error, or 'bug', immediately without having to wait for users to discover and report it. White box testing also enables developers to optimise their code by going through it and removing any superfluous or replicated code. However, white box testing can be costly because it takes time to test all the code thoroughly and requires highly skilled developers or programmers. Using automated white box testing will not work if the code is changed frequently during testing. White box testing does not take into account any required features that have been missed out or not fully implemented as it tests only the code that is already present.

In black box testing, the developer and tester are independent. The developer does not have any influence over the tester, so the tests and the results of the tests are not biased. Tests are carried out from the point of view of a user so are focused on the features and functions of the whole product. However, black box testing may not cover every input possibility, feature or function of the product so testing may miss potential errors and faults. Also, not all specifications are detailed enough for the tester to be able to properly test every feature. Another disadvantage of black box testing is that the causes of any errors or faults may not be obvious, easy to determine or describe, especially as the tester has limited, or no, knowledge of the workings of the code.

16.5 Implementation

The **implementation** stage follows the development and testing stages. In this stage the new system is prepared and placed in its working environment in the real world. Implementation involves installing new hardware, if required, and the new system, and the training of the end users. Where a new system replaces a previous system, it is called system changeover.

Implementing a new system to replace a current system has to be carried out carefully and methodically to avoid disruption to the company having the new system and to avoid corruption or loss of their data. The method chosen to change over the systems depends on the specific scenario and is usually detailed in the specifications written at the beginning. Whichever method is used, the main tasks are:

- to replace the current hardware with new hardware as necessary to ensure that the new system is supported and will run properly
- to alter, or plan to alter, any business procedures to match the new system, as these amendments may be part of resolving problems with the current system
- to train the end users in the use of the new system.

The IT technicians in the company will also have to be trained to manage and maintain the new system. There may also be a need to convert the data used by the current system to allow it to be used in the new system. Sometimes, different data types or formats are to be used in a new system so these must be converted before the old data can be used. If the current system is not computer-based but is a manual system, for example if a manual accounting system using books and ledgers is being replaced with a software accounts application, then it is necessary to digitise and enter data from the current system into the new application.

Parallel running, direct changeover, phased implementation and pilot implementation are methods of system changeover.

16.5.1 Parallel running

The **parallel running** method of system changeover has both the current system and the new system running at the same time until the new system is shown to be working as expected with few or no risks to the data or business activities. For example, a shop changing its accounting software would run both the current and new accounting software at the same time until the shop owner was confident that the new software was working as expected. When the installers and clients are confident that the new system is working as expected, the old system is removed.

Advantages and disadvantages of parallel running

Parallel running enables comparisons of the output of the new system with that of the old system to ensure that the new system is working as expected and the results of its processing are reliable. The new system can be amended and optimised for enhanced performance while the old system is still in use. Also, if the new system fails or has defects, then the old system can still be used until the new one can be used again or the defects are corrected. End users can be trained in the use of the new system while retaining the familiarity of having the current system still in place.

However, running two systems side by side can be expensive in terms of the costs of users' time to input data twice, once into the old system and once into the new, and to compare the data to look for errors. Also, there is a great deal of effort required to accurately enter two copies of the data into two different systems. The two systems have to be kept in synchronisation with each other so that the outputs can be reliably compared and this may be difficult where the two systems run at different speeds or require different methods of data entry. Where a new system takes up considerable space, for example because new hardware is needed, the company may not have sufficient room to easily house the new hardware.

16.5.2 Direct changeover

With **direct changeover**, the current system is completely replaced with the new system all at once. Sometimes this is called a 'big bang' changeover because the changeover is instant and every user moves to the new system at once, for example a bank changing from its current stocks and shares trading system to a new system overnight. Bank employees stop working with the current, old system at the end of their working day and start with the new one at the opening of business on the following day.

Advantages and disadvantages of direct changeover

Direct changeover means that the new system is fully available immediately to all end users. The full installation takes less time than the other methods, but there is no option to go back to the old system if the new system fails or has defects.

16.5.3 Phased implementation

Phased implementation is a method that replaces the parts or modules in a current system with new parts or modules in stages. Over a period of time, as more parts or modules are replaced, the whole system is changed from the current system to the new system. For example, in a supermarket that is changing its EPOS system, first the checkouts are replaced in the bakery section and fully tested. When they are shown to be working properly, the checkouts in the meat section are changed over, and so on until all the checkouts have been replaced. Then the supermarket can start changing other parts of the whole system. In this method, some current system parts work alongside the new ones. Phased implementation can be considered a combination of parallel running and direct changeover; while some of the older parts are still running alongside new ones, some parts have been completely replaced.

Advantages and disadvantages of phased implementation

Phased implementation allows the end users to be trained on each new part or module and so become gradually used to using the new system. However, while parts of the old system are still in place, there is no option to go back to the whole of the old system as some parts may have gone forever. There is, therefore, often no backup option for these parts and any data in the new parts may be lost if that part fails. However, in some cases where a module fails, it may be possible to reinstall or use the old one while the new one is corrected. It can take a very long time to fully implement a whole new system under phased implementation compared to other methods, but it does allow for the testing and evaluation of parts of an installation before moving on to the next. However, it is not possible to separate some systems into parts that can be replaced individually.

16.5.4 Pilot implementation

Pilot implementation involves replacing the current system in one part of
a company or business but leaving the rest of the company using the current, old
system. The new system is installed and tested in this part of the company, and
only when it is shown to be working correctly are other parts of the company
changed over. For example, a travel agent with several branches could change its
booking system in one branch and test it there and only change the system in its
other branches when it is satisfied that the first branch changeover is successful.

Advantages and disadvantages of pilot implementation

Pilot implementation allows all the functions and features of the new system to
be tried and tested by end users before the company changes completely to it.
Also, any defects or problems with the new system can be rectified before the
full implementation across the whole company. End users can be trained to use
the pilot system, and they are then able to train or support other end users when
they have the new system installed. There is no backup or fallback option for
the part of the company that is using the new system if the new system fails, but
other parts of the company can carry on with the old system.

16.5.5 Advantages and disadvantages of each implementation method for given situations

The method of implementing a new system depends on the particular situation.
The advantages and disadvantages of each method have to be considered when
choosing the most appropriate method for the situation. For example, if a new
and improved installation of a different method of internet connection to several
villages in a remote area is required, pilot implementation is a useful way of
determining whether a full changeover to a new system is actually worth the
cost, time and effort. A pilot implementation to one village would determine the
problems, difficulties and improvements of the new connection method. Parallel
running would also be possible, but very costly, to run both methods at once
and compare them.

Parallel running of systems is used when the system is critical to the operations of
a company. For example, changing the database management system that stores
and manages the accounts of customers of an online bank is critical to the bank's
workings. Running the new system alongside the current, old system for some
time is vital to the bank and to the customers to ensure that all customer details
are managed correctly and that they can be retrieved if the new system fails.
The cost of running both systems, the time and effort to compare the system's
performance cost, the time taken for training the bank employees and, possibly,
the cost of employing more staff for both systems are drawbacks of parallel
running. However, the critical nature of the tasks undertaken by the system may
make parallel running the only sensible choice for the bank. If the bank chooses
to use the direct changeover method for its DBMS and it fails or produces
incorrect customer details for the accounts, the bank will be unable to fall back
to the old system, will lose customers and may go out of business. Possibly, in the
case of a modular DBMS, phased implementation may be possible and a better
choice, but this may not provide all the functions and features required of
a completely new system.

Direct changeover is used when a company, or client, is satisfied that the
new system will work reliably and perform as expected from the moment it is
installed. Often, companies will choose direct changeover where the system

is not critical to the operation of the company. For example, changing to a different software application for word processing letters is not critical to the company operations, whereas changing the accounting software application could well be. In the case of a change of accounting software, where bills have to be sent or paid and taxes accounted for, parallel running of the two applications would be used so that the processing and output can be compared, even though running two systems at once is costly. Direct change may take less time than other methods but, in the case of accounting software of a bank or other financial institution, it would be very risky as data could be lost or incorrect output could be produced if the new system had defects.

Phased implementation is used when it is possible to replace part of a current system while leaving the rest in place. A database management system, or a software app for a smartphone, may consist of several modules that carry out specific tasks. A phased replacement of the modules allows each module to be tested and optimised alongside the others, but it does take a long time to completely replace a system.

Activity 16i

In each of the following scenarios a new system is to replace a current system. Choose the most appropriate method of implementation: parallel running, direct changeover, phased implementation or pilot implementation. For each, explain why that method is the most suitable.

1 Replacing a stock control system in a supermarket.
2 Replacing the user accounts system of an online bookshop.
3 Changing from local, hard disk storage of company documents to cloud-based storage.
4 Replacing the workflow system for taking and fulfilling orders in a mobile (cell) phone store.
5 Replacing the booking system for rail tickets in a travel company with many branches.
6 Replacing the operating systems of desktop computers in the offices of a multinational company.
7 Replacing the manual counter services in the branches of a bank with automatic teller machines in the foyer.
8 Installing 5G mobile services in cities.

16.6 Documentation

Documentation for a new system is important because it describes and explains how to use the system and how it was created and developed. There are two main types of documentation that are produced for a new system. The user documentation provides end users with the necessary information that they need to install and start the system, and how to access and use the system. The technical documentation describes and explains how the system was created and developed. It provides the information required by those who maintain the system and for developers who may work on the system in the future.

16.6.1 Requirements documentation

Requirements documentation contains all the system and user requirements. The documentation is also called a requirements specification and is important because it collects together and records all the information that describes how a new system should function and carry out its intended purpose. It shows all stakeholders how the new system should look and work and contains functional and non-functional requirements. Functional requirements are descriptions of the features of the new system, how it will respond to specific commands and what users expect the new system to be able to do. Non-functional requirements are, for example, the limitations of the new system, its reliability and its security features. Requirements specifications can also be used when finding developers to create the new system because it tells them what is to be produced.

16.6.2 Design documentation

Design documentation must include the design specifications, but it can also include other documents that describe the intended end users, product features, project deadlines, details about implementation methods that will be used, and any other decisions about the design that have been made and agreed by the stakeholders.

16.6.3 User documentation

User documentation is necessary so that end users can carry out the tasks for which the system was created. The documentation contains all the information necessary from the point of the installation and starting of the new system, through data entry, to the output of the system's results in the format required by the user. The features and functions of the system are included, along with guidance on how to get the best out of the system and, often, some simple tips or advice on its use and ways to resolve common problems.

The introduction to the user documentation usually contains an explanation of the purpose of the system so that the user understands what it is to be used for. A user should be able to determine what the system can do from this description of the purpose. The hardware requirements and the supporting software requirements, for example the operating system needed, are also shown in the documentation. This is important because end users need to be confident that the new system will run on their hardware and software. If necessary, the users may have to change or upgrade their computing environment to ensure that the new systems will run. Having clear information about the minimum, and optimum recommended, hardware and software requirements in the user documentation is essential to avoid end users being unable to use the system.

Installation instructions are included for systems that can be installed by end users, for example a new smartphone app or desktop software, but often the installation is carried out by technicians working for companies. For system installations in companies, such as installing a new DBMS in a bank, these instructions may be omitted or appear in other documentation.

Instructions are provided on how to start the new system or how to load it into the computer. The login and logout procedures are explained to end users. Administrative information about the creation and management of user accounts may also be included.

Most of the content of user documentation is about how to use the features and functions of the new system. The instructions may be accompanied by screenshots of the most common activities that end users carry out so that these procedures are easier to understand and follow. The documentation contains the information necessary for an end user to use the system for its intended purpose.

Assistance in resolving common problems is also included. A user will encounter errors and problems when using the system, so the user documentation will include a troubleshooting guide, instructions on how to deal with system errors and explanations of error messages. This helps the user to understand what caused the error or problem and how to resolve it.

A frequently asked question (FAQ) section is included to help end users in the use of the system. A list of sources for additional help, guidance and support is also included to answer or resolve problems that are not covered in the documentation.

With the increased availability and use of networks and the internet, user documentation is often available online. Not only can user manuals and guides be downloaded from the developer's website, but the documentation can be accessed online in a web browser. This allows online user documentation to include video and audio instructions and help, as well as the traditional text and images. Online video tutorials can be used to show users how to carry out tasks or resolve problems in software or systems that might take lengthy written explanations.

Advantages of user documentation include reducing the need for training end users on all of the features and functions of a new system and enabling end users to learn to use a new system in their own time and at their own pace. There are very few, if any, disadvantages but the necessity of producing and distributing new user documentation whenever a system is changed or updated may sometimes be costly.

16.6.4 Technical documentation

Technical documentation details and explains every part of the system so that advanced users, technicians, future analysts and future developers can know and understand exactly how a system was created, how it works, how to maintain it and how to improve it.

The purpose of the system section of the documentation explains what the system does and the problems it is designed to solve. In the technical documentation, this is explained in great detail and uses the documents from the analysis and design stages of the development.

The design specifications for the system are included in the technical documentation. All of the designs (for example for input screens, output print or display formats and user interface screens) are included as well as the system flowcharts and data flow diagrams.

The program language that was used to develop the system is specified and the program code itself may be included. Annotated listings of the code can help when the technician's system requires maintenance and when future analysts and developers attempt to improve it.

The technical details of the input formats describe what types of data are accepted by the system, the output formats describe how the data leaves the system, and the processing and storage of the data in the system are described. File structure details are given so that technicians and developers know where and how the data is stored. If the system is a database management system, technical documentation includes details such as the tables used in a relational database and their relationships, which are essential for an understanding of the workings of the database. Similarly, descriptions of validation rules and the error messages they generate are included so that problems can be resolved.

Technical documentation also contains the test plans and test documentation, showing the test data with the results of testing and the actions taken. This is included for all parts of the system. Also, any known problems that could be rectified in the future are described so that future developers can attempt to resolve them.

The advantages of producing technical documentation include a reduction in the calls to support services because customers or users may be able to solve problems themselves. The documentation can provide safety and security warnings about the product and include disclaimers against damage to the product by end users or technicians. Disadvantages include the resources, such as time, required to produce a complete technical description of the product and the cost of distributing the documentation. Also, publishing technical documentation may reveal industrial secrets about the new system.

16.7 Evaluation

The **evaluation stage** determines whether or not the new system meets the specifications set out by the analysts and designers. A new system is evaluated to determine whether it is working as it was designed to do and is behaving as expected. Also, the evaluation seeks to determine whether the end users are able to access and use the system as required. An evaluation shows whether the system has any problems and how these may affect its workings. An evaluation should also determine the opportunities for developers to add more features or functions in the future.

16.7.1 Methods of evaluating a new system

The evaluation of a new system is carried out by the systems analyst. The analyst compares the new system with the requirements, system and design specifications that were produced at the beginning of the system life cycle. If the new system meets the requirements and criteria set out in these documents, the analyst can show that the development has been a success and has produced the required new system.

A new system should also be tested or evaluated to determine whether it carries out its intended use as expected and in accordance with current regulations. This ensures that the new system can actually carry out the expected tasks as described by its developer and seller and is, for example, safe to use. This is important because the developers and sellers may be subject to financial claims if the system fails to perform to an accepted standard, is used for purposes other than what was originally intended, or causes problems or damage to other systems. This evaluation requires thorough and complete testing of the new system and a review of the test results.

The end users' experience of using the new system is collected by gathering data using questionnaires, interviews and observations, in a similar way to how data about their use of the old system was gathered. The responses are analysed and used to determine whether or not the users find the new system easy to use, whether the new system makes their jobs easier or more difficult, and whether they found any problems or errors in the new system.

When evaluating a new system, analysts examine it, considering its usability, sustainability and maintainability. These can be used to evaluate whole systems, software applications or additions to systems. Each aspect can be assigned specific criteria and even given a score to quantify the evaluation.

An evaluation of the usability of a system or application includes determining whether the user documentation is well structured and easy to understand and whether it includes all the information that users need. It will also examine how easily the new system can be installed and loaded or run, and how easy it is for end users to understand and learn all of the features and functions necessary for their work. A questionnaire can be used to collect information from end users about these aspects, which asks yes/no questions followed by the option to comment. This allows the responses to be quantified and analysed. This evaluation is used to provide feedback to the developers to resolve problems or improve the new system.

Sustainability evaluates the management of the system or software application in terms of who owns it, who can license it for use and how future use of, for example, the system code is controlled. This is important because the system has to be deployed and maintained, so an evaluation determines how easy this is to do without infringing copyright laws, for example.

Evaluating the maintainability examines how easily the system can be updated or upgraded, tested and maintained. An evaluation of the new system will determine how easy it is to obtain or download updates and upgrades, and how long these updates or upgrades will continue to be made available by the developers to give an estimation of how long the system will be supported. How easily the system or application can be tested and updates or upgrades applied is also included in the evaluation.

The interoperability with existing systems and related software is evaluated to determine how well the new system has integrated into the existing processes.

An analyst can find limitations in the new system from an evaluation and should produce a report detailing these for the client. Limitations are features or functions detailed in the requirements specification that the new system is unable to provide or carry out. These limitations form the basis for the descriptions of improvements that should be made to the new system.

16.8 Maintenance

Maintenance of a system involves checking that the system is functioning correctly, replacing or updating modules or components that are defective or nearing the end of their planned life, and the routine activities that ensure the system keeps functioning properly.

The purpose of system maintenance is to ensure that the system keeps working as expected throughout its intended lifespan. Errors are corrected, features and functions are added or removed, and the performance is improved during the lifetime of the system. Any activity that addresses these tasks is part of system maintenance. Any changes to the system that are made during maintenance activities are tested, recorded, evaluated and documented so that the details can be added to the technical documentation.

System maintenance is planned during the analysis and design stages so that it can be included in the budget for development and in the plans for the implementation and deployment of the system, and can show when the maintenance will occur and who is responsible for carrying it out. Details of who is responsible are usually shown in a Service Level Agreement (SLA), which forms part of the contract between client and developers negotiated at the start of the development process.

16.8.1 Perfective maintenance

Perfective maintenance is carried out to improve the performance of a new system during its intended lifespan. This maintenance introduces improvements or changes that are not important enough to require a completely new system but, for example, make the system easier to use, or produce additional report formats or reduce the amount of storage space required for files by amending the database table structure.

Perfective maintenance can be carried out by applying software updates that remove, amend or add code to the system.

Advantages and disadvantages of perfective maintenance

Perfective maintenance can increase the useful lifespan of the hardware components and software in systems, which reduces the time and costs involved in replacing them with new versions. It reduces the number and occurrence of failures and errors in systems, which means that productivity can be kept at a maximum. It does, however, require the downtime or interruption to systems and processes, which can cause a loss of productivity and increased costs due to the time taken to carry out maintenance and testing. Perfective maintenance may need to be carried out when the systems are not being used so there may be increased labour costs involved.

16.8.2 Adaptive maintenance

Adaptive maintenance amends, modifies or updates the system when problems that were reported during the evaluation stage begin to affect the end user experience of the system and need to be resolved, or if the end users require the system to work with new hardware or software or there are changes to regulations or laws that affect the operation of the system. For example, changes in regulations that govern the lending of money require the amendment of accounting systems in banks.

Adaptive maintenance can be carried out by using updates or altering parameters within the system. For example, an update could enable the use of different hardware but a setting could be altered to apply a different tax rate.

Advantages and disadvantages of adaptive maintenance

Adaptive maintenance ensures that the lifespan and use of existing hardware, software and systems can be continued or extended when user requirements change, legislation and regulations change, or new hardware or software is installed without the costs and disruptions of installing completely new system components. It does, however, incur costs and loss of productivity while the maintenance, such as an update, is carried out, and during testing of the changes.

16.8.3 Corrective maintenance

Corrective maintenance identifies and corrects problems or errors so that the system is returned to its fully functioning condition. Corrective maintenance is either carried out immediately the error is discovered if the error is serious or is deferred if the error is considered minor. For example, an error that prevents data being output to all printers is significant and must be rectified quickly, but the correction of an error that prevents output to only one of many printers can wait.

Corrective maintenance is carried out in steps. When an error or fault occurs, the error is first diagnosed and the cause of the error located. This means that technicians have to determine what the error actually is and where in the code, or which module or setting, is causing the error. The error is then corrected and the solution is tested to ensure that the problem is solved. A record of the problem and its solution is kept for future reference.

For example, if a new system presents a problem with printing its output, a diagnosis will include whether it refuses to print any output or just refuses to print to specific printers, and the cause of the error would be traced to either the system software or problems with the printers themselves. The error is then corrected by modifying the system software or settings or correcting the problem at the printers. Testing the output from the system to all of the printers checks whether the problem is solved and the system can be fully used again. The error and its correction are recorded in a maintenance log or record. Often, errors in systems can be corrected by applying a 'software patch', which is used to replace some lines of code or add extra code to the system.

Corrective maintenance is also used when the new system does not work in exactly the way that was intended. There may be a faulty logic in the flow of data that was not discovered at the testing stage, so this must be corrected, for example by rewriting some code in a module to return the system to full functionality. This type of error is more difficult to correct as it may require considerable reprogramming.

Advantages and disadvantages of corrective maintenance

Corrective maintenance saves managers and technicians time because it requires less planning. It is a one-off process based on the immediate needs caused by the fault or failure, allowing technicians to focus on other tasks until required, and it can reduce the downtime of, for example, a system compared to the regular downtime of other types of maintenance. In the short term, corrective maintenance can reduce costs because work needs to be done only when required. However, carrying out maintenance only when there is a fault or failure can mean that hardware and software is not checked or monitored during its lifetime. Relying on corrective maintenance may mean that problems go undetected, which can result in deteriorating performance, increased unpredictability and, in the case of hardware, safety concerns.

16.8.4 Preventative maintenance

Preventative maintenance is the planned inspection, checking, testing and servicing of hardware and software to try to ensure that it does not develop significant faults. The aim is to prevent breakdowns or malfunctions so that hardware and software is used for the maximum amount of time at maximum potential with minimal maintenance costs. Planned inspections and tests of hardware can help identify potential problems caused by wear and tear, neglect, misuse and accidental damage. Preventative maintenance can be scheduled by how much use the hardware is getting, how often the hardware is used, how long the hardware has been in use, the status of the hardware or combinations of these and other factors, for example the way users work with the hardware. Action is taken in good time to prevent faults in the hardware having an impact on the workings of, for example, a network. Preventative maintenance of software aims to keep the software working as long as possible by addressing small issues or 'glitches' (often called 'latent faults') that are usually unnoticed by users and have little impact on normal use but may become of greater significance if left unattended. These are often corrected by upgrades, patches or minor changes.

Advantages and disadvantages of preventative maintenance

Preventative maintenance follows a schedule so that it can be planned, which reduces the impact on the user's productivity, reduces the costs related to errors and faults interrupting work, and helps to monitor the performance and keep track of hardware and software. It reduces the risks of failure by lowering the probability of unexpected faults, and the lifetime of the hardware and software can be extended. However, more IT technicians may have to be employed to regularly check and maintain the hardware and software, which may increase costs. There may be more interruptions due to the requirement for regular checking, which may reduce productivity.

Activity 16j

A software app has developed a problem and is now not displaying its user interface correctly on smartphones. A software engineer has been asked to fix this problem.
1   What type of maintenance will the software engineer carry out?
2   Describe the steps that the engineer must take in order to fix the problem.

Activity 16k

Explain the difference between adaptive and perfective maintenance.

16.9 Methods of software development

Developing a new system or new software involves, as described earlier in the chapter, finding out what is required of the software and creating the user requirements and system and design specifications. These are used by software developers, engineers and programmers to construct and test the new system or software application. During the construction process, different methods of development can be used with differing dependencies on specification documents, but all are meant to produce a final product that satisfies the client's requirements.

16.9.1 The agile method of software development

The **agile method of software development** is based upon:

» full involvement with the software developers by clients and other stakeholders
» increased collaboration and transparency between the developer team members
» rapid and predictable delivery times for the product
» predictable and manageable costs and timings
» a focus on the ability to adapt and change the product during development to meet changing demands and requirements from the client and others (for instance customers).

The agile method is designed to involve the client at every stage by providing clients with working software to try out and test in a short time. This is because the delivery of working versions for the client and end users is considered to be of the highest importance. When the client, and the end user, has reviewed the software and given feedback, the software is amended and given back to the client for review again. This process continues until the client and end users are satisfied that the product meets their requirements. The timescales, sometimes called 'sprints', for the production of versions and the client reviews are kept short, typically only a few weeks, to allow changes and amendments to be made quickly. Each sprint may focus on one feature or function of the new software so that each is completed and made ready for inclusion in the whole product.

For example, when developing a new app for a smartphone using the agile method, the client who wishes the app to be developed writes an initial requirements document that may not be very detailed but should include what the app is supposed to do (for instance create and edit a text message), what features it needs to have (such as a choice of font colours) and, possibly, some features that it will not have to begin with but that may be added later (for example the ability to use emojis). A meeting between the client and developer may occur to review the requirements and change them. Developers often want to have several meetings to review requirements to avoid 'feature creep' later on.

The next step is to gather a team of programmers or engineers and decide on, for instance, the programming language and the software tools to use so that the requirements can be met. A series of meetings would probably be necessary to review and amend the decisions if new features are added.

The app code is divided into sections for the various features and functions that can be programmed and tested quickly. These are completed and shown to the client for review. The code is repeatedly tested and reviewed to ensure that it is working correctly. The testing and amendments are repeated as many times as necessary according to the test results so that problems and errors are corrected. At each stage, the client and developers review the app code and make amendments and additions, adding features if required and testing the new features, until the app is finished. Errors and problems are detected early in the development of the app.

The client and developer may go back to the requirements document and add features to, or remove from, it at any stage. The agile method allows for this as it is focused on the production of a satisfactory end product, the software app, that delivers what clients, and maybe end users, want.

The agile method of software development works best when there is no definite limit to the time needed or to the budget for the development. Changes in the requirements are expected to occur during development. The development team can readily adapt to the changes in the requirements and the client needs the end product to become available quickly.

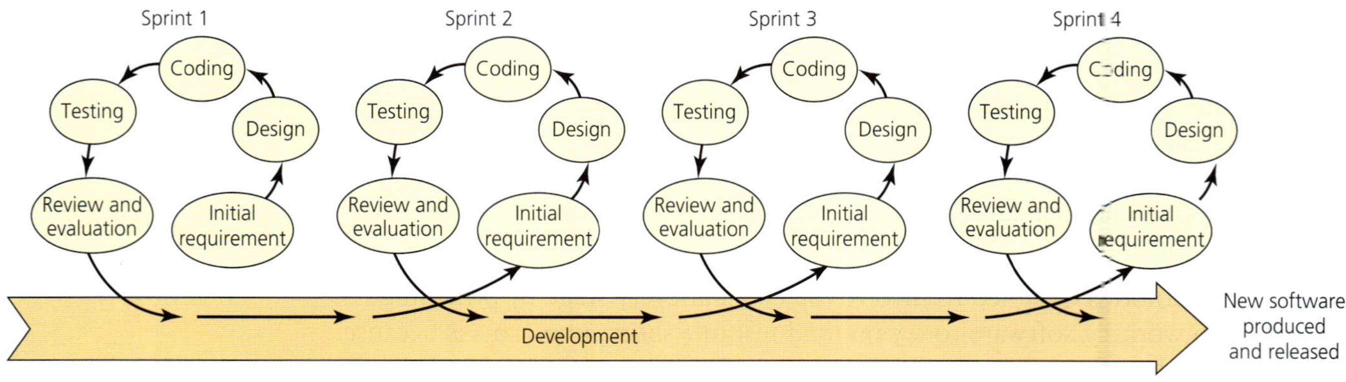

▲ **Figure 16.9** Agile method of software development

Advantages and disadvantages of the agile method

The division of the development process into sprints increases the transparency to the client, who collaborates more closely with the developers. This allows errors and problems to be discovered and eliminated quicker as each sprint is limited to a short amount of time and there is a shorter time between the coding and the testing and review. The final version is produced more quickly than with other methods and can be made available to the clients and their customers faster. This means that clients can start to make profits on their new software more quickly. However, the great amount of interaction and collaboration between the developers and stakeholders requires skilled management and can take a lot of time and effort.

The agile method does not allow an accurate assessment of how long the development will take or how much it will cost. This is because there is no definite set of requirements at the start of the development. The lack of detail in the initial documentation can make it difficult to replace or add team members to the development because they have no detailed requirements documents for reference.

Using the agile method to develop software applications can result in the development process taking much longer than expected because clients may keep adding more functions and features. This extends the development time, adds to costs and may mean that the software application is never fully completed.

16.9.2 The iterative method of software development

In this method of development, software developers follow the designing, coding and testing, deployment and evaluation stages to create a new software application that is not complete or perfect when released. Feedback from the client and end users is used to improve the software by following the stages again and a new, improved software application is released. This is carried out repeatedly until the software application meets all the initial requirements properly. Repeatedly going through the development process stages is called 'iteration', which gives the method its name.

The **iterative method of software development** is often combined with the incremental method, where a new software application is first released without all the intended features and subsequent releases are used to add features. Each new release adds a feature or function that the end user can make use of. Further, combining the incremental and iterative methods can also be considered to be an agile method because the new software can be altered between releases to meet different requirements.

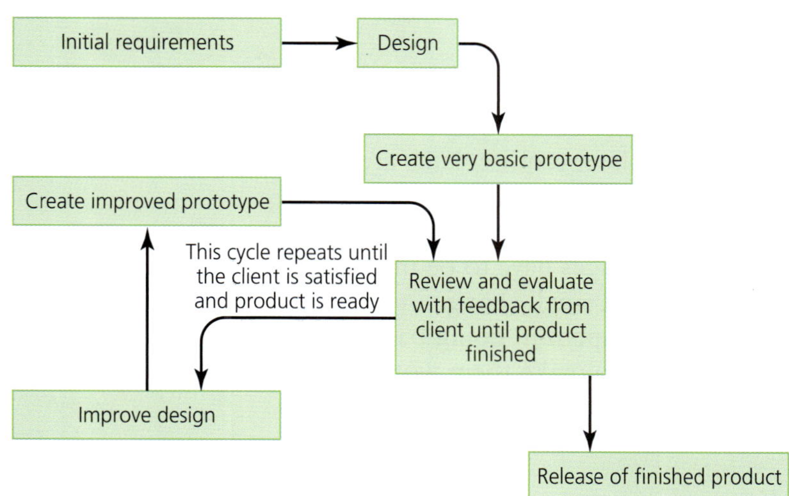

▲ **Figure 16.10** Iterative method of software development

Advantages and disadvantages of the iterative method

The advantages of the iterative method are that less time is spent on the documentation and more on the designs and programming. An incomplete product is created quickly and shown to clients and end users who discover defects and, using feedback, make suggestions for improvements. An end product with many desirable functions and features can be created quickly but, because all the requirements were not detailed at the beginning, any problems with the underlying system processes or architecture may not be discovered until later. These can be expensive to resolve in terms of time and money if the software is nearing completion.

16.9.3 The incremental method of software development

In this method of software development, a new system or software application is divided into sections, each of which represents a function or feature and is developed in the same way as a whole new product would be. The initial idea for the function or feature (for example the ability to use different font sizes), the section requirements, the designs, the programming and testing, and the reviews and evaluation are carried out on each section.

The client is involved in every phase of every section, and the development of the next section can occur only after the client has approved the section under development. Each of the phases in the section development often has a time limit to ensure that the section is delivered on time. This also produces the final new software application in a shorter time compared to some other methods, such as the waterfall method.

Usually, the requirement that has the highest priority for the client is developed first. When that requirement is met, the requirements for that component are not open to change and are frozen for the duration of the development of the other components. This is because changes to finished components can mean that components are forever being revisited and the development is never-ending.

The **incremental method** is best suited to developments where very highly skilled programmers are not available or they lack the training to deal with constant demands to amend or change functions and features.

As each function or feature is completed, it is added to the next version of the software until all the requirements are met and in place.

Advantages and disadvantages of the incremental method

The main advantages of the incremental method are that the new software application is created quickly and that it can also be changed at any point in the development process. However, because not all the client requirements are available, or thought of, at the beginning, problems can occur at any stage. While changes can be made at any time as the client reviews a component, changes in one component or module may mean that changes have to be made in other components or modules to accommodate them. A small change in one module may mean that a great deal of time has to be spent in going back to amend the code in other, already completed, modules to make them work together. This can also add to the overall costs.

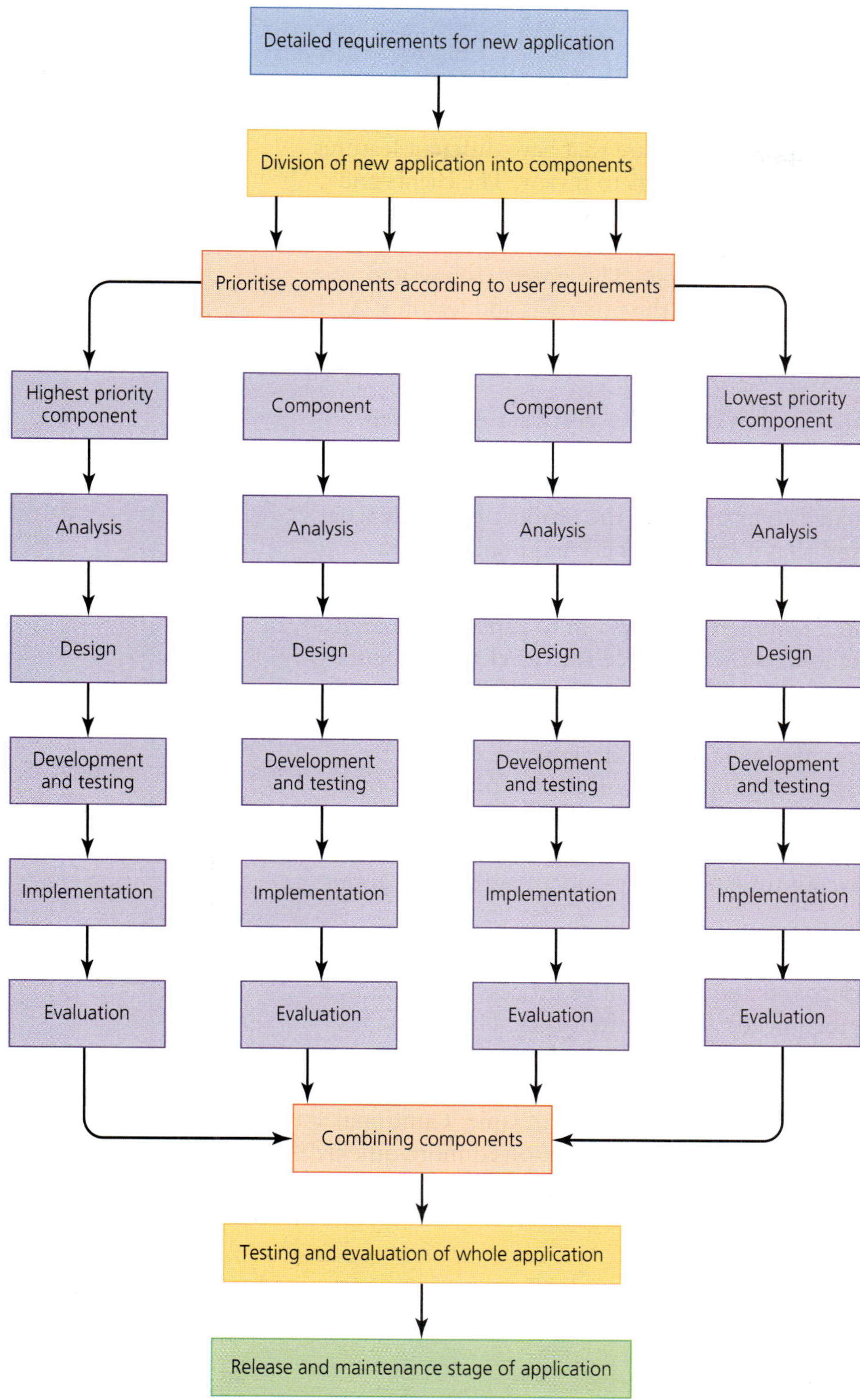

▲ **Figure 16.11** Incremental method of software development

16.9.4 The rapid application development method of software development

The **rapid application development (RAD)** method has similar stages to the other methods. There is a stage where the requirements are set out, a **prototyping** stage where functional versions of the new software application are quickly created, a reviewing stage using feedback from clients and end users, and a final stage where the application is completed and released.

In the stage where the requirements are decided, a detailed specification is not usually produced. Instead, an outline of what is required is produced. This allows the detail to be added at a later stage and to be altered to suit changing demands.

The prototyping stage produces versions of the software that have different features or functions and can be shown to clients and end users to review. The clients and end users provide feedback to the developers about the features and functions that they like or dislike or which work or do not work. The feedback is used to correct, amend or enhance the next versions of the prototypes. This process is repeated until the clients and end users are satisfied with all the functions and features. At this point, when all the stakeholders agree on the functions and features to be included, the final product is created by the developers.

In the next stages of RAD, the final version of the new software is reviewed and more feedback received. The final version is reviewed by the stakeholders so that all the features and functions, as well as the visual appearance and other aesthetic considerations, are evaluated to ensure that the product is exactly what the client and end users wanted and that it includes every requirement that appeared during development.

RAD can also be used to develop a complete new system to replace a current system. In the initial stages, stakeholders (for instance the developers, clients who could be company managers, end users and IT technicians) meet to decide upon the needs of the business, the scope of the system to be developed and the requirements. When these are decided, the designs are created by systems analysts and users to show all the inputs required, the processes that the system has to carry out and the resulting outputs. This is an iterative process with the stakeholders in constant contact with each other. At the end of this stage, the stakeholders have an understanding and an agreement about what is to be created.

During the programming stage, the system is constructed. There is continuous interaction between the stakeholders and another iterative process to produce a system that meets all of their agreed needs and requirements. The last stages are implementation and deployment when the system replaces the current, old system. In comparison to other methods, RAD can take less time from initiation to deployment because the stages are compressed in time. Continual iterative production and feedback mean that problems are resolved more quickly and client and end user requirements are met more quickly.

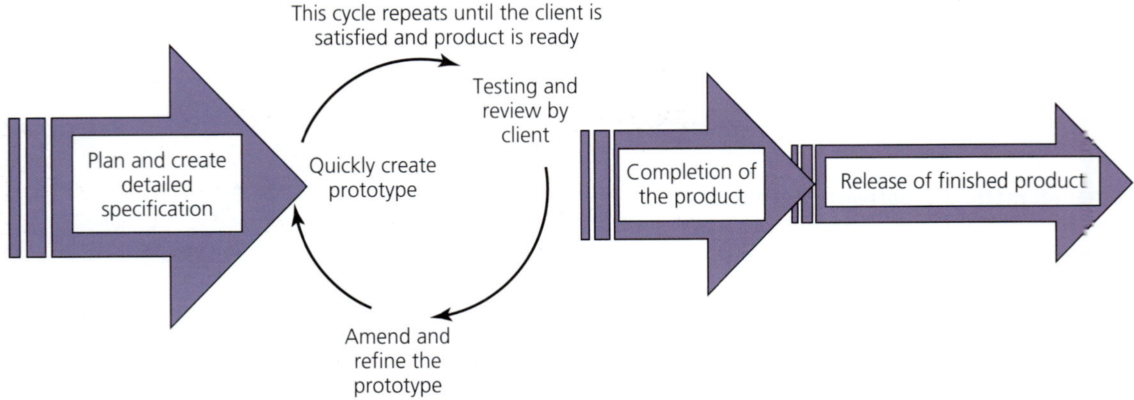

▲ **Figure 16.12** Rapid application development (RAD) method

Advantages and disadvantages of the rapid application development method

RAD has the advantage that the client and end user requirements are a priority, which means that these can be changed at any time during the development. However, the continual need to ask stakeholders to review and provide feedback requires good collaboration skills from all members of the team. This is difficult to achieve and manage with very large teams working on the development, and RAD is usually more successful with small teams working on software that needs a short development time.

The involvement of clients, and usually end users, is necessary throughout the development and, while this usually results in a better final product, it can be more complex and time-consuming to manage than when there are fewer people involved. The close involvement of clients and end users can move the focus of the development away from the underlying functionality and more towards features that are visible or of aesthetic importance only, for example not focusing on the control of data input but on how the colours of the input screen look to the eye. This can also be a problem if developers continually have to make minor alterations to some features and do not spend enough time working on how all the components work together.

RAD does not work well with the development of very large systems. The large number of people, and the large number of software components or modules that are involved, mean that interactions between developers and stakeholders, the iterations of the many stages in the development of components and the analysis of all the reviews and feedback are complicated to manage.

16.9.5 The waterfall method of software development

The waterfall method of software development consists of a series of stages that follow on from each other. The stages in the waterfall method are:

» a full analysis, design and production of the specifications before development work begins
» programming the code for the application
» testing and review by the developers
» review by the client and end users
» deployment and maintenance.

Each stage, from the starting initiation, through analysis, design, programming and testing, to the implementation, deployment and on into maintenance, depends upon the successful completion of the deliverables from the previous phase. Progress in developing a new software application can be considered to flow 'downwards' in a linear sequence from start to finish. This is why it is called the waterfall method.

In the waterfall method, the development can move on from one stage to the next only when all the reviews have been carried out and any problems resolved. If problems are discovered later in the development, such as errors in the code, then a return to the previous stage is required. It may be necessary to return to the design stage if defects are found at later stages.

Advantages and disadvantages of the waterfall method

The main advantage of the waterfall method is that the time and costs of the later stages can be reduced because most problems are detected at the initiation, analysis and design stages when the requirements are decided. Most of the

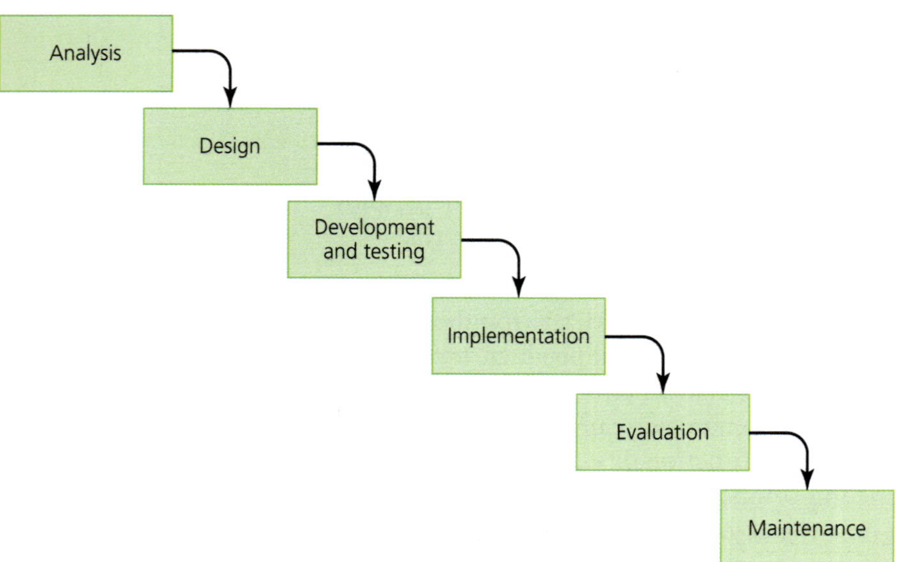

▲ **Figure 16.13** Waterfall method of software development

planned development time is used in the initial stages so that the resulting in-depth documentation should highlight and resolve potential problems before the actual programming work begins. The development process is structured and controlled throughout so that deviations from, or alterations to, the functions and features are kept to a minimum to ensure that the client requirements are properly met. The structured control of the development process also allows clearly defined, distinct and recognisable milestones to be set and used to monitor progress.

A disadvantage of the waterfall method is that the clients may not know or fully understand their needs or requirements at the beginning. They may need to see a functional prototype before they can be sure of exactly what they want. In the waterfall method, the requirements are set out at the beginning, making it difficult to change them if the client asks for alterations or more features as the development progresses.

16.10 Prototyping

Prototyping is used during the development of a new software application. A prototype is an early version of a new software application that can be used and tested. A prototype is usually functional but may not have all the functions and features of the final product. Although a software prototype may look and feel different from the final product, a prototype allows the testing of all the critical functions and features before the final implementation.

Prototyping a new software application allows the developer and client to check whether the new application meets the specifications and whether the development is proceeding as originally agreed, for example whether it is likely to be finished on time. During the prototyping, end users may be asked to try out the new application and provide feedback that is used to improve it, the inputs and outputs are checked to determine whether or not the application is working as expected, and the opportunity is taken to review and revise the specifications so that improvements or additional functions or features can be added.

The methods of prototyping new software applications are evolutionary, incremental, throwaway and rapid prototyping. All methods require:

» an understanding of the specifications
» the design and creation of a working version
» the prototype
» the application
» the testing of the prototype, for example by a small group of end users
» the gathering and analysis of feedback from the testers
» a review of the feedback
» the revision and amendment of the prototype.

The process is then repeated until the developers and client are satisfied with the application and the final product is ready.

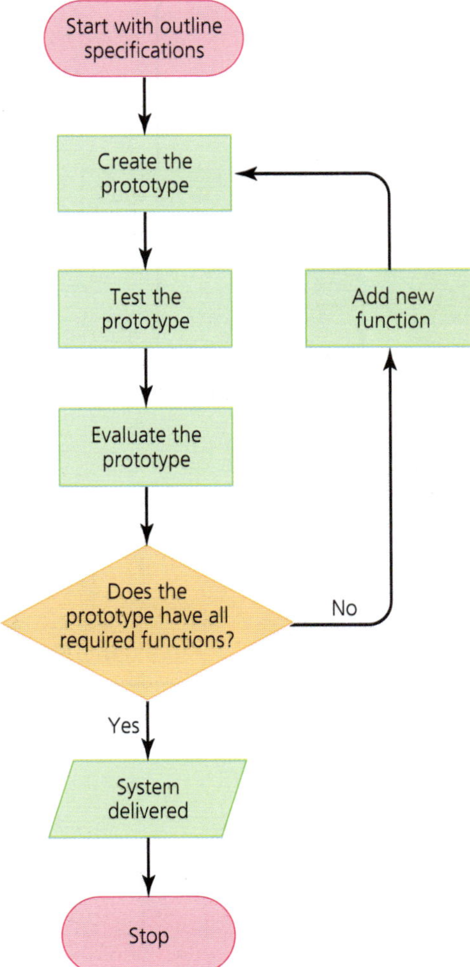

▲ **Figure 16.14** The prototyping process

16.10.1 Evolutionary prototyping

In **evolutionary prototyping**, a working prototype is first developed that has only the basic functionality. Once the basic version has been released, additional functions and features are added and tested. When these are fully functional, more functions and features are added to the basic prototype version and tested. The progressive adding of new functions and features builds on the existing application until all of the required functions and features have been added to the application. The new application 'evolves' during development from the basic version on to the completed version.

Advantages and disadvantages of evolutionary prototyping

The team developing the new software application starts by creating the parts of the application that they fully understand (there is no need for them to fully understand the whole new application). The team concentrates on creating a properly working basic version. It does not risk creating a full product that has all the functions but does not function satisfactorily because the team does not fully understand the requirements. The team can add the other functions or features to the software as development proceeds. The developer can add new functions or features that the client did not mention, or even think of, in the specifications as they develop the software. There is continuous communication between the developers, clients and end users to allow the new application to change and evolve during the development process. Also, the client is more likely to receive a final product that matches the requirements, and any problems experienced by the end users are more likely to be resolved before the final product is made available.

Evolutionary prototyping creates a product that works but may not have all the functions and features that were required. This means that a client may decide that all the original functions are no longer required and that the partially completed prototype is good enough. The full product is

▲ **Figure 16.15** Evolutionary prototyping

never completed. Also, some of the originally required functions may never appear in the final product because the client wanted different functions during development. In cases where the client or the developers keep adding additional functions or amending the application, the development process may go on for a very long time or may never properly end, resulting in budget overruns, a loss of motivation by the client or developers, or confusion among end users.

16.10.2 Incremental prototyping

With **incremental prototyping**, a new software application is divided into its component parts and each part is developed into a prototype. All of the prototypes are tested and refined into fully working parts of the application. When all of the prototypes are fully tested and functional, they are combined into the complete application.

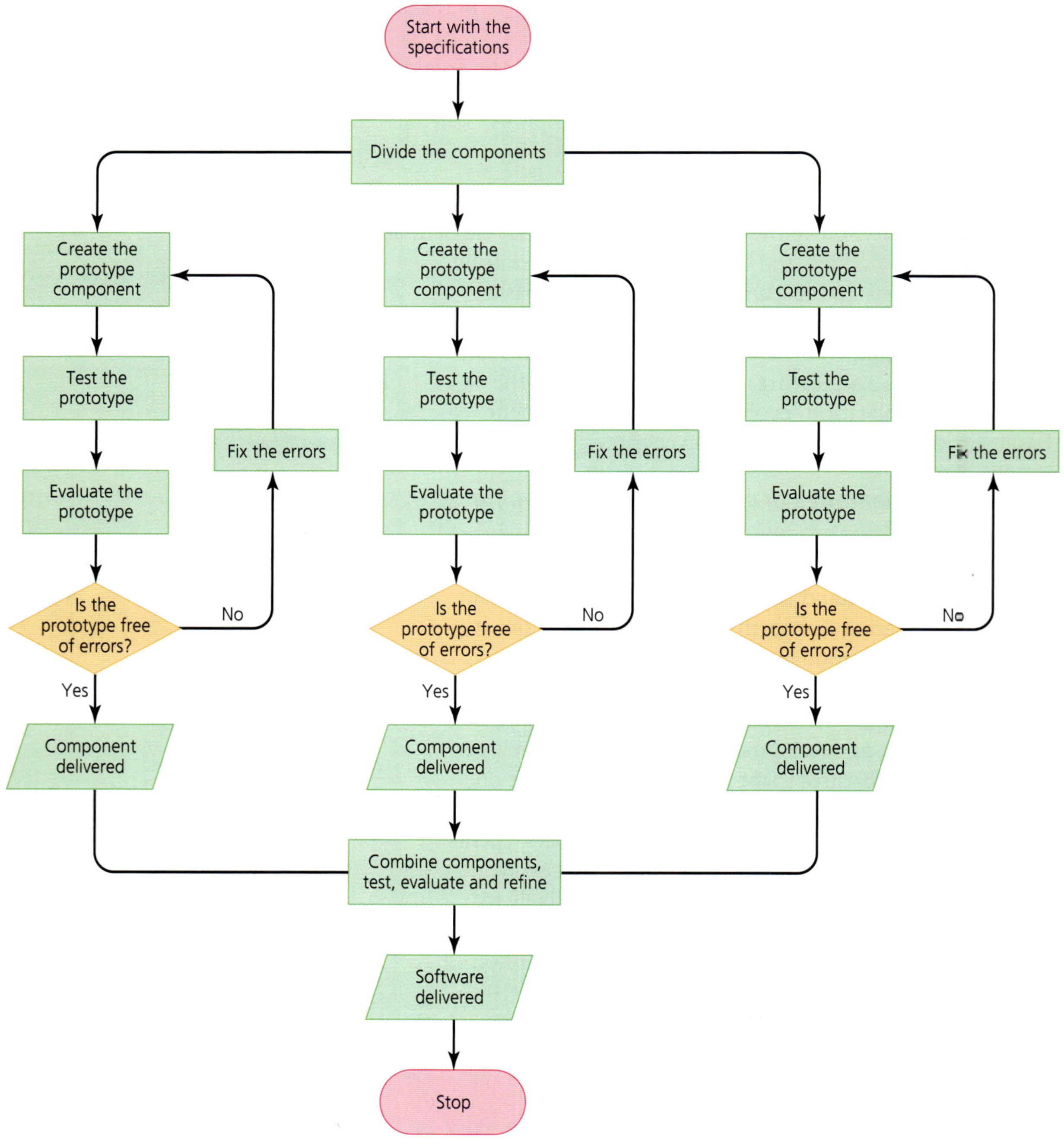

▲ **Figure 16.16** Incremental prototyping

Advantages and disadvantages of incremental prototyping

One advantage of using incremental prototyping is that the specifications detailing the requirements and designs are fully considered during the development. Other advantages are that each part of the new application is fully tested before being combined into the whole, the testing and refining of each part is easier and more convenient before it is combined into the whole, and delivery of the complete application can be achieved in a shorter time because parts are developed at the same time as others.

The main disadvantage of incremental prototyping is that it raises costs because the development of different application parts are carried out in parallel, requiring more programmers. There may be extended testing of the complete application to ensure that all the parts work correctly together and this may increase the costs of development.

16.10.3 Throwaway prototyping

Throwaway prototyping creates a working version of the new software application for testing and review. When the testing and review are finished, the prototype is discarded (thrown away), and a new prototype, based on the feedback from the earlier prototype, is created for more testing and review. This is repeated until the fully working, complete version is produced. Throwaway prototypes are also used to allow clients and end users to gain more insight into their actual requirements by trying out versions of the software.

Advantages and disadvantages of throwaway prototyping

Throwaway prototyping is carried out so that the client and end users can quickly see what the new application will be like and the developers can receive feedback. Problems with the new application can be detected at an early stage in the development and throwaway prototyping may cost less than the other methods because there is no need for more than one group of programmers as parts are not being developed concurrently. Changes and refinements can be made early in the development process and incorporated into the next prototype. This can reduce the time and costs of testing the final product.

A disadvantage is that because the prototypes are discarded, so the time and effort used in creating each prototype is not invested in the final product. Sometimes, useful refinements are discarded when a prototype is thrown away and do not appear in the final product.

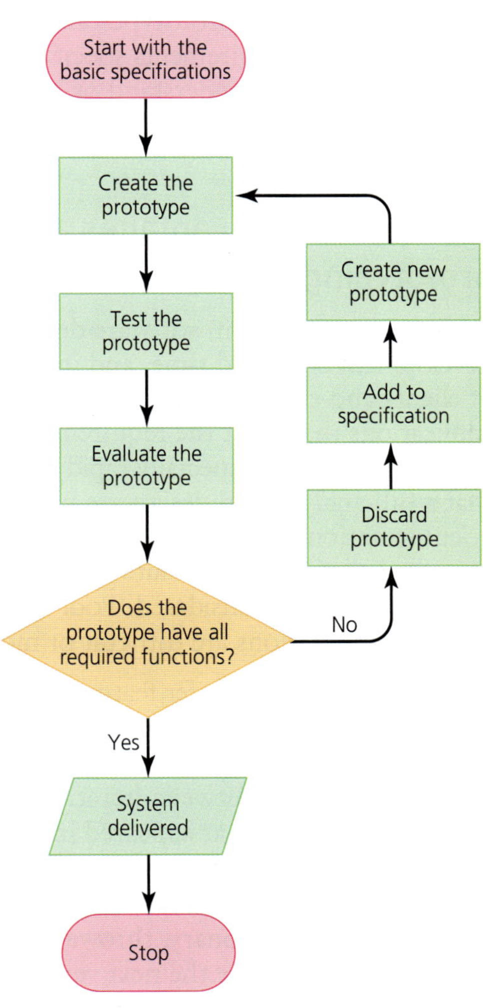

▲ **Figure 16.17** Throwaway prototyping

16.10.4 Rapid prototyping

Rapid prototyping creates a working version of the new software application in a very short time. The prototype is used for preliminary testing by developers, clients and end users, for example as a proof of concept, which means to show that the application can actually be developed. It is also used to enable clients to clarify their requirements and what they expect the application to be able to do. The prototypes are discarded and new ones created with the refinements.

Advantages and disadvantages of rapid prototyping

Rapid prototyping is carried out so that the client and end users can have a working version of the new application to try out and so that the developers can receive feedback in a very short time. The reduced time allows for the final product to be completed quickly. Rapid prototyping requires good communications between the developers, client and end users throughout the development, and the quick, repeated prototype–feedback cycle results in a product that meets all, or nearly all, of the requirements.

The disadvantage of rapid prototyping is there is often difficulty in managing the development process because there is no clearly defined plan. The development follows the results of the testing of the prototypes and may not stick to a pre-determined requirements specification. In many cases, the client may not fully understand or be clear on the requirements until development starts and may change these as the process continues.

16.10.5 Other advantages and disadvantages of prototyping in software development

Overall, prototyping in software development reduces development costs and time, and increases the involvement of clients and end users with developers, which results in applications that meet the requirements better. On the other hand, prototyping, especially rapid or throwaway, can mean that a full analysis and design are not carried out, and full documentation is not produced, before the work starts. This can mean that desirable solutions or features are sometimes overlooked or not considered. Poor solutions can mean that software applications are difficult or more expensive to maintain in the future.

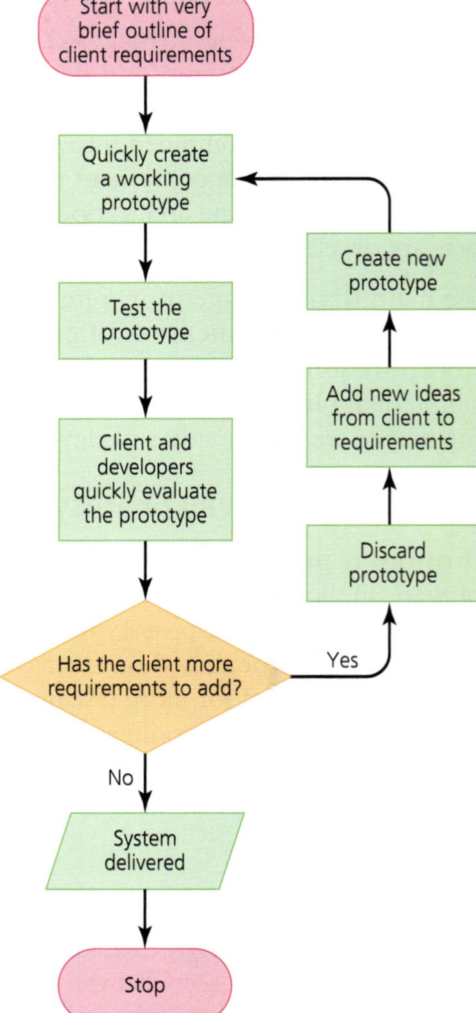

▲ **Figure 16.18** Rapid prototyping

Some end users may prefer features in the prototypes that the client does not want and so become confused when they use the final product. This is especially so with throwaway or rapid prototyping where functions and features are tried, added or removed between iterations. Also, clients may require that every function or feature that appeared in the prototypes be included in the final product. This may lead to disagreement between developers and clients because other than with incremental prototyping, detailed and binding specifications are not used. Evolutionary, throwaway and rapid prototyping use versions of prototypes to create the next version depending on the feedback received, whereas developers using incremental prototyping follow the specifications to create and test each required part of the new application.

Activity 16l

1 Explain why incremental prototyping would be used when developing a gaming app for smartphones but throwaway prototyping would be used for an app that allows users to make short written notes.
2 Explain why rapid prototyping requires good communications between clients and developers.

Practice questions

1 Use the data flow diagram that you drew for Activity 16f to draw a
   level-1 DFD for the coffee shop system. Use the symbols shown in
   this textbook.   (8)

2 An accounting system has been used in a bank for several years. IT
   technicians have worked on the system and corrected a few software
   errors, improved and updated it over the years to keep it working properly.

   a What types of maintenance have the technicians carried out?   (3)

   b Explain why some errors in the system were not fixed during the
      development of the system.   (4)

17 Data analysis and visualisation

> **In this chapter you will learn:**
> ★ about transforming, cleansing and combining data
> ★ about displaying the results of analysis.

> **Before starting this chapter, you should:**
> ★ be familiar with:
>   – spreadsheets
>   – relational databases
>   – database management and file concepts
>   – Chapters 8 (especially 8.4) and 10 (especially 10.1 and 10.4) of the AS syllabus and in *Cambridge International AS Level Information Technology*
> ★ be able to analyse, interpret and display data to communicate information visually.

17.1 Data analysis and visualisation

17.1.1 Transforming and cleaning data to extract meaningful information

Data transformation is the changing of data from one format, structure or value into another. The original source of the data may not be in the condition that is required for processing or, for example, creating reports. Data transformation can be carried out by specialist software or with scripting languages such as SQL. Data can be transformed in several ways:

» Constructive data transformation adds, copies and replicates data, for example data can be combined with other data, such as customer information from sales databases with that from marketing databases.
» Destructive data transformation deletes fields or records, for example it can simplify data for analysis such as anonymising information by removing names and changing specific ages into age ranges to ensure that individuals are not identified during data mining.
» Aesthetic data transformation changes data to a standard form, for example using the same standard format for dates or names.
» Structural data transformation reorganises data by renaming or combining entities in a database.
» Normalisation of data for use in relational databases is also a type of data transformation.

Cleaning data is the process of removing incorrect, incomplete, irrelevant, improperly formatted, replicated or corrupt data from a **data set**. If data is not cleansed, the faulty data can slow down processing, cause inaccuracies in calculations and may cause inaccurate, unexpected or biased results to appear.

At the end of data cleansing, the data should be:

» of high quality and valid, for example be what is needed for the analysis that will use it
» accurate, for example be the true values, such as a person's name spelled correctly every time
» consistent, for example be the same where the data is recorded twice, such as a person's name or address
» complete, for example have all the necessary data
» uniform, such as using the same units of measurement.

Data cleansing involves auditing the data set using specialist software or commercial database management software to check for and remove anomalies and contradictions. Any anomalies that the software cannot reconcile are reported and have to be removed manually. Data cleansing can be expensive because it can take up a large amount of computing resources, and is time-consuming to check and carry out manually.

17.1.2 Getting data from different sources

Combining data from different sources is called **data integration**. Where the data comes from is called a 'data source'. When collecting data, primary data sources are the original data, for example from interviews, from observations, on data capture forms or readings from sensors. Secondary data sources are those that have derived data from the original, primary sources. In computing and IT, a data source usually refers to the location where the required data is to be found, which could be a text file, a spreadsheet, a database file, XML or other stored data. Data sources are used to combine data when merging, searching or analysing spreadsheets or databases, or when creating mail-merged documents. Websites responding to user requests (for example when searching for information, during online shopping or banking) combine data from different sources, such as customer, stock and financial databases.

Activity 17a

Files for data sources

A CSV file can be used as a data source.
1 Describe the structure of a CSV file.
2 Describe one other use for CSV files.

Reports can be created in word-processing software using data that is dynamically linked to spreadsheet or database files. Embedded charts or tables can be used for illustration. Linking allows information in a report to update automatically if the **source file** is updated because only the location of the source file appears in the document. Embedding a section, or object, from a spreadsheet does not allow for automatic updating because the data becomes part of the new document and is no longer part of the source file.

Combining data in spreadsheets

The simplest way to combine the data from two or more spreadsheets is to copy and paste the data from the original into the others. This method is useful only if the data being copied is unlikely to change. If the data does change, the copied data will not change automatically so it will have to be updated manually.

If there are several sources of data that are required to be linked, for example when summarising or using data from several spreadsheet in another, the original data in the source spreadsheets can be linked to the summary spreadsheet. In this case,

17

when the source data is amended, provided that the summary spreadsheet is still linked and can still locate the original data and automatic calculation is enabled, the data in the linked spreadsheet will update. The links can be broken so that the data is no longer dependent on the source data, but then it will also no longer update automatically. Links can be created using formulas or by hyperlinks. The link information contains references to the spreadsheet name and to the cell where the source data is stored. In the case of using a different, separate spreadsheet file for the source, the filename is also included.

In Microsoft Excel®, cells can be linked to cells in different workbooks (files) by opening both workbooks, or all if there are more than two to be linked, and typing the = (equals) sign in the cell where the data is to go, selecting the cell in the other workbook from which the link is to come and then pressing Enter, as shown in the image.

In Excel, the data in the linked workbook cell usually updates when the data in the source workbook is changed. A similar method is used in LibreOffice® Calc® but the data may not automatically update.

Spreadsheet applications also include tools to import and combine data from different sources. In both Microsoft Excel and LibreOffice Calc, there are several ways to import and combine data from different sources. Excel has numerous options for importing from files such as other Excel workbooks (files), CSV and XML as shown here.

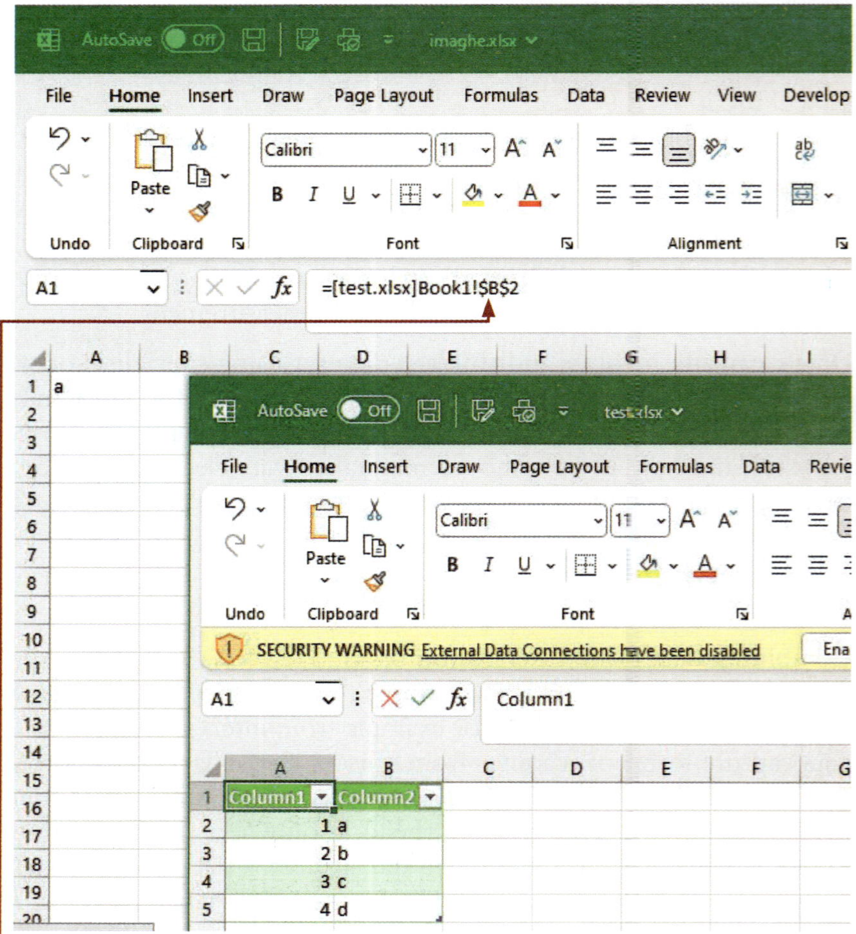

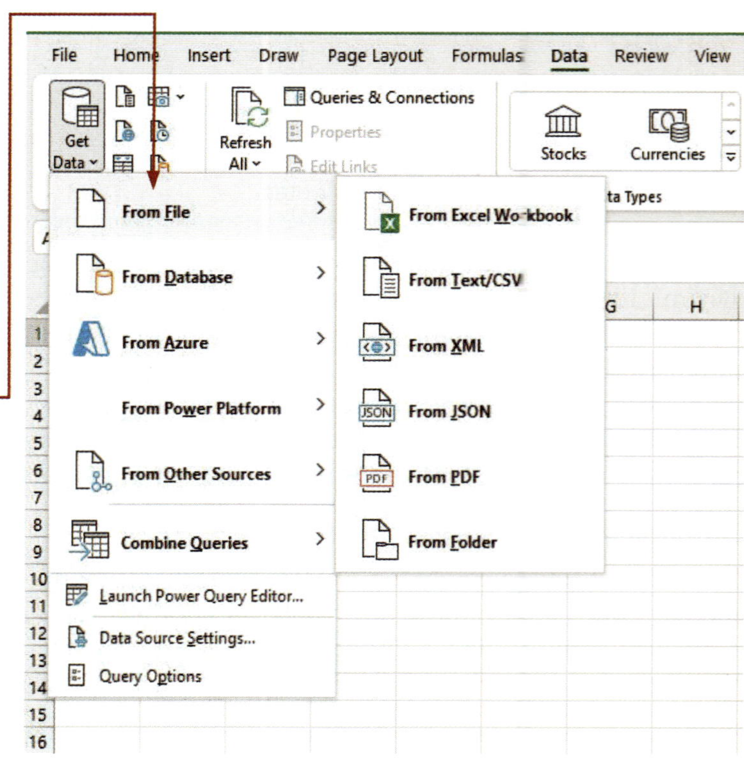

Calc uses the **Open** menu option as shown here.

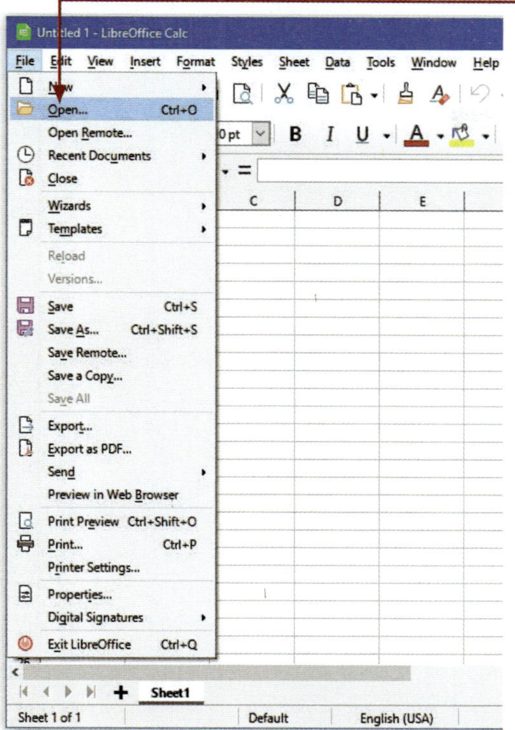

When a file has been chosen for import, both spreadsheet applications provide options for transforming the data into their own format before loading it into the cells. In Microsoft Excel it looks like this:

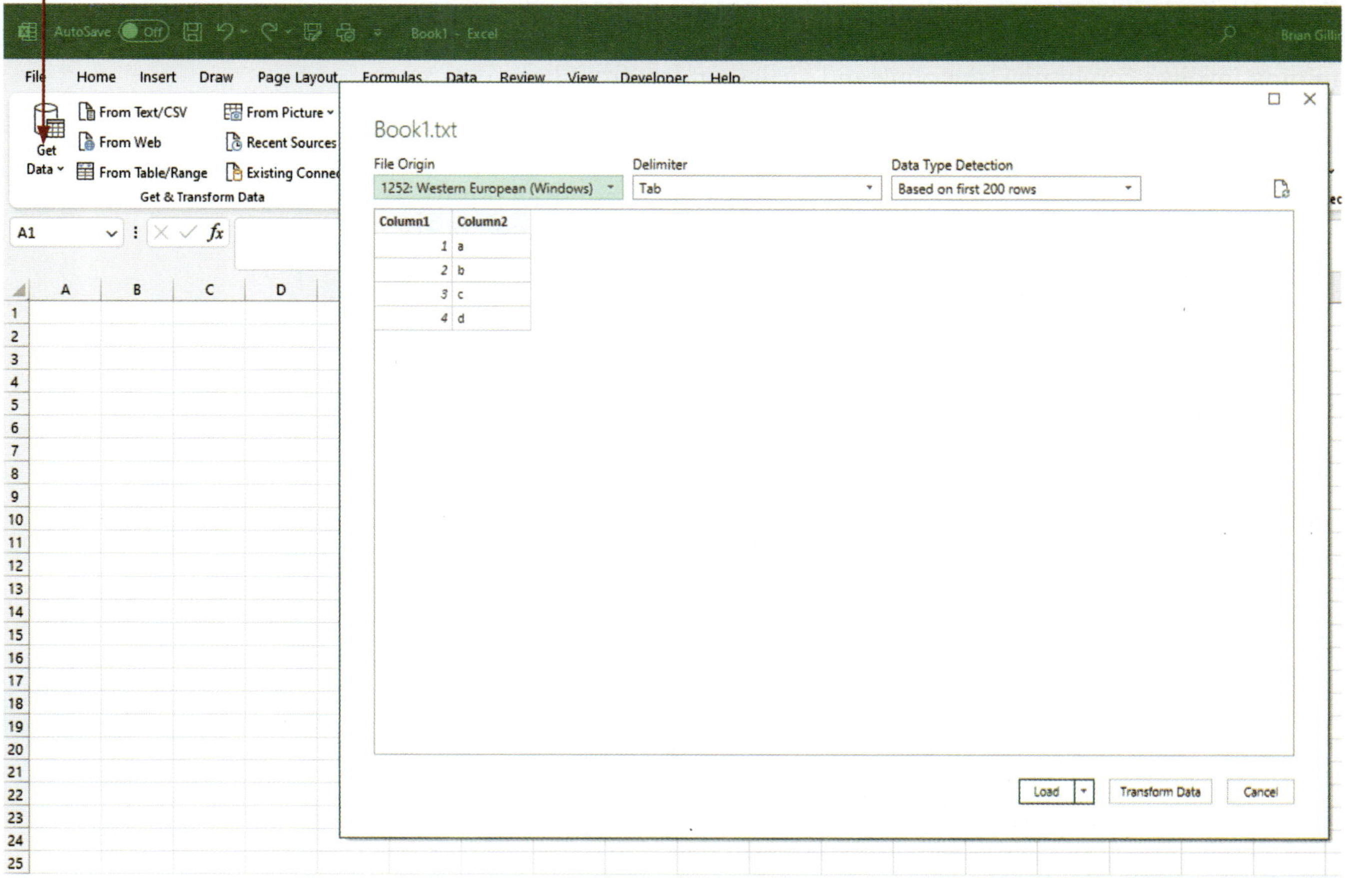

Importing data from files in LibreOffice Calc looks like this:

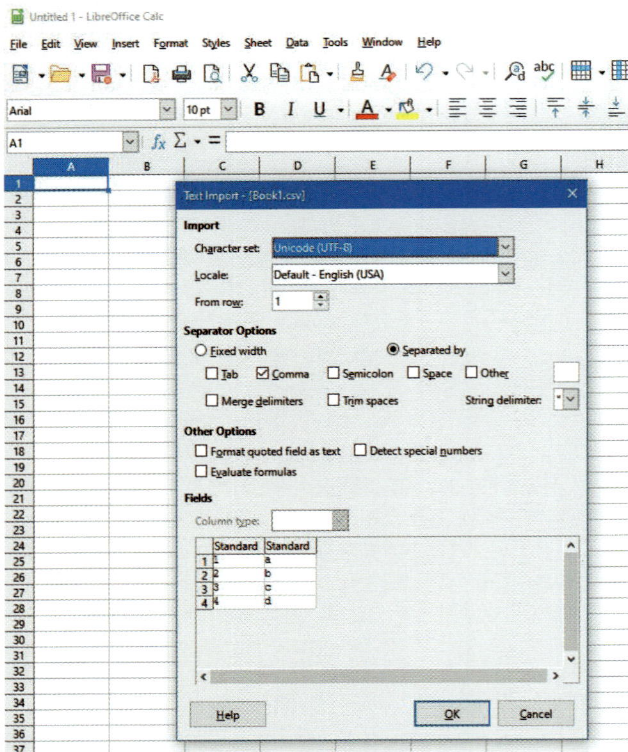

Combining data in databases

Combining data from several different databases allows complex queries, or searches, to be carried out across vast amounts of data. Businesses can query their databases (for example customer, stock, order and financial) for the purposes of marketing research or creating reports for stakeholders. Data mining involves combining, querying and analysing data from large data sets from many databases.

Relational databases store data in different tables within the database. Many tables can be linked so that data can be extracted. There are several ways to join tables in Microsoft Access® or in LibreOffice Base®. Structured query language (SQL) can also be used to select and extract data across tables in relational databases.

Comparing and consolidating data from two data sources

In spreadsheets, formulas can be created using the functions provided by the spreadsheet application to compare and consolidate data from different sources. Comparisons between data (for example equals to, greater than, less than) can be made in spreadsheets. Complex comparisons and decisions based on the results of the comparison can be used in, for example, nested **If** and various types of **Lookup** statements. In-built functions, such as **Concatenate**, can be combined into complex formulas to bring together the contents, or various parts of the contents, of cells from the same worksheet, different worksheets or different workbooks.

When working with data stored in database tables, comparing data to find out whether or not the data in the fields match can be very useful. For example, for marketing and further sales opportunities, database tables of customers and their purchases and of salespeople can be compared to find out which customers have been sold which items by which salespeople.

To compare two database tables to find matching data, queries can be used. A query is created either to make a new join between fields, which must be the same type, or by using an existing join between the tables, to extract only the data that matches. Fields that hold similar data but are of different types, for example a field that has a number data type but is to be compared with a text data type, cannot be joined. However, a query can be created to use a field in one table as the criterion against which a field, or fields, in other tables can be compared. A select query is created to include the tables, the fields to be displayed are included and the field to be used as the criterion is selected. The type of comparison is chosen, for example equals to. Complex queries can be set up across several tables.

Data consolidation takes data from different sources, cleanses it and combines it. This enables the processes of transformation, analysis and reporting to be carried out more quickly and easily. In businesses and organisations, data is consolidated in a data warehouse. A data warehouse is a large database that takes and stores data from other databases and upon which analysis is carried out. A specialised program, or script, is used to extract and load (EL) data from the source and place it in the data warehouse. This is done by running queries to extract the required data, creating tables to receive the data and loading the data into the data warehouse. Specialised EL tools (ELT) are also available commercially. This is inexpensive, but for smaller businesses and individuals, Microsoft Access and LibreOffice Base have the ability to consolidate data.

Microsoft Access can work with several different file formats and has a powerful query and SQL system that can consolidate data in many ways according to the requirements of users.

Splitting data into discrete fields

It may be necessary to split data in one field into several new fields. For example, importing customer details from a poorly designed spreadsheet or database may have all the address details in single field. To enable searching, for example of just the street or city, the data would need to be split. Ideally, this would be done before the data is imported as part of the transformation process but, if not, database queries can be set up to extract the data from the field and place it in new fields. Microsoft Access has functions, such as **Split()**, which can be used in queries to extract data from a string in a field for use in other fields or reports. The **Split()** function has the syntax:

```
Split(string,delimiter,limit,compare)
```

Where **string** is the contents of the field, **delimiter** (or **separator**) is where the split should be and this is taken to be a space if nothing else is stated here, **limit** is the number of splits returned and this is set to -1 by default, which means all of the splits are returned to the system for use in other fields, and **compare**. **Compare** is optional but can be used to make comparisons based on the string contents.

In LibreOffice Base, the syntax of **Split()** is:

```
Split(string,delimiter,number)
```

The first two parameters are the same as for Microsoft Access but **number** is how many strings are to be returned.

Activity 17b

Splitting data fields

Split() is a function for splitting strings. Individual or groups of characters can be extracted from strings in database fields for use in other fields. Research and make notes on the other functions that could be used to extract characters from a string.

Merging and combining data into required fields

While merging tables with fields of data is sometimes required, it is not usual to permanently combine data from several fields into one field. It is usual database practice to keep data at the lowest possible level of detail, or as atomic as possible, which is why relational databases often have many tables with joined fields.

However, merging and combining fields at runtime; that is, only as and when required and not altering the underlying data structure, can be used to give an overview of the data, for example in reports. In Microsoft Access, two fields can be combined in a query using the **&** character, which will concatenate the chosen fields with a space separating each field. In both Microsoft Access and LibreOffice, SQL scripts can be used to select and combine data in different fields.

17.1.3 Displaying data to communicate information

Data from spreadsheets and databases is useful only if it can be understood. Raw data is meaningless, but when data is processed, put into context and understood, it becomes information. How data is displayed to a user can help with the communication of the information.

Graphs and charts can be created from data. Graphs and charts present data visually, which makes it easier to read and understand. Graphs and charts enable viewers to visualise trends and relationships between data with minimal mathematical or analytical skills. There should be a clear, meaningful title and the axes should be clearly labelled. The chart or graph should be uncluttered and with no unnecessary data or information, the units of measurement should be clearly visible and appropriate, and the axes should not be distorted or disproportionate. Where appropriate, the data source should be included.

Activity 17c

Revise the different types of graph and charts and their uses.

Pivot table reports

Pivot tables are used in spreadsheets to summarise large amounts of data. They can also be used to look for patterns and trends and can include links to external data sources.

It is important to ensure that the data you wish to analyse with a pivot table is correctly formatted and suitable, meaning it has been transformed and cleansed ready for use. For example, each column in the spreadsheet should have a header (or title), there should be no blank columns or rows and there should be no rows with totals. Table 17.1 shows a simple set of data about the sales of computing items that can be used to create a pivot table in Microsoft Excel.

▼ **Table 17.1** A simple set of data used to create a pivot table in Microsoft Excel

Product ID	Type	Item	Total sales
156	Printers	Inkjet	564.21
201	HDD	1TB	1200.67
768	SSD	2TB	452.69
769	SSD	4TB	325.45
236	Graphics	Scanner	120.50
125	Network	Router	245.35
158	Printers	Laser	452.10
222	HDD	500 Mb	76.90
784	SSD	250 Mb	46.75
127	Network	WAP	87.90

The data is entered into a Microsoft Excel spreadsheet and prepared for use, as shown below.

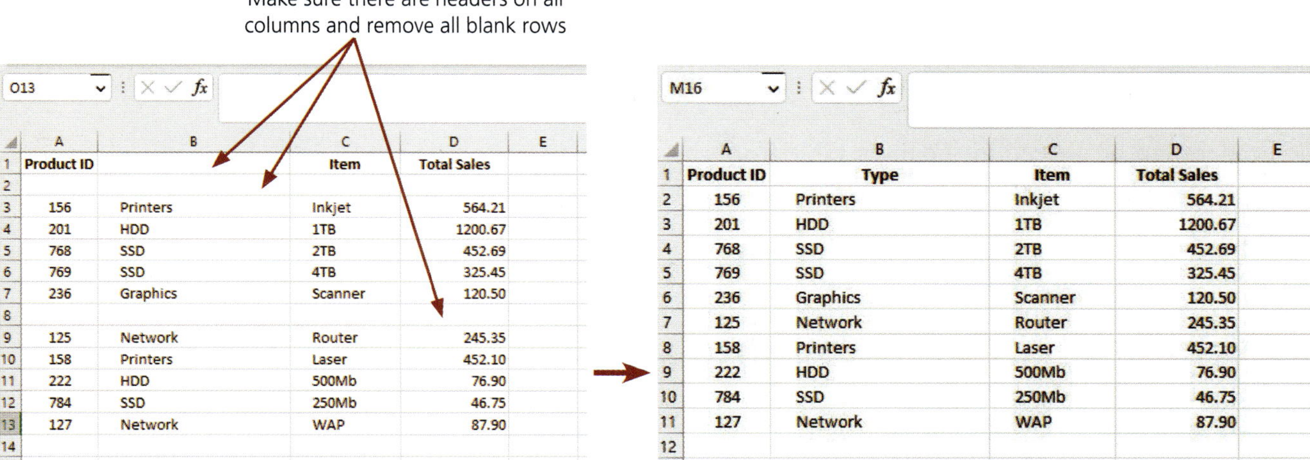

When the table is prepared, a pivot table report can be created from the **Insert**, **Pivot table** menu options. The pivot table report is placed in a new sheet in the spreadsheet workbook, as shown below.

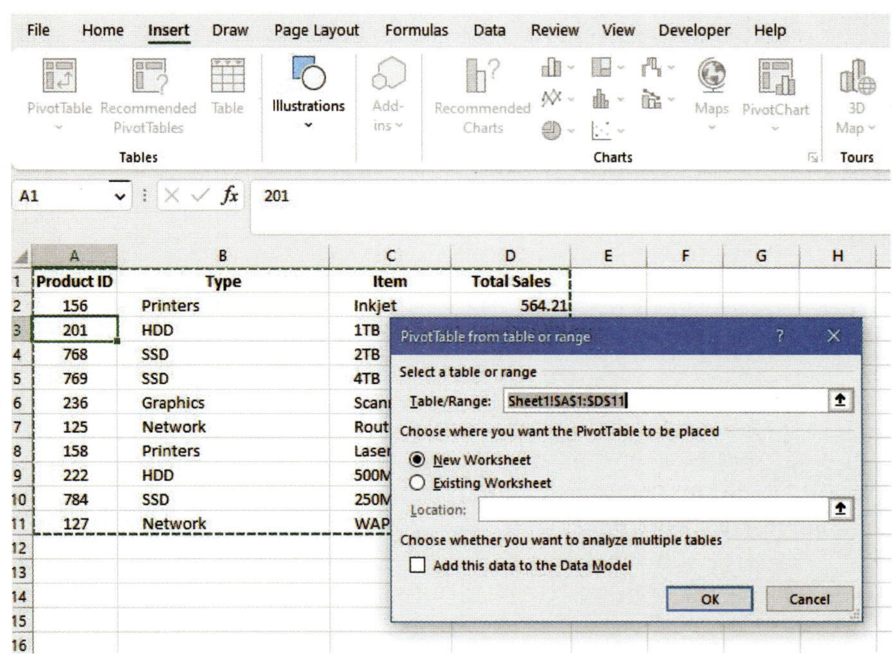

The data to be included in the pivot table report is chosen from the dialogue box that appears when the table is created. The **PivotTable Fields** dialogue box can be opened at any time by a right-click on any cell in the pivot table and choosing the **Show field list** option, as shown here:

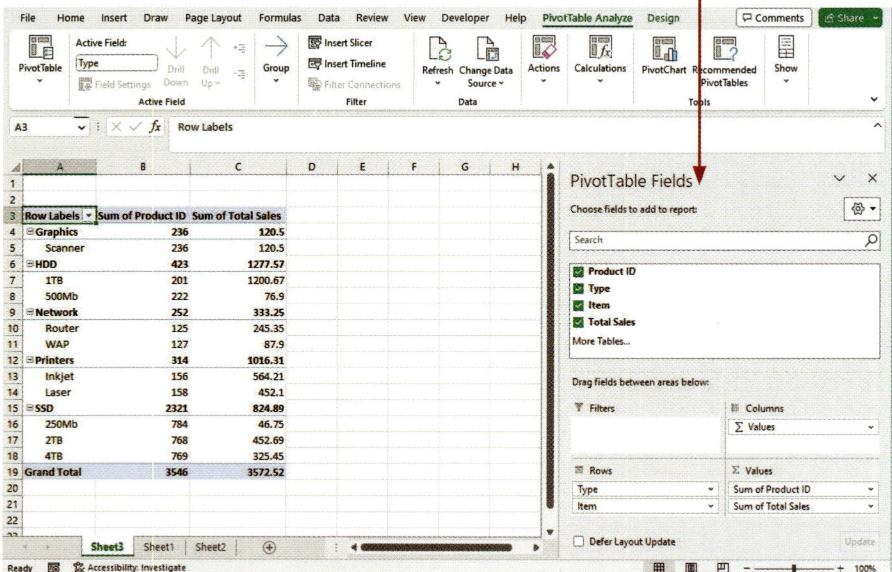

The headers can be altered in the pivot table from the default to customise the table as required. There are options to search, sort and filter the table contents, as shown here:

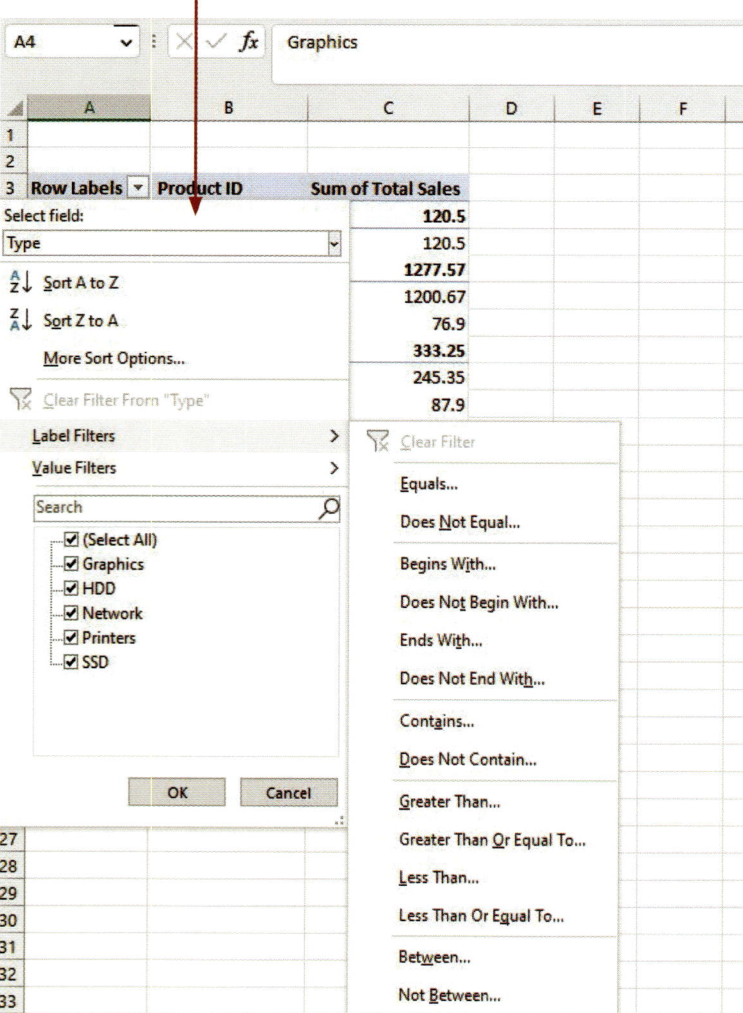

Extracting and viewing data from the contents of the original sheet is made much simpler by using the pivot table to report. For example, to find out and display the total sales of items, open the **PivotTable Fields** dialogue box and move it by dragging the field **Sum of Total Sales** to the filter box, as shown here:

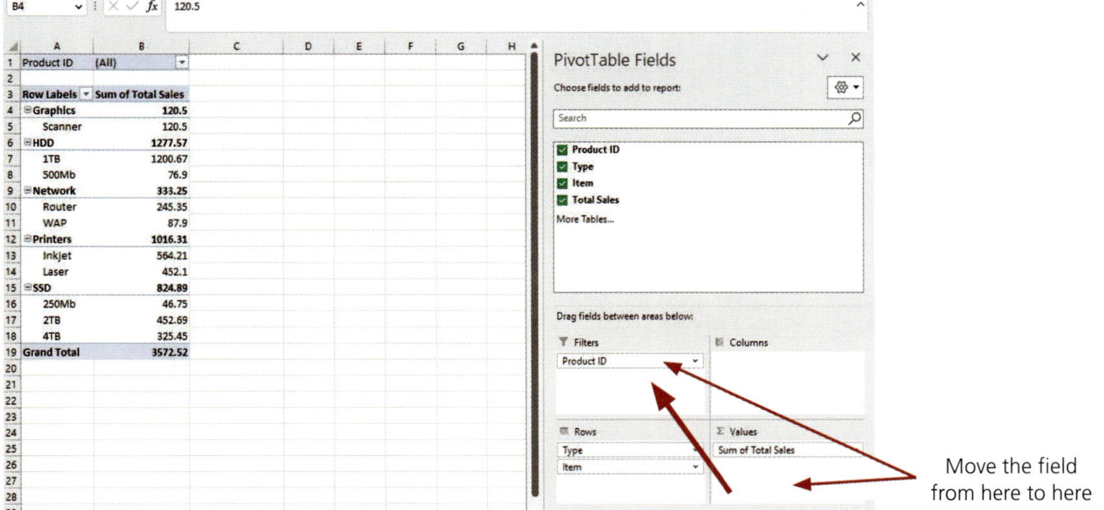

By choosing the **Product IDs** for the items from the drop-down list, the sales for these can be displayed, as shown here:

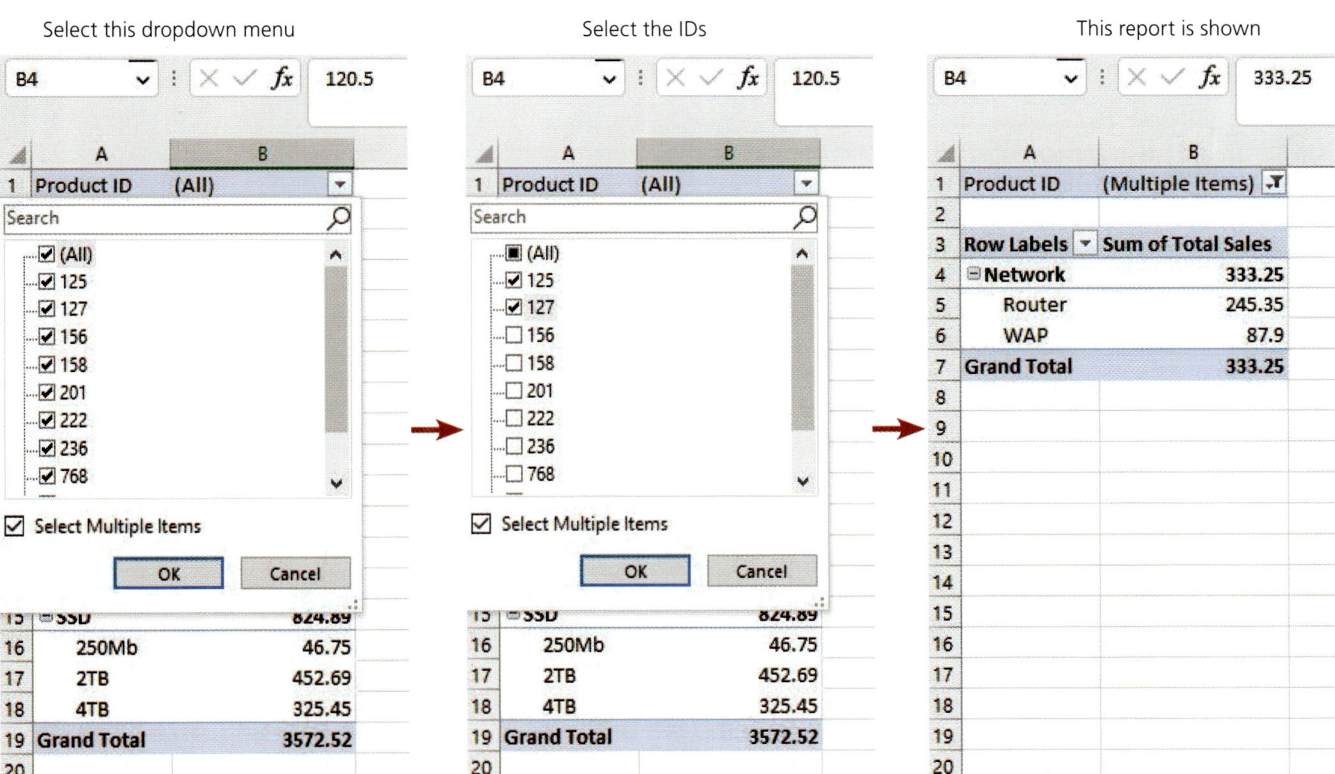

17

In LibreOffice Calc, pivot tables are created from the **Data, Pivot table** menu options. The layout of the data fields is carried out by dragging the column titles as required, as shown here:

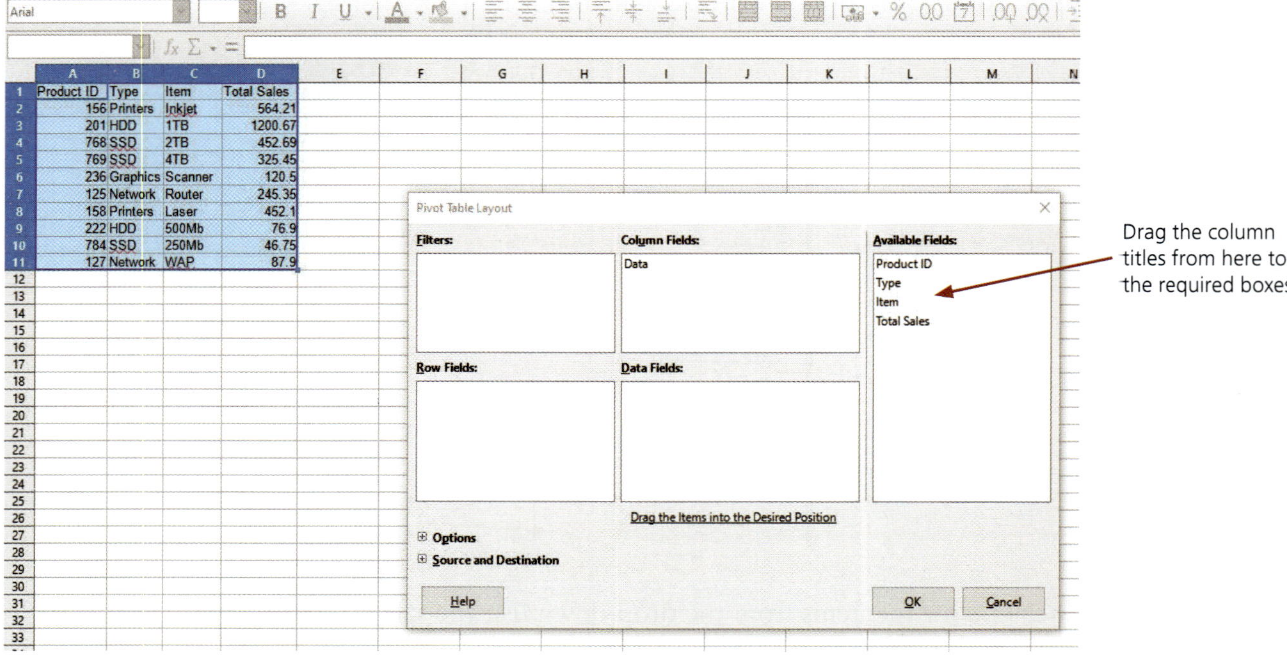

Drag the column titles from here to the required boxes

The same end result can be achieved in Calc as in Excel. The dropdown menus also offer the same sorting, filtering and searching of data, as shown here:

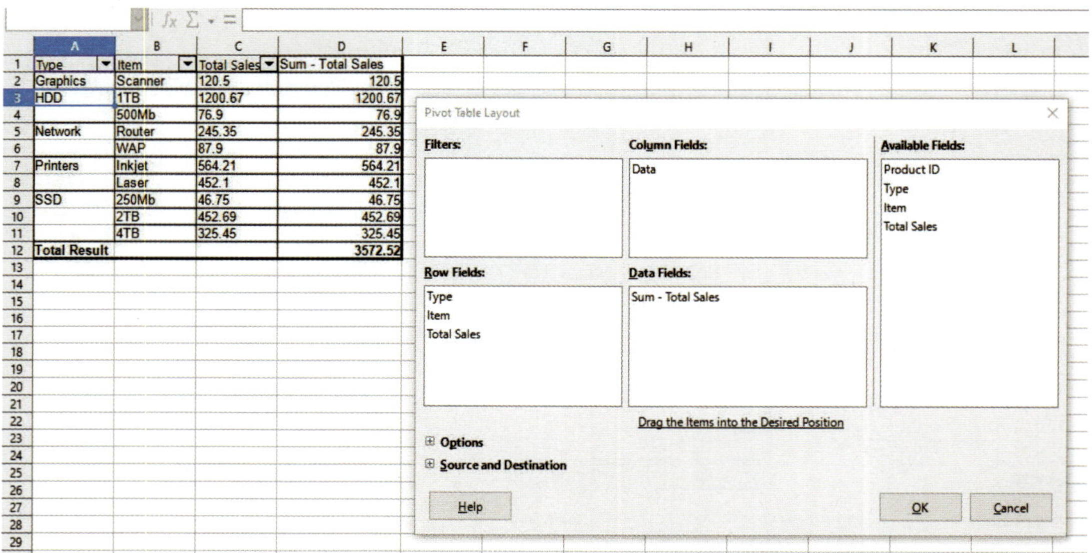

Pivot tables allow the data to be explored to create many different reports so it is worth spending time exploring their use with your own data.

If the data in the source sheet is altered, then the pivot table must be manually refreshed, from the menu options, with the new data.

Pivot charts

Pivot charts are used to display the data in a pivot table. Pivot charts display the data in much the same way as charts made from data in ordinary worksheets. A pivot chart can be edited to change elements like the titles, legends or colour, but the data itself cannot be edited in the chart. This is because the data depends on that in the sheet on which the pivot table is based.

In Microsoft Excel, pivot charts are created from the **Insert**, **Pivot chart** menu options, as shown here:

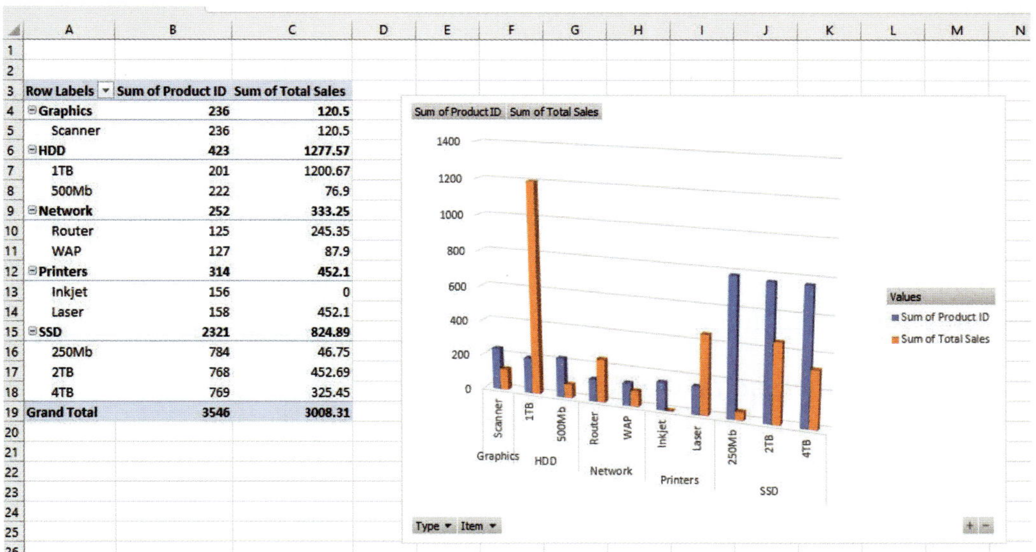

Similarly, in LibreOffice Calc, select any cell in the pivot table, and from the **Insert** menu choose to insert a **Chart**, as shown here:

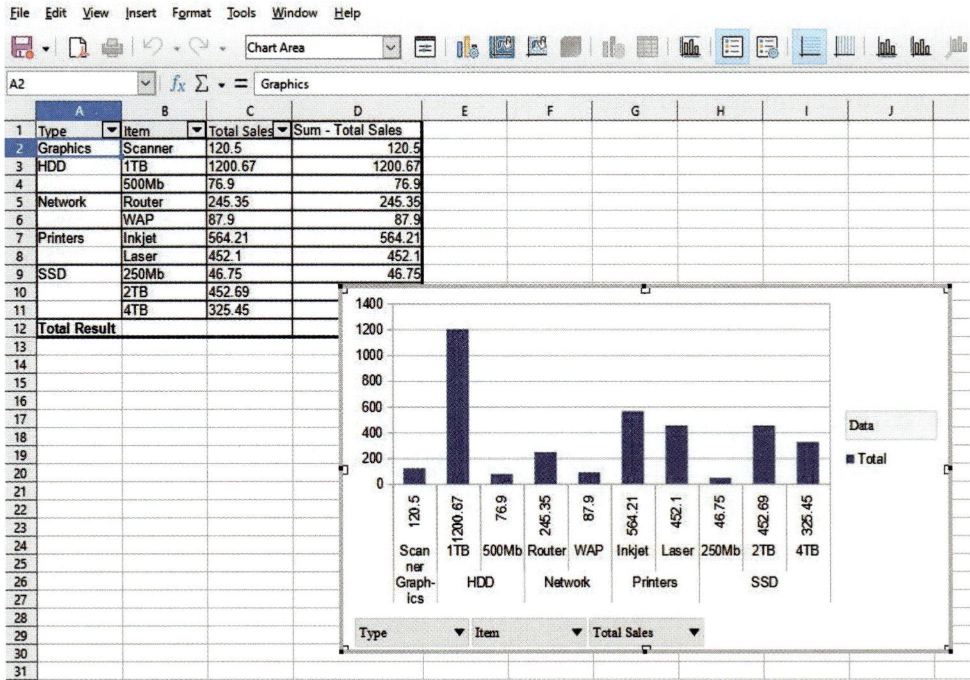

The format and layout of the charts can be customised from the menus and dialogue boxes. As for pivot tables, any changes to the original data source are not reflected in the chart until it is manually refreshed. However, in the same way as for pivot tables, the content of pivot charts can be filtered.

17.1.4 Dashboards

Instead of generating a new report from the data each time, different visualisations of data can be presented in one view and examined together in a single visual representation called a dashboard. To create a dashboard, data must first be imported, or entered, into a spreadsheet application, for example Microsoft Excel, and if necessary transformed so that its structure is suitable, such as no missing rows or columns, each row representing a unique record and the data correctly formatted as a spreadsheet table. The data can be manipulated, summarised and analysed using formulas, filters, charts, pivot tables and pivot charts to extract the required information for the user. The resulting views are grouped together to form the dashboard. Dashboards usually look best if the gridlines are not showing. Excel's toolbars and ribbons can be hidden too if necessary. Dashboards should be kept simple and easy to understand and navigate. Shapes and colours can be used but should not be crowded. Very bright, contrasting colours should be avoided unless these are to draw attention to particular details. Users can interact with the dashboard to manipulate the display using filters available in the dropdown menus on the charts.

Creating an interactive dashboard in Microsoft Excel

A dashboard showing summaries and analyses of the data in Table 17.1 (in Section 17.1.3 'Displaying data to communicate information') can contain shapes, tables and pivot charts. This is done by creating a new worksheet and giving it an appropriate name, for example Dashboard. It is often preferable to move this sheet to be the first tab in the list at the bottom of the workbook.

Using shapes to display information from the data

A shape, inserted from the Insert tab on the ribbon, from Illustration and Shapes, can be used for a simple display of a value. Note that formulas that do calculations cannot be placed in a shape so use references to cells that contain the formula. Put these on a different sheet. Insert a shape, click inside the shape and type the equals sign (=) to start a formula and go to the different sheet, click on the cell with the calculation and press Enter. The value of the calculation will appear in the shape. For example, a formula to calculate the value of the total sales of printers, from the data table, could be found using =SUMIF() in cell K3 on sheet1. So the formula in the shape would be =Sheet1!K3. The shape can be formatted and have labels, in text boxes, as required:

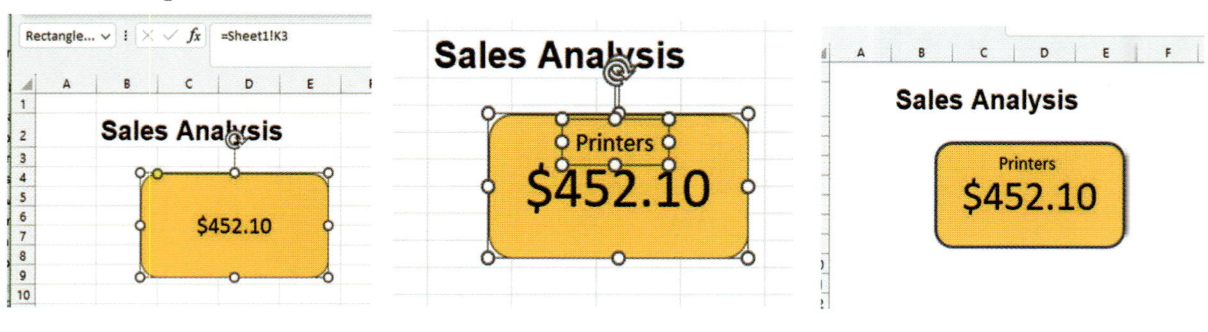

| Inserting the formula | Labelling and formatting the shape | Finished shape |

Grouping the shape with its label boxes enables easy repositioning as required. The value in the shape will automatically update if the source data changes so the dashboard will show the latest data. The shape shows the format of the value in the source cell, for example currency shown as $ will show as $ in the shape. Shapes can be linked to the data source so that, for example, a click takes the user to the original data.

Using tables to display information from the data

A table on the dashboard can be used to display information using references to the data either from the data source or from a pivot table made from the source data. As for shapes, functions such as SUMIF() can be used to extract meaningful data but, unlike shapes, formulas using functions can be placed directly in the tables. Also, the =GETPIVOTDATA(data_field, pivot_table, [field1, item1, field2, item2], ...) function (the data_field is the field in the table where the data is to come from, pivot_table is the pivot table that holds the data to be retrieved and the field and items describe the data to be retrieved) can be used to extract data directly from a pivot table.

Using pivot tables and charts to extract data for display

The pivot charts from the other sheets are copied and pasted onto the dashboard sheet and arranged as required. Suitable legends and titles can be added and formatted using the tools in Excel. The border and colours can be changed using the 'format shape' options.

A simple dashboard using shapes, a table and a pivot chart created from the data in Table 17.1 is shown here:

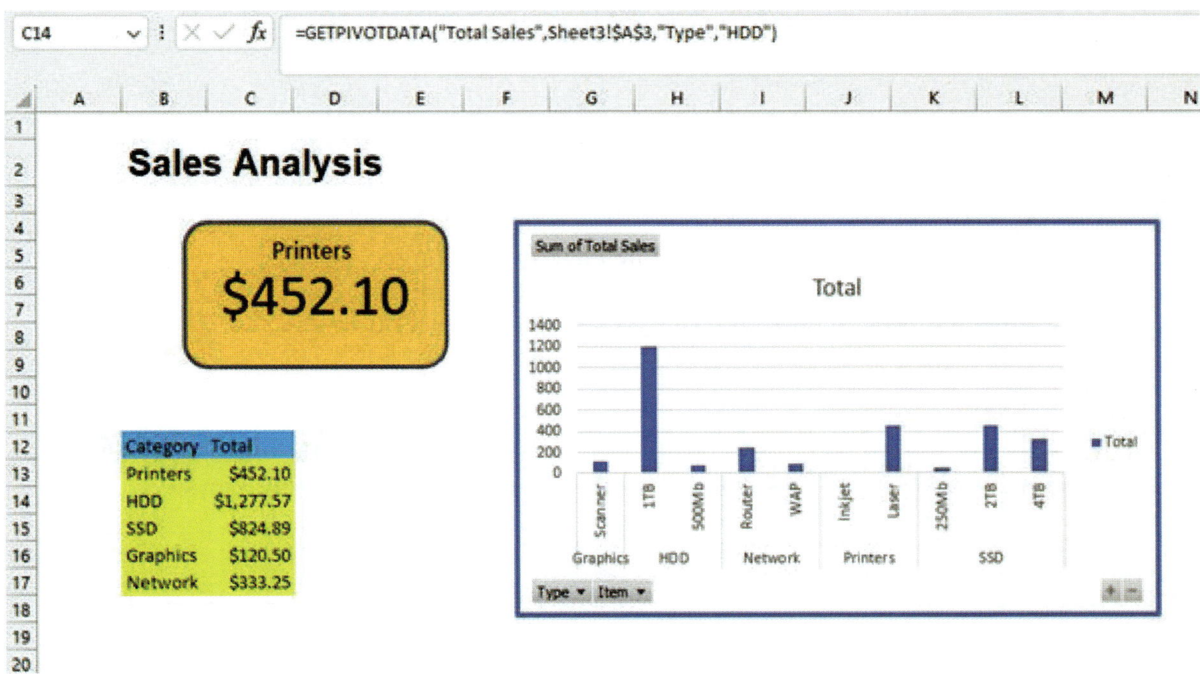

The chart can be manipulated by using the filters that appear on the bottom left of each of the pivot charts, as shown here:

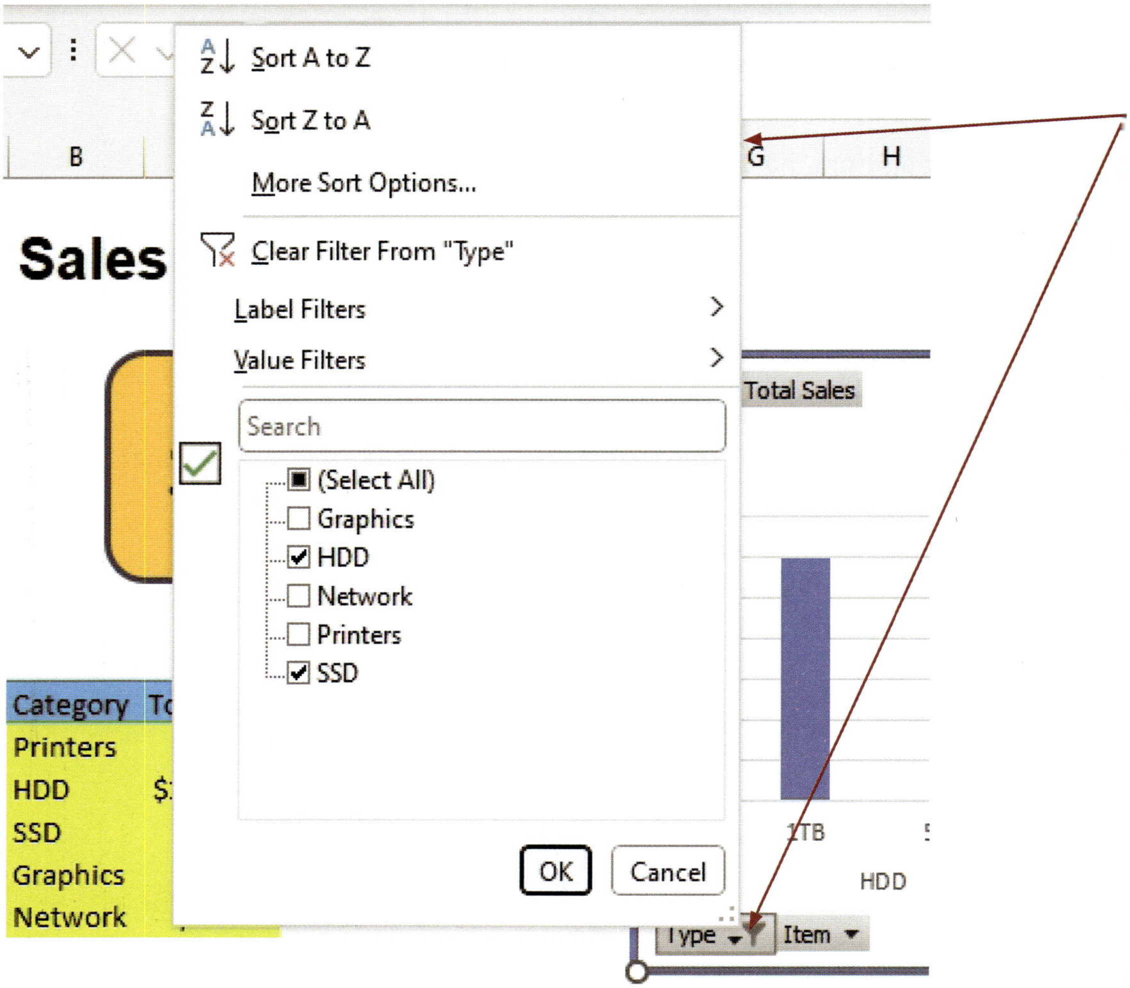

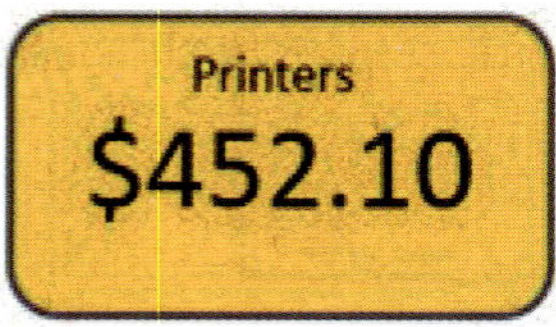

Filter options are accessed from the dropdown menus

When the filters are altered on the right-hand chart, the chart will change to reflect the new filters. For example, selecting only HDD and SSD in the Type filter changes the right-hand chart in the dashboard to this when OK is clicked:

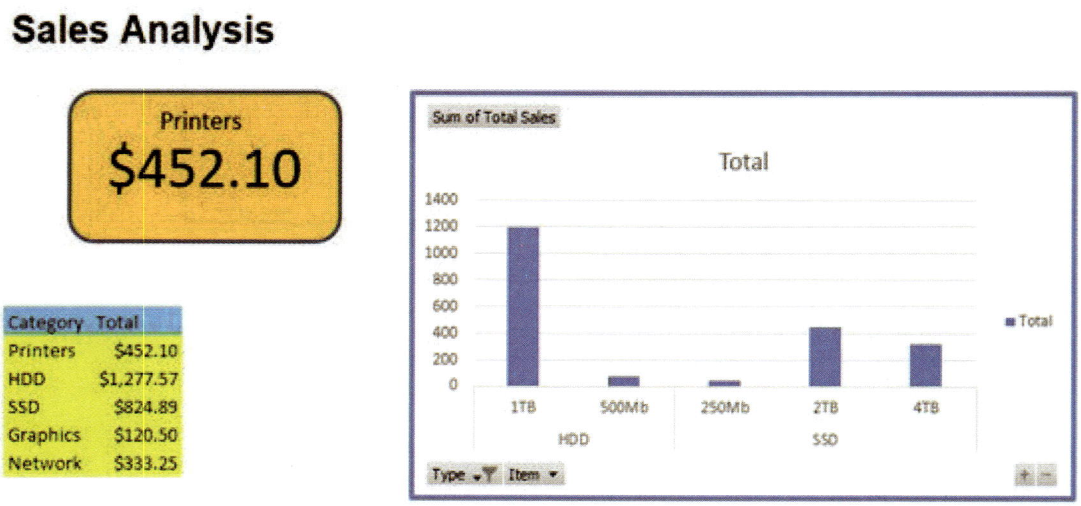

Sales Analysis

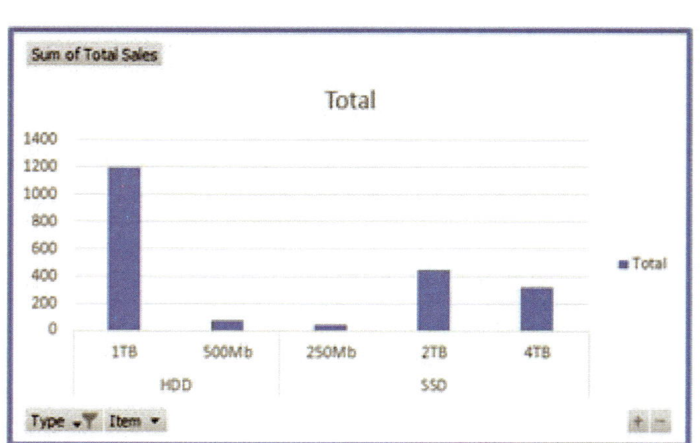

The user can view and analyse the data using the filters from the dashboard without seeing, or even knowing about, the data table and pivot charts. Care needs to be taken when using a pivot table for more than one display on a dashboard because changes in the filters on one display may cause unexpected, unwanted or unsightly changes to other displays on the dashboard.

Data displayed in pivot tables and pivot charts does not automatically update when the source data is changed so a dashboard using them will not do so either. The tables and charts used for a dashboard can be manually refreshed from the 'PivotChart Analyze' tab where there are options to refresh one or all of the charts. This tab only appears when a pivot table or chart is selected. There is also an option to have the pivot tables and charts updated from the source data whenever the file is opened.

Practice questions

1 Explain why data should be cleansed before it is analysed. (4)

2 Pivot tables can take a long time to set up before they are ready to use. Describe two other problems that analysts using pivot tables might come across. (4)

3 The table shows some figures from the sales of car parts. Enter the data into a spreadsheet.

Product ID	Type	Part	Total sales
1	Door	Handle	236.89
2	Seat	Cover	451.23
3	Engine	Cam belt	562.45
4	Windows	Seal	129.89
5	Door	Seal	135.76
6	Windows	Motor	257.21
7	Seat	Bolts	110.9
8	Door	Lock	156.86
9	Screen washer	Liquid	56.78
10	Engine	Oil filler cap	23.59
11	Lighting	Turn bulb	103.45
12	Lighting	Main bulb	99.99
13	Screen washer	Bottle	45.56
14	Lighting	Interior bulb	25.19

Prepare the data ready for analysis. Use it to create a pivot table report and chart to show the sales for each of the sections of a car: Door, Seat, Engine, Windows, Screen washer and Lighting. (10)

4 This question requires access to the data set in the file **cheese_sales_data.csv**, which can be found at **www.hoddereducation.com/CambridgeExtras**.

Using the cheese sales data, create a dashboard as shown here. The dashboard should use shapes, tables and data from pivot tables and pivot charts.

Cheese Sales Analysis

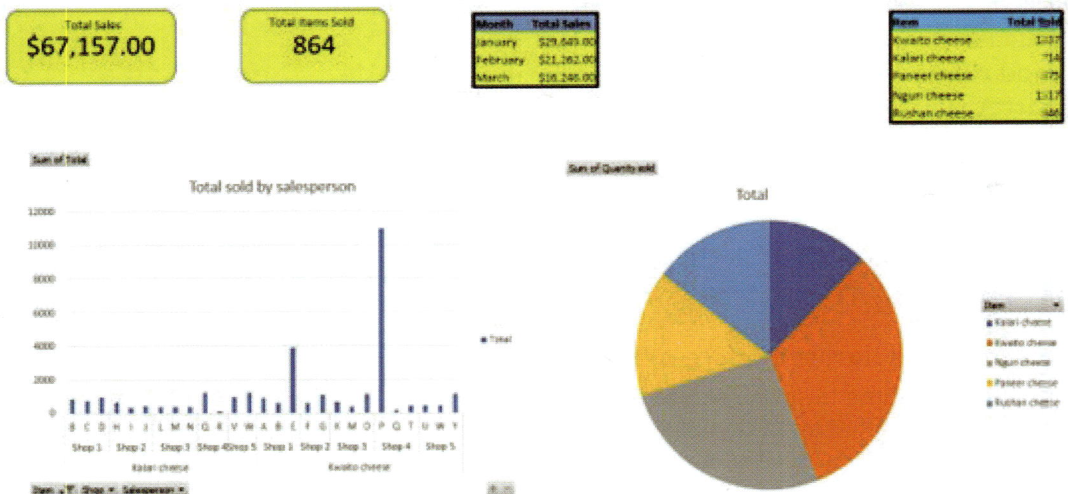

The dashboard should have:

» a title–'Cheese Sales Analysis'–in Arial font, bold and font size 20pt.

» a rectangular shape with rounded corners. It should be approximately 5 or 6 columns wide and 8 rows deep, and have a yellow fill, with a 3pt thickness for the line. The shape should display the Total Sales in Arial font, font size 36pt.

» a shape with the same size and formatting to show the total of the items sold.

» suitable titles for the shapes, centred and in Calibri, font size 20pt.

» column titles in bold and in a different colour from the values for the tables of monthly sales and total cheese sales. The tables should be surrounded with a thick border. Formulas should be used to insert the relevant data.

» a pivot chart showing bar charts of the whole data, 10 columns wide, 20 rows deep and a 2pt black border, with a suitable title.

» a pivot chart showing a 2D or 3D pie chart of the totals of the items sold, 7 or 8 columns wide, 15 rows deep and with a 2pt blue border.

» the shapes and chart displays locked so they do not resize when rows or columns are altered.

» hidden gridlines. (25)

18 Mail merge

In this chapter you will learn:

- ★ to create, edit and use a source file containing data for mail merge
- ★ to create, edit and use a master document with a structure suitable for linking to a source file
- ★ to specify rules for managing the document content and for selecting the recipients
- ★ to set up fields for automatic completion and manual completion
- ★ to set up fields for calculations
- ★ to use software tools and manual methods to ensure error-free accuracy
- ★ to perform mail merge.

Before starting this chapter, you should:

- ★ be able to create and edit documents using a word processor
- ★ be able to create and edit tables and linked charts in spreadsheets
- ★ be able to create and edit source files in a text editor, spreadsheet and database management system.

You may have already learnt about mail merge; however, this is included in the A Level content of the syllabus to practise your problem-solving skills.

18.1 Mail merge

Mail merge is used for automatically making labels, printing addresses on envelopes for mailing letters or addressing emails to large numbers of recipients. It is also used for combining source materials, such as text and data, from different sources into letters and documents such as reports. Complex combinations of source materials can be made. The output of documents can be customised to individual recipients by adding content that is specific to the individual. **Placeholders** are used to determine where the specific content appears in the document.

Mail merge automates the creation of personalised documents from a standard document. The standard document is the template and is called the **master document**. The personalised, or customised, information that is inserted comes from a source file. Figure 18.1 shows a label template with a data source of addresses to illustrate the concept of mail merge.

18.1.1 Using mail merge

In the following worked example of how to use mail merge, a letter about study courses is to be sent to the students attending a college. A master document will need to be created for the letter. A source of data for inserting into the letter will also need to be created.

A master document can be used many times and only the source data needs to be updated. Editing the source data can be carried out by opening the source file in its native application and changing the data. Usually, the master document must be closed before the data can be edited because the source file is locked while in use. However, some word processors will allow the source data to be edited while in use and some, such as MS Word, will even enable the editing from within the master document itself.

The label template is the Master Document:

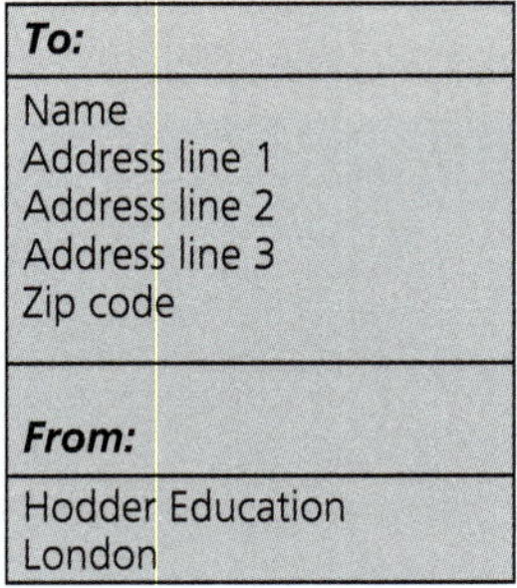

The list of recipients is the source document:

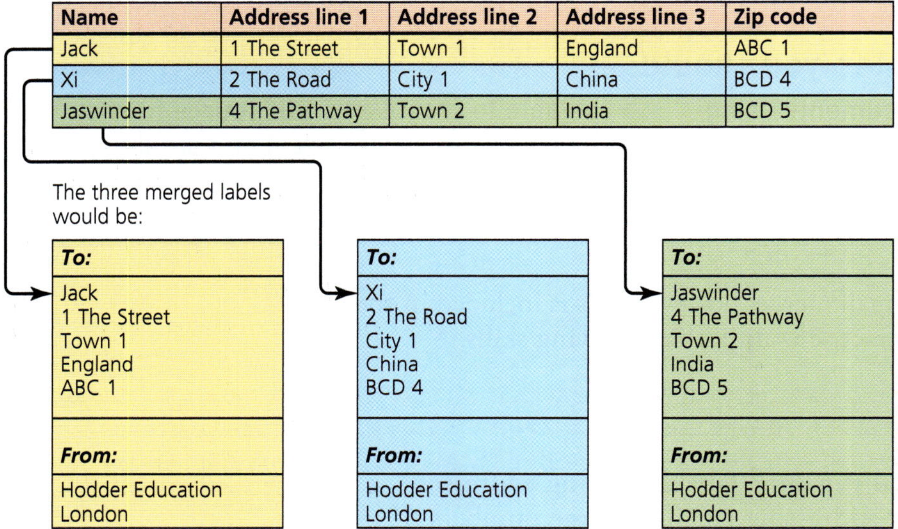

The three merged labels would be:

▲ **Figure 18.1** Mail-merged set of labels

Creating master documents

A master document can be a word-processed document such as a letter or a set of labels to be printed. The printed document is sent physically using postal services. The master document can also be an email template that is sent electronically to recipients. The master document has positions with variables called 'placeholders' or 'fields', where the personalised content is inserted during the merge process. In the worked example, it is a letter.

Creating source data files

A source file is used to provide the data to be inserted. Source files can be text files, database tables or queries, word-processed tables or lists, contact lists in email software or spreadsheets. The personalised content can be anything from names, addresses and dates to customised information for the recipient.

The files used for mail merge must be stored locally on your computer or in a shared folder accessed from your computer. The data in the source file to be inserted into a master document is separated into discrete items in the source file by a delimiter, such as a comma or space, in text files or in tables, rows or columns.

The types of files that can be used are:

» Text files in which the data items are separated by specific characters. The separating characters are called delimiters. The most common delimiters are commas when the file is called a comma-separated value (CSV) file and tabs in a tab-separated value (TSV) file. Text files are useful as they can be opened in any text editor or word processor.
» Spreadsheet files where the data is formatted in columns and rows
» Databases from which tables and queries can be used
» Lists in word-processed documents, as long as the data is arranged appropriately. A table of data in a document is quick and easy to create. It can be linked to a master document.

New lists can be created within a document. For example, in MS Word, **Mailings**, **Select recipients**, **Type a new list** opens the dialogue box where a data source list can be created, as shown here:

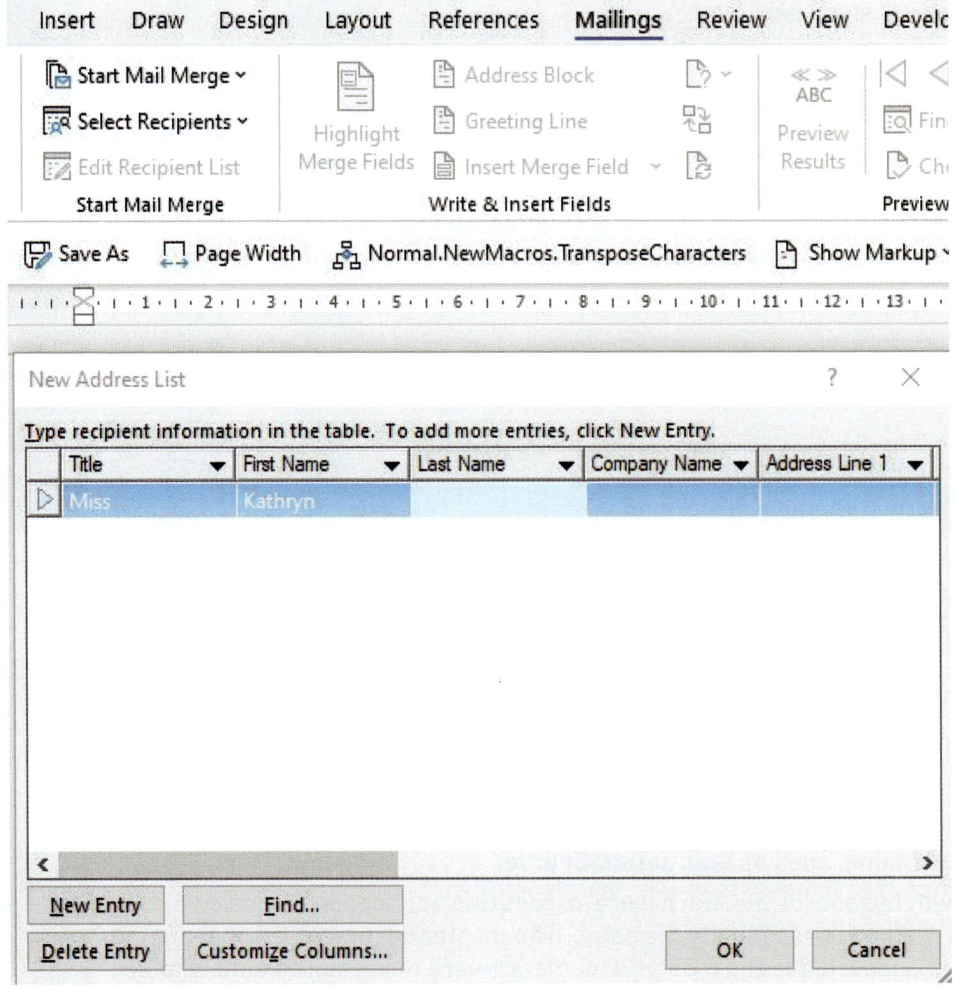

Multiple data sources

It is usual to have one source file in use by a master document in mail merge. However, there are occasions when data is required from several sources, for example different spreadsheets or databases. When a mail-merge task requires data from more than one source to be inserted into a master document, the data is selected in each of the sources, extracted and combined into a new single source file. Chapter 17 describes how this can be done by applying spreadsheet and database skills to extract the required data for the task and make it available for the master document to use. The new single source file is used in the mail merge.

18

Task 18a

A worked example of mail merge

1 Create the letter below in a word-processing application. This is the master document. Ensure that the text, apart from the heading at the top, is left justified. Save the letter with a suitable filename, such as **StudentCourses**.

College Study Courses

The student's full name will be inserted here
The address of the student will be inserted here

Date to be inserted here

Dear *given name of student will be inserted here*,

Here is a list of three courses you will study next term:

Course 1 will be inserted here
Course 2 will be inserted here
Course 3 will be inserted here

We would like to contact you with more details of these courses. Please confirm that we have the correct contact details for you. Our records show that your mobile (cell) number is *number to go here* and your email address is *email address to go here*.

Yours sincerely,

The Principal

2 Create a data source for the letter. Make a table in a spreadsheet like this:

Given name	Family name	Address	Mobile (cell) number	Email address	Course 1	Course 2	Course 3
Jaswinder	Patel	India	079898989	jas@winder.com	IT	Science	Physics
John	Welsh	Scotland	089876784	john@welsh.com	English	Geography	History

Add three more students and their details to make a total of five students in the source file.

Save the file with a suitable filename, such as **CourseDataSource**.

3 Link the master document with the source file. Each word-processing application has its own mail-merge process but the method is essentially the same. The master document is connected to the source file. A placeholder is inserted in the master document where the required data is to be placed. The placeholder is the link to the field in the source file. When the mail merge is performed, the placeholder extracts the data from the source file field and puts it in the master document. The mail-merge process works through the source file until all the entries have been placed into separate documents.

Task 18b

LibreOffice

Open LibreOffice Writer® and create the master document. Save it as **StudentCourses**. Use LibreOffice Calc to create the data source file and save it as **CourseDataSource**.

» Step 1: Link the master document to the source file. Open the **StudentCourses** document and from the **File** menu select **Wizards**. Select **Address data source** and choose **Other external data source**. Clicking **Next** takes you to **Connection settings**. Click **Settings** and choose **Spreadsheet**. Browse to the **CourseDataSource** file, test the connection and click **Finish**. The **Field assignment** wizard opens.

» Step 2: Assign the fields by choosing the appropriate field in the **Field assignment** dropdown lists, as shown here:

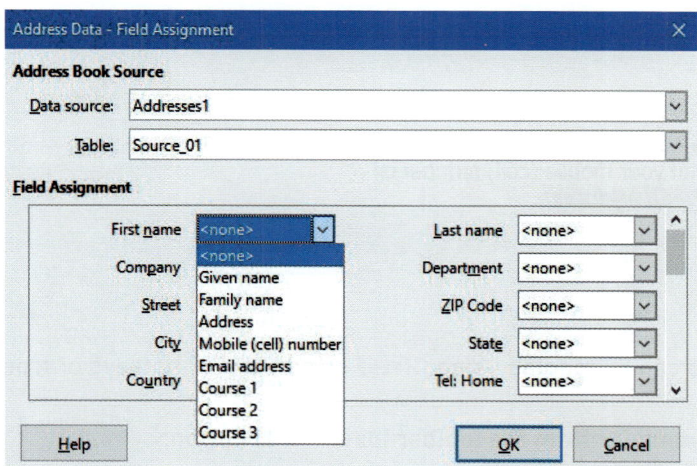

In this example, not all the fields will be assigned, so choose ❮none❯ for unassigned fields. The course options can be assigned to the **User** fields. Clicking through to **Finish** stores the assignments with the document. Save the document to keep the assignments.

» Step 3: Insert the fields as placeholders in the master document, as shown below.

Highlight the area or text where the field is to go and from the **Insert** menu, choose **Field**, **More fields**. This opens the **Fields** dialogue box. It can also be opened from the button on the toolbar or by the **Ctrl+F2** hotkeys. Select the **Database** tab and **Mail merge fields** in the **Type** column and the fields from the linked spreadsheet will appear in the **Database selection** column.

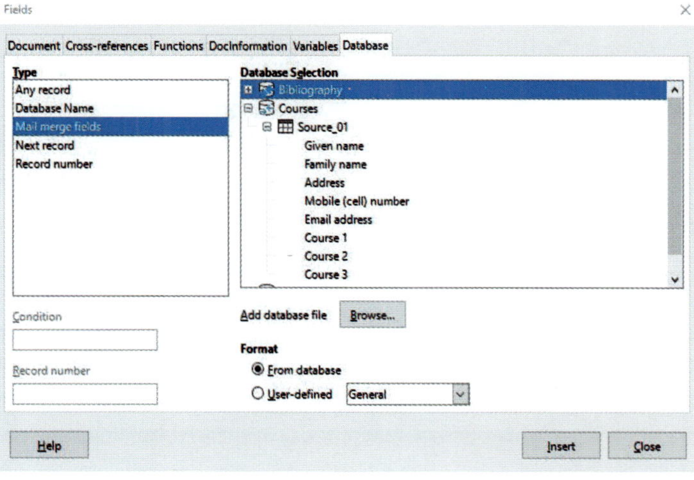

By repeating the insertion of fields in the appropriate positions, and not forgetting the space between the **Given** and **Family name** fields, the master document is completed and now looks like this:

<u>**College Study Courses**</u>

<Given name> <Family name>
<Address>

Date to be inserted here

Dear *<Given name>*,

Here is a list of three courses you will study next term:

<Course 1>
<Course 2>
<Course 3>

We would like to contact you with more details of these courses. Please confirm that we have the correct contact details for you. Our records show that your mobile (cell) number is *<Mobile (cell) number>* and your email address is *<Email address>*.

Yours sincerely,

The Principal

The shading of the fields and the full field names can be toggled using **Ctrl+F8** and **Ctrl+F9** hotkeys or from the **View** menu.

The date field is inserted from the **Insert**, **Field** menu or from the toolbar **Insert field** button.

Note that fields can be quickly placed in the master document by viewing the data source with **Ctrl+Shift+F4** and dragging the field name into position, as shown here. This is an easy way of adding fields to a master document.

Drag the field header from the data source into the master document

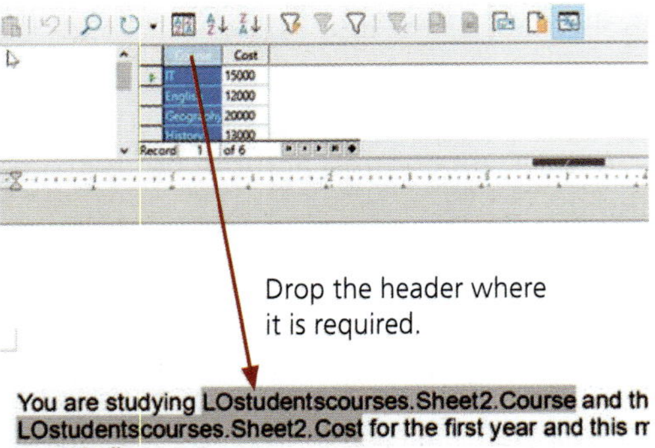

Drop the header where it is required.

You are studying LOstudentscourses.Sheet2.Course and th LOstudentscourses.Sheet2.Cost for the first year and this m your studies.

Note that the formatting of the inserted data can be changed by altering the format of the inserted field, for example you can remove the bold or italic format if you wish by amending the master document or by amending the merged document.

To preview the mail-merged documents, choose **Toolbars** from the **View** menu, and ensure that **Mail merge** is ticked. Use the arrows to move through the documents and **Ctrl+F8** to remove the shading. **Ctrl+F9** shows the field names again. ➜

The **Mail merge** toolbar enables the editing, saving and printing of the merged documents, as shown here:

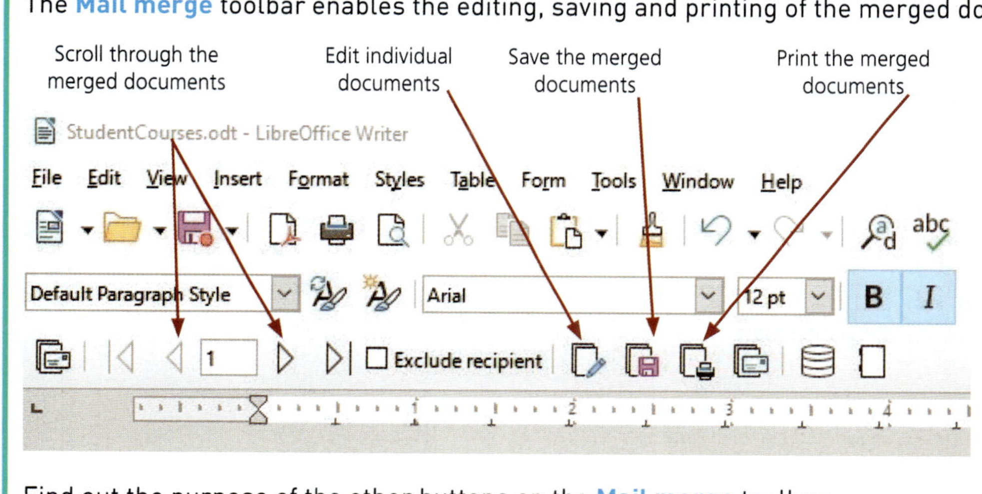

Find out the purpose of the other buttons on the **Mail merge** toolbar.

Task 18c

Microsoft Office

Create the master document in MS Word and save it as **StudentCourses**. Create the data source file in MS Excel and save it as a workbook called **CourseDataSource**.

» Step 1: Link the master document to the source file. Open the master document in MS Word. From the **Mailings** toolbar, choose **Select recipients** and select **Use an existing list**, browse to the **CourseDataSource spreadsheet** and select it. Click **Open**, leave **First row of data contains column headers** ticked, and click **OK**.

» Step 2: Insert the **merge fields** as placeholders. Highlight the place where the field is to go, and from the **Mailings** toolbar, choose **Insert merge field** and the list of field names appears like this:

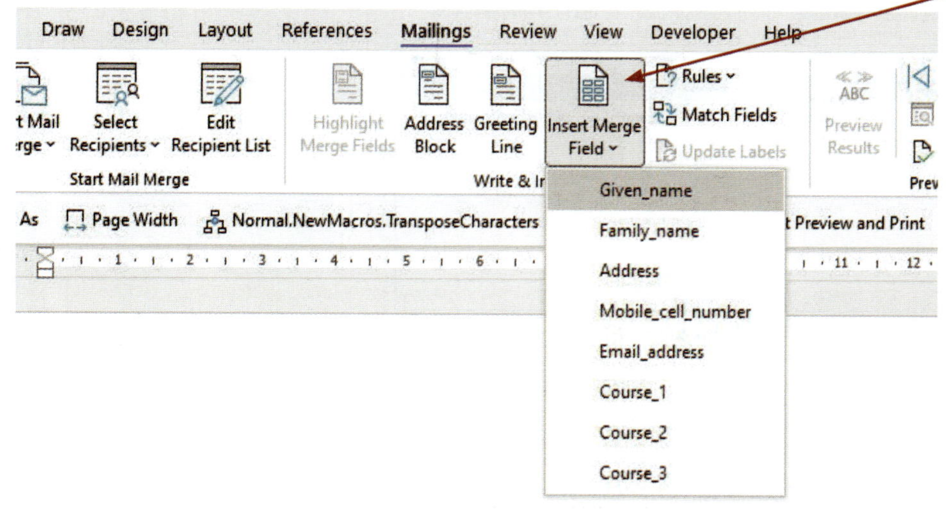

By repeating the insertion of fields in the appropriate positions, and not forgetting the space between the given and family name fields, the master document is completed and now looks like this:

College Study Courses

«Given_name» «Family_name»
«Address»

Date to be inserted here

Dear *«Given_name»*,

Here is a list of three courses you will study next term:

«Course_1»
«Course_2»
«Course_3»

We would like to contact you with more details of these courses. Please confirm that we have the correct contact details for you. Our records show that your mobile (cell) number is *«Mobile_cell_number»* and your email address is *«Email_address»*.

Yours sincerely,

The Principal

The list of 'recipients' can be seen by using the **Edit recipient list** button, as shown here.

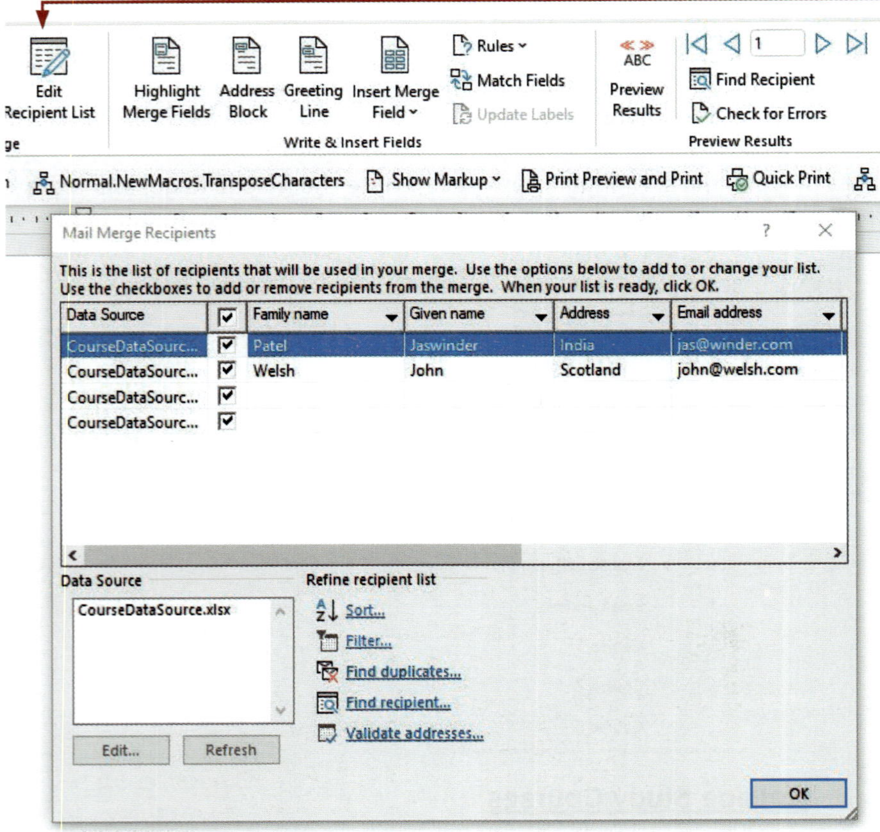

The Principal

The date is inserted from the **Insert**, **Date and time** menu. The format of the date can be chosen and it can be set to update automatically or not. If it is not set to be updated, the current date is inserted and does not change. Dates can be inserted into any document, not just mail-merged ones.

» Step 3: Perform the mail merge. The **Mailings** toolbar provides the means to preview and print the mail-merge documents, as shown opposite.

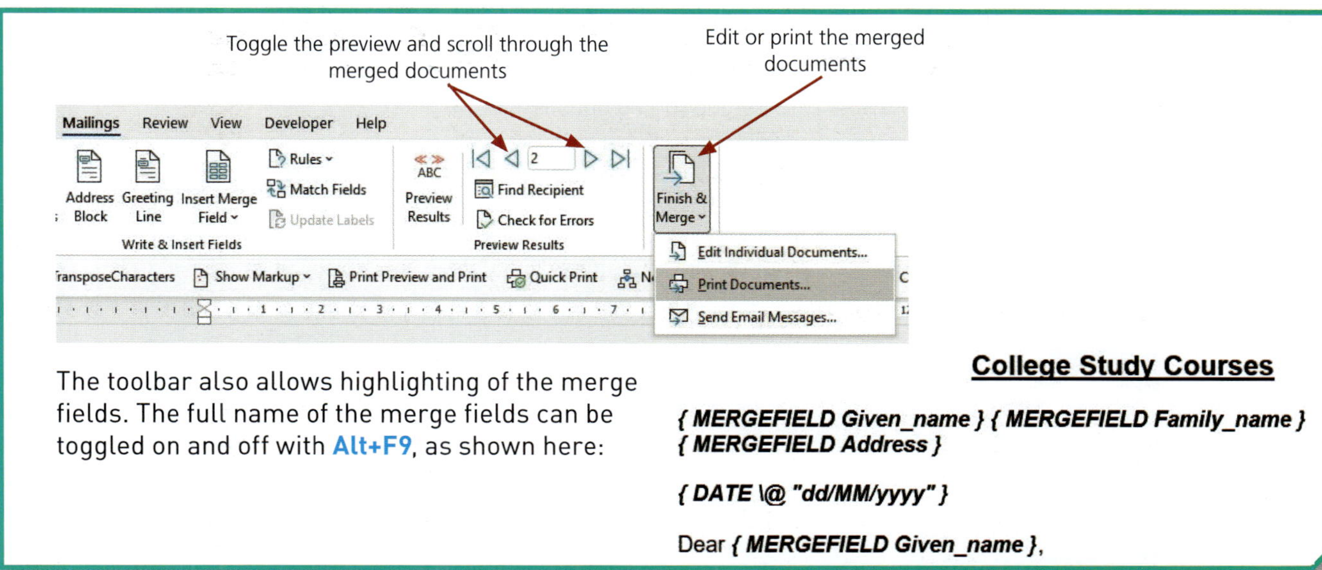

The toolbar also allows highlighting of the merge fields. The full name of the merge fields can be toggled on and off with **Alt+F9**, as shown here:

Entering merge fields into documents

Merge fields are usually entered from the menus. However, there are alternative ways of accessing and entering merge fields that may be useful. For example, using copy and paste is a quick way to duplicate fields as necessary.

In Apache OpenOffice®/LibreOffice, the data source can be viewed by selecting, from the **View** toolbar, the **Data sources** option. This opens the data source in a window above the document. It can also be opened with **Ctrl+Shift+F4**. It is closed with the same actions. A merge field from the data source can be inserted into the master document by dragging its name to the required place on the page.

In Microsoft Office, **Ctrl+F9** inserts the curly brackets that enclose a merge field. If you know the exact name and correct syntax for a field, it is possible to manually type the field details. However, it is much easier and more accurate to use the supplied menu options in MS Word.

Alt+F9 toggles the view of the merge field names.

Sorting recipient lists

In OpenOffice/LibreOffice, the sort options are available from the **Mail merge** toolbar. Use **Ctrl+Shift+F4** to view the data source and then choose the sort options: like this:

1 Choose Sort order from the Mail merge toolbar

2 Set the sorting options and click OK

Or use the Ascending or Descending order buttons

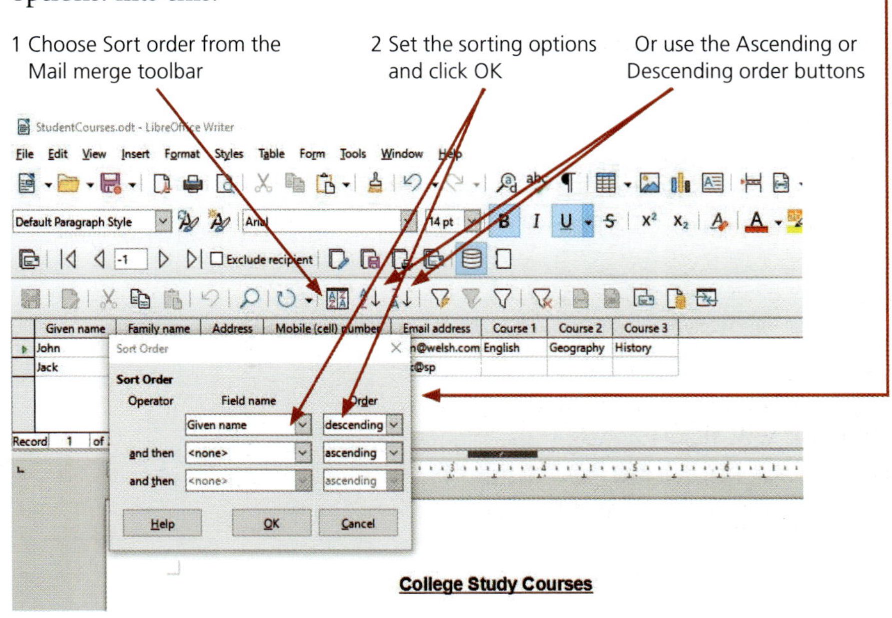

In MS Word, the recipients list can be sorted using the **Mailings**, **Edit recipient list** menu options, like this:

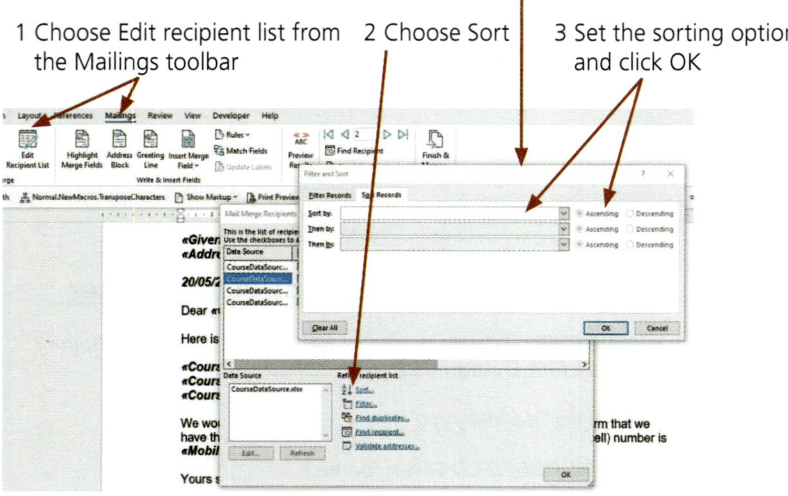

The sorts can be reversed by choosing <none> in the **Sort by** dropdown lists.

Creating labels

Commercial labels come in many shapes and sizes and are printed on adhesive paper that can easily be fixed to letters and packets. There are many suppliers of commercial labels, and word-processing applications such as OpenOffice/LibreOffice and Microsoft Word® have built-in facilities to use these.

Creating labels in OpenOffice/LibreOffice

In OpenOffice/LibreOffice, the label options are found from the **File** menu under **New**, **Labels**, as shown here:

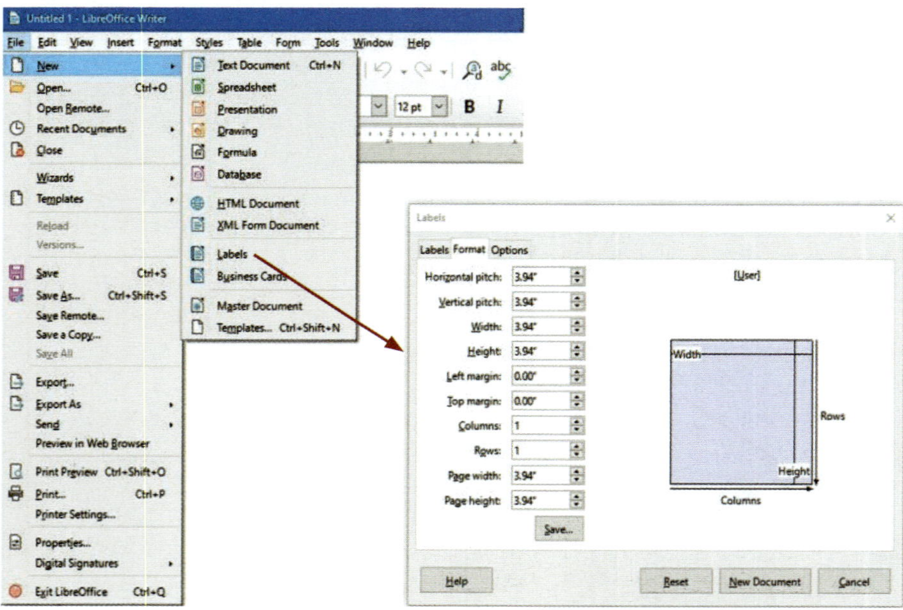

The fields to be used in the label are chosen from the **Labels** tab, as shown here:

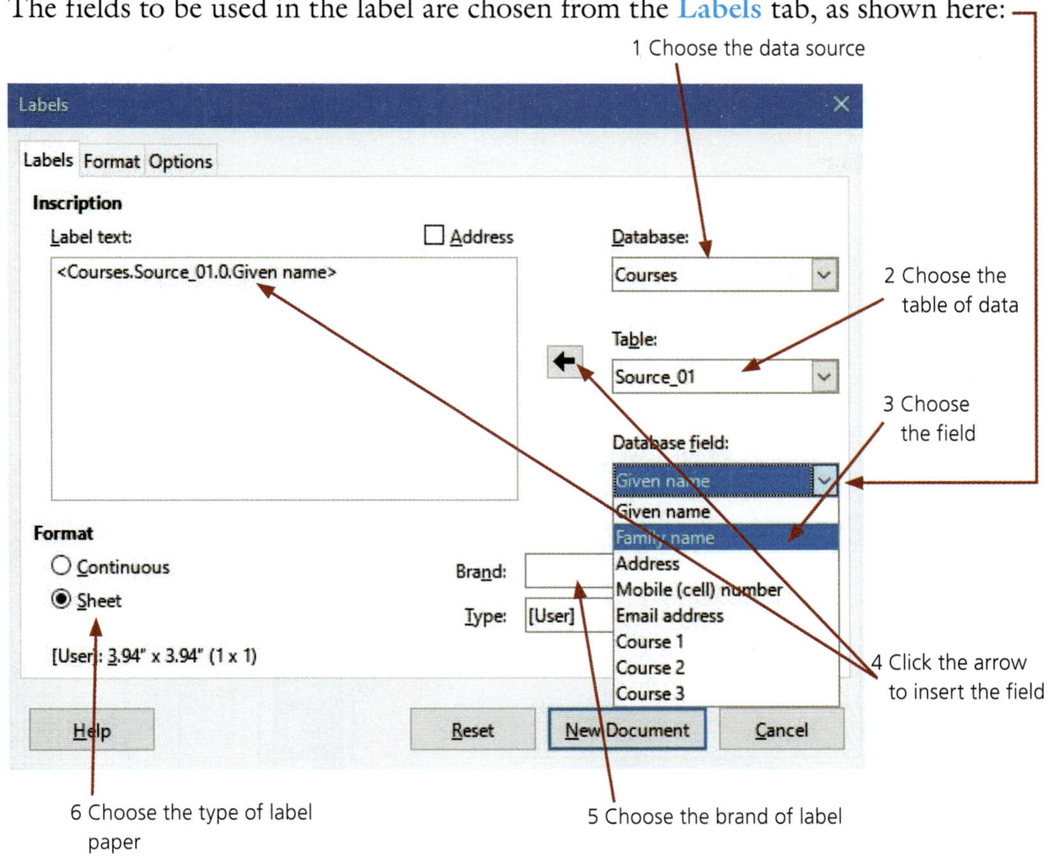

The size of the label is chosen from the **Format** tab, and the layout on the page is chosen from the **Options** tab, as shown here:

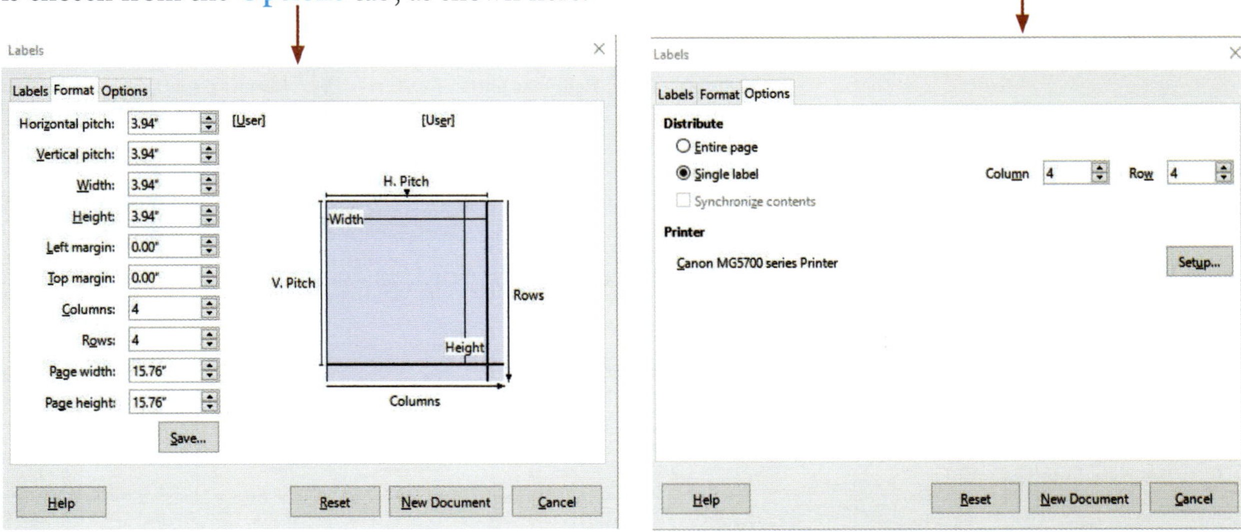

When the **New document** button is clicked, a page of labels is created ready for mail merging. This page of labels can be saved, previewed or printed.

Creating labels in MS Office

In MS Word, labels can be created from the **Mailings** toolbar. Open a new document and, using the **Select recipients** button, link it to an existing data source file of names and addresses. Choose the **Start mail merge** button and select **Labels**. Set the options you wish to and click **OK**. Click **Update labels** to see the layout on the page.

Select the first label on the page and use the **Insert merge field** button to choose the fields you want in the label. Click **Update labels** to replicate the fields into all the labels.

Use the **Preview results** button to see the completed labels. If necessary, adjust the dimensions of the labels.

In MS Word, from the **Mailings** menu, many options can be set, for example for label supplier or vendor, number of labels per page and the label sizes, as shown here.

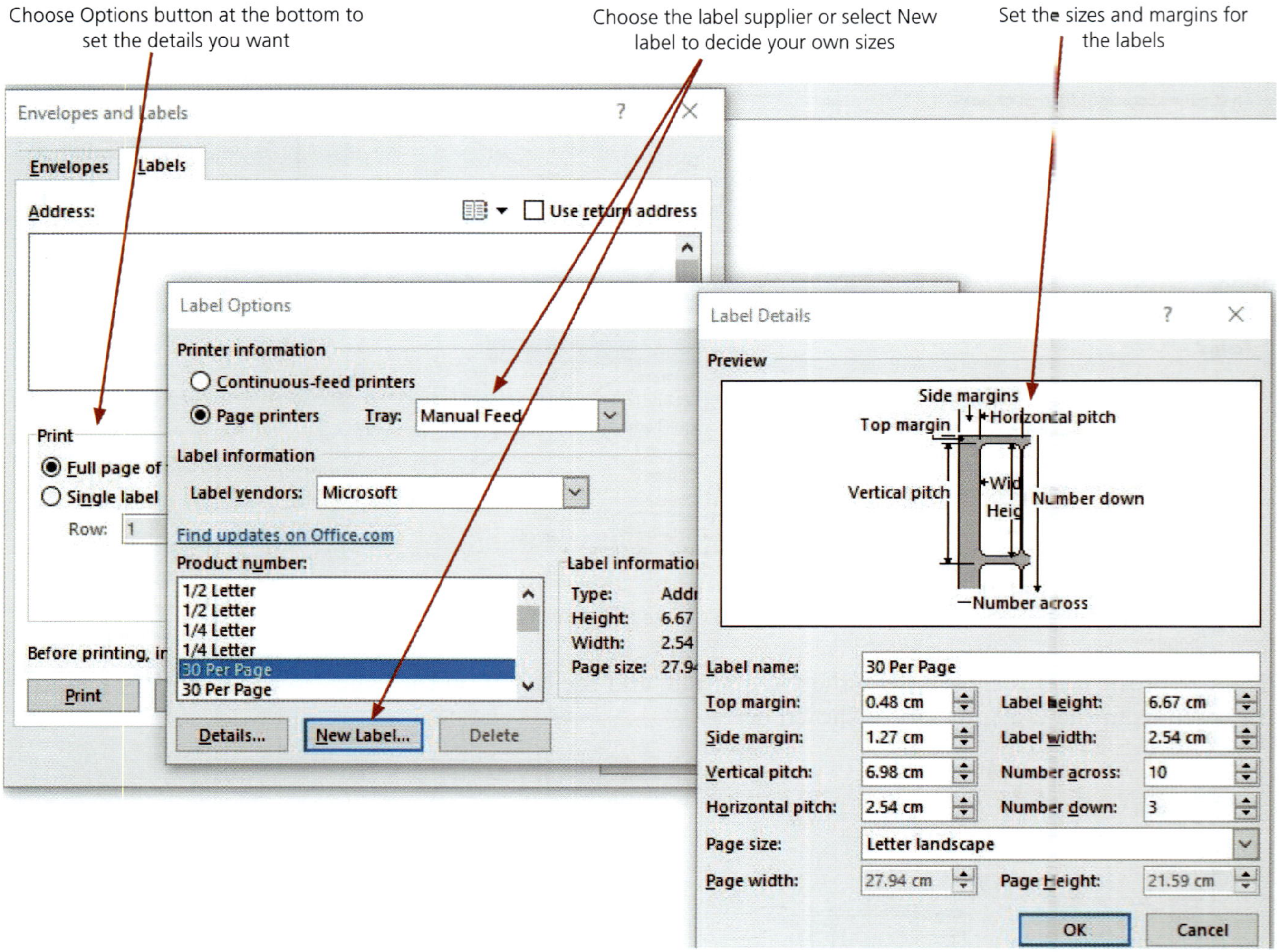

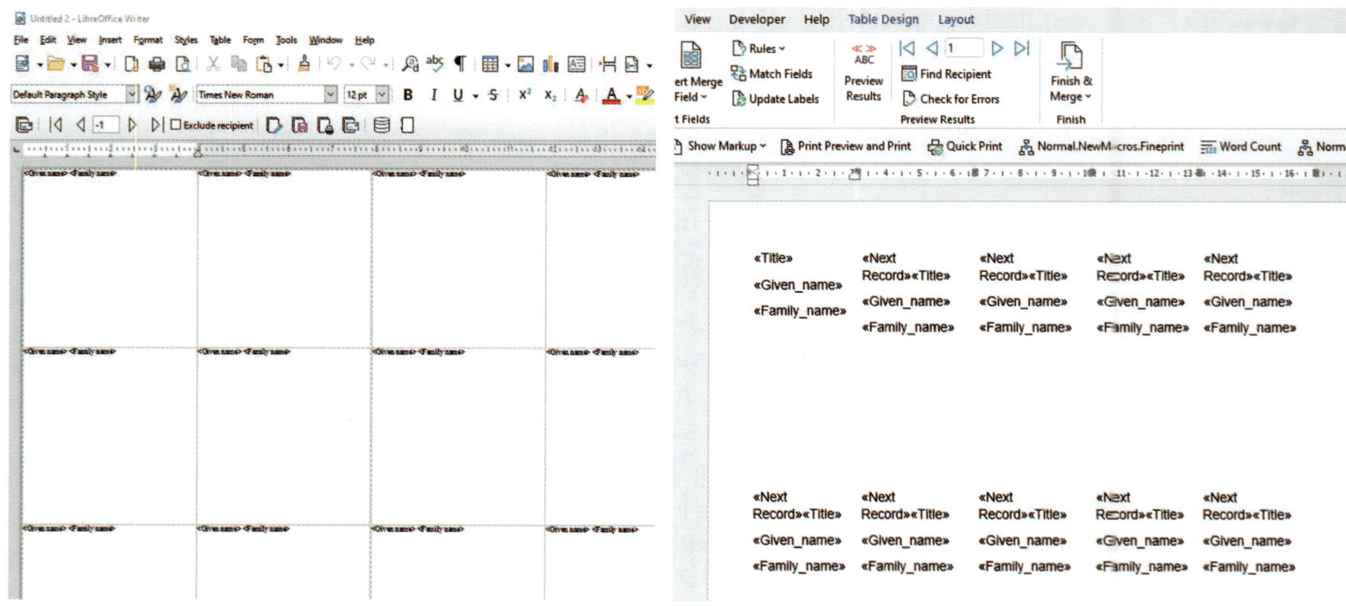

Similar page layouts are created in OpenOffice/LibreOffice and MS Word, as shown here:

Task 18d

Creating labels

1   An online store sells a variety of goods, packages them and ships them to customers by courier. Create a master document for a packet label in a word-processing application. Save the label with a suitable filename, such as **Label_master**. The label should look like this:

> ***Ship to:***
>
> Title Given name Family name
> Address line 1
> Address line 2
> Address line 3
> Town
> Country
> Zip code
>
>
> ***From:***
> My Online Retailer
> London

2   Use the information in this list to create a source file in a spreadsheet application and save it as a CSV file with a suitable filename, such as **Customer_details**. The list shows three customer details to include.

Field	Customer 1	Customer 2	Customer 3
Given name	John	Audrey	Amrita
Family name	Wales	North	Banerjee
Title	Mr	Miss	Mrs
Address line 1	1 The Avenue	45 Elizabeth Way	56 The Road
Address line 2	Crundale	Parnell	Alipur
Town	Haverfordwest	Auckland	North Delhi
Country	United Kingdom	New Zealand	India
Zip/postal/area code	SA62 7DT	4414	110007

In the source file, the field names should make up the first row and the customers the next three rows.

3   Perform the mail merge to create the labels.

Performing the mail merge

When the mail merge is carried out, the letters, labels or other documents such as business cards can be sent to files for saving and using later or sent directly to a printer. To carry out the mail merge and create the new documents in OpenOffice/LibreOffice, the destination is chosen from the **Mail merge** toolbar. In MS Word, the **Finish & Merge** button is used.

Directory mail merge

A **directory mail merge** creates a list or document using specified fields from a data source, such as a spreadsheet or database.

Task 18e

Directory mail merge

Explore the use of directory mail merge to create a list. You will need a data source, for example of friends or made-up people, with given and family names, addresses, phone numbers and their email addresses in a spreadsheet. Practise by creating a list, for instance of family name, given name and email address.

Links

Links and fields should be updated after amending the contents of the data source or altering the master document, or else a mail merge may not work properly. In OpenOffice/LibreOffice, the update is performed from the Tools menu by selecting Update and then the option required. F9 will update the fields directly. In MS Word, the update happens after making edits to the data source from within the master document when the user is prompted to update it. It is good practice always to allow the update.

Embedding tables and charts

Charts and tables can be embedded into word-processed documents. The charts and tables can be from external sources such as spreadsheets or databases. Strictly speaking, embedded tables and charts that are expected to update when the source data is changed must be 'linked'. If they are not linked, they will not update and will reflect the data only when it was first copied. The charts and tables can be updated, either manually or automatically, when the data in the external source changes. Documents that have embedded data from external sources must 'know' where the source files are stored so they can access the data. Moving files can cause the links to be broken. If links are broken, they must be remade or else the data cannot be shared between the sources and the documents.

For this example, a survey of the number of different types of trucks passing through a small village has been recorded and entered into a spreadsheet. A table or chart showing, for example, the number of each type of truck on each day can be created from the data, like this:

	A	B	C	D	E	F	G
1							
2		Truck survey					
3							
4			Monday	Tuesday	Wednesday	Thursday	Friday
5		Pickup	2	2	2	2	1
6		Tanker	3	4	7	1	2
7		Tractor-trailer	1	2	5	2	2
8		Van	4	6	11	8	3
9							

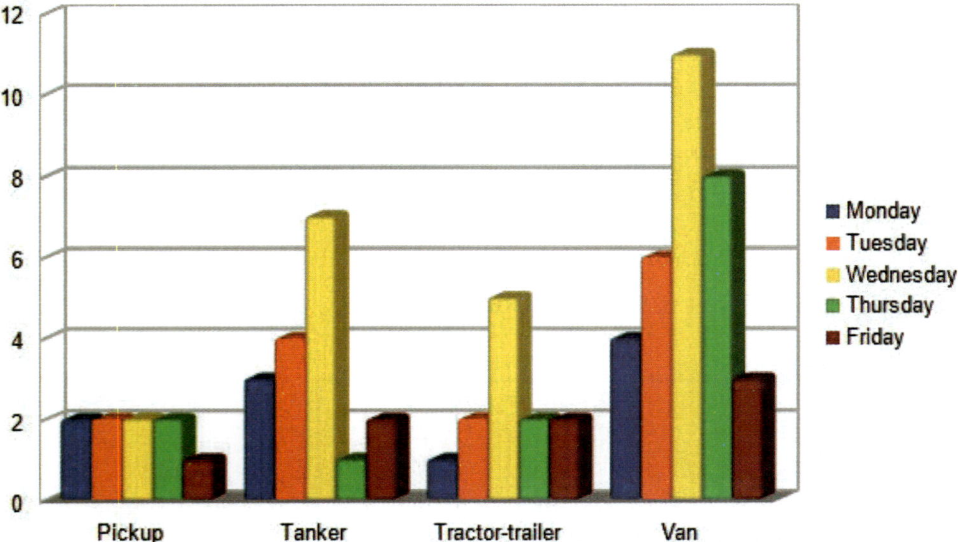

Embedding tables and charts in OpenOffice/LibreOffice

The spreadsheet table and a chart are created in Calc. These can be embedded and linked in a Writer document to be used in, for example, a letter, report or summary leaflet, so that the latest data is always shown. To do this, both the spreadsheet and the document are opened. It is not sufficient to copy and paste the table or chart because, while this displays the data, it pastes a copy of the data only as it was at the time of copying. To insert a 'live' copy, carry out these steps:

» Step 1: Choose the table: In the spreadsheet, highlight the table of data to be inserted and copy it to the clipboard using **Edit**, **Copy** or **Ctrl+C** in the usual way.
» Step 2: Insert it into the document: Put the cursor at the place in the document where the table is to be inserted. From the **Edit** menu, choose **Paste special** or use **Ctrl+Shift+V**. This opens a sub-menu, from which **Dynamic Data Exchange (DDE link)** should be chosen:

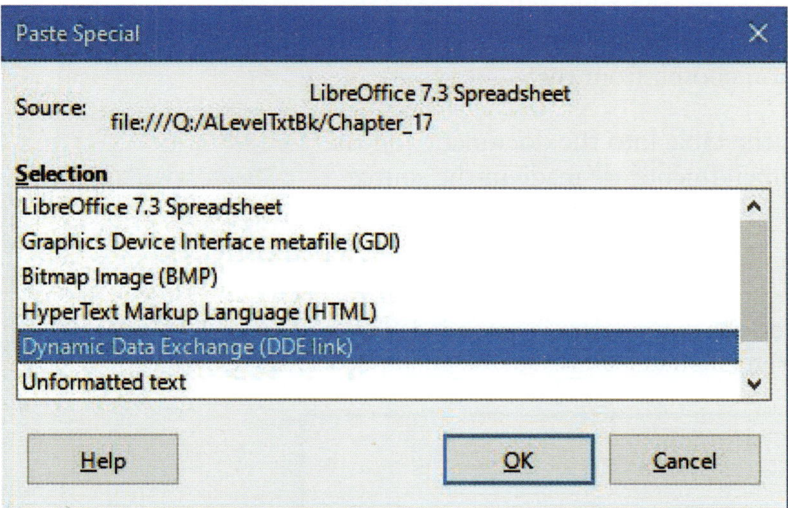

When **OK** is clicked, the table is pasted into the document. Any changes to the data in the spreadsheet will be shown automatically in the table in the Writer document. The updated data will appear in the Writer document when the user returns to it from amending the spreadsheet – a mouse click anywhere in the document is usually enough.

The table can be formatted as required in the Writer document, but the data should only be edited in the source spreadsheet. While it is possible to edit the data in the document, these edits will not be copied back to the spreadsheet. However, edits made in the spreadsheet will overwrite changes made in the document.

Charts cannot be embedded and linked in the same way as tables. A chart that is to be updated must be created in the Writer document itself. To create a chart that can be updated, carry out these steps:

» Step 1: Repeat the steps from above to insert the table of data from the spreadsheet into the document. Ensure that it updates when the data is changed in the source spreadsheet.
» Step 2: Insert a chart into the document: Highlight the embedded table in the Writer document and from the **Insert** menu, select **Chart**. Choose the type and layout of the chart and click **OK**. The chart can be resized and moved to wherever it is required in the document.

When data is amended in the source spreadsheet, updated data will appear in the table in the Writer document, but the chart is not automatically updated. To show an updated chart, the user must select **Update** from the **Tools** menu and choose either **Update all** or **Update chart**. The chart will then reflect the new data.

Embedding tables and charts in MS Office

The spreadsheet table and chart are created in MS Excel. It is not sufficient to copy and paste the table or chart because, while this displays the data, it only pastes a copy of the data as it was at the time of copying. To embed 'live' data from an Excel spreadsheet in a document created in MS Word, carry out these steps:

» Step 1: Choose the table: In the spreadsheet, highlight the table of data to be inserted and copy it to the clipboard using **Copy** from the **Home** toolbar or **Ctrl+C** in the usual way.

» Step 2: Insert it into the document: Put the cursor at the place in the document where the table is to be inserted. From the **Home** toolbar, choose the **Paste** button and select one of the options that includes **Link &….** As the mouse pointer is moved across the options, the table will appear in the document. Alternatively, the **Paste** options can be shown by clicking the right button of the mouse and the option chosen from there.

Choosing the option to **Link &…** inserts the table into the document and the data will automatically update when any amendments are made in the source spreadsheet. If any option not including **Link** is used, the data will not update.

Charts are embedded and linked in the same way as tables. Copy the highlighted chart from the Excel spreadsheet and paste it into a Word document using an option that includes a **Link**. When the source data is amended, the chart will automatically update to reflect the changes.

> ## Task 18f
>
> ### Embedding tables and charts
>
> Create a table and chart in a spreadsheet using the truck survey data. Create a report about the trucks going through the town and embed the table and the chart into the report. Change the data in the spreadsheet to see how the table and chart change in the report.

Setting up fields

Master documents are designed to be used again, but sometimes there is data that has to entered every time a new mail merge is performed. This data can be entered manually or automatically. Manual data could be, for example, the name of the person sending the letters, but an entry showing the date of the letters could be entered automatically.

Manual completion

A prompt is placed in the master document and the user is expected to overtype it with the required information. If the user does not overtype the data, the prompt is printed in the merged documents.

In LibreOffice, open the dialogue box from **Insert**, **Field**, **More fields** and select the **Functions** tab and **Input field**. **Ctrl+F2** also opens the box. A prompt is typed into the **Reference** box and the default text to go in the document is entered like this (see opposite):

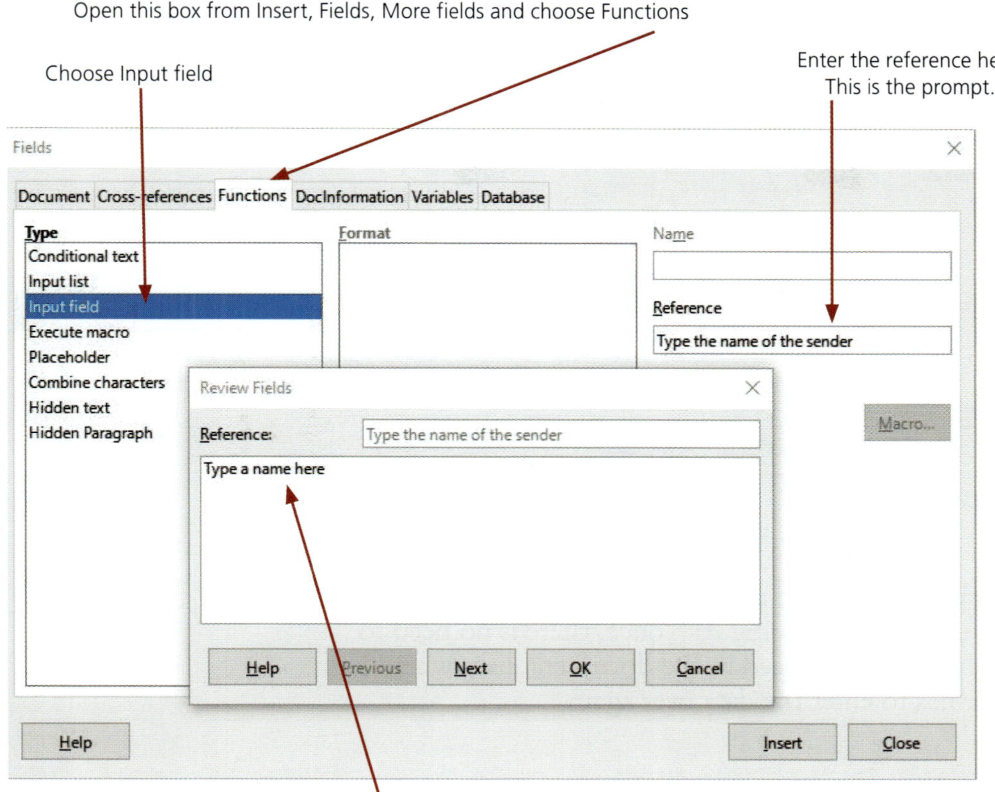

Enter the default text here. This appears in the document.

When **OK** is clicked, the input field is placed in the document. Click **Close** to remove the dialogue box.

When the mouse is moved over the default text, the prompt appears as seen here.

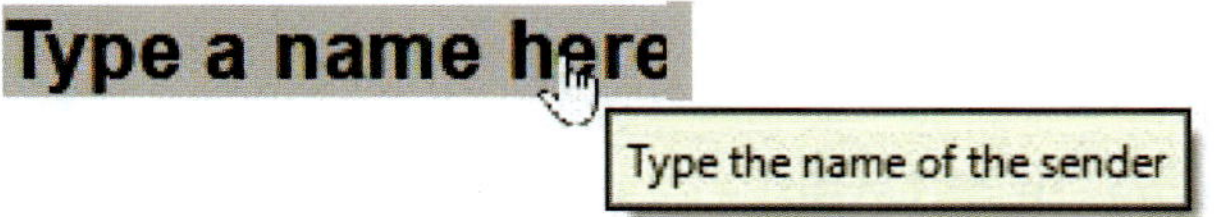

Double-clicking on the text opens a dialogue box in which the user can type a name. The entered name appears on all the merged documents when the merge is performed.

In MS Word, **Fill-in** is used and the dialogue box to set it up with a prompt is accessed from the **Mailing**, **Rules** menu. The details are entered and the user can choose whether nor not the prompt appears once only. The prompt appears to the user when the merge is performed.

The merge field **Ask** is used to enter different data into each individual document that is output from a mail merge. When a merge is performed, the user is prompted to complete a dialogue box before the merge. To use **Ask**, a bookmark for reference is created first by using **Ctrl+F9** to insert a pair of curly brackets. Inside the brackets, a name of a bookmark is entered, as shown here:

{ TypeinName }

This is placed wherever it is required in the document. It can be inserted as many times as needed.

From the **Mailings**, **Rules** menu, **Ask** is chosen, which opens the dialogue box shown on the next page.

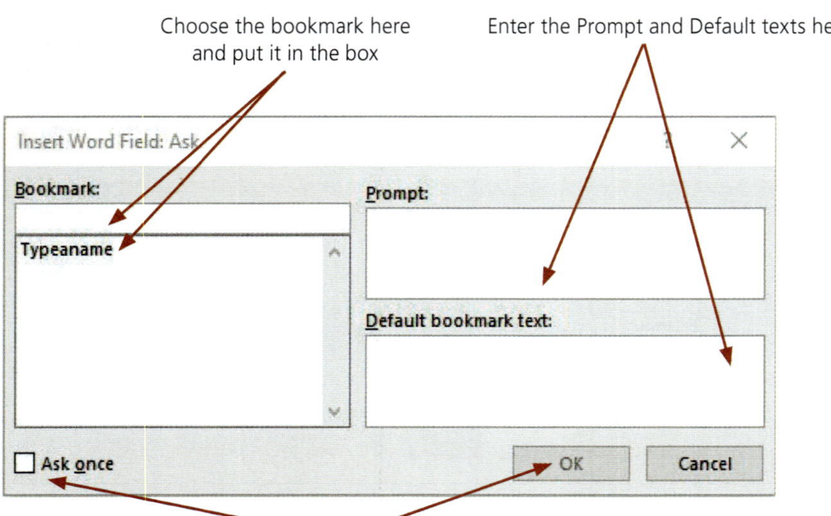

Choose whether or not to ask once and click OK

The advantage of using **Ask** over **Fill-in** is the ability to reuse the prompt by placing bookmarks in the document. By choosing **Ask** once, there is no need to enter the data repeatedly. **Fill-in** has to be placed in the document everywhere the data is needed and the user has to enter the data over again.

Automatic completion

Some fields can be set to complete automatically when a mail merge is carried out. The insertion of a date field was described in Task 18b and a time field can be inserted in the same way. The date and time format can be chosen and the field can be set to update automatically.

Document properties such as the number of pages in the document, the page number and other information about the document can be inserted directly or in fields. These can be placed, depending on the property, in the main document body or in headers and footers.

In LibreOffice, the fields can be inserted from the **Insert** menu. In MS Word, page numbering options can be found in the **Insert** toolbar. Fields may need to be manually updated using the **Tools** menu.

These and the other properties can be found in the in the **Header/Footer** toolbar. To access these, select the header or footer in the document and the toolbar appears. From the toolbar, the date, time and document properties are available. Some fields may need to be manually updated. For example, the **Page number** field will update automatically when in a header or footer but may need to be manually updated when placed in the body of a document.

Calculations and the use of formulas

Calculations can also be carried out in mail-merge fields to manipulate data from a data source file for inclusion in personalised documents. The calculations use numbers and **arithmetic operators**. Arithmetic and other operators are described in Chapter 21. For example, the cost of tuition for courses can be included in a letter to students and calculations can be carried out on this field to show expected increases or additional charges in future years.

Task 18g
Calculated fields
Steps 1–3 are the same in both LibreOffice and MS Office.

1   Create a table in a spreadsheet with this imaginary data:

Course	Cost
IT	15000
English	12000
Geography	20000
History	13000
Physics	14000
Science	21000

There is no need to include the currency because this is set up in the master document.

2   Create a master document and link it to the data source in the spreadsheet. The document should contain text, with Course and Cost replaced with merge fields, such as:

You are studying **Course** and the **Cost** of tuition is Cost for the first year and this must be paid before you are allowed to start your studies.

Ensure that the merge is working correctly so that the cost of the course matches the course in the merged documents.

3   Add these lines to the master document:

In the second year, the cost will rise by 10% so **XXXX** will be added to make a total of **YYYY**.

In the third year, the cost will be the second-year cost reduced by 2.5% plus $500, to cover the cost of additional items, so this will be **ZZZZ**.

Merge fields are placed at **XXXX**, **YYYY** and **ZZZZ** to carry out the calculations and insert the results using formulas that contain the merge fields to be used. Choose the appropriate currency symbol and place it before the merge field. Alternatively, the currency symbol can be added by editing the field and entering a symbol to be placed before the field.

Task 18h

LibreOffice

Show the data source with **Ctrl+Shift+F4**. Highlight and delete **XXXX** and open the **Fields** menu with **Ctrl+F2**. Choose the **Variables** tab and select **Set variable**. A variable must be set so that it can be used in the formulas in the merge fields. Using the variables makes complex formulas possible because LibreOffice will not allow more than one field header to appear in a formula. In the name box, enter **CourseCost** and in the **Value** box drag the header of the **Cost** column from the data source to the box. Click **Invisible** so it does not appear in the documents and then click **Insert**. Now choose the type **Insert formula** and the variable will appear in the **Select** column.

Click in the **Formula** box, go to the named variable **CourseCost** and double-click it. It appears in the **Formula** dialogue box to represent the values in the **Cost** column of the data source and can be used in calculations. Add the arithmetic operator for multiply and **0.1** to calculate 10 per cent. Note that the % sign will not automatically appear so will have to be added to the document text, or the format of the formula result could be set here in the dialogue box.

The formula now reads **CourseCost*0.1** and clicking **Insert** places it in the document text as a **calculated field** and shows the result when the merge is run.

In the same way, enter the formula **CourseCost+(CourseCost*0.1)** into the document at **YYYY** to calculate the second-year total cost. Calculating the third-year cost needs a more complex formula to be entered, in the same way as above, at **ZZZZ**:

(CourseCost+(CourseCost*0.1)-((CourseCost+(CourseCost*0.1))*0.025))+500

Care must be taken with the syntax and the order of operations. If the syntax is not correct an error will occur, and if the order of operations is not correct the result will not be as expected. Changes such as correcting errors can be made by double-clicking on the field to open the dialogue box. Copying and pasting the formula sections can make constructing complex formulas easier.

Task 18i

Microsoft Office

» Step 1: With the cursor at **XXXX**, from the **Insert** toolbar choose the **Formula** menu, open the dialogue box and enter the formula. At this stage, enter **XXXX**, the arithmetic operator for multiply and **0.1** as shown below. The format of the number should also be entered.

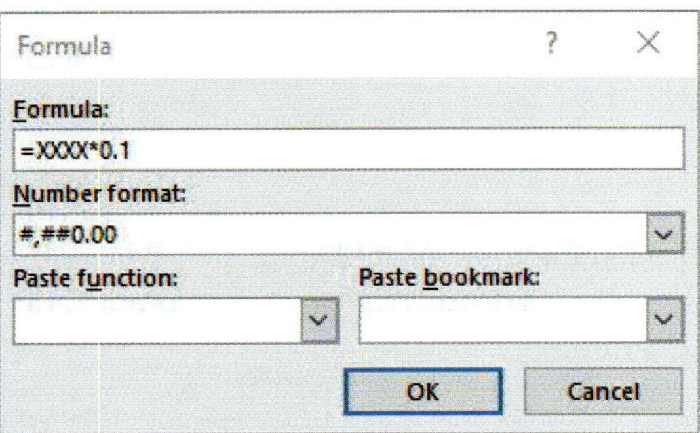

When **OK** is clicked, an error message about an undefined bookmark is shown because the formula has to be edited to include the merge fields.

» Step 2: Insert the merge fields. View the field codes with **Alt+F9**. Highlight and copy the merge field for **Cost** and paste it into the formula in place of **XXXX**, as shown here:

Replace XXXX here with the merge field so it becomes:

{ =XXXX*0.1 \# "#,##0.00" }

{ ={ MERGEFIELD Cost}*0.1 \# "#,##0.00" }

Hide the codes with **Alt+F9** and preview the documents to see the results of the calculation in the merge documents.

The process is repeated at **YYYY** for the total cost. A formula is inserted to add the 10 per cent to the original cost. The merge field for the total cost at **YYYY** is:

{ ={ MERGEFIELD Cost \# "#,##0.00"}+{=

{ MERGEFIELD Cost \# "#,##0.00"}*0.1 } }

Again, the currency symbol can be in inserted as appropriate.

Calculating the cost at **ZZZZ** is more complicated because the order of the arithmetic operations has to be considered. In this example, the addition is carried out after the multiplication so there is no problem. However, **operator precedence** must always be considered. This is described in more detail in Chapter 21. To ensure that the operations are carried out in the required order, brackets should be used as is usual in formulas. The merge field to calculate the cost at **ZZZZ**, leaving out the number formatting for clarity, is shown here:

{ =({ ={ MERGEFIELD Cost }+{={ MERGEFIELD Cost }*0.1 } } -({ =

{ MERGEFIELD Cost }+{={ MERGEFIELD Cost }*0.1 } }*0.025))+500 }

The simplest method of accurately creating such complex merge fields is to construct the formula and then use copy and paste for inserting the merge fields.

Specifying rules

Selecting recipients

It is not always desirable, or necessary, to include all recipients from the data source in a mail merge. Perhaps only a few letters or labels to specific people need to be created for printing. Sometimes, letters need to be sent only to certain recipients in the list because, for example, only a few are required to attend a meeting, they live in different countries, do different jobs, are in a specific age group, or for some other reason. Selections from recipient lists can be made by filters or by setting up rules to be used during the mail merge.

Filtering recipient lists

Filtering can be used to pick out specific entries in the data source and merge only these.

In OpenOffice/LibreOffice, the filter options are available from the **Mail merge** toolbar. The field is chosen, the condition set and the value to be used in the evaluation is entered. The condition is the comparison to be made and the value is what to compare the field with. This is shown here:

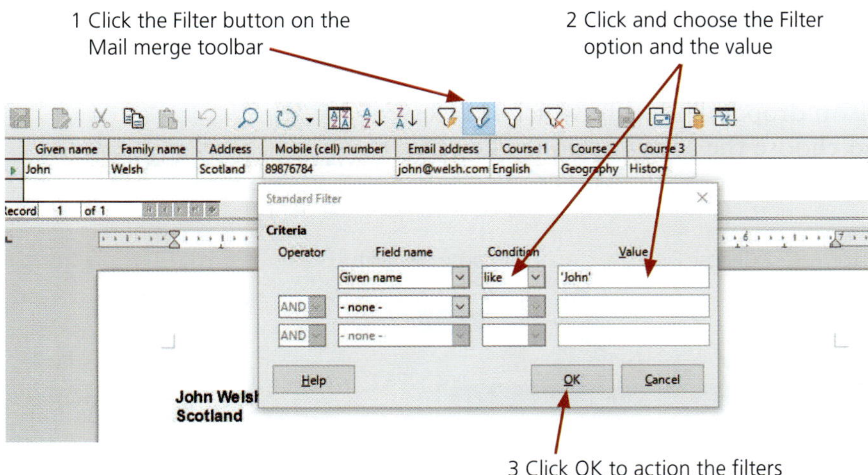

The condition can be 'like', which means the same as the value, or 'not like', which is not the same as the value. In the example, 'like' would filter so that only those with a **Given name** of **John** are used, whereas 'not like' would use all those whose given name was anything but **John**. The condition can also be 'null', where nothing is entered in the field, or 'not null', where something is present. Null filters only those with no entry, whereas 'not null' filters all those with an entry. In the example, 'null' returns only those with no **Given name** entry whereas 'not null' shows every entry that has a **Given name**. Up to three fields, with different conditions, can be used to set up complex filters.

In MS Word, the filter options are accessed from the **Mailings** toolbar, as shown on the next page.

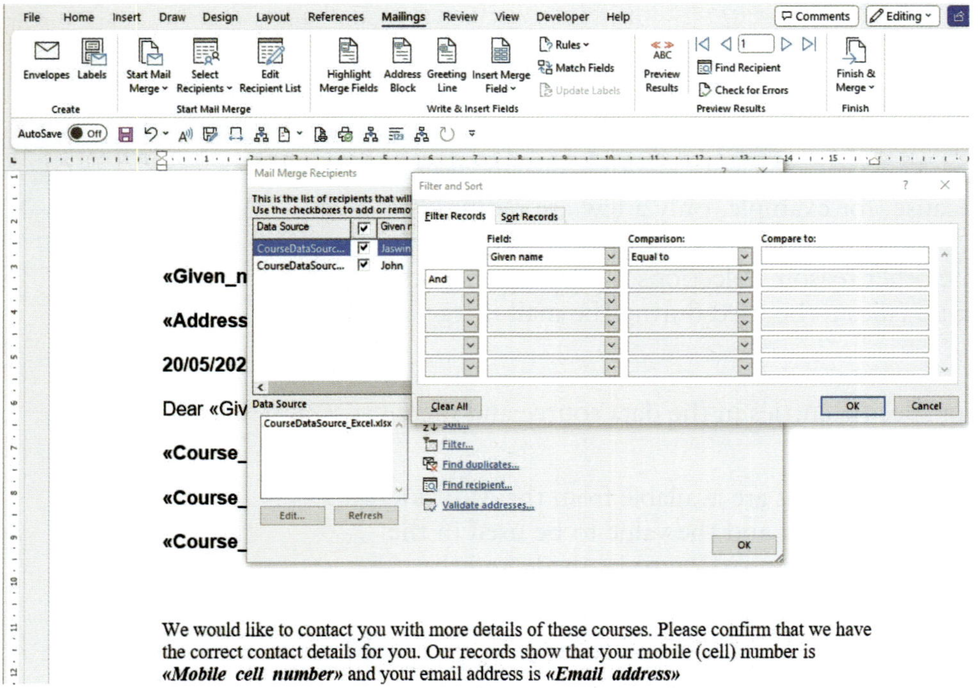

The conditions are set in the **Comparison** dropdown menu, which is accessed by: **Mailings**, **Edit Recipient List**, and choose the **Filter** option. The value to be compared with is entered in the **Compare to:** box, shown here:

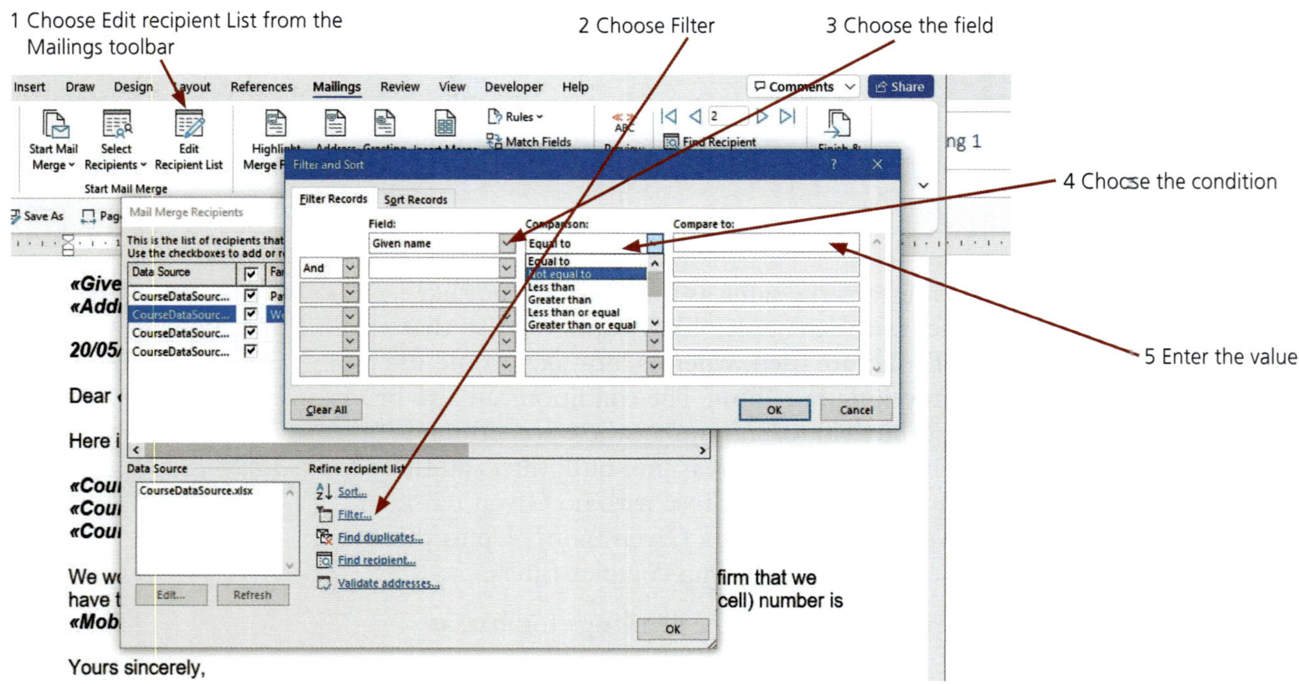

The conditions can be chosen from the **Comparisons** menu, where they are self-explanatory, and are:

- » Equal to
- » Not equal to
- » Less than
- » Greater than
- » Less than or equal
- » Greater than or equal
- » Is blank
- » Is not blank
- » Contains
- » Does not contain

Combinations of these used on several fields allow complex filters to be set. When the **OK** buttons are clicked, the recipient list will update to show the effects of the filter or filters. The comparisons are not case sensitive, so the value **JOHN** matches John, jOHn or any other combination of cases in the letters.

Using If…Then…Else

If…Then…Else enables content to be inserted in merged documents depending on the outcome of a comparison test. It can be used to select specific recipients, and exclude others, by setting conditions by which they are selected. A condition to be tested is set up and the alternatives chosen. When the merge is carried out, the result of the comparison selects the text to be inserted. For example, when merging a student's marks into a letter, a condition could be set up to choose text to be inserted depending on the marks.

In LibreOffice, a dialogue box opened from **Insert**, **Field**, **More fields** allows **Functions** to be chosen where the condition and actions can be entered, as shown here.

Open this box from Insert, Fields, More fields and choose Functions

Choose Conditional text

Enter the condition, Then and Else parameters here

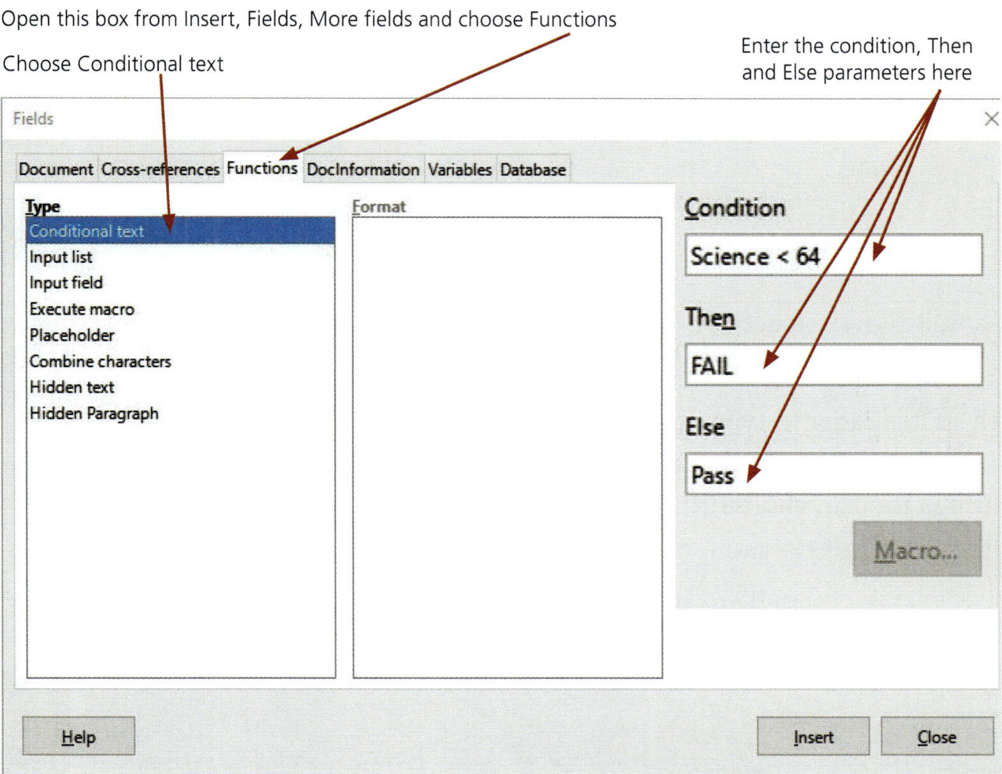

Clicking **Insert** places the field into the document. If a student has a mark in the field called **Science** of less than **64**, **Fail** will be placed in the document; otherwise **Pass** will appear.

In MS Word, this is done from the **Mailings**, **Rules** menu in the same way except that the comparisons are stated for you.

In another example, when sending letters to students about their course, it may be that some courses are taught at the main campus and others elsewhere. For those in the main campus, the sentence **This will be taught in the main campus.** is required, but for all others the sentence **You will be notified later of the campus to attend.** has to be inserted. This can be carried out using **If…Then…Else** and testing the contents of a merged field during the merge.

In LibreOffice, place the cursor where the text is to be inserted, open the **Field** dialogue as before and choose **Functions**. The field in the condition must be inside square brackets, state the full file path and the text must be in quotes. An empty field can be tested for by using " " with nothing in between. To ensure that the syntax is correct, the simplest way to enter the field in the **Condition** box is to drag it from the data source header into the box. You can view the data source using **Ctrl+Shift+F4**. The **comparison operator** has to be entered manually; for the same as it is **==** and for not the same as it is **!=**. This is shown here:

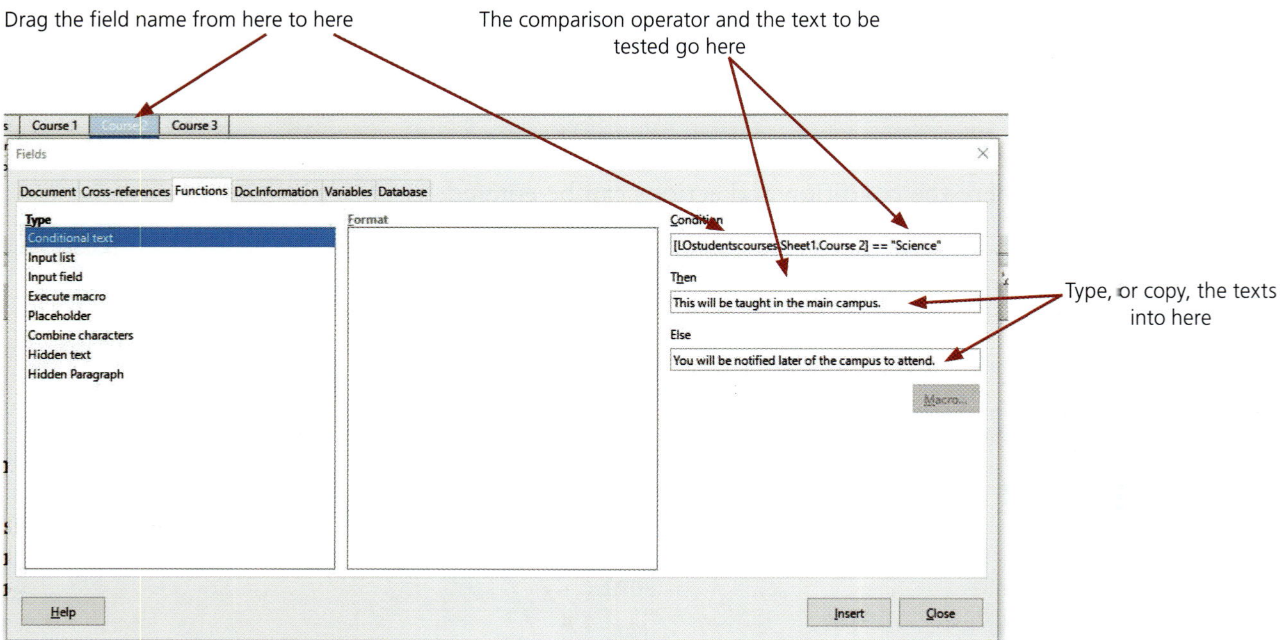

The text can be of any length so it is easier to type it, check it and then copy it into the dialogue box.

In MS Office, from the **Mailings** toolbar, choose **Rules** and select **If…Then… Else** from the menu, like this:

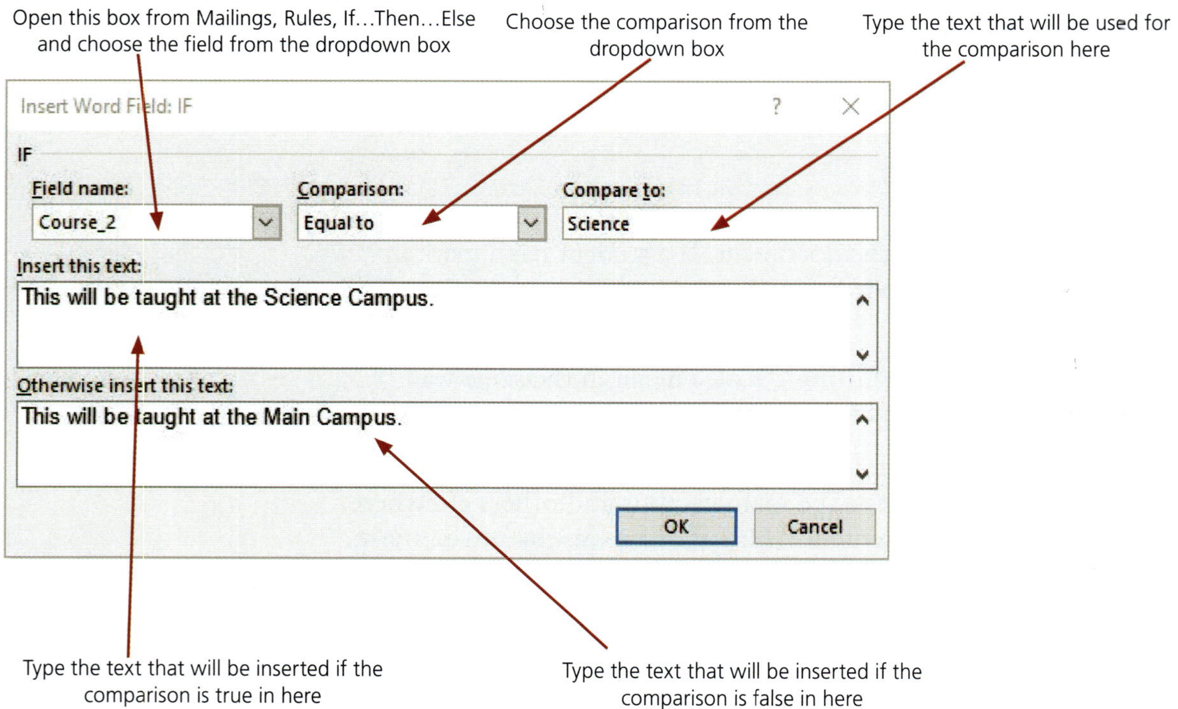

The comparisons that can be used are in the dropdown menu.

When the mail merge is carried out, the comparison is used to insert the chosen text. For example, in the Student Courses letter, when used on the **Course 2** field, the following text is produced.

IT
Science   This will be taught at the Science Campus.   OR   **English**
Physics   **Geography**   This will be taught at the Main Campus.
                                                                        History

If either text box is left empty then, depending on the type of comparison, nothing may be shown when the comparison is made.

The **If…Then…Else** statement can be altered, for example to change the wording of the texts, by using **Alt+F9** to view the statements and directly changing the contents. Copy and paste from other text sources can be used, as shown here:

{ MERGEFIELD Course_1 }
{ MERGEFIELD Course_2 }   { IF { MERGEFIELD Course_2 } = "Science" "This will be taught at the Science Campus." "This will be taught at the Main Campus." }
{ MERGEFIELD Course_3 }

If…Then…Else can be extended to deal with multiple comparisons by 'nesting' statements. In MS Word, nested statements are created by viewing the fields with **Alt+F9** and then entering the mail-merge codes and text manually, for example the nested statement shown below could be inserted in the Student Courses letter:

{ IF { MERGEFIELD Course_1 } = "IT" "You will study IT first." " { IF { MERGEFIELD Course_1 } = "English" "You will study English first." "You are not studying IT or English as your first course." } "}

If IT is the first course, the document shows the text **You will study IT first.** Else it checks whether English is the first course. If it is, it shows **You will study English first.** If neither is the case, the document shows **You are not studying IT or English as your first course.** Many statements can be nested. However, the spaces, quotes and brackets must be absolutely correct for them to work as expected. Do not forget that the curly brackets must be entered with **Ctrl+F9**. An easy way to create nested statements in MS Word is to create each statement separately from the toolbar and menus and then cut and paste them into each other.

Using complex **If…Then…Else** fields with long sentences means that it is essential that thorough checking of the document is carried out before a mail merge is performed.

Task 18j

The selection of recipients for an invitation to a conference could be made on the basis of where the recipient lives. Create a spreadsheet with the names of ten different recipients, five of whom live in one country, two in a different country, two in another different country and one in another. Create a master document of an invitation with mail-merge fields that selects the recipient depending on their country.

Using Skip Record If

Skip Record If can be used in a merge field to 'skip over' records or entries in the data source that are not to be included in the mail merge. **SkipIf** compares two expressions and if the result of the evaluation is **true**, the merge moves on to the next merge document. The syntax of the **SkipIf** merge field is:

```
{ Skip Record If field-to-be-evaluated conditional-
operator value-to-be-compared-against }
```

The conditional operators are the same set of comparisons, without the **Contains** or **Does not contain** options, that can be used with filters.

In MS Word, with the cursor at the place of insertion, **Skip Record If** can be inserted from the **Mailings** toolbar using the **Rules** menu, as shown here:

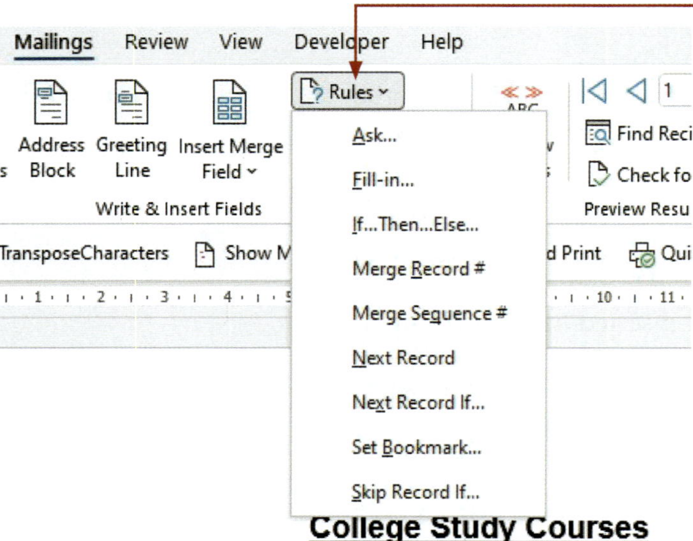

From the lists in the dropdown menus, the choice of field and operator and the value can be entered like this:

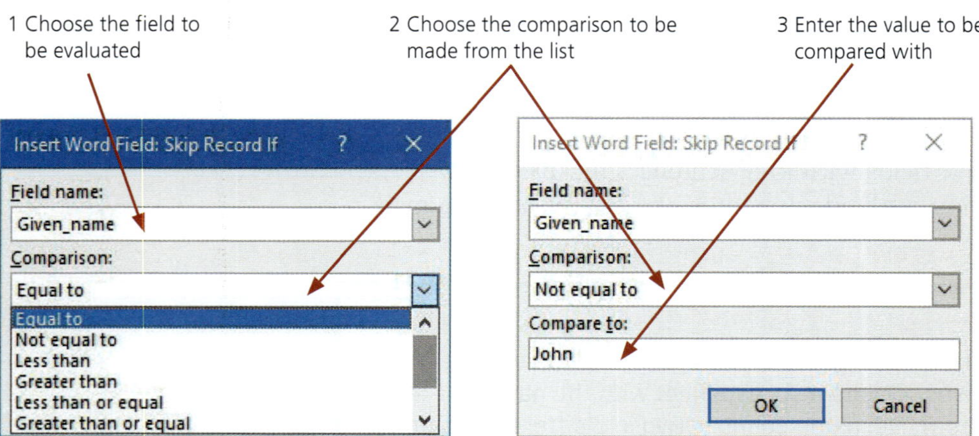

When **OK** is clicked, the merge field is inserted into the document at the cursor. In this example, when the merge is carried out, any records with a **Given name** that is not **John** will be passed over and not included, so only those records with **John** as a **Given name** are merged.

The **Compare to** box is left empty if the comparison is set to **Is blank** or to **Is not blank**.

Task 18k

Moving to next records

Find out what the merge fields **Next record** and **Next Record If** are used for. Set up a new mail merge or use the ones from previous activities to try them out.

Practice questions

1 Describe the circumstances in which each of these would be used in a mail-merge master document:

   a **If...Then...Else** (2)

   b **Skip Record If** (2)

   c **Next Record If**. (2)

2 Describe how mail merge can be used when sending emails in marketing. (6)

19 Graphics creation

In this chapter you will learn:
- ★ about digital graphics and how they are constructed
- ★ how to use image editing software to manipulate bitmap and vector images.

Before starting this chapter, you should:
- ★ have a working knowledge of geometry
- ★ have access to image manipulation software such as GIMP for use with bitmap images and Inkscape for vector images.

Digital graphics is the processing, storage and display of digital images. The processing can include compression and the display can include printing images to make hardcopies. Digital images consist of data that represents the visual characteristics of physical or imaginary objects. Computer processing turns the data into a form that can be made visible to humans. The two forms of digital graphics are bitmap and vector.

Bitmap graphics

Bitmap graphics create images made up of picture elements called pixels that are arranged in rows and columns and stored in computer systems in arrays. Digital images made up of pixel arrays are called raster images. Each pixel holds data about its colour and the brightness of the colour. The greater the number of pixels and the higher the amount of data stored per pixel, the greater the resolution of the image will be. Consequently, high-resolution bitmap images consist of vast amounts of data, so compression techniques are used to reduce the amount of data to manageable levels. Bitmap graphics are used, for example, for digital artworks and digital photographs.

Vector graphics

Vector graphics create images from data that specifies geometric shapes. Vector graphics are used in computer-aided design (CAD) and 3D animation systems and in graphic designs and illustrations.

19.1 Common graphics skills

Image manipulation software provides the tools for creating and editing digital graphics. These tools allow the user to manipulate every aspect and feature of an image.

19.1.1 Layers

The individual components, such as the shapes, of a digital image are called elements. Complex digital images are created in **layers**. A layer can be visualised as a transparent sheet holding separate image elements. The layers are laid or stacked on top of each other so that, when viewed, all are seen together at the same time and appear as one image. A layer may be larger than the image **canvas**, so parts of the layer may lie outside the actual image area.

When using graphic manipulation software to create an image, especially a complex image with many elements, each component, or **element**, of the image should be created and edited on a separate layer so

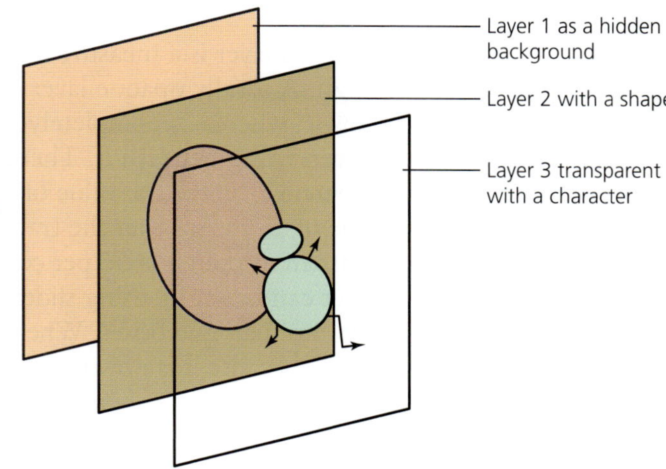

▲ **Figure 19.1** Layers in an image

that it can be manipulated, edited or deleted without affecting the other layers and their elements. This means that any changes or effects added to one layer do not affect the contents of the other layers. Layers can be added or removed without affecting the other layers, and the properties of each layer can be individually set and amended. When a specific layer is being worked on, it is called the 'active' layer. When working with layers, it is good practice to give each layer a sensible name so that its contents are easily recognised. This is especially important if an image contains many layers as, often, they are not all visible at once. Naming a layer allows the graphic designer to access a specific layer quickly.

Ordering layers

The order in which layers are placed in the 'stack' determines which elements appear in front of and which elements appear behind other elements in the finished image. Moving a layer lower down in the stack will make its elements appear behind those on layers higher in the stack, and vice versa.

Grouping layers

Layers can also be placed or **grouped** for further convenience. Layers, and groups of layers, can be made invisible (hidden) so they do not interfere with the editing of other layers. The visibility of layers, and layer groups, is easily toggled between visible and hidden from the software menu or with keyboard shortcuts.

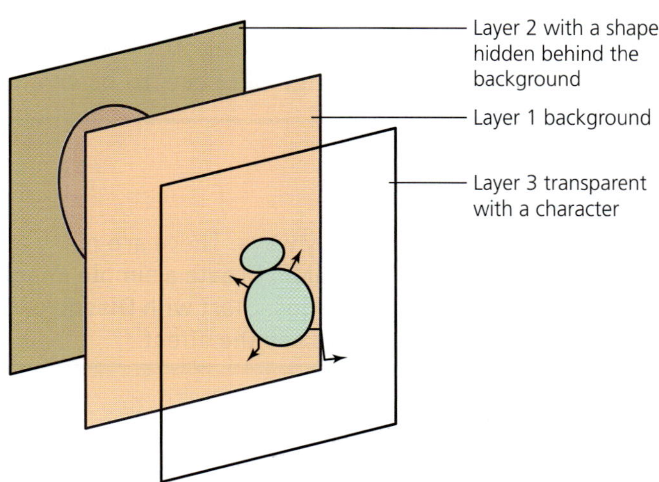

▲ **Figure 19.2** An image layer moving up or down the layer stack

Locking layers

Layers can be locked to prevent further editing and so protect them from accidental alteration. Layers can be partially locked where, for example, only some pixels are locked and cannot be edited, or only the transparent areas are locked so nothing can be drawn in these areas, or the position of the layer can be locked so it cannot be unintentionally moved. A layer can be completely locked so no changes at all can be made until it is unlocked. Unlocking the layer allows changes to be made. All the layers in a group can be locked so they can be protected together. In some image manipulation software, locking layers can be used to move a group of layers together so that all the elements can be moved around the image canvas while retaining their positions relative to each other.

Layer opacity/transparency

The **opacity/transparency** of a layer is a measure of how little, or how much, the layer can be seen though. A totally opaque layer cannot be seen through and hides everything below it, whereas a completely transparent layer is totally 'see-through' and shows everything beneath it. There is an inverse relationship between opacity and transparency, where the value of one is 100 minus the other, for example an opacity value of 45 means that the transparency value is 55 or if the layer is 34 per cent transparent then it is 66 per cent opaque. The opacity/transparency values in a layer can be set by using sliders in the software tools or by entering the values directly into dialogue boxes. When the opacity of a layer is set to 0, the layer will be invisible because it is now totally (100 per cent) transparent. Individual colours in a layer can also have their opacity/transparency altered, enabling only selected parts of the layer to be made transparent rather than the whole layer.

The opacity/transparency of a layer is handled by its 'alpha channel', which may be automatically included when the layer is first added. This channel specifies which areas of the layer are transparent and which are opaque. Usually these are black and white pixels but any colour can be used as the transparent colour.

Blending layers

The **blend** setting of a layer determines how the layer appears when viewed with the other layers. There are several different blend settings available, depending on the software. Here are some examples. A normal blend setting allows only the top layer to be seen. A dissolve blend shows a lower layer through a higher layer through random pixels to give an appearance similar to two video sequences dissolving into each other at a transition. There is also a range of blend settings based on the colour, brightness and opacity of individual pixels in the layers that allow a wide range of effects to be created.

Task 19a

Blend settings

Layers can be blended in many ways. There are many blend settings available for layers in Photoshop and GIMP. Create a simple image with several layers and experiment with the blend modes. Start with **Dissolve**, **Lighten only** and **Darken only** and then try the others to see the effect.

Flattening layers

When an image is flattened, all the layers are combined, or merged, into a single layer. During the **flattening** process, alpha channels are removed and any areas in layers that were set as transparent show the elements from layers below. An image is usually only flattened when it is exported to a format that does not support transparency. Flattened images do not allow the individual elements to be independently edited.

Merging layers

Layers can be merged into one layer while retaining the properties of the active layer. Merging combines the elements of the layers into one, editable layer. The merge can be set with various options such as discarding any hidden layers during the merge, **cropping** away parts of layers that lie outside the image area or expanding layers to match other layers.

19.1.2 Transform tools

Transform tools are used to alter the image or elements within the image.

Resize

Resizing an image changes its actual, physical size by changing the number of pixels. For example, changing the image size from 1920 pixels by 1080 pixels to 1344 px by 756 px changes the number of pixels in the image from 2 073 600 to 1 016 064, losing over a million pixels from the image. Resizing an image to a larger size adds pixels. Removing pixels can reduce the quality of the image by removing detail when pixels are lost, and adding pixels blurs the image as the new pixels are inserted by the software's interpolation of the properties of the new pixels.

Scale

While **scaling an image** may have a similar visible effect as resizing, these are not always exactly the same. Scaling changes the viewed size of the image but may not change its physical size. For example, scaling an image to 50 per cent of its size for display on a web page means that it is rendered at half of its real size, which still has the original number of pixels. Using stylesheets to specify the size of an image on a web page determines the size of the displayed image but does not change the number of pixels in the image. In this way, a single image can be displayed in many different sizes without being physically altered.

However, in several image manipulation software packages, the image scaling tool is actually as image resize tool.

Shear and skew

Shear applies a horizontal or a vertical slant to an image. The slant, or **skew**, can be applied to whole images or to individual elements within an image.

Shear is often used on whole digital photographs to reduce the distortion produced when taking the photograph from an angle such as looking upwards and sideways.

Reflect

Reflect, often called flip, produces a mirror image of the original image by rotating it through 180° either horizontally or vertically. Only the direction of the image is changed; all other properties remain the same.

Rotate

Rotate moves the image, layer or element in a circular motion around a central fixed point. The amount of rotation can be entered as values in a dialogue box displayed by the **Rotate** tool or carried out by dragging by the desired amount.

▲ **Figure 19.3** Shear applied to an image

▲ **Figure 19.4** An image reflected

▲ **Figure 19.5** Image rotated about its centre

Move

The **Move** tool is used to move an element to a new position on the image canvas. The element to be moved is selected and then dragged to its new position.

Envelope

In image manipulation, an **envelope** is an object that is used to distort or change the shape of other, selected objects. Envelopes can be created from user-created objects or from the tools supplied by the software, such as a mesh or a warp shape. Different image manipulation software packages implement envelopes in different ways and may call them by different names, for example 'warp distort', but the principles are the same. A mesh envelope overlays the object to be distorted with a grid that is then dragged with the mouse pointer into the desired distorted shape. A shape envelope warps the object within to take on the shape of the enveloping object.

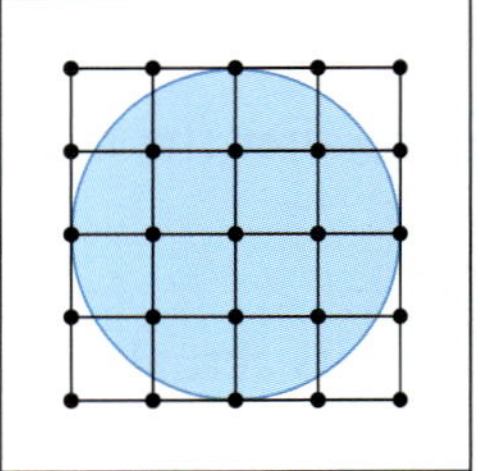
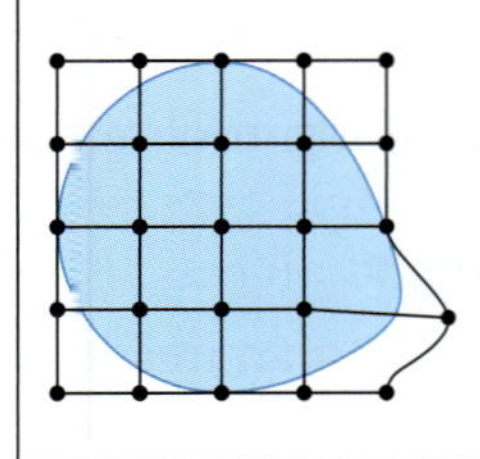

Mesh grid envelope

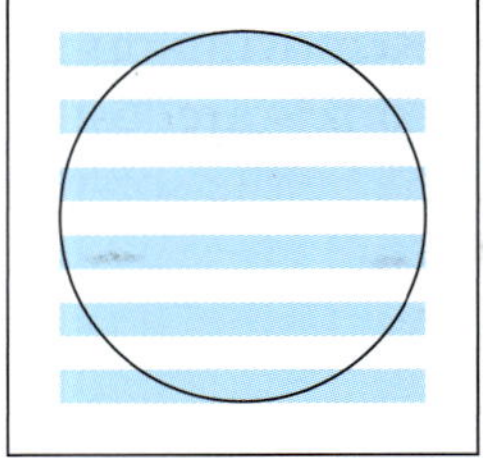

Envelope created from another object

▲ **Figure 19.6** Using envelopes to distort images

Perspective

Perspective gives flat, two-dimensional images the impression of being three-dimensional by making near objects appear larger than those further away from the viewing point. In animations, objects that become smaller in a sequence can be seen as moving further away from the viewer and vice versa. In still images and drawings, perspective techniques are used to convey the sense of distance and 3D structure in the image. One-point perspective keeps the vertical lines or edges parallel, but the horizontal lines or edges converge on a single point in the image. Two-point perspective has two sets of horizontal lines converging on two separate points in the image, giving the impression that the viewer is at a corner of a building, for example. With three-point perspective, the viewer appears to be looking up or down at the object, and the vertical and horizontal lines converge on points at the top or bottom of the image.

The perspective tool is often used to correct or remove extreme perspective distortion in digital photography. For example, a digital photograph taken while standing at the base of a building and looking up will appear distorted – an illustration of three-point perspective – and this can be corrected so that the photograph is more pleasing.

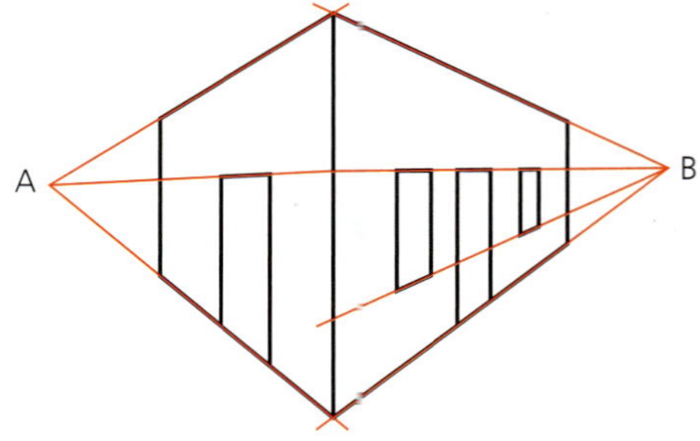

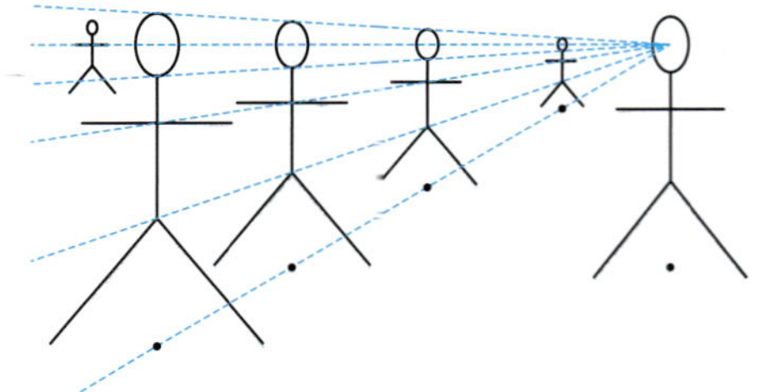

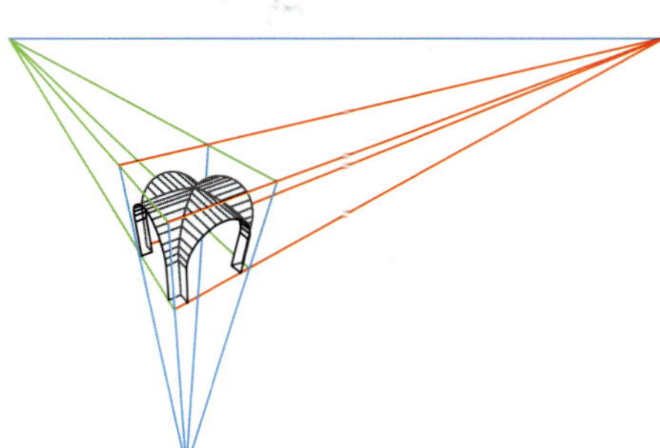

▲ **Figure 19.7** One-, two- and three-point perspective

▲ **Figure 19.8** The perspective tool used on a digital photograph

19.1.3 Grouping or merging tools

Grouping and merging objects, elements and layers allows them to be treated as a single entity. However, while grouped objects can easily be ungrouped, merged objects usually cannot be unmerged.

Group and ungroup

Grouping several layers, objects or elements together allows them to be moved, resized and edited as if they are one. The properties of a group can be edited and the changes are applied to the whole group. Ungrouping a group of objects separates them into individual items again.

Combine/join

This is often used to combine digital photographs together. Combining and joining images can be carried out in several ways. Images can be superimposed on each other, selections in the images can be superimposed or individual elements in the images can be combined to create a composite photograph.

Images can be joined side by side or vertically stacked, or however the user wishes to display them.

Add/subtract

Usually when a shape is added to an image, it will be drawn on a new layer so it can be edited separately. But if it is required on the same layer as an existing shape, the **Add shape** option is used. The added shape or shapes can be used to change the shape of an existing shape by combining two or more shapes together on the same layer. If the added shape is used to remove part of an existing shape, for example to **cut** part of a shape away, then this is referred to as a 'subtract'.

Intersect

When two or more shapes overlap and only the overlapping areas are made visible, this called an intersect.

19.1.4 Alignment and distribution tools

Alignment

Elements, objects and layers can be placed almost anywhere on an image canvas by selecting them and dragging them into position. Often, multiple elements, objects or layers need to be aligned with each other and this can be tedious and time-consuming. **Alignment tools** make the precise positioning easier and much more accurate. The tools are especially useful when positioning or aligning multiple elements or objects with each other. The element, object or layer to which the others are to be aligned is called a 'target', and this is the first to be selected. Other elements, objects or layers are then selected, usually by selecting while holding down the **Shift** key, and the alignment tool is used to align these to the target.

The alignment tool has various options for the alignment. Alignment can be with another layer, object or element or with the whole image. When aligning with a target, **Left align** moves all the selected items so that their left edge aligns with the left edge of the target, and **Right align** aligns the right edges. In a similar way, **Top align**, **Bottom align** and **Centre align** aligns the top, bottom or centre of the selection with that of the target. There may also be an **Align middle** option, which aligns items vertically to the middle of the target.

Aligning with the whole image works so that the **Align** tool aligns the selection with the left, right, centre, top or bottom of the whole image area. For example, **Centre align** places the entire selection in the centre of the actual image rather than aligning the centres of the selections.

Distribution

Distribution refers to how elements, objects and layers are spaced and placed around the image canvas. Vertical distribution places them one above each other from top to bottom in an image, while horizontal distribution places them side by side from left to right in the image. The distribution tools have several options that can be applied to the distribution, for example by aligning against one of the four edges or to the centres of the selection, layer or whole image. Setting offset for the selected items places each item a set distance from the previous and can create an overlapping effect. An offset can be from any of the edges. Positive or negative offsets can be used to determine in which direction the offset occurs.

Order

The order of layers in the image determines which layer takes precedence over the others. The layer at the top has the highest precedence and any elements in that layer will be seen first and will obscure, depending on the opacity settings, the elements in the layers below. A 100 per cent opaque top layer will obscure all the layers below, whereas a 100 per cent transparent top layer will allow the elements on the layer immediately below it to be seen. Adjusting the opacity/transparency values of layers and objects within layers controls how much of the lower layers are seen.

Layers can be raised or lowered in the 'stacking order' as required by the user to achieve different effects. Raising a layer gives it precedence over layers below, while lowering a layer in the stack means that the visibility of its contents depends on the settings of the layers above it. Layers can be moved up and down the stack either from the **Layer** menu options or by dragging and dropping the layer to be moved to its new place in the stack.

Bring to front/send to back

These tools only move objects within a layer. Bringing an object or element to the front places it on top and it will obscure those behind it. Conversely, sending an object to the back places the object behind the others in the layer.

19.1.5 Layout tools

These tools are used for accurately positioning and aligning elements and objects in an image when creating and editing. They also make accurately measuring pixels and the spacing between them much easier.

Rulers

Rulers are usually placed on the left side and on the top of the images. The default unit of measurement of a ruler is usually pixels, or px, but this can be changed, for example to inches, millimetres, points, picas or other units that the user might wish to work with. Rulers can be turned on or off as required.

Grids

A **grid** is an overlay of squares on the image. It looks like the graph paper used to hand-draw mathematical and scientific graphs, and acts in a similar way by providing reference points to accurately align elements both horizontally and vertically. The grid size can be set in different units in much the same way as with rulers. The appearance of the grid, for instance solid line or dotted line, can also be set by the user.

Guidelines

Guidelines can be created from the rulers to temporarily place lines across the image to allow easier placement of elements when moving or creating elements.

Snapping

To make the placement of elements easier and less time-consuming, elements can be made to **snap** to the nearest guidelines or grid point. When an element is moved close to a guideline or grid point, it is automatically placed (snapped) onto the point. A default distance from the point for the snapping – often 8 px – is set, but this can be changed by the user. The snapping option can be turned on and off as desired.

19.1.6 Colour

Colours are used in graphical design and in images for several reasons. Colours can, for instance, express the mood of a scene, enhance the meaning of objects and elements, code objects using different colours (such as red for warnings), improve the recognition of objects and elements by viewers, and improve the accessibility and usability of objects, for example icons. However, because the ability of humans to perceive and interpret colours differs widely between individuals, the choice of colours must be made with care to ensure that everyone is included.

Colour wheel

Colour wheels are a circular representation of the available colours. They are designed to make choosing colours easier. The colour values are often shown alongside the colour wheel.

Colour systems

How humans perceive colour is not fully understood but, in very simple terms, humans are sensitive to three main colours: red, blue and green. Combinations of varying amounts of each of these colours in the light captured by the human eye is interpreted by the human brain as the full range of colours that we can see. The colouring of objects and elements in graphical designs and images makes use of this to display the range of colours as close as possible to the colours that the viewer would see on a real object. When viewing a real object the viewer sees the light reflected from the object, whereas when viewing a digital image of the object on a computer screen, the viewer sees light that is transmitted from the screen. When viewing a copy of the same digital colour image that has been printed on paper, the viewer is again viewing reflected light. Colour systems have been developed to try to make all the colours in these appear to be the same to the viewer.

Colours are arranged in a circle and can be selected easily. This colour wheel from GIMP also displays information about the colours.

▲ **Figure 19.9** A colour wheel

RGB colour system

The **RGB** (red, green, blue) system is additive, in that colours are produced by combining, or adding, varying amounts of red, green and blue. Red, blue and yellow are called the primary colours and cannot be created from other colours; green is created from blue and yellow. Combination colours made from the primary colours are called secondary or tertiary colours. Secondary colours are made by combining two primary colours, and tertiary colours are made by combining a secondary colour with another primary colour.

Each of the primary colours can be varied in intensity from none (completely off) to full (completely on). With all three RGB colours fully on, the resulting colour is white, and with all off the colour is black. The values can be represented in percentages, where 0 per cent represents none and 100 per cent represents fully on, but is usually represented as from 0 to 255, which is based on the number of bits in a byte used to store the value. The 0–255 scale is used when specifying colours in image editing and in style sheets for web pages. For example, **rgb(255,0,0)** in a style sheet specifies that red be set to 255 and green and blue both be set to 0, resulting in a pure red colour. A value of **rgb(255,255,255)** will produce white. RGB values can also be shown in hexadecimal format, for example **rgb(255,255,255)** can be shown as **#ffffff** (or **#FFFFFF**) instead.

In image editing, the colours can be set by inputting the values directly in the colour tool or by dragging sliders. Alternatively, colours can be picked from a colour wheel, or palette, and the values automatically assigned by the software.

The RGB system is used to store digital images in memory or on storage devices, where the RGB values for each pixel are stored in the usual binary form, and it is used to generate colours on computer displays. In the older cathode-ray tube (CRT) displays, red, green and blue phosphor dots were lit up by varying intensities of beams of electrons. Modern displays have an LCD, plasma, LED or OLED red, green and blue light source for each individual pixel. These are so small and so close together that they are indistinguishable at normal viewing distances and appear as solid colours to the viewer. Screens with huge amounts of pixels can be manufactured to provide very large, high-resolution displays.

HSL colour system

The **HSL** (hue, saturation and lightness) system is a method of representing colour where the hue (H) is the amount of colour ranging from 0 to 360, from red through yellow, green, blue and back to red, on the colour wheel. As for the RGB system on which it is based, HSL is used to adjust colours in digital images and when displaying colours on computer displays. The saturation (S) has a value from 0 per cent, which represents none of the colour set in hue is present (in other words it is grey), to 100 per cent where the colour is fully present. The lightness (L) value can be compared to mixing a fully saturated colour with white. This is on a scale that ranges from 0 per cent, the darkest possible (fully black or no white), to 100 per cent, the lightest possible (all white). The tools in image editing software allow each of the values to be set in the usual ways.

CMYK colour system

The **CMYK** system uses four colours: cyan, magenta, yellow and key (or black), to produce different colours. It is a subtractive colour system and is used in colour printing. Colour prints are viewed by looking at the light reflected from the paper, or other medium, on which the image is printed. With white paper viewed under white light, if all the light is reflected and viewed, the image appears completely white. Different colour inks are used to print areas of the image to stop different colours being reflected; that is, to subtract colours from the image. For instance, printing an area with black ink stops all light from being reflected from it so the area appears black. Therefore, inks are used to subtract colours from an area, for example yellow ink stops all light except yellow from being reflected so the area would appear yellow.

In the CMYK subtractive system of producing colours, using all colours produces black and using no colours produces no colour. In the case of 'no colour', the actual colour seen is the colour of the paper. The key (black) colour is used because if only C, M and Y are used then the resulting colour is more of a 'muddy' colour than a proper black. Using the additional key (black) ink makes up for any deficiencies in printing processes and, more importantly, provides the fine details in the printed image and saves ink. Less black ink is used to produce black than if black were produced using a combination of three colour inks.

Inkjet printers used in homes usually have four ink cartridges that apply ink in one pass over the paper to create the printed image, but large-scale commercial printing may use a different approach, such as screen printing. The image is divided into separate images for each colour, so that one separation will show only the cyan areas, another will show only the yellow areas, and so on. These separated images actually appear, when viewed on screen, to be in greyscale because it is the ink that determines the colour during printing, so it is important that each image is correctly named with its colour. Registration marks are placed on each separation to ensure that the final printed image has the colours in exactly the right place. During the printing process, printing plates are made for each colour and each plate in turn prints its colour on the paper.

Colour-management system

It is not possible to make the colours appear exactly the same on every device, screen or printed copy because of the different methods of producing colours for display on screen. Also, the colours seen in a printout do not always appear exactly as they did on screen. There are complex colour matching, colour profile and configuration methods to try and ensure that colours match what the graphic designer intended. If the final image is designed to be displayed mainly on computer screens, then graphic designers should use the RGB or HSL colour systems, but if the final image is destined to be printed (for example in a magazine, a book or as a poster) colours can be set up in image editing software for use with CMYK and colour separation systems.

Colour-management systems (CMS) use a standard set of reference colours to create colour profiles for each device, screen or printer. The colour profiles are used to alter, or translate, colours when the images are displayed on the different devices, screens or printers to try to make them the same.

Colour-picker tools

The colours of the elements, the foreground and background can be changed or set in different ways. **Colour-picker tools** allow colours to be chosen and set by conveniently clicking on the required colour from a colour chart or wheel. When a colour is clicked, the colour values are automatically selected and can be used to colour objects or elements.

Colours can be directly picked from an existing colour in the image. A tool resembling an eyedropper is used to 'suck up' a colour from the image and this colour becomes the colour used by the paint brush or for filling objects. This is useful for picking a colour when correcting spots or marks on digital photographs.

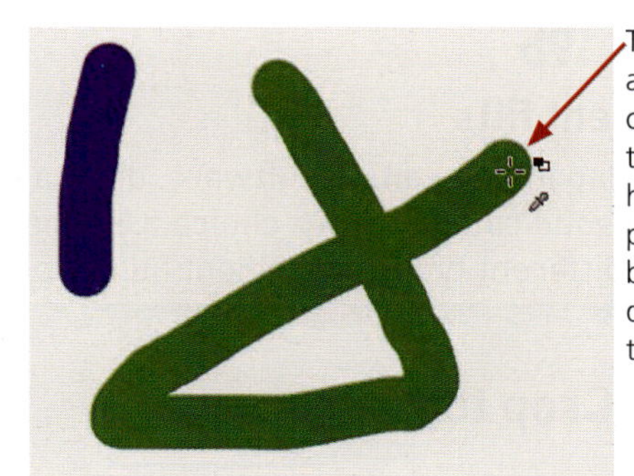

▲ **Figure 19.10** Colour-picker tool in GIMP

Colour fills

Objects can be filled with colour in several ways. The colour to be used in a **fill** is chosen or selected from the colour menus before the fill tools are used to apply it. The fill tool, often called a bucket fill tool, fills a selected object or area with the chosen colour, but how it fills can be controlled with various options. It can, for example, fill only a selected area, fill only the background, or fill or replace one colour with another. The type of fill can be solid or gradient.

Solid fills

A solid fill applies a solid, full colour to the entire selected area and, while its opacity can be changed, the colour is the same throughout the fill.

Gradient fills

A gradient fill applies new colours as a blend across the filled area. Usually, gradient fills are created in their own layer and these need to be adjusted to match the specific areas where the user wants the gradient fill to appear. The blend can be the same colour, starting from a light and ending in a darker shade of the colour, or it can be a mixture of two or more colours that gradually changes from one into the others. The suddenness of the gradient can be configured from smooth to very sudden changes across the area.

Gradient fills can be either linear or radial. In a linear gradient, the start, or origin, of the gradient is on one side of the area and the end is at the other side. The direction, or angle, of the fill gradient is set by the user. There are many other options available to the user, such as zooming in or out of the gradient to change the visibility of the ends or edges. A radial gradient fill has a circular shape and starts from a central point with one colour and ends with another colour at its edge. Again, there are many options that can be used. For example, the colour can start in the centre and end at the edge, or this can be reversed, with the colour fading towards the centre; the opacity can be altered so other layers show through the gradient; or the fill can be repeated to give the appearance of waves emanating out from the centre. A spiral gradient fill arranges the fill gradient around the centre in spiral fashion, a square gradient around a square, and so on.

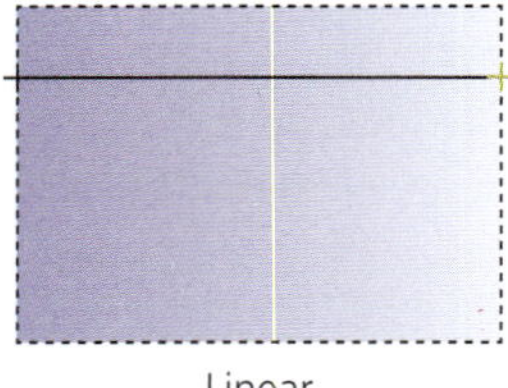

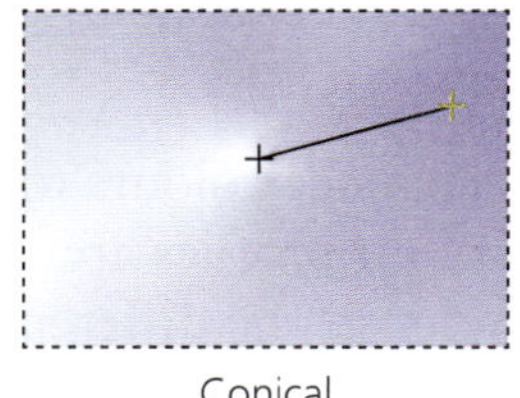

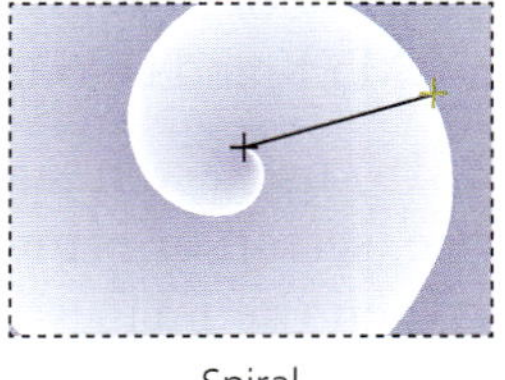

▲ **Figure 19.11** Gradient fills in GIMP

Task 19b

Gradient fills

In bitmap image editing software, draw a shape, for example a rectangle. Fill it with a solid colour of your choice. Use the gradient fill tool to experiment with as many different types of gradient fill as you can.

19.1.7 Crop tools

Cropping an image removes unnecessary objects or parts of the image to leave only those that the user wishes to be seen. An area can be marked out by selecting it with the **selection tools** or by using the cropping rectangle. When the image is cropped to this area, all the rest of the image is removed from visibility and the visible area changes so that the unwanted parts are not seen.

19.1.8 Resolution

The **resolution** of a digital image is how much detail it has, and this depends on the number of pixels present in the image. The absolute number of pixels in an image is shown by its width and height in pixels, for example 1920 px by 1080 px. The greater the number of pixels in the image, the more detail can be seen. However, it not simply the absolute number of pixels that is important, but also how many pixels there are in a given area and how they are arranged in the image. The more pixels in a given area, the higher the resolution and vice versa.

Resolution is measured and described as the number of pixels per inch (ppi) or, for printed images, dots per inch (dpi). The ppi of an image is usually set when it is first created as a new image, but it can be altered. Changing the ppi does not change the number of pixels; instead it changes the physical size of the image. To change the number of pixels, the Scale or Resize tools are used. Low-resolution images have few pixels per inch while high resolutions have many. High-resolution images look to be clearer, sharper and crisper and appear as higher quality with more detail.

When displaying images on web pages, compromises may need to be made. The need for higher-quality images with high resolution has to be balanced with these images usually having a large file size even when compressed. Large files take time to download, process and display in web browsers. Some users may become frustrated if they have to wait for images to appear and may move to another, competing, website. The most common resolution chosen for images for display on web pages is 72 ppi. However, because modern computer monitors, and the best screens on smartphones, are capable of displays in excess of 100 ppi, images for web pages may need to be of higher resolution to avoid appearing as low quality and unappealing to viewers. Displaying images on web pages is further complicated since web images use pixel dimensions rather than resolution, so images can have differing resolutions but display at same size and look the same.

When images are to appear high quality in magazines or books, the resolution needs to be high; at least 600 ppi. This is especially important if high grade, glossy paper is used, as any lower resolution will not show the detail in the image. An acceptable quality for most magazines and books is around 300 ppi, and for use in newspapers it can be much lower as quality is often seen as less important than the content. If an image is to be displayed on a large billboard as a poster, it will usually be viewed from a distance so it can be of low resolution; around 150 ppi. Because the human eye will not be able to make out the detail when far away, there is no need for the expense of printing high-quality images in such large sizes.

When displaying images on computer monitors, the characteristics of the monitor affect the appearance of the image as viewed. A low-resolution monitor will not do justice to a high-resolution image and a low-resolution image will not look good enlarged on a high-resolution monitor. Images that have been designed for use in presentations should have a resolution suitable for projection or display on large TV screens. Given that audiences are often some distance away, images of at least 72 ppi, with a pixel count of 1024 px by 768 px, are suitable. This fits well with the resolutions of television systems.

19.1.9 Exporting an image

Each image editing software package saves its files in its own proprietary file format. These formats allow the files to hold the image data and all the extra data about its content such as the layers, the object's elements and the editing history. While this is essential when creating images, the drawback of these formats is that the proprietary format cannot usually be imported into other software applications and the extra data is unnecessary for viewing the image. All image editing software applications allow for the exporting of images in a format that can be imported and used in other software applications, which overcomes this drawback. There are several formats that are used, each with its own characteristics and uses.

Image file formats

Numerous file formats are available for storing digital images. Some formats are used when creating, editing or exchanging digital images, such as TIF and BMP, and some are used when the images just need to be displayed, for example JPEG, PNG and GIF. Some formats allow the file size to be reduced by applying compression when the file is saved. There are different formats for using bitmap images and vector images.

Scalable Vector Graphics

Scalable Vector Graphics (SVG) is an open standard, meaning that anyone can use it, for storing vector graphic images. It is supported by almost all web browsers. It is a markup language that describes two-dimensional images in an XML-based text format. It can be opened and edited in any text editor if required and can be searched, scripted and indexed. Its scripting capability makes it very suitable for use on interactive web pages and web-based applications. It does not directly support compression but SVG files, because they are text, can be compressed using third-party compression tools. A SVG file contains all the data required to construct a vector file including the coordinates of point, the length and angle of lines, the type of lines, and the size and shape of objects. The properties of each shape are also stored, such as its colour, the type of line used to surround it, and the colour and weight of the line. Images stored as SVG can be zoomed in and out for viewing, resized and scaled without any loss of quality.

Windows Bitmap

Windows Bitmap (BMP) format is lossless (see Section 19.3 'Compression'), so files can be very large. While it can be used with most applications, it is most commonly used by the Microsoft Windows operating system and Microsoft applications as it was created by Microsoft for this purpose. As its name suggests, it is a bitmap image file format, and is mostly used when creating images or transferring images between applications.

Joint Photographic Experts Group

Most commonly referred to as **Joint Photographic Experts Group (JPEG)**, there are several file extensions associated with JPEG images, for instance jpg, jpeg and jif. This file format uses **lossy compression** (see Section 19.3 'Compression') to reduce the file size. The amount of compression applied to the image during saving can be altered to compromise between loss of quality and file size. Applying large amounts of compression will significantly reduce the file size but can greatly reduce the quality and introduce intrusive 'compression' artefacts that spoil the image. Repeatedly saving, loading and resaving in jpg format introduces 'generational' degradation because of the losses during compression. Maintaining high quality can mean that the file size is very large. Images in jpg format store image data in 24 bits, with 8 bits being used for each of the three primary colours. Greyscale images use just 8 bits. It can handle 16 million colours but the JPEG format does not support transparency or animation.

The benefits of using JPEG formats are that they are almost universally acceptable in image editing software, can be used with all web browsers, software applications and apps, can have a small file size so are quickly transferred over networks and the internet, and download times to web browsers are short. However, the lossy compression can reduce the quality of the image, with edges of objects becoming blurred, detail in the image being lost, and the transitions between colours and objects in photographs becoming sharp and looking unreal with a posterisation effect. Compression artefacts such as dots or 'noise' can appear.

Portable Network Graphics

The **Portable Network Graphics (PNG)** format supports **lossless compression** so compressed images retain their quality. The format uses 24 bits to store colour data so can handle 16 million colours. It may also have an additional 8 bits to store an alpha channel with data for transparency. The PNG format supports transparency but not animation. However, a development offshoot of PNG called Animated Portable Network Graphics does so.

PNG is an open format so anyone can use it, meaning it can be used in almost any application or app and is supported by all web browsers. It does not lose image quality because it uses lossless compression so is not subject to generational degradation. However, lossless compression does not significantly reduce the file size so PNG files are usually larger than, for instance, JPEG files when storing the same image. This means their download times are longer than for the JPEG files. To compensate for this, the format allows for the progressive display of an image so that the viewer can see the image before it is wholly downloaded. The PNG format is designed for viewing on computer screens, for example in web pages, and converting PNG images to CMYK for printing often results in the colours being inaccurate on paper.

Graphics Interchange Format

The **Graphics Interchange Format (GIF)** is mainly used for basic images that are displayed on the internet, for example in web browsers. It supports 256 colours using 8 bits per pixel. Basic animations can be created and stored as GIFs by combining several images in the file.

The format uses lossless compression so image quality is retained. The small number of colours available means that file sizes are small so GIFs transfer and load quickly over the internet. The creation of simple animations in GIFs requires little skill or technical knowledge so it is a popular format on social media. However, the low number of colours compared to other formats means that images are not very realistic so it is not suitable for digital photographs. Editing GIFs with more than one image in the file, such as an animation, can be difficult and often not possible at all.

Tagged Image File Format

The **Tagged Image File Format (TIFF)** has a file extension of .tif or .tiff and is supported by all operating systems and many applications. Some versions of the format can use lossy compression in an attempt to reduce file sizes, but most use lossless compression to store images in order to retain the high quality of the image. The high quality associated with this file format means it is often used to store digital photographs taken by digital cameras. TIFFs are suitable for high-quality digital photographs because they can be of high resolution and have extra image data included.

However, the extra image information to be stored in the file means the file size of TIFFs can be very large, which may limit the number of very high-quality photographs that can be stored on the memory card in a camera or other storage devices. The large file size of TIFFs means that the format is rarely used when sending or sharing images over networks or the internet. TIFS are not used on web pages because their large size means that download times can be long. TIFFs are often produced by scanners when documents are scanned for optical character recognition purposes or when scanning detailed photographs. TIFFs can also be used to store a collection of other images in one file, so that a group of lower-resolution JPEGs created by one designer could be stored in one TIFF image. This single image can then be transferred to another graphic designer.

Portable Document Format

The Portable Document Format (PDF) is designed for use with documents. It is designed to be completely independent of hardware and software so is supported by all operating systems. PDF documents can be imported into and exported from almost all software applications, making documents 'portable'.

PDF files can contain text, images, formatting information, interactive forms, video and other media. As well as the content, PDFs contain a complete description, including details of the fonts used for text and of how the document should appear when displayed on screen and on paper. PDF files consist of objects and streams that hold the instructions and content of a PDF document. Streams of content are encoded to save space and may also be encrypted to keep them secure. A typical PDF file will, in simple terms, have a header that holds data (for example about the PDF version used to create the document) and a file body containing dictionaries (collections of objects) and streams (including text and information about fonts and images). There is also a cross-reference table with information about the objects in the file and a file trailer. The trailer holds information about where all the content in the document is to be placed when a PDF file is opened; this is the first area that the software application looks at. The trailer is also used when searching the document. Raster images are stored as a dictionary describing the image and a stream of image data, whereas vector images are stored as instructions to recreate the image.

The benefits of using PDFs are:

» They can be used on almost all computing devices and platforms.
» They can be of small file size (although if many different fonts and images are used this is not always the case).
» The quality of the document is high, with all the formatting, detail and resolutions preserved.
» Any type of image can be included.
» The content is difficult to edit and alter so the integrity and security of the document is high.
» Documents can be protected with passwords.

There are some drawbacks when using PDFs. They can be difficult to read on computer screens if they are in the A5, A4 or A3 layout used for most documentation meant for printing on paper. For example, user manuals are often placed as PDF files on web pages and as such are difficult to follow as they were intended for reading from paper copies. Editing PDFs can be difficult, especially if a paper document has been scanned to PDF. Such PDFs are usually one large image of the document, and extracting text from them is difficult. Different software applications may interpret PDF instructions in different ways so the document may not always appear the same. Not all web browsers can interpret all the data in PDF files so some documents may not be rendered accurately. This can cause problems when the PDF document contains interactive forms that require user input. The form may not accept the user input properly or may simply not work at all.

19.2 Compression

Compression is the process of encoding information in fewer bits of data than are used in the original representation of the information. Compression reduces the bit rate of an image. Compression is used to reduce the amount of data that is stored in a file so that file takes up less computer memory and less space on storage media, as well as to reduce the time taken for it to be transmitted between devices, such as over the internet. Compressed files are decompressed when they are loaded into software applications. The compressing and decompressing can require considerable processing and may take some time. Lossless and lossy compression are used.

19.2.1 Lossless compression

Lossless compression retains all the image data so that, when it is decompressed, the image is perfectly restored with no loss or change of detail or colour from the original. Lossless compression relies on there usually being a lot of repeating data in an image that can be represented by a type of 'shorthand' or coding. For example, a large square block of dark purple colour may be made up of 2 000 000 pixels, each of which has data describing its colour and brightness. This is a very large amount of data to store and transmit, but much of the data is repetitive and, as such, is redundant. Storing data that instructs the computer or software application to draw the square image and fill it with pixels of dark purple results in a much smaller amount of data to store or transmit, but the resulting image is the same. This is called 'run-length' encoding and is the most common method used. It encodes what is there and counts of how many there are.
An index file of the image data is required to reconstruct the data into an image.

Lossless compression only works well with images that contain redundant data. Using lossless compression to act upon repetitive and redundant data can reduce the file size of images by over 50 per cent of the original while still retaining the original quality. Lossless compression is used for storing images that need to be edited, for archiving images, for storing artworks and for technical drawings where the original quality must be preserved.

19.2.2 Lossy compression

Lossy compression of images takes into account the perception of detail and colour by humans. In a similar way to the compression of video, some details and colour data can be removed without affecting the overall visual perception of the image. However, as with video and audio files, once the data is removed it cannot be put back so the final image is not exactly the same as the original. There are several methods of lossy compression, for instance removing much of the colour data and storing only a representation of the main colours, storing more data about the brightness of colours (the luminance) compared to the data for the actual colours (chrominance) because humans perceive changes in brightness more than changes in colours, and using some parts of an image often being similar to other parts so there is no need to store all the data. The latter method is called fractal compression because the compression algorithms make use of fractal codes to recreate the image. Lossy compression is suitable for digital photographs where the loss of data does not significantly affect the overall visual appearance.

19.2.3 Effects of different methods of compression on images

All compressed images have a lower bit rate than the original, meaning there are fewer bits to transfer in a given unit of time, for example one second. Bit rate is more commonly used when referring to video than to still images but can refer to the number of bits used to represent an image.

Lossless compression has little, if any, visual effect on the final image compared to the original as no image data is lost. However, while the size of the file is significantly reduced, there is also an index file of the image data that needs to be stored, so the reduction may not be as much as would be expected. Lossless compression can also remove unnecessary image metadata (data about the data in the image).

Lossy compression can have a significant visual impact on an image but often this is tolerated in favour of the reduction in file size. However, careful use of lossy compression can make a considerable reduction in file size without being obvious, or even perceptible, to the human eye. Lossy compression removes image data so it results in a loss of resolution or detail, reduces the number of colours and reduces **colour depth**. Compared to the original image, an image that has been compressed in this way often looks less vibrant, and less detailed and appears to be lower in quality or may appear 'blocky'. Opening and saving images using lossy compression results in repeated compression, which loses even more data so the images are repeatedly degraded. Lossy compression can introduce artefacts such as visible blocks of pixels of the same colour appearing because the data for the pixels has been reduced to much the same as the surrounding ones so they all look the same. Other artefacts could be edges of objects or colours being blurred or distorted in shape, halos appearing around objects, or banding or striping appearing across colours because the gradations in the colours have been reduced.

19.3 Text

Text is added to an image using text tools. The text tools are accessed in different ways in different applications so you will need to explore their use. They allow text to be entered and positioned, the font and size to be chosen, and the colour and formatting to be set. The tool also allows the text to be edited later if necessary, provided the text has not been converted to pixels and remains as a text object.

19.3.1 Selecting font style

Fontstyle is the size, weight and formatting, colour and overall shape of the characters that make up text. The size is how large, referring to height, the characters appear on screen or when printed and is measured in points (pt). There are about 72 points per inch, or 2.54 cm, so a 72 pt letter would be about an inch tall. In images, font sizes are measured in pixels (px) or pica, denoted by p in printing but by pc in cascading style sheets (CSS). One pica is about one-sixth of an inch, but this differs slightly in different areas of the world. Mostly, text sizes in images are specified in pt or px. The weight and formatting determine the prominence of the text and can be, for example, bold or italic. If the text is in its own layer, the opacity can be used to affect the weight of the text. Text colours can be chosen from the whole range available. The shape can be altered in the software in a number of ways, for instance along a path, so is not limited to the font itself.

Fonts can either be serif, which have the easily distinguishable small strokes that finish the ends of the characters, or sans serif (without serifs). Serif fonts are more difficult to read on screen than sans serif fonts so are often used for titles and headings, while large blocks of text are in a sans serif font. In images, the serifs might be more difficult to make out on some backgrounds so sans serif might be preferred. However, this is a matter of choice for the author.

The choice of font is made from dropdown menus in the usual way. Pre-installed fonts may be used or new fonts can be designed and added by the graphic designer.

The readability of the text can be improved by carefully adjusting the amount of white space between letters. Kerning is the adjustment of the spacing only between selected characters so that there are different spacings between each letter of a word. Image editing software has tools for adjusting kerning. Tools are also provided for adjusting the spacing between all the letters of a word at the same time. Line spacings can also be adjusted using the tools.

19.3.2 Fitting text to a path or shape

Fitting text to a **path** allows the graphic designer to make text appear to flow across an image in different ways. The path can be an open path such as a line or a closed path, or shape, such as a square, triangle or circle. Text can be typed along a pre-existing path or it can be written first and then fitted to a path.

When typing along a path, the path is created first, the text cursor is placed at the start of the path and the text is typed. When fitting to a path, the text and the path, consisting of a line, are written and drawn as separate objects. For this action, the whole text is treated as a single object. Once the text is fitted to the path, its appearance along the path or its distance from the path can be adjusted as required.

Text can be added using text tools that allow the characters to be entered directly. The properties of the text, for instance the choice of font style, colour, size and weight, can be set or altered from the dialogue boxes.

Text can be fitted easily to the lines of shapes in the same way as to open lines. However, when working with bitmap images, tools for fitting text inside or around shapes may not be available in some image editing applications. The text itself has to be distorted and shaped by the graphic designer. Vector image editing software usually has tools for shaping text in and around shapes.

Text can be fitted to a curve, in a similar way to fitting it along a path, for example to create a flowing banner across an image.

19.3.3 Converting text to editable vector shapes

Text tools enable the entry of text and 'vectorise tools' can be used to convert it. After conversion, the whole word or sentence can be edited together or each individual character can be altered separately.

19.4 Vector graphics

Vector graphics are made up of points, lines, curves and polygons arranged in geometric shapes. A point (called a **node**) is an exact location on the image canvas, a line connects two points (one at the start and one at the end of the line) and a polygon is made up of several lines connected together. A shape can be made of one or many polygons. Vector images are stored in files as a set of instructions that define all of the points/nodes, lines, curves and polygons in the image.

19.4.1 Using software tools for the creation and editing of vector graphics

Vector drawing tools

Vector graphics software provides many tools for creating and editing vector graphics. Lines and shapes can be drawn freehand or created from pre-set shapes. In the Inkscape software, lines are considered as paths. An 'open' path has two ends and is what is usually considered as a line, a 'compound' path is made up of two or more paths, and a 'closed' path has no end, such as a circle or a square shape.

Like bitmaps and animations, and for the same reasons, layers can be used to separate lines, curves, shapes or objects from others when creating a vector graphic.

Freehand drawing

The pen tool allows users to freely draw paths anywhere on the canvas. Any bends in the path have points (nodes) placed automatically by the tool as the path is drawn. The tool is selected, the pointer is clicked where the start of the path is to be and then dragged around the canvas with the left button held down, drawing as it goes. The end is marked by a further mouse click.

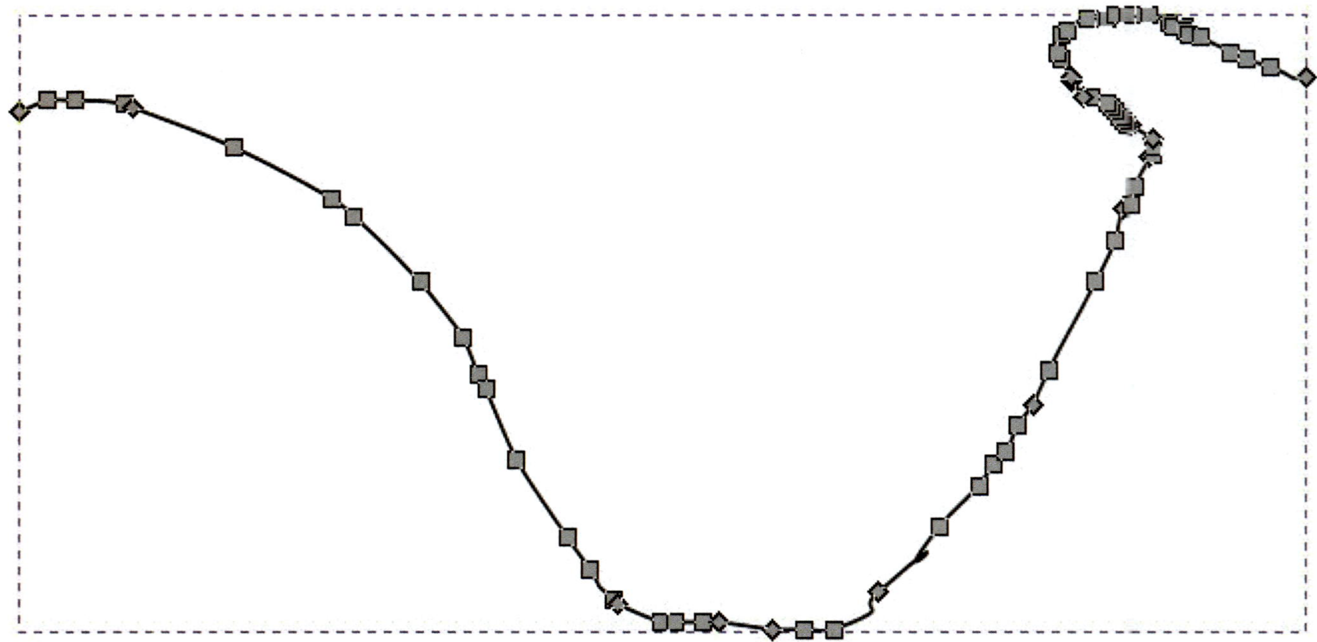

▲ **Figure 19.12** Freehand path showing nodes drawn in Inkscape

Straight lines

A straight line (a straight path) is drawn with the same tool. The start of the path is chosen with a mouse click and the pointer moved to the end location without holding down the left button. A further mouse click marks the end location. A straight path with nodes (points) only at the start and end is drawn by the software. To create a path made up of more than one straight line, the Bezier curve tool is used.

▲ **Figure 19.13** Single straight-line path drawn in Inkscape

Bezier curves

Bezier curves, or paths with multiple straight lines, are drawn with the **Bezier curve and line** tool. To draw a path with more than one straight line, select the **Bezier line** tool, position the cursor and click on the canvas. Move the pointer to the end location of the first line and click, then move to the location of the end of the next line and click again. Repeat this until you reach the final end point of your multiple-line path and then double-click the mouse.

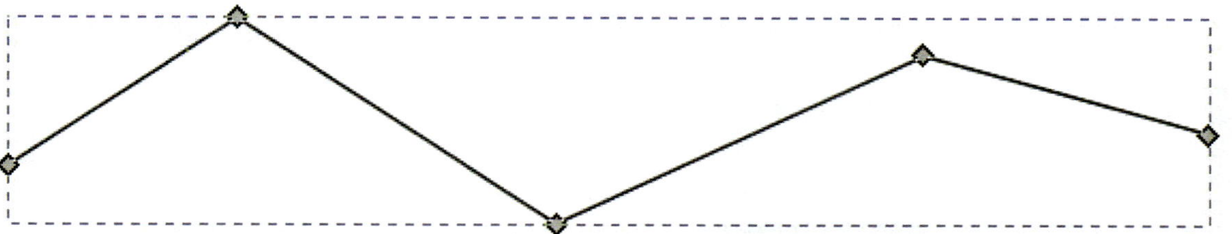

▲ **Figure 19.14** Multiple straight-line path drawn with the **Bezier line** tool in Inkscape

A curve is drawn using the **Bezier curve** tool by positioning the mouse cursor at the start location and clicking the mouse button. Move the mouse cursor across or around the canvas to where you wish to place a node and click once, then move the mouse cursor and a curve will appear. Node handles will appear centred on the node. Repeat these steps to place more nodes and draw curves as desired.

The **Bezier curve** tool cannot be used to alter the curve but the nodes and the handles can be moved by deselecting the **Bezier curve** tool, clicking on the dots and dragging the node or handle. There is also a **Node-editing** tool available to edit the curve using the nodes.

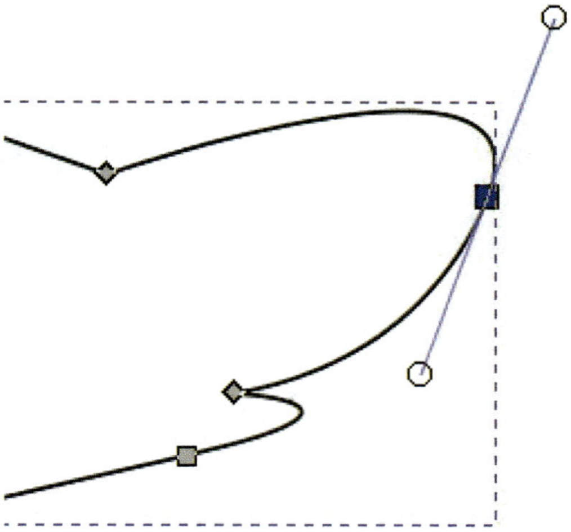

Shape tools

Rectangles, ellipses, circles, arcs, stars, polygons, spirals and 3D shapes can easily be created using the tools provided by the vector graphics software.

▲ **Figure 19.15** Curve path drawn with the **Bezier curve** tool in Inkscape showing the node handles

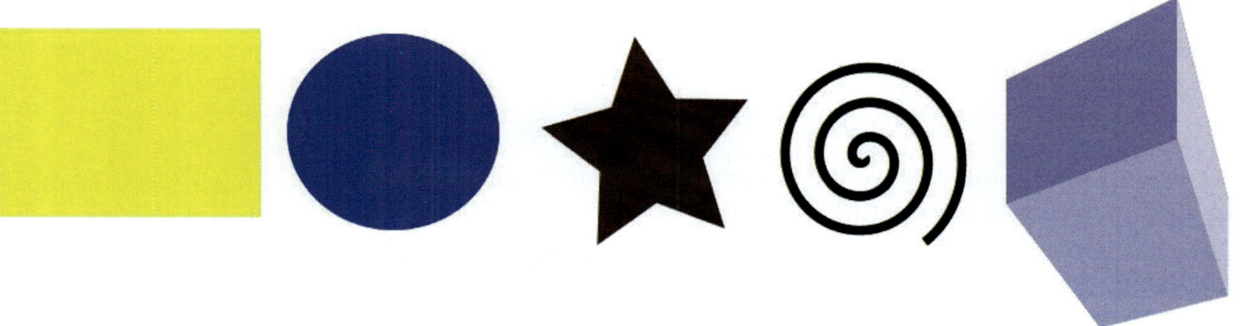

▲ **Figure 19.16** Shapes created with tools in Inkscape

When a shape is selected, its properties are displayed, for example colour, opacity, line type and corner type, and these can easily be set or edited using the available tools.

Envelopes and perspectives

An envelope in vector graphic editing places a shape around the object, such as another shape or text, and allows it to be distorted to the envelope shape. The shape of the envelope can be altered by dragging its nodes into new positions and the enclosed shape or text will follow the distortion. Ready-made tools are included in commercial vector graphic editing software applications, but in Inkscape the use of envelopes is more difficult. In Inkscape, shapes that are paths are manually distorted. Text should be converted to a path. The object is selected and the menu option **Paths**, **Path effects** is chosen. From the right-hand panel, the **+** icon is selected and the **Envelope deformation** effect is chosen. This provides options to manually 'bend' the top, bottom, left and right edges of the selected object into whatever position is required.

▲ **Figure 19.17** Distorting an object with the **Envelope deformation** effect tool in Inkscape

In a similar way, the **Perspective envelope** effect can be added to the path effects in Inkscape and used to alter the perspective of selected objects.

▲ **Figure 19.18** Distorting an object with the **Perspective envelope** effect tool in Inkscape

Selection tools to select and manipulate parts of a vector graphic

Nothing can be edited unless it is selected, so the ability to select elements in images is a fundamental feature of all image editing software. Selection tools act like a 'hand' to choose and 'grab' elements for editing. The basic editing tool is the arrow pointer, also called the pick tool, which is used, for example, to select elements and to grab them for moving or editing by dragging handles or nodes. Pressing the **Space bar** after using a drawing tool may also select the shape or object. Multiple selections can be made by holding down the **Shift** key while using the **Pick** tool or by drawing around the chosen elements while holding down the left mouse button. The selected element or objects can then be edited as a group, for example moved, resized, copied or deleted.

Convert to curves

Straight paths, or lines, can be converted to curves. For example, a sawtooth path can be converted to a wavy path.

▲ **Figure 19.19** Converting to curves

This is carried out by selecting the path, making the nodes visible by choosing the **Edit path by nodes** tool (or pressing **N** in Inkscape), selecting a node and pressing **Shift+A** in Inkscape. The path is changed to a curve and the handles can be used to adjust the curve.

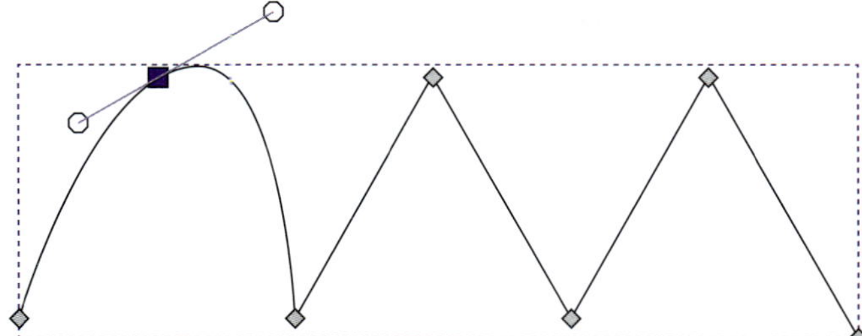

▲ **Figure 19.20** The node and handles used to adjust the curve

Replication

Replicating elements or objects produces multiple copies of the original. Elements and objects can be copied and pasted in the usual way, either by using the menu options or by the **Ctrl+C** and **Ctrl+V** shortcuts. There is also a **Duplicate** option, which creates a duplicate of the original. Copied or duplicated objects are placed on top of the original by default so may have to be moved. In some vector graphic software, the distance from the original of the duplicate's location can be specified. Copies and duplicates are independent of the original and must be edited separately.

Cloning creates copies of the elements or objects. Clones are also placed on top of the original and may have to be moved. However, editing the original also changes the clones at the same time.

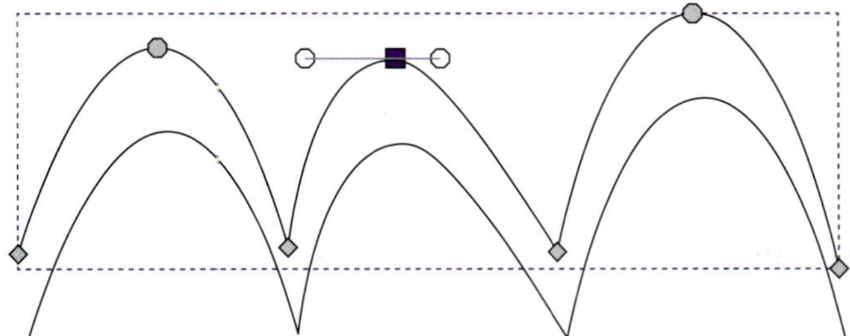

▲ **Figure 19.21** Changes to the original curve also affect the clone curve

While their size and rotation can be changed independently, most properties of clones cannot be edited directly as they depend on those of the original. Clones can be separated, or unlinked, from the original and this allows them to be edited independently.

Transformation tools

Transforming an element or object means to change its size, shape or orientation. Transformations include the scaling, resizing, moving, rotating and shear of individual elements or of objects. For rotation, the element or object is rotated about its own centre. Transformations can be made manually, or freehand, by dragging the shape using edges, nodes and handles or by using the tools provided. Tools allow the **transformations** to be made precisely and accurately. Guidelines and grids can be used for snapping the changes to a precise position.

> ## Task 19c
> ### Using transforming tools
> Draw a 3D shape and experiment with the transformation tools to change its shape.

Fill tools to colour elements

These tools act to fill shapes with solid and gradient colour in a similar way to the tools used in bitmap graphic editing software. Colours and gradients are selected from the tools and their dialogue boxes.

Node and path editing

The position, direction and length of a path are determined by its nodes, or control points. A node can only be connected to one or two other nodes. It is impossible to connect more than two nodes to a single node, so a Y-shaped path has to be created from two paths each with nodes at the same location at the point at which the branch occurs. A node has a defined position on the *x*- and *y*-axes of the image canvas and nodes are used to define the start and end points and the direction of the path. A node has a central point with handles, which can be selected and dragged to change the position and direction of the path. Moving the handles of a node changes the direction of the path around the node and so alters the curve.

Moving a node alters the position of the path on the image canvas. Nodes can be added to allow more precise changes to the shape of the path or deleted to remove unnecessary nodes. Removing excess nodes can make the path appear smoother and simpler and can make editing quicker. Nodes can be joined, which merges the nodes but attempts to preserve the path's shape. Breaking a path creates two new ends so the nodes at those locations have only one handle. End nodes can be joined together so that multiple paths can be made into one.

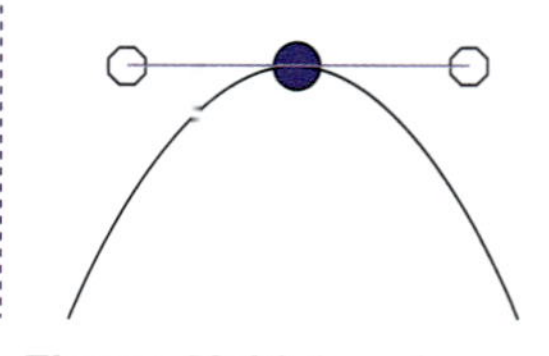

▲ **Figure 19.22** A node on a path

Symmetrical nodes

Symmetrical nodes have handles that are always 180° to each other and as one handle is rotated, the other mirrors its movement. The handle lengths are always the same as each other and cannot be changed independently so any adjustments made to one are mirrored in the other.

Asymmetrical nodes

These are the same as symmetrical nodes except the length of the handles can be altered independently to change the curve shape on either side of the node.

Cusp nodes

In this type of node, the handles can be rotated independently so they are not always 180° opposite each other. The handles can also have different lengths. Cusp nodes can be used to make extreme, sharp changes to a curve.

Smooth nodes

Smooth nodes have handles that are on a straight line going into a curve and are used to make gentle, smoother changes in the direction of the curve.

Align and distribute nodes

For precise positioning of paths, tools are provided to align nodes vertically and horizontally to the centre of the selected nodes or to pre-set locations. Nodes can also be evenly distributed vertically and horizontally using the options in the tool dialogue boxes. When the options are used, the end result can look much neater than if the nodes are positioned manually.

Task 19d

Nodes

In vector imaging software, draw a Bezier curve and experiment with the types of nodes to discover how they change the shape of the curve.

19.4.2 Converting bitmap images into editable vector shapes

Bitmap images are converted into vector images by the process of vectorisation. The pixels that make up areas are converted into lines, curves and other geometric shapes that can be described by mathematical instructions, as for any other vector image. Automatic conversion by software can be used, or the objects within a bitmap image can be manually traced to create nodes. A combination of both is usually used. Once converted, the image can be edited as for any other vector image.

Advantages and disadvantages of converting bitmap images into editable vector shapes

The advantages of converting bitmap images into vector images are that file sizes may be smaller so saving on storage space or transmission times, and the images can be scaled without the quality being affected. Vector graphics are more suitable for text-based images because the shape, size and path of the text can be more precisely controlled. Scaling text in a bitmap can make it blocky and less readable than scaling text shapes in a vector image. The elements, shapes and objects in an editable vector image can be adjusted precisely and accurately and more easily than editing pixels in a bitmap image. Bitmap images are not suitable for artwork that includes fine lines or sharp edges, but an editable vector image allows these to be created and altered with precision.

The disadvantages of converting bitmap images into vector images are that vector images do not display the continuous, fine gradations of colours that bitmaps do. Digital photographs do not appear realistic when converted to vector formats. The processing of vector images can require more computational power because the device has to be able to process the instructions and make the calculations necessary to display the image.

19.5 Bitmap graphics

Bitmap graphics are made up of pixels. Each pixel represents a point or dot in the image and has information about the colour and the properties of the colour at that location in the image. Different bitmap file formats have different structures, but they all store the information about each pixel in memory or on storage media as a series of bits arranged in an array, from which the name 'bitmap' is derived. The file usually consists of a header defining the type of image format, followed by the image data itself and lastly any other extra data (metadata) about the image.

19.5.1 Using software tools for the creation and editing of bitmap graphics

Selection tools to select, remove or hide parts of a bitmap image

Selection tools are used to mark out, or select, areas in the active layer of an image so that the area can be edited separately. Unselected areas are not affected during the editing. Areas can be selected in different ways depending on the editing that is to be carried out. Basic selection tools mark out areas such as squares and ellipses. The areas enclosed by the square or ellipse are available for editing until deselected.

Lasso tool

This tool is a free selection tool that is used to mark out an area by freehand drawing around the area that is to be edited. The pointer is placed at the start of the area, the mouse button is clicked and the area drawn out with the button held down until the pointer is back at the start point. When the starting point is reached again, a further mouse click marks out the end point of the area. Straight lines for bounding the area can be drawn and the area can be resized by clicking on the end point and dragging. The **lasso tool** does not allow precise selections to be made easily, but it is a convenient method of making a preliminary selection that can be refined with the use of other tools.

Magic wand tool

The **magic wand, or fuzzy, select tool** uses the similarity of colours of contiguous (next to each other) pixels to mark out areas of the active layer. It works well when selecting objects or shapes with sharp edges as the pixels next to each other will have very similar colours. When using the tool, a pixel is selected by clicking on it and the tool selects all the adjacent pixels that are of a similar colour. The effect is similar to a flood fill in that the selection spreads out across the image. The result is one contiguous selection of similarly coloured pixels. Pixels that are not touching these are not selected. The threshold, or sensitivity, can be adjusted to control how close the similarity has to be when the tool selects the pixels.

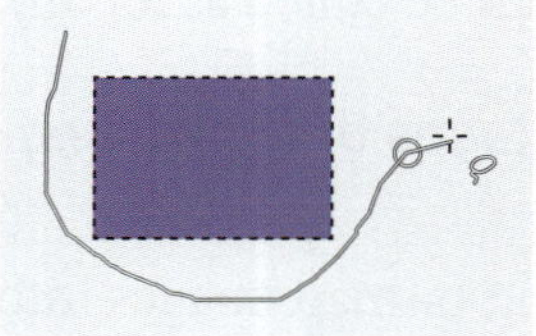

▲ **Figure 19.23** Lasso and Magic wand (fuzzy select) tools in GIMP

Colour select

This tool selects areas of an image based on similarities in their colour, in a similar way to the magic wand/fuzzy select tool. However, this tool selects the pixels of the selected colour whether or not they are contiguous, meaning that it selects the pixels of the colour everywhere in the active layer. As soon as the selected pixel is clicked, the selection starts and spreads and, in the same way as for the magic wand, the threshold can be adjusted to control the selection of colours.

Cut out

An image cut from an existing image can be used in the creation of a new image or added to an existing image to make a composite image. The Magic wand tool is used to carefully draw around the area to be cut out and the Cut option is chosen from the Edit menu. Other selection tools, such as the Lasso tool, can also be used to select the area to be cut. A new image can be created from the Paste as new image option or the area can be pasted, preferably as a new layer, into an existing image. The pasted area can be resized, rotated and edited as required.

Crop

The Crop tool is used to remove unwanted areas from around the area that is to be retained. The Crop tool marks out, shown by the crop overlay lines, a rectangular area in the active layer. Handles on the corners allow the rectangle to be adjusted. When Enter is pressed, the surrounding area is removed and the interior of the rectangle left on the layer. There are advanced options available with the Crop tool, for example to set an aspect ratio or alter the opacity.

▲ **Figure 19.24** Image crop and cut out

Masking tool

Masking tools can be used to change the opacity of a layer. The opacity of the whole layer can be changed or the opacity of selected areas can be altered. The masking layer is on top of the other layer and the opacity settings adjust the alpha channel (which represents the degree of opacity of a colour) of the masking layer. Different colours can be used as the mask, which allows the designer many options when creating images.

Working with colour

Colours can be selected, applied and edited in several ways. The Colour select tool, the colour picker and the Magic wand tools can work with colour.

Solid and gradient fills

As described previously, a solid fill applies a colour uniformly across an area, whereas gradient fills fade one colour into others across the area. Linear gradient fills start with the foreground colour and change into the background colour along a single line drawn by the user, bi-linear gradient fills spread out in opposite directions from the starting point, and radial fills create a circle starting with the foreground colour at the centre and ending with the background colour at the edges. There are more gradient fills, such square and shaped, available in some graphic editing applications and all start with the foreground colour in the centre and end with the background colour. The size of the shape and the rapidity of the transition between colours is determined by the gradient line drawn by the user. Conical gradient fill can be symmetric, where the gradient is even all around, or asymmetric, which has the appearance of a ridge where the line is drawn.

Convert to black and white or greyscale

Converting a colour image to black and white or to greyscale can make the image appear more dramatic. The simplest way to convert an image is to remove the colours by desaturating them or changing them to greyscale. A tool is usually supplied to do this automatically with a single click.

However, more control over the appearance of an image when converted can be achieved by choosing how the colours are removed. This could be done by adjusting the RGB or HSL layers or by adding a layer and using it to convert areas, or the whole image, to greyscale. In Photoshop, using a black and white adjustment or greyscale layer retains the original colours but overlays them with a layer that can be adjusted to darken or brighten areas of colour to change how they appear in black and white.

Adjust brightness, contrast, colour balance, shadows and highlights

Brightness is how dark or light, or intense, pixels in the whole image or in areas of the image appear. Contrast is the difference between the maximum and minimum intensity, or brightness, of the pixels in the image. These can be adjusted with the software tools, but care must be taken as brightness is relative and the apparent brightness can alter depending on the viewing conditions.

Colour balance refers to the relative intensities of the colours in an image. Usually these are the three primary colours of red, green and blue. Adjusting the colour balance alters the amount of colours in the image so it is mostly used for colour correction. This is necessary to ensure that neutral colours like white, grey and black appear as they should, for example in digital photographs, and that white is white and not tinged with another colour. This is described as 'white' balance, but sometimes the term 'grey' or 'neutral' balance is used instead. Deliberately altering the white balance by varying the intensity of the component colours can be done for artistic effect.

In photographs with backlighting or washed-out areas due to very bright light sources, shadows may be too dark or highlights may appear too light. The details in these areas may not be visible. Shadow and highlight tools can improve the visibility of the details not by lightening or darkening the whole image, but by selectively lightening or darkening the pixels in the affected areas by comparing them with surrounding pixels.

Brush, pencil and pen tools

These tools are used for creating lines or shapes by 'drawing' or 'painting' on the canvas. Selecting a tool and moving the pointer on the screen creates 'brushstrokes' similar to painting on paper or canvas with a real brush or writing with a pen or pencil on paper. The benefit of using image editing software and its tools is that the shape of the brush or pen, its weight (or size in pixels) and its colour and opacity can all be adjusted with the options dialogue boxes, giving the graphic artist full creative control. The tools work on objects and layers.

Pre-set and customised options

All graphic editing software has pre-set options, which are the most commonly used settings. The settings produce an effect that the software writer deems to be the best for the tool or task so that the user can get started quickly. Nearly all tools, and some actions and settings, can be customised according to the requirements of individual users. There is usually a **Preferences** option that allows the software to be extensively customised so it behaves as the user wishes. For example, the visual appearance of the software on screen, the options for each tool and the type of file to export as default, such as JPEG or PNG, are among the settings that can be altered to suit the user.

Activity 19a

Pre-set and customised options

Explain why image editing software has pre-set settings but allows users to customise them.

Tools/filters to alter parts of an image

Many tools are provided to alter objects, areas and shapes in an image. The tools are often combined with other tools. The effects of these tools can also be made available as filters.

Distort

Distorting parts of an image changes its shape, scale, size or perspective. Tools for distorting are often grouped under **Transformation tools**.

Clone

In bitmap image editing, the **Clone tool** is used to copy an area of an image and place it elsewhere in the image. A simple clone of an area can be achieved with copy and paste but the **Clone** tool gives more control. After using a selection tool, for example the lasso tool, to choose an area to copy, the **Clone** tool can then be used to control how the area is pasted into the receiving image or layer. The brush used for pasting the cloned area, using a pattern to tile over the existing image, how the edges are drawn, and other options can be set. Unlike cloning in vector image editing, changes to the original cloned area do not affect the pasted clones.

Erase

The **Erase** tool removes areas of colour from the active layer. The erased area will have a level of opacity depending on the setting of the alpha channel for that layer. If no alpha channel is present, the background colour will show. As with most tools, the options for the **Erase** tool are many. The brush, its size, the opacity and many other settings can be adjusted to customise the erase action.

Blur

Blur can be used in many ways, for example to change the focus, give an appearance of motion or add artistic effects, for instance to a digital photograph. Blurring a background draws the attention of the viewer to the foreground objects. Like other tools, blur has many available options to choose from, such as the brush and its size, opacity and spacing. Blur can be applied to any area of an image, but to blur large areas it is usual to create a new layer for the blurring. Blurring can use considerable computing resources so users may notice a 'lag' when blurring large areas of an image. There are various types of blur available, for instance Gaussian blur that smooths edges by removing the pixels at the extremes of the intensity range, selective blur that can reduce graininess in photographs by the adjustment of the threshold for the blur, and linear and circular blurs. Blur is also available as a set of filters and is often combined with other tools, such as sharpen or smudge.

Smudge

The **Smudge** tool produces a smearing effect similar to wiping a finger across an area of real paint that is not fully dry. Again, there are numerous options that can be set for smudging, such as the level of opacity of the smudged pixels and whether or not the pixels are erased as the smudging occurs.

▲ **Figure 19.25 Smudge** tool effects in GIMP

Sharpen

Sharpening an image or a shape in an image increases the contrast at the edges of shapes so they appear more prominent to the viewer. Sharpening can be carried out using the **Sharpen** tool or by applying a **Sharpen** filter. Sharpening alters the intensity of pixels with intermediate values to those with high or low intensity so the gradations are removed.

Red-eye removal

Red eye in photographs is caused by the light being reflected from the blood vessels in the back of the retina of the eye. It occurs when the light source is close to the camera, for example as in a smartphone using flash. While it is best to avoid red eye by moving the light source, this is not always possible, so tools are provided in some image editing software to remove the red eye. Selecting the **Red-eye removal** tool in Photoshop and clicking on the red area of the eye usually removes the red colour and replaces it with a more natural-looking dark colour. In GIMP, a selector tool such as the **Ellipse-shape selector** tool is used to manually select the area and then **Red-eye removal** is chosen from the **Enhance** option in the **Filters** menu.

19.5.2 Resizing the image and canvas

The size of the image and of the canvas are adjusted and set independently. The image and canvas size are usually measured by the number of pixels in the height and width. Other units can be used, for example centimetres or picas. It is usual to set the image and canvas size for new images when the image is first created, but digital photographs may need to be resized or scaled to suit the user's needs.

Changing the canvas size allows more space for adding objects to an existing image or photograph.

19.5.3 Changing colour depth

The colour of a pixel is determined by the intensity of each of the primary colours. This, in turn, is determined by the number of bits used to hold information about each colour. This is referred to as the 'bit depth' or 'colour depth'. A pixel with a bit, or colour, depth of 1 can have only two values, one representing black and the other white. A bit depth of 8 gives a possible 256 values. An 8-bit greyscale pixel can have 256 shades of grey. Pixels with 256 bits for each of the three primary colours in the RGB system can have 16 777 216 colours. Three colour channels, each with 8 bits, is often called 24-bit colour. Extra channels can be added, for instance for more colour information or for transparency values.

When all the bits for all the colours in all the pixels are placed in a file, the file size can become very large, even with compression. Reducing the number of bits used in the image for the pixel colours can reduce the file size.

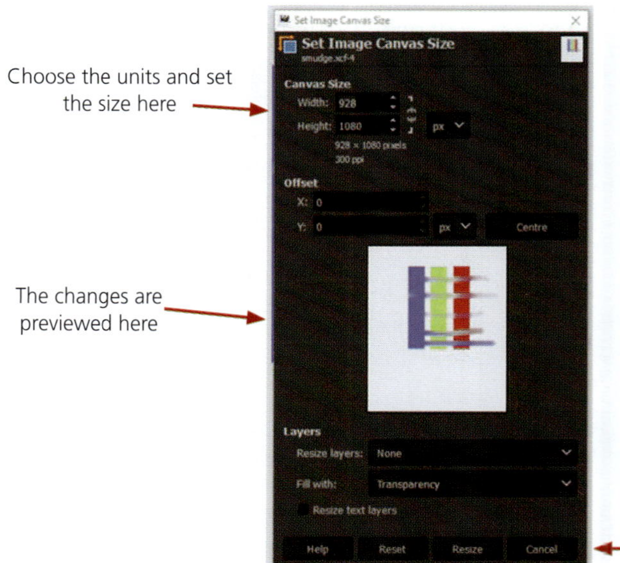

▲ **Figure 19.26** Changing the canvas size in GIMP

However, the more bits are used, the higher the quality of the image appears due to the greater range of colours, and using fewer bits can produce images that do not look as realistic.

The number of bits used can be altered from menu options. In Photoshop, the **Image** tab and the **Mode** menu provides the options. In GIMP, the option is found in the **Image** tab, then the **Precision** menu.

19.5.4 Changing resolution

Image resolutions can be altered from the menu options. As previously described, low-resolution images have less detail than high-resolution images. The required resolution of an image is chosen to suit its intended purpose.

Practice questions

1 Layers are used when creating a computer graphic.
  a Explain why layers are used. (4)
  b Explain why layers are usually merged in the final image. (3)

2 a Describe how rulers, grids and guidelines can assist in the placing of objects when creating digital images. (6)
  b Describe the differences between the effects on an image of crop and cut out. (4)

3 Colour wheels are used in graphic image creation as an aid to choosing colours.
  a What is a colour wheel? (2)
  b Describe two limitations of using colour wheels. (2)

4 Why are colour-management systems (CMS) needed? (6)

5 Explain why bitmap images may become pixelated and suffer a loss of quality when resized or zoomed into but vector images do not. (6)

6 Explain why lossy compression can have an adverse effect on image quality but lossless compression usually does not. (6)

7 Colours in bitmap images are stored as bits.
  a Explain why a bit depth of 8 allows only 256 colours to be used in an image. (2)
  b Why do some 256-colour bitmap images have 32 bits? (2)

20 Animation

In this chapter you will learn:
★ the types of animation
★ the basic principles of computer animation
★ how to use computer software to create animations.

Before starting this chapter, you should:
★ have knowledge of bitmap and vector images
★ have access to animation software, for example the free Synfig Studio application.

In computer animation, still images are digitally arranged and manipulated in a way that makes them appear to move. The illusion of movement is created by using computers to generate sequences of digital images and display them on screen at pre-determined intervals. Each digital image in a sequence is slightly different from the previous image, with the position or arrangement of objects moved very slightly, so when the sequence is played the objects appear to move around the screen.

20.1 Different types and methods of animation

20.1.1 Cel animation

Cel animation is named after the technique of manually drawing and painting images onto transparent celluloid sheets. When an object is to be shown moving on screen, many images are created with each image showing a small amount of movement of the object. Each image is individually photographed onto film or digital video. A background with objects that are not intended to move may be used and this is placed behind the cels as they are photographed. When the cels are displayed in sequence by playing the film or video, the objects appear to move against the background.

Today, cel animation is rarely used to create complete animations. When an animation production team wishes to make their animation appear like a traditional cel animation, their drawings and backgrounds can be hand-drawn and digitised by scanning into a computer system. Software is then used to create the animation by linking the drawings, adding effects such as colours or transitions where necessary, and storing the final animation as digital video. Computer software can create sequences of images in a shorter time and with greater detail than most human animators. The large teams of skilled human animation artists are no longer needed and this reduces the costs of creating an animation. When computer software is used, the animation is already in a digital format so there is no need for expensive photography. The errors in, for example, exposure and alignment of cels that sometimes happen when photographing many different cels do not occur when using computer software.

20.1.2 Stop motion

Stop motion creates a similar illusion to time lapse. Traditionally, stop motion involves physically altering or moving real objects very slightly between individual photographs of the object. The photographs are displayed in sequence to show an illusion of movement of the object. Usually, the objects are made of clay, plasticine or other modelling material that can be easily shaped, but any object can be used. Stop motion is considerably cheaper to use than CGI and the appearance of the objects can be more realistic. Computer systems can be used to control the movement of objects, such as puppets, in stop motion and this can result in smoother and more realistic-looking motion.

20.1.3 Time lapse

In **time lapse**, photographs or **frames** in a film or video are captured at a lower frequency, or rate, than the rate at which they are subsequently played back or displayed. The result is that any movement or changes captured appear to the viewer to be happening at a much faster rate than they would normally be seen.

Slow-motion (high-speed) photography creates the reverse effect, where the movement of objects appears to be much slower and is often used to study the movements of fast-moving objects.

20.1.4 Flip book

Flip books have been used for a long time to create the illusion of movement using still images. A booklet of pages with images that differ slightly from each other are flipped or flicked so that each page is viewed quickly in sequence. Viewing the images in quick succession creates the illusion of movement in the images in the same way that watching a sequence of movie or film frames does. Software applications are available that can convert digital video files into a flip book format ready for printing.

20.1.5 2D and 3D animation

Two-dimensional (2D) objects appear flat whereas **three-dimensional (3D)** objects appear to have depth and volume. While all animations are displayed or shown on a flat surface or screen, the illusion of depth and volume adds realism.

2D objects in animations are drawn with the dimensions of height and width around the x- and y-axes and are viewed from only one angle, usually the front. The lighting of 2D objects appears to be from one direction only and they are not realistic when placed into scenes with 'live' action or when used in computer games.

20.1.6 CGI

CGI is computer-generated imagery that creates two-dimensional (2D) and three-dimensional (3D) objects that appear to move on screen. The objects can be anything that a computer can generate in an image, for example the environment in which live-action is set, the characters or actors in a movie or game, or text that overlays other objects. The images are used in movie productions, TV programmes, commercials and computer games. CGI is also used in simulations for training purposes.

3D objects in animations are drawn with the dimensions of height, width and depth around the x-, y- and z-axes, allowing them to appear, unlike 2D objects, to be able to rotate though 360° and be lit and viewed from different camera angles. 3D objects can appear to have solidity and textures, which in computer games or when they are placed into 'live' action movie scenes gives them a more realistic appearance.

2D animation is often used in presentations and demonstrations where realism is not important, but viewers prefer to see more realistic 3D animations in movies or computer games. 2D animation focuses on manipulating images, whereas 3D animation involves computer-generated or augmented worlds in which characters and objects move about. For movies or games, motion capture is used to record movements of characters or objects. The recorded movements are converted into digital data. Computer software uses the data to create 3D objects in motion.

Computer-aided animation digitises hand drawings whereas computer-generated animation is created solely in software. Both can create 2D animation, but 3D animation is computer-generated.

Activity 20a

Different types of animation techniques

Create a table to show at least one advantage and one disadvantage of each type of animation.

20.2 Components of an animation

Animations are made of sequences of **frames** that contain objects. An object in computer animation is a shape made up of point locations in space called vertices (a single point is a vertex), lines (or edges) joining two points, and polygons (three or more edges making up a surface). The surface is called a face. Objects can depict almost anything, such as text, characters and features in backgrounds. Digital photographs can also be included in computer animations.

Animations have several components. The primary components are the objects in the scene with which they interact, such as the main characters, other characters and important features. The focus of attention for the viewer is usually on the primary components. Secondary components are used to bring a scene to life so that the enjoyment of the viewer is enhanced. Examples of these might include: minor characters, additional shapes or lines, sounds or lighting effects. Secondary components can also be used to emphasise the motion of primary characters or objects and to draw the viewers' attention to details of the primary components, such as a movement of an arm, finger or eyebrow or the turning of a switch.

20.2.1 The principles of animation

The twelve principles of animation

Based on the work of Walt Disney, it is generally considered that there are twelve principles of animation. All of these, although drawing is usually done in software, are applicable to computer animation.

Squash and stretch

This adds an appearance of 'give' to an object when it has external forces applied to it. It makes the movement of an object more convincing to a viewer. For example, a leaping animal would be drawn so that it appears to squash down before jumping and stretch out when leaping upwards.

Anticipation

Anticipation prepares a viewer for an action, for example by a character in an animation. For instance, an escaping character about to scale a wall might bend at the knees before climbing. An athlete might crouch before starting a race. These would induce anticipation in the viewer in preparation for a major action.

Staging

The position of a character in a frame of a scene, the camera angle and lighting effects can all be used to 'stage' a scene so that the viewer is in no doubt about what is of the highest importance in the scene. Staging makes it clear to the viewer where to look and what to concentrate on.

Straight-ahead action and pose-to-pose

Straight-ahead action creates a smoother illusion of movement by drawing a scene frame by frame from beginning to end. This can introduce errors such as small changes in object or character shape and proportion, but is considered better at creating realistic fast action sequences.

Pose-to-pose uses a technique of first creating the **key frames** in a scene and then filling in the missing frames. This technique is considered better for dramatic scenes where the composition of the scene is important. The addition of the missing frames is carried by computer software and eliminates shape and proportion errors.

Follow-through and overlapping action

Follow-through action means that moving parts of an object or character keep on moving after the main part has stopped. For example, a person's arms may keep moving after the character has stopped running or jumping. This is more realistic than a complete stop of the whole person because it makes the object seems to obey the familiar laws of physics and actions that viewers are used to.

Overlapping action means that different parts of an object are made to move at different rates, just like in real life. The torso of a human runner would move differently from the arms, legs or head during an action sequence. Exaggerating the movements of different parts can be used to convey different effects.

Slow in and slow out

Moving objects take time to accelerate and time to slow down, so animations look more realistic if there are more drawings at the start and at the end of a movement than in the middle. The movement is shown to be slower at the start and end than in the middle of the action.

Arc

Movements of natural objects, such as fingers and arms, or thrown objects, follow an arc so animators follow this when drawing movements to make them more realistic.

Secondary action

Secondary actions emphasise the main action. They can express emotions, add dramatic sounds and add movements to help bring a scene to life.

Timing

This is the number of drawings or frames that make up a particular action in an animation. Timing determines the speed of the action. Timing is critical for setting the mood of a character, its emotion or reaction to another action. Incorrect timing may make the animation appear to defy the laws of physics and is not realistic.

Exaggeration

This is used to artistic effect in, for instance, cartoons, where character features and actions are often exaggerated so that they do not appear as a dull copy of reality.

Solid drawing

This is a technique where an object is drawn in such a way as to give it the appearance of volume and weight in a three-dimensional space. Traditional animators who drew by hand were skilled artists and painters and took many years to learn how to accomplish this. Modern-day animators using computers to draw do not need this high level of skill but must still understand the principle and how, and when, to apply it.

Appeal

This is artistic effect that determines how a viewer reacts to a character or an action. It is the skill of the animator that makes a character appear real and interesting.

The basic principles of animation

Frames

A frame consists of one individual, complete image or drawing within a series of complete images in an animation, video or film. The image that makes up a frame can contain any number of objects, such as images, shapes and text.

To make an animation, a series of frames is played to create the illusion of movement. The number of frames played in one second is called the frame rate. It is expressed as frames per second (fps). There must be enough frames displayed to avoid the perception of flicker when viewed. A frame rate of over 10 or 12 frames per second will be perceived by most people as movement instead of individual images. Film projection systems use 24 fps. Modern television standards have a rate of around 30 fps but these differ depending on the country, system and video standards in use.

Due to the large number of drawings needed when done by hand, animations are sometimes drawn 'on twos', which means that only one drawing is shown for every two frames. A frame rate of 24 fps on film would need only 12 drawings per second. This is sufficient for most movements, but fast action would require the use of drawing 'on ones', meaning that there is a separate drawing for every frame. Using both techniques ensures that the viewer perceives smooth, uninterrupted movement but reduces the total number of drawings needed and keeps production costs down. Cheap cartoons were often drawn 'on threes' or even 'on fours'. The time and costs of drawing large numbers of frames is much less of a concern in computer animation so these are 'on ones'.

Key frames

The start and end points of an element, movement or action in an animation
are defined by **key frames**. For example, a character jump would be defined by
a key frame just at the start of the jump and by another at the end of the jump.
The first key frame would set the position, and state the colour and shape of the
objects in the frame at the start of the sequence. The end, or next, key frame in
the sequence sets the new positions and state of the objects. An animator would
create these in software, for instance by dragging objects or moving objects and
changing their colours and size as required. When the two key frames are set
up, the computer animation software is used to create the frames in between,
by a process called 'tweening', so there is a smooth flow of movement from
one key frame to the next. The time between the two frames, determined by
the number of intermediate frames, determines the speed of the movement or
action. A frame can only be associated with one key frame.

Property key frames

Key frames hold the settings for specific properties of the objects in the frame.
The properties can be the opacity, volume, weight, scale or rotation of an object
and the background colours or text size. The settings for the properties can be
set and changed from menus in computer animation software or by dragging
objects on screen within the frame. **Property key frames** can be placed between
other key frames to affect only some properties or some objects within a given
sequence. For example, to speed up the movement of a flag in a sequence but
not of a bouncing ball, a property key frame could be inserted between two key
frames at the point where the change was required.

Timings

Timing is the speed at which an action or movement takes place. The speed
of an action is determined by how many frames it takes to happen. The more
frames it takes, the longer the action appears on screen and the slower it appears
to be. Faster action is achieved by having a shorter time on screen, which is
produced by using fewer frames for the action. The timing of an action depends
on the context and properties of the moving object within a scene. For example,
different types of ball behave differently when bounced.

The spacing of objects in a frame also affects the timing of their movements in
a sequence. For example, the smaller the spacing between a ball drawn in a series
of frames, the slower the movement appears. Conversely, the larger the spacing
between the drawn balls, the faster they appear to move. Spacing can be used to
follow the 'Slow in and slow out' principle, described above. For example, if the
overall timing of a sequence of a bouncing ball is 30 frames then there are 30
drawings. The spacing of the drawn balls in the first few frames and in the last
few frames is less than of those in the middle frames, so the ball appears to start
and end its bounce slowly.

Animation software contains a **timeline** where the animator can see or decide
when changes or actions occur in a sequence.

Coordinates

The shape and position of objects in a frame are determined by their coordinates.
The Cartesian system of mutually perpendicular axes is used to define the
position of a point in an animation frame. In 2D animation, x- and y-axes are
used and in 3D animation there is an additional z-axis. Coordinates are

specified as x,y for 2D and x,y,z for 3D. The starting point or origin is defined as 0,0 in 2D and 0,0,0 in 3D.

Coordinates can be used to define the shape of an object and, in this case, the origin is at the exact centre of the object. They may also be used to position an object on screen, where the origin is usually at the bottom-left corner. Different animation software may have different settings for origins and some software even has different arrangements for the axes (for example Blender starts with the z-axis vertical and the x-axis pointing out of the screen), so make sure you know the arrangement for the software you are using.

Inbetweening

The properties of the objects in key frames are defined by the animator to specify the start and end points in an animated sequence. When these have been completed, the frames in between are drawn by the computer software. The process of creating the 'inbetween' frames is called 'tweening', a shortening of the term **inbetweening** used to describe the manual drawing of frames in traditional hand-drawn animations. Tweening creates all the frames between two key frames so, for example, a sequence of 100 frames would have two key frames and the 98 in between frames would be computer-generated using the animation software's tween function.

Animation software has tools for controlling how the motion, shape, size and colour of objects change from one key frame to the next. Careful and precise control of the tweening process by skilled animators can create enthrallingly realistic animations.

Morphing

Changing the shape of an object by making one shape appear to transition smoothly or dissolve or distort into another is called **morphing**. Animation software then distorts one image into the other while matching the points. The process involves very precise measurements and manipulation of the position and number of pixels in the images to be morphed.

Key points are marked on the original drawing or image and similar points are marked on the final drawing or image. Each pixel in the original image is mapped to an equivalent pixel in the final image in a process called forward-mapping. Reverse-mapping is the process of sampling the pixels in the final image so they can be used as an 'equivalent' pixel for the forward-mapping. When all the pixels in the original image are mapped to those in the final image, the animation software creates a sequence of intermediate images that show a gradual distortion and transition from one to the other. The original image is warped, distorted and faded out at the same time as the final image is faded in.

Animation software can now morph any shape into another automatically with little input or effort from the animator, so the process is now quite common in animations.

Layers

A layer is the equivalent of the traditional, transparent sheet of celluloid. In the same way as for layers in still computer graphics, objects drawn and animated on one layer are separate from objects on other layers. Objects can be edited and manipulated without affecting objects on other layers. When an animation is rendered into the final movie, the layers are combined into one.

20.3 Using software tools for the creation and editing of computer animations

20.3.1 Worked example of creating an animation

A typical practical examination task will set a brief with a specification for the animation. The worked example is carried out in the free 2D-animation software package Synfig Studio. Synfig is a powerful application that requires careful study and learning to utilise it fully, and the worked example uses only a few of its options and some of its capabilities. You should take time to investigate the various panels and tabs available.

The brief

A short animation is required to advertise a coffee shop. The shop serves many different coffees and some light snacks. The owner of the coffee shop wishes the animation to have these features:

- an aspect ratio of 16:9 or to be set to 512 by 288 pixels
- a yellow background at the start
- the text **Our Coffee Shop** to appear one letter at time starting 2 seconds into the animation
- the text to be in red and in a 48 pt, or 32 pixels by 32 pixels, sans serif font
- each letter to be on screen for 0.5 seconds before the next one appears
- the text to disappear once all the letters have been displayed
- an image of a coffee cup to appear in the bottom right of the screen
- A white rectangle to appear in the middle of the screen
- in the rectangle the sentence **All our coffee is freshly brewed** to show for 4 seconds
- the text of the sentence to be black, 16 pixels by 16 pixels and in a sans serif font
- the text to fade in and fade out
- the rectangle and image to disappear
- a suitable logo for a coffee shop to appear in the centre of the screen
- the logo to grow to fill the screen and rotate as it grows
- the animation to be set to loop at least four times or be set to repeat.

Creating the animation

The time taken for the animation to run will depend on the number of frames per second set for playback. This is set in the **Canvas** dialogue box in the **Time** tab. The length of the animation is set in frames in the dialogue box. A simple calculation of multiplying the number of frames per second by the number of seconds will give the total number of frames needed for the animation.

The length is entered here:

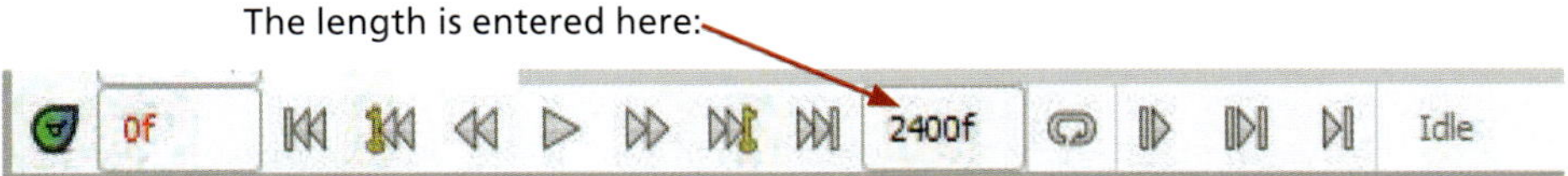

The length can be changed here at almost any point during the creation of the animation.

When making any changes to the parameters of objects or layers, it is important that Synfig is in editing mode. This is shown by the canvas having a red outline and the icon at the bottom right of the canvas appearing as running.

This icon can be used to toggle editing on:  and this icon to toggle editing off: 

Configuring the canvas size

The canvas size is the area in which the animation is drawn. It can be set in the settings by specifying the number of pixels on the *x*- and *y*-axes. The aspect ratio should be set to **16:9** as this is the usual format for video shown on TVs and monitors. Most animation software will also show the canvas size with rulers at the edges. The view of the canvas on screen can be adjusted with **Zoom** options, but this does not affect the animation canvas size.

The canvas size is set in Synfig Studio by choosing **Properties** from the **Canvas** tab or by pressing **F8**. The pixel sizes are entered directly into the dialogue box. Choosing **Apply** and closing the box sets the size. Setting the canvas size to **512 × 288** also sets the image size ratio to **16:9**, as shown in the dialogue box.

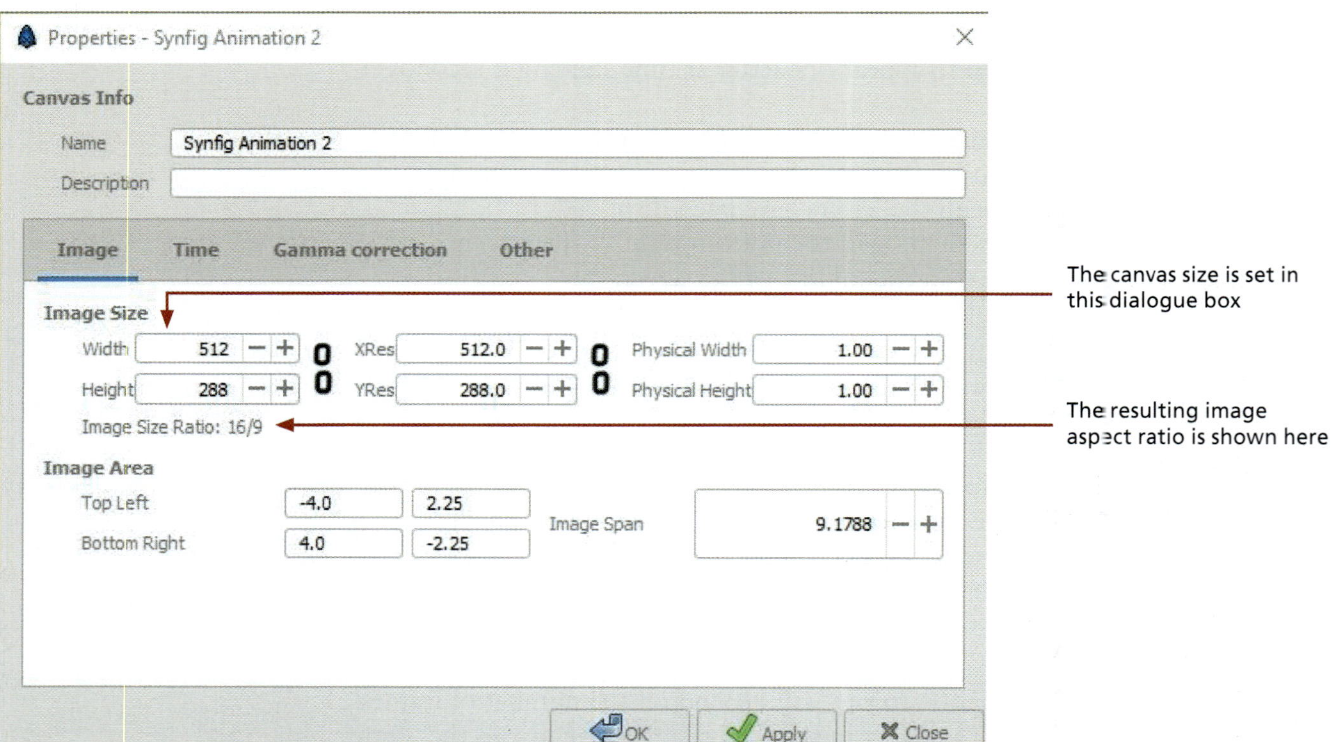

The canvas size is set in this dialogue box

The resulting image aspect ratio is shown here

Setting the background colour

To set a background colour, a new layer is added and its colour set, for example to yellow. Choose a **New layer** from the **Layer** tab and select **Geometry, Solid Colour** as the new layer.

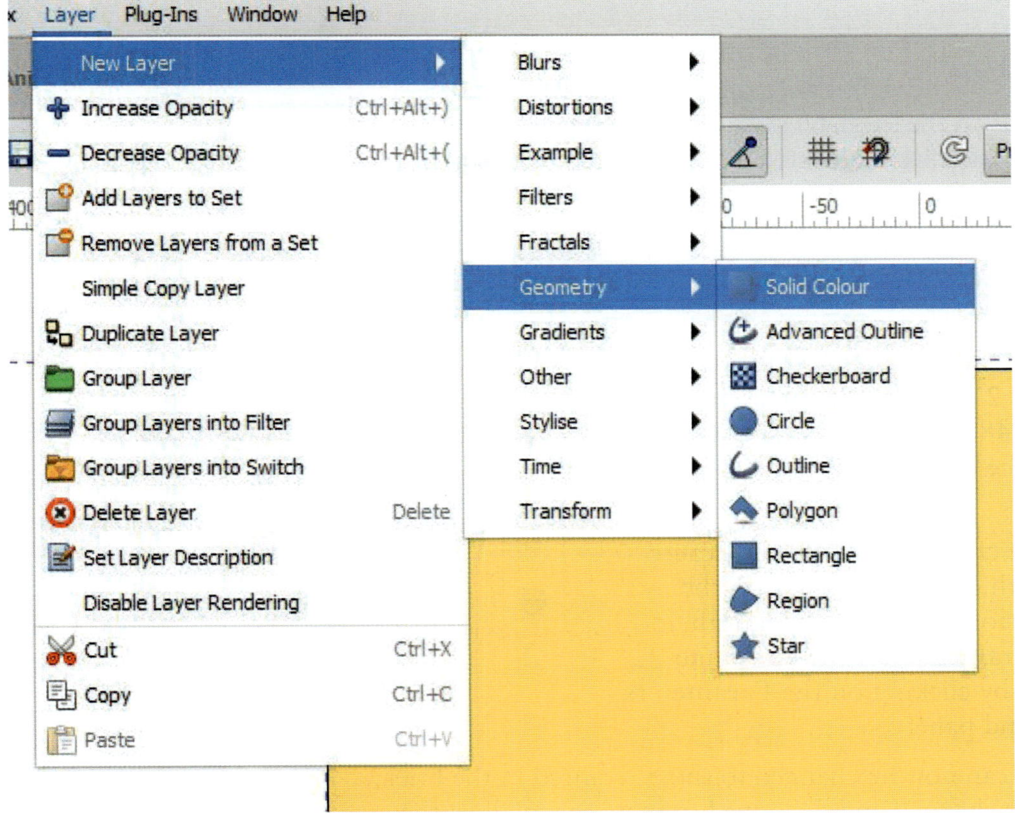

The colour of this layer is chosen by selecting the layer from the right-hand panel, selecting a colour from its parameters and adjusting the sliders to the required colour.

The RGB values can be set using the sliders or directly entered. In this example, yellow is set as **Red 255**, **Green 255** and **Blue 0**. Once these are entered, select **Set as fill** and **Close**, as shown below.

Using the grid

The canvas can have a grid overlaid on it to aid the placement of objects. The grid can be turned on and off and the size of the grid boxes can be adjusted as required. Also, when it is enabled (from the options or button on the toolbar), a **Snap to grid** option allows handles to be snapped to the nearest grid point for convenience and precision during the placing and adjustment of objects.

The layer is selected in the right-hand panel:

and the colours chosen:

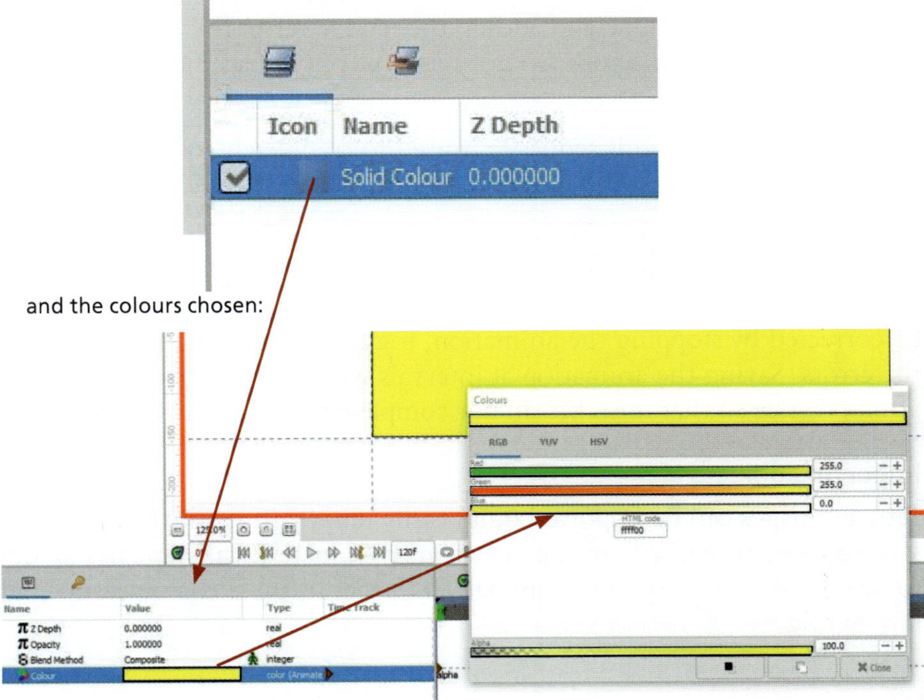

Making the text appear

Create a new **Text layer** from the **Layer** tab by selecting **Layer**, **Other**, **Text**. The new layer appears in the right-hand panel and can be selected to make its properties appear on the left-hand side. In the properties, select the **Value: text** layer by double-clicking and the dialogue box shown here appears.

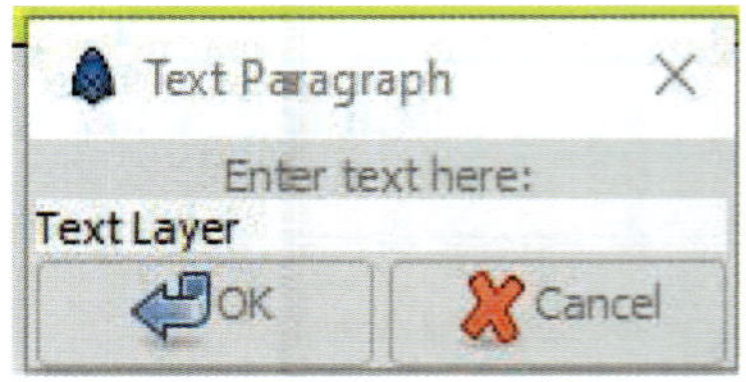

Remove the words **Text layer** and click **OK**. This is now a text layer that can be used for making the text **Our Coffee Shop** appear in the size and colour and at the intervals specified in the brief.

The text must appear after 2 seconds. At 24 fps, 2 seconds of animation takes 2 × 24, or 48, frames, so move the animation to the 48th frame by entering **48** in the **Frame count** box at the bottom left of the canvas window, as shown here.

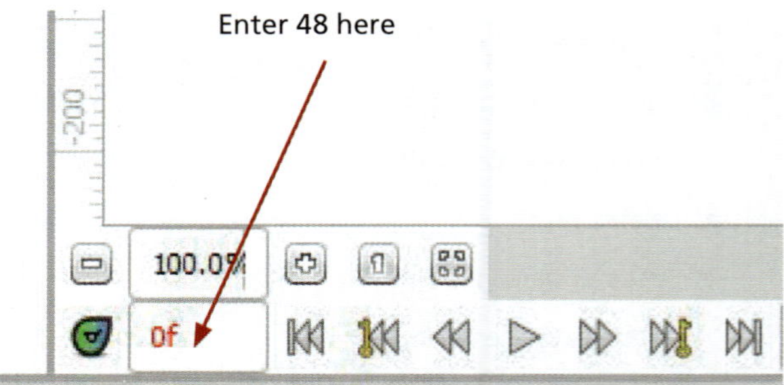

A new key frame should be created at this point because the layer is to be edited with changes, so click on the **Key frame** icon in the bottom left-hand panel, right-click and select **Add key frame**. Select the key frame. Go back to the properties by clicking on the **Parameters** icon in the bottom left-hand panel.

The properties of the layer can now be changed to those required in the brief. Click on the property to be changed and dialogue boxes appear to allow the new values to be input. Change the colour to **red**, the text to **O** and the size to 32.000000 px by 32.000000 px, while leaving the other parameters alone.

Move the frame along by half a second (12 frames) to frame 48 + 12 (that is, 60), create and select a new key frame as before and return to the layer properties. Note that actual seconds can be entered in the dialogue box, for example 2 s, instead of the number of frames. Change the text to **Ou**.

Move the frame along by another half a second (12 frames) to frame 60 + 12 (that is, 72) and create and select a new key frame as before and return to the layer properties. Change the text to **Our**.

Repeat the steps of moving on by half a second (12 frames), creating and selecting a new key frame and adding the next letter of the phrase for all the letters in **Our Coffee Shop**. Do not forget to put a space before the uppercase letters **C** and **S**.

The last key frame in this sequence is used to completely remove the phrase so that it disappears, as required by the brief.

It is good practice to test the animation at intervals. Do this by toggling the edit mode to off and moving to the first frame. Selecting the **Play** button will run through the frames and show you whether there are mistakes. These can be corrected by stopping the animation, toggling to edit mode and correcting the errors. Saving the animation at intervals with different version names is also good practice so that mistakes do not completely ruin all of your work.

Importing objects

Bitmap images can be imported quite easily into Synfig, but vector images saved as .svg format will need conversion. Fortunately, Inkscape has an option to save its images as Synfig in animation format so these can be easily imported into Synfig.

The brief does not specify the exact moment of the appearance of the coffee cup, so it is assumed that it appears when the text has disappeared. Source a copyright-free image of a coffee cup. In the practical examination, the image would be supplied for you. Create a new layer for importing images from the **Layer**, **New layer**, **Other**, **Import image** tabs. Choose your coffee cup image and import it. Most bitmap formats are supported, but choose an image that has a small file size. There is no need for a high-resolution image for this animation.

The size and location of the image are adjusted using the handles that appear on the canvas. Experiment with these to find out what each does and position the coffee cup image in the bottom right-hand corner of the canvas. Be sure to make the final image quite small so as to allow space for the white rectangle and text that will be added next.

The timing of the appearance of the image can be adjusted in several ways. One of the simplest, but more time-consuming, ways is to start at the first frame and set the opacity of the layer to 0 in its properties. Setting the opacity to 0 effectively makes the image invisible. Set the opacity to zero at each key frame until the one at which you wish it to appear. In this case, the key frame will be the same as the one where the coffee cup frame disappears. You may to have to manually adjust the opacity to 0 in the frames before this key frame if Synfig attempts to fade in the image, or you may wish to allow it to fade in. The brief does not specify whether or not a fade is required so the choice is yours. The brief does not specify how long the image should be visible either, so it can remain on screen for the remainder of the animation. Use this as an opportunity to experiment with the property settings for layers and images.

Importing and creating vector objects

A hand-drawn image that has been digitised or an existing bitmap image can be converted into a vector object in Synfig. Import the image on to the canvas, resize it as required and then use the **Spline** tool from the toolbox to trace around the shapes in the image to create vector objects. At the end of the trace the spline is joined to complete the outline of the object. The objects can be resized, coloured and moved as required in the animation frames.

> ## Task 20a
> ### Creating a vector object
> Hand-draw a simple, outline cartoon character on paper. Digitise it (scan or photograph it) to JPEG format. Import it into Synfig and use the **Spline** tool to create vector objects. Colour the objects. Use the objects to create a simple animation based on your drawing.

Adding the white rectangle

Rectangles can be added using the **Rectangle** tool or by adding a new layer from the tabs. Add a rectangle and use the handles to position and size it to fit in the middle of the canvas. Change the colour to white. Use the opacity parameter at the key frames to make the rectangle appear at the same time, or just after, the coffee cup. It should be set to disappear at the same time as the text (which will be placed within it in the next section) disappears.

Adding the text in the rectangle

The brief is specific about the size, timing and fading of the text so these instructions must be followed precisely. Create a new text layer as before and enter the text **All our coffee is freshly brewed** from the brief. To make it fit in the white rectangle, a carriage return will need to be placed after the word **coffee**. The text will automatically be placed in the centre of the canvas so there is no need to adjust its position. Set the size to 16 × 16 and the colour to black. Select the sans serif font. Set the opacity to 0 until you wish it to appear, then set the opacity to 0.7 for 12 frames so that the text appears to fade in. The text has to show for 4 seconds so needs to be visible for 96 frames. Use the opacity setting to fade it out. This is a simple but tedious way to fade the text in and out. For more complex animations, you should explore more sophisticated methods. Do not forget to ensure that the white rectangle and the coffee cup image disappear too, as specified in the brief.

Task 20b

Waypoints

Most animation software has more than one way to adjust parameters. Waypoints and graphs can be used in Synfig to adjust the opacity of layers, for example. Waypoints can be shown by selecting the parameter and displaying its graph (by selecting the **Graph** icon) on the timeline.

Using waypoints can be much faster when adjusting many values. Explore the effects of the use of waypoints in Synfig to adjust the opacity of a layer. Drag the triangles (waypoints) on the graph around in editing mode to see the effect on the opacity values. The effect in the animation can be viewed in playback mode.

Adding a suitable logo for a coffee shop

Source or create a suitable logo for a coffee shop. As previously, in the practical examination you would be provided with a logo to use. In this example, the brief specifies that the logo should appear in the centre of the screen, rotate and grow in size to fill the screen. When it has filled the screen, the animation should end.

Create and select a new **Import image** layer. Import the logo and resize it to be small. It will automatically be imported in the centre. Create a new key frame just after the rectangle and text have disappeared. A suitable time would be a second or two (24 or 48 frames) later. Set the opacity of the layer to 0 in the frames before this new key frame. At this key frame, the opacity can be set to 1 to make the logo appear. Set the opacity parameter in the layer settings or use **waypoints** and alter the graph to adjust it.

To make the logo rotate, create a new key frame 2 seconds later. The brief is not specific but 2 seconds is a suitable time for the growth of the logo. The number of rotations is not specified so you can choose how many times the logo spins as it grows. Here, just one rotation is described. At the new key frame use the handles on the image to horizontally flip the image. Next, use the handles to 'grow' (resize) the image to fill the screen. The animation software will create the frames in between the two key frames so that the image of the logo appears to rotate and grow from one key frame to the next.

Looping the animation

One method of **looping** the animation by the four times specified is to use a Time loop layer. Create the layer from the Layer tab. If it does not appear as first in the right-hand list of layers, right-click on it and raise it to the top of the list. Being at the top of the list ensures that it affects all the layers below. In this example, set the parameters of the Time loop layer so that the Link time is 1, the Local time is a frame value several seconds after all your animation has completely run, such as 500, and the Duration is the same as the Local time. In Synfig, the Link time is the frame at which the repeated loop starts and the Local time is the frame at which the looping starts; in other words the frame on which to go to the Link frame. The Duration is the number of frames, counting from the Link time, to be included in the loop. In this example, the looping starts at frame 500, goes back to frame 1 and runs to frame 500. It will do this until the animation reaches its final frame, so ensure that the total length of the animation is set to at least four times the animation that you have created. This allows the animation to be repeated four times before it stops.

> ## Task 20c
>
> **Looping and repeat**
>
> Explore the options in Synfig, or your chosen animation software, to loop or repeat complete animations or just sections of an animation.

Rendering the animation

To allow the animation to be run without the animation software, the animation has to be rendered. **Rendering** converts and saves the animation in a format, such as AVI or MP4, that can be played on many different computer platforms. In the practical examination, you will be told what formats you can use so that the examiners can view your work. To render your animation, choose the Render option and set the parameters for your animation. Rendering make take some time but will eventually produce a standalone animation that you can easily run.

20.4 The use of animation variables when creating animations

Animation variables control the positions of objects in an animation. Sets of animation variables in each successive frame determine the positions of the objects. If the variables change from frame to frame, the objects appear to move. Animation variables can be generated from motion capture images or entered and adjusted manually. Manually adjusting the coordinates of an object can be used to refine the positions of objects or to move them to pre-determined locations on the canvas. In Synfig, the position of an object can be altered in its properties.

20.4.1 Object properties in animations

The properties of objects in an animation can be controlled and adjusted as required from their property settings or by dragging handles on the object on the canvas. The size, position and orientation settings can be precisely set by entering values, but they can be easily altered by dragging to create animation sequences.

20.4.2 Masking an object

The opacity (transparency) of the object can be used to determine when it appears or disappears in an animation sequence. However, if one object is animated to move behind or in front of another object, masking must be used. There are two types of mask used in animations. A hiding mask is used so that everything covered by the mask is hidden, and a revealing mask shows everything in the layers below it.

For example, in Synfig, masking is controlled by the **Blend method** property. A mask can be created on any existing image or drawn using the **Spline** tool to make a shape to be used as a mask. The blend method property of the shape is changed to 'alpha over' for a **Hidden** mask, and when placed in the required position, it allows objects under it to show through, making it appear that the masked area is hidden. A **Revealing** mask can be created using the blend method property of **StraightOnto**.

20.4.3 Fill and stroke

Fill sets the colour within the boundaries of an object. The **Fill** tool allows the colours of a selected object to be changed as required.

Stroke sets the colour of the line that surrounds an object. When creating a new outline shape or object, the current outline colour is used.

20.4.4 Setting paths

An object can be made to follow a set path during an animation. A path is drawn to determine where the object must go and then the object is linked to the path. When animated, the object moves precisely along the set path. This is useful as there is no need to redraw the object as it moves in every frame, the movement is precisely where the animator requires it and the path can be altered if changes are needed later.

In Synfig, the path is drawn across the canvas, in a layer as usual, using the **Spline** tool. The object is drawn, in its own layer, anywhere on the canvas. When the object is completed, the object layer is selected from the right-hand layer list, and, while holding down the **Ctrl** key, the path layer is also selected using the **Left** key. Select the dot in the middle of the object then right-click anywhere on the line. Choose **Link to spline** from the menu and the object will attach itself to the line. The object can now only be moved to a position on the line and when animated moves only along the line.

Practice questions

1 Explain why it may be necessary to trace a bitmap image and convert it to a vector image in an animation. (6)

2 Explain the difference between straight-ahead action and pose-to-pose action. (4)

3 Describe how an animator can determine the speed of movement of an object across a screen. (6)

21 Programming for the web

In this chapter you will learn:
- ★ the structure and syntax of JavaScript code
- ★ object-based JavaScript programming techniques and terms
- ★ how to use JavaScript to:
  - add interactivity to web pages
  - interact with, and change, HTML elements, content and styles
  - display data in different ways
  - react to common HTML events.

Before starting this chapter, you should:
- ★ have basic mathematical skills
- ★ have a working knowledge of HTML and CCS
- ★ be familiar with the terms: element, interactivity, text, string.

21.1 Using JavaScript

JavaScript is a scripted programming language for use on the web. Uses of JavaScript include creating web pages, web applications and games. Most JavaScript code is **client-sided**, meaning it is executed on a computing device by a JavaScript engine in a web browser and the scripts are embedded in, or run from, HTML code. JavaScript cannot display its own data, so it uses the display elements of HTML or the web browser to show its output to users or viewers of the page.

21.1.1 Adding interactivity to web pages

JavaScript scripts can be in the head or body, or both, of the HTML page and any number of scripts can be inserted. JavaScript has its own set of rules, or **syntax**, which must be followed if the code is to run. JavaScript does not need or use line numbers, but sometimes they are shown in printed examples so as to be able to reference individual statements.

JavaScript enables users to interact with web pages by allowing the web page to react and change according to the actions of the user. The entering of text or numbers, clicking of buttons, movement of the pointer over an image or a touch on part of a page can be used by JavaScript to change the behaviour of the page and the content displayed on screen.

21.1.2 Inserting JavaScript in HTML

JavaScript is embedded in HTML code between **<script>** and **</script>** tags. Task 21a, with comments, makes an **alert box** with some text appear on a web page. When **OK** is clicked, the alert box closes.

Task 21a

Preparation

This is a worked example to get you started with JavaScript.

Coding in JavaScript is best carried out in a plain text editor. Most operating systems provide a basic text editor, but it is useful to use one that has a few extra features. In these examples, Notepad++ running under Microsoft® Windows has been used, but there are alternatives available for Windows and other operating systems that have similar features. It is not sensible to code JavaScript in a word-processing application as these add their own commands, markup codes and formatting to documents.

Some HTML code is required to run JavaScript.

Open a text editor and type in, with indents, this code:

```html
<html>
<body>
<!-- the next HTML line displays some text on the page -->
<h2>This HTML code runs my JavaScript</h2>

    <script>
            // Your code will go here
    </script>

</body>
</html>
```

Save this code with a filename you will remember and in a folder where you can find it again. Test it out in a web browser to make sure it works. The JavaScript you write can be placed inside the `<script>` tags; with no JavaScript it will produce a white page with a single line of text, which you can customise if you wish.

Task 21b

Type this JavaScript code between the **script** tags:

```html
<script>
    alert("This text is displayed from JS");
// The next line creates and displays a box using the alert keyword
</script>
```

The alert box requires **OK** to be pressed to close it. **Alert** may also be written as **window.alert()**.

The complete code for the page, with HTML code shown in red and JavaScript code in blue here, should be as shown opposite.

```html
<html>
<body>
<!-- the next HTML line displays some text on the page -->
<h2>This text is displayed by HTML code</h2>

<script>

    alert("This text is displayed from JS");
// The next line creates and displays a box using the alert
keyword

   /*

      The alert box requires OK to be pressed
      to close it. alert may also be written as window.alert()

   */

</script>

</body>
</html>
```

21.1.3 External scripts

JavaScript code can be stored in external files and this **external script** inserted
into HTML when needed. The files are plain text, created in any text editor
and saved with the .js extension. External files are used so that the code can be
reused by inserting it into several web pages, and it is easier to edit one external
file than many web pages. External files are inserted into HTML using the
HTML arc attribute placed in the HTML where the writer wants the code to
be run, for example inserted into the body of the HTML.

Task 21c

Create this HTML file and save it:

```html
<html>
    <body>
        <script src="myjscode.js"></script>
    </body>
</html>
```

Save the JavaScript code shown in blue from Task 21b, between, but not including, the **<script>** tags, as
a file with the .js extension, for example as **myjscode.js**, in the same folder as the HTML file.

Run the HTML file to display the alert box as before.

It is important that the file path to the JavaScript file is known to the browser. This is usually by relative file locations. The external files can be any JavaScript code that would be placed between the `<script>` and `</script>` tags. However, because these tags are already in the HTML statement, they are not needed in an external file.

The advantages of using external files are that the length of the page code is reduced because only the location of the JavaScript file is specified, and the time to display the web page is reduced because there is less code to load when it is first opened. Also, the file can be cached by the web browser for reuse and this reduces page loading times even more. The code only has to be written once because it can be reused by many web pages and it is easier to edit and amend the code as there is only one copy of it. The readability of the code is increased because it is not embedded in lines of HTML and web authors can work on the HTML code and JavaScript code separately. This reduces the development time.

The disadvantages of using external files instead of code embedded in the HTML include the web browser having to make more requests to the web server, which can cause delays. If the code is reused an error in the file will affect many pages, which can impact on the whole website. Also, a change in the JavaScript in the file will affect many pages and some may have unexpected problems when interacting with other scripts, so it is necessary to check every page that uses the file. This is time-consuming.

21.1.4 Displaying the data from JavaScript

JavaScript does not have its own means of displaying data to users or viewers of web pages. It makes use of HTML features, dialogue boxes or the browser **console**.

Using document.getElementById()

The **document.getElementById()** method of displaying data works by getting an element whose ID matches the supplied **string**. The ID of the element has to be set before the method is used. An ID is set in the HTML code of a web page with, for example:

```
<p id="name _ of _ id _ goes _ here"></p>
```

or:

```
<div id="name _ of _ id _ goes _ here"></div>
```

Using innerHTML

JavaScript can use the **document.getElementById(id).innerHTML** method to write data into an HTML element. The element is defined by the attribute **id** and the content by **innerHTML**. The contents of innerHTML are displayed on the web page by the browser. This is the most common way to output data from JavaScript into a web page. In the script in Task 21d, the **id** is defined as **example**, referenced by **document.getElementById("example")** in the script and the content of **innerHTML** changed to **CIE**, which is displayed on the web page.

Task 21d

Create and run this script:

```
<html>

<body>

<h2>Using innerHTML and document.getElementById("id")</h2>

<p id="example"></p>

    <script>

        document.getElementById("example").innerHTML = "CIE";

    </script>

</body>

</html>
```

Using document.write()

JavaScript data can be output to a web page using **document.write()**, but using **document.write()** after the page has loaded, for example in response to a button click, will delete all the HTML and display only **IT Examination**, as shown here:

Task 21e

Create and run this script:

```
<html>

<body>

<h2>Using document.write() as the page loads shows this line too</h2>

    <script>

        document.write("IT Examination");

    </script>

</body>

</html>
```

If a script should not be run immediately, buttons and JavaScript **functions** can be used to specify when scripts are run. It is mainly used when testing scripts and in most other situations it is better to use **innerHTML**.

Task 21f

Create and run this HTML code to show how the page is cleared:

```
<html>
<body>
<h2>Using document.write() after the page loads clears the rest of the page</h2>
<p>HTML has written CIE on this page</p>
<p>but is all cleared when IT examination is written to the page</p>
<button type="button" onclick="document.write('IT Examination')">Click here</button>
</body>
</html>
```

Using an alert box

The alert method is used when data or information such as a message is to be shown to a user. The data is written into the alert box using **window.alert()**. This is described later in this section.

Using console.log()

Web browsers have a console that allows users to check on JavaScript for debugging and diagnostic purposes. Data, such as values, **variables** and **arrays**, can be written into the browser console from JavaScript by using **console.log()**. The data is not normally shown in the web page but can be viewed in a browser tab by accessing the console from the browser menu or by pressing **F12**. This example outputs the text **"This is a textbook"** to the console and, while the web page remains blank, the text can be viewed in the console, as shown here:

```
<html>
<body>
    <script>
        let txt = "This is a textbook"
        console.log(txt);
    </script>
</body>
</html>
```

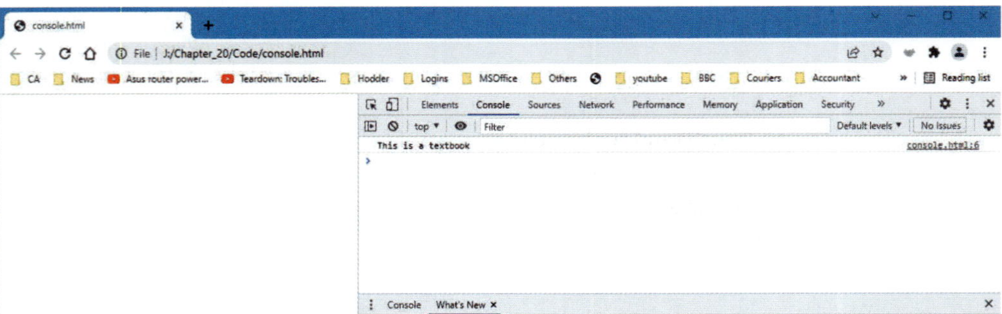

21.1.5 Changing HTML content

JavaScript makes use of the HTML Document Object Model (DOM) to access the elements of a document written in HTML. It creates dynamic HTML by being able to react to, change, add or remove any and all of the elements, attributes, events and CCS styles in a web page.

Text and string manipulation

Text is stored and manipulated by JavaScript in strings. A string is any number of characters, including none at all, inside quotes. Single or double quotes can be used. Quotes can be included in a string so long as they are different from the quotes that enclose the string. For example, the statements assigning the string **Brian** to the variable **a**:

```
let a = "Brian";
```

and

```
let a = 'Brian';
```

work in the same way. To assign **Brian says "hello" when meeting Katey** to **a**, the syntax is:

```
let a = 'Brian says "hello" when greeting Katey';
```

If the quotes inside the string are the same as those enclosing it, the assignment does not work properly or may not work at all. Quotes can be included in strings by using the backslash escape character. Putting \ before a 'problem' character turns it into a character that can be placed in a string. For example, this works fine:

```
let a = "Brian says \"hello\" when greeting Katey";
```

Another method is to use backticks to define a literal string, which can then include either or both single and double quotes. In this example, the backticks are before B and after y:

```
let a = `Brian says "Look, it's Katey"`;
```

JavaScript can manipulate the characters in strings to create new strings by removing or adding characters or joining strings together. It can also extract characters from strings. It is important to note that JavaScript creates a new string and does not change the original string.

To illustrate some of the many actions that can be performed on strings, we will use an example of assigning a string to the variable **strgtxt**:

```
let strgtxt = "Cambridge Assessment Education"
```

Some of the actions that can be carried out on the string are shown in Table 21.1. The value in the variable **strgtxt** can be manipulated by using several methods. The syntax is to attach the method to the variable name and assign the new value to a new variable. The value of the new variable can be used or passed to the HTML for display on the web page.

▼ **Table 21.1** Some of the actions that can be carried out on the string

Action required	Method	Syntax example	Result
Find the length of a string.	Use the length method by assigning the length to a new variable.	`let x = strgtxt.length`	The length of **strgtxt** is returned and assigned to **x**. In this example, **30** is stored in **x**.
Extract part of a string by specifying the first and last character positions to be extracted. *Note that the first character position in a string is denoted as 0 and that the character in the end position is not included in the return.*	Use **slice(start, end)** The first character to be extracted is at the start position in the string but the last is the character before the end position.	`let x = strgtxt.slice(3,9)`	The third to the eighth characters are extracted and assigned to **x**. In this example, **bridge** is stored in **x** because the end, ninth, character is not included in the return.
	The parameters can be negative numbers, which are counted from the end of the string.	`let x = strgtxt.` `slice(-27,-21)`	Counting from the end, the 27th to the 21st characters are extracted and assigned to **x**. Again, in this example, **bridge** is stored in **x**.
	Note that if the second parameter is not included, then the whole of the string starting from the first parameter is returned.	`let x = strgtxt.slice(10)`	In this example, **Assessment Education** is returned.

Join strings together.	Use **concat()** to join two strings together.	To join strings, they are assigned to variables and the result assigned to a new variable:
		`let strgtext2 = "A Level examination"`
		`let newstring = concat.strgtxt(" ",strgtext2)`
		The inclusion of **" "** places a space between the words.
			In this example, **Cambridge Assessment Education A Level examination** is returned to the variable **newstring**.

Find the position of a string inside a string.	Use **strg.search()**	The string is assigned to a variable and the position to another variable:
		`let strgtxt = "Cam"`
		`let pos = strg.search("a")`
			In this example, the value **1** is returned to the variable **pos**.

> **Activity 21a**
>
> There is an alternative method to **slice(start, end)** for extracting part of a string.
> It also specifies the first and last character positions.
> 1 What is the alternative method?
> 2 Describe how this method is used.
> 3 There are many more methods that can be used on strings in JavaScript.
>   Strings can be searched, converted between upper and lower case, trimmed
>   or split or their contents matched with a stated value.
>
> Create a table of at least six methods, other than those in Table 21.1, that can be
> used on strings in JavaScript. Create your own examples for each and show the
> result that would be returned.

Regular expressions

Regular expressions are used when replacing and searching for characters in
strings. The sequence of characters that makes up the pattern to be replaced
or searched is called a 'regular expression'. Regular expressions can be one
or more characters in length. For example, replacing International with
Assessment so it becomes Cambridge Assessment Education means that
International is the regular expression being used. The JavaScript code to do
this is explored in Task 21g.

> **Task 21g**
>
> Type in this code and click the button:
>
> ```html
> <html>
> <body>
> <!-- the next HTML line displays some text on the page -->
> <h2>This HTML code runs my JavaScript</h2>
> <button onclick="ReplaceFunction()">Click to run the JavaScript to change
> the text</button>
> Cambridge International Education
>   <script>
>     function ReplaceFunction() {
>     let title = document.getElementById("Textid").innerHTML;
>     document.getElementById("Textid").innerHTML =
>     title.replace("International","Assessment");
>     }
>   </script>
> </body>
> </html>
> ```

Searching Cambridge International Education for International returns the
position of the first character of the regular expression. Note that the count
starts at 0, so the b in Cambridge is at position 3.

Activity 21b

Write a script to find and display the position of **International** in **Cambridge International Education**. Don't forget to give the element an id and use **innerHTML** to display it.

Numbers and calculations

Calculations are carried out using the arithmetic operators. When writing code to perform calculations, operator precedence applies, as described later in this chapter. JavaScript does not distinguish numbers into integer or decimal, and stores all numbers as floating point.

JavaScript uses 64 bits to store the **floating point numbers**. The details are beyond the scope of this syllabus, but it does put a limit on the size of the numbers that can be stored. Also, arithmetic using floating point storage may not always be perfectly accurate. For example, adding **0.1** to **0.25** or to **0.45** produces the expected results of **0.35** and **0.55** respectively, but when adding **0.1** to **0.35** with:

```
let x = 0.1 + 0.35;
```

the expected result is for **x** to be **0.45**, but JavaScript does not produce this result. Because of this, care must be taken to thoroughly test the code. Such inaccuracy can be overcome by careful coding or by additional arithmetic on the numbers.

Task 21h

1 Create and run this script to add **0.1** and **0.35** to show the inaccuracy.

```
<html>
<body>
<h2>JavaScript numbers</h2>
<p>Floating point arithmetic is not always 100% accurate.</p>
<p id="demo"></p>
    <script>
        let a = 0.1;
        let b = 0.35;
        let x = a + b;
        document.getElementById("demo").innerHTML = "0.1 + 0.35 = " + x;
    </script>
</body>
</html>
```

2 Amend the script as opposite so that the correct result of **0.45** is returned. While multiplying the values by 10 before the addition and then dividing the result by 10 works in this example, such a simple approach may not work in other examples.

```
<html>
<body>
<h2>JavaScript numbers</h2>
<p>Floating point arithmetic is not always 100% accurate.</p>
<p id="demo"></p>
    <script>
        let a = 0.1;
        let aa = a*10;
        let b = 0.35;
        let bb = b*10
        let x = (aa + bb)/10;
        document.getElementById("demo").innerHTML = "0.1 + 0.35 = " + x;
    </script>
</body>
</html>
```

Images

Images can be included in web pages by accessing the image files using the
<img src="filename"> tag. The filename is the name of the file as it is stored,
for example **myimage.jpg**, and the HTML tag in the code of the website in the
browser finds the file and displays it on the web page. The filename can include
the path to the image file if the image is not stored in the same place as the
HTML code. JavaScript can be used, for example, to change the image to
a different one when an event occurs, such as a mouse click, or when the mouse
pointer moves over the image or to access the image element properties.

The image to be manipulated is given an id by, for example:

```
<img id="myselfie" src="photographofme.jpg"
```

To change the image to a different one, **document.getElementById()** is used
to access the **src** property:

```
document.getElementById("myselfie").src = "adifferentphotoofme.jpg";
```

This can be placed in a function to be called when an event occurs or, for
example, after a set interval of time.

Other properties of the image can be manipulated in a similar way. The width
and height can be set and changed using JavaScript code with
document.getElementById() by adding the property to it, for example:

```
document.getElementById("myselfie").width = 1000;
```

This sets the width to 1000 pixels. The image can be set back to its original width with:

```
document.getElementById("myselfie").naturalWidth;
```

In a similar way, the height can be set with:

```
document.getElementById("myselfie").height = 1000;
```

This can similarly be reset with:

```
document.getElementById("myselfie").naturalHeight;
```

21.1.6 Changing HTML styles with JavaScript

The styles of HTML elements, such as background colour and text colour, font, size, alignment and many more, can be set or changed using JavaScript. The **HTML styles** are changed using:

```
document.getElementById(id).style.property = new style;
```

where **property** is the style to be changed and **new style** is what it is to be changed into. For example, this changes the style of the element with the id **bg1** to the Courier font:

```
document.getElementById(bg1).style.fontfamily = "Courier";
```

> ## Task 21i
>
> 1 Put this script into the body of an HTML page. Type your name where **XXXXXX** is and then run it in a web browser to see how the script can change the font display.
>
> ```
> <h1>Changing font styles with JavaScript</h1>
>
> This is my text and this is my name XXXXXX.
>
> <button type = "button" onclick = "fontdisplay()">Change the size and
> font family of my text</button>
>
>     <script>
>
>         function fontdisplay() {
>
>             document.getElementById("mytextID").style.fontFamily =
>             "Arial,serif";
>
>             document.getElementById("mytextID").style.fontSize = "larger";
>
>         }
>
>     </script>
> ```
>
> 2 Experiment with different font families and sizes by amending the values.

21.1.7 Showing and hiding HTML elements with JavaScript

HTML elements can be shown or hidden using the syntax:

```
document.getElementById(id).style.visibility = "visible";
```

or

```
document.getElementById(id).style.visibility = "hidden";
```

The display property of style can also be used:

```
document.getElementById(id).style.display = "none";
```

This will hide the element from view. The difference is that when the **display** property is set to **none**, it leaves no trace of the element on view, whereas the **visibility** property **hidden** leaves blank spaces where the element once was.

21.1.8 Reacting to HTML events

Programmers can use JavaScript to react when **HTML events** take place in a web page. The events are used to call a script or a function written in JavaScript. Events that can be used by HTML to trigger a script include, among many others, when a script is loaded, using keys on a keyboard, mouse clicks, mouse movements and the dragging and dropping of elements on a page.

Common events and examples of how they can be used are shown in Table 21.2.

▼ **Table 21.2** Common events and examples of how HTML events can be used

Event	Syntax	Example of use
A web page has finished loading into a browser.	`<HTMLelement onload="JSscript">`	`<body onload="myJS()">` Here **onload()** is within the body of the HTML and starts a script called **myJS** when the page has finished loading.
A mouse button is clicked.	`<HTMLelement onclick="JSscript">`	`<button onclick="myJS()">X</button>` Here **onclick()** calls **myJS** when a button is clicked. Text prompting the user to click it can be placed at **X**.
The mouse is moved over an element. This is often used when the mouse pointer is moved over an image.	`<HTMLelement onmouseover="JSscript">`	The **JSscript** changes the image source, size or other attributes when the mouse pointer is over the image.
The mouse is moved away from an element. This is often used when the mouse pointer is moved out of an image area.	`<HTMLelement onmouseout="JSscript">`	Often used with **onmouseover()** to change an image back to its original condition when the mouse is no longer over the image.
The press of a key on a keyboard.	`<HTMLelement onkeydown="JSscript">`	Often used where a user inputs a character into a form element such as a text area or textbox. It can be used, for example, to run a script to produce an alert or to change the text colour when a key is pressed.
The value of a selected element is changed and the element is no longer selected or the focus.	`<HTMLelement onchange="JSscript">` The HTML element can be, for example, an input field or the **select** element.	When a new value is typed into an input box to replace what is already there and the user moves on, a script/function **JSscript** is called, to create an alert to inform the user that the value has been changed.

User interaction

There are several methods that enable users to interact with JavaScript code. Alert boxes display only data to the user, while the prompt and confirm statements enable user interaction with the dialogue boxes.

Alert()

The **alert()** statement is used when information such as a message is to be shown to a user. It displays a box on screen with the information and an **OK** button that the user is forced to interact with. The focus is taken away from the window in current use and the user cannot access any other parts of the web page until the alert box is closed. If alert boxes occur often, users can get distracted and annoyed and may move to other websites.

Alert boxes are often in a function that is called in response to an HTML event, such as when the mouse is clicked or the pointer moves over an image. For example, in this code an alert box opens when the **Click me** button is clicked. The box is closed by clicking the **OK** button:

```
<button onclick="openmyalertbox()">Click me</button>

<script>

    function openmyalertbox() {

        alert("Hi, this is my alert box!");
The text to be displayed is enclosed in brackets and quotes

    }

</script>
```

The script produces the output shown here:

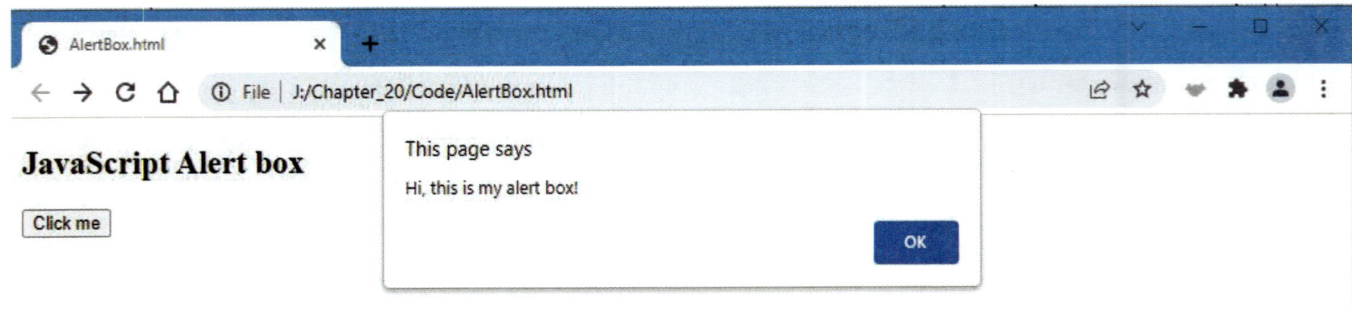

Prompt()

The **prompt()** statement is used to open a box that requires a user to input
a value. This type of box is referred to as a dialogue (or dialog) box. The
value from the user is returned for use by the code when the user clicks **OK**.
A **Cancel** option is provided to close the dialogue box and, if the user does
not supply a value, null is returned. A default value can be specified and this is
returned instead of null. For example, this produces a box that asks "Who are
you?" but the entry line is blank:

```
let name = prompt("Who are you?");
```

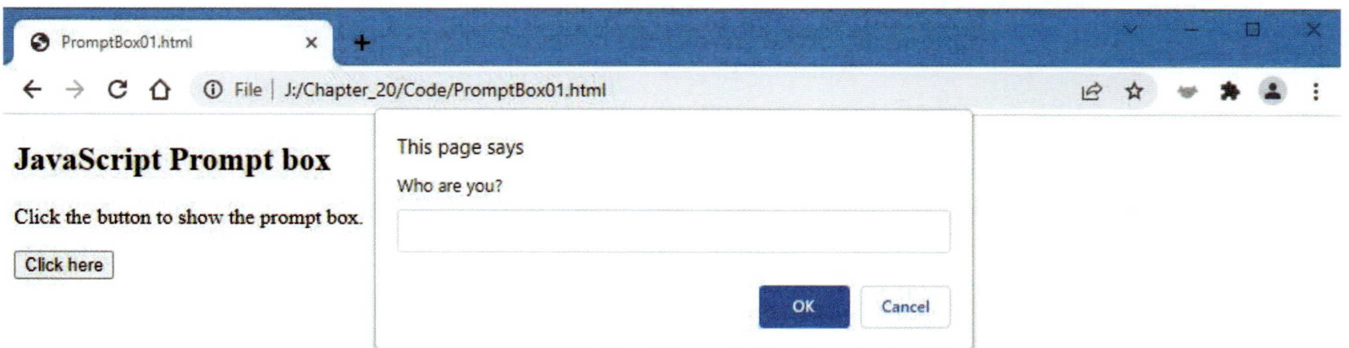

This produces a box with the word **Name** inserted in the entry line:

```
let name = prompt("Who are you?","Name");
// Message and default entry inside the brackets
```

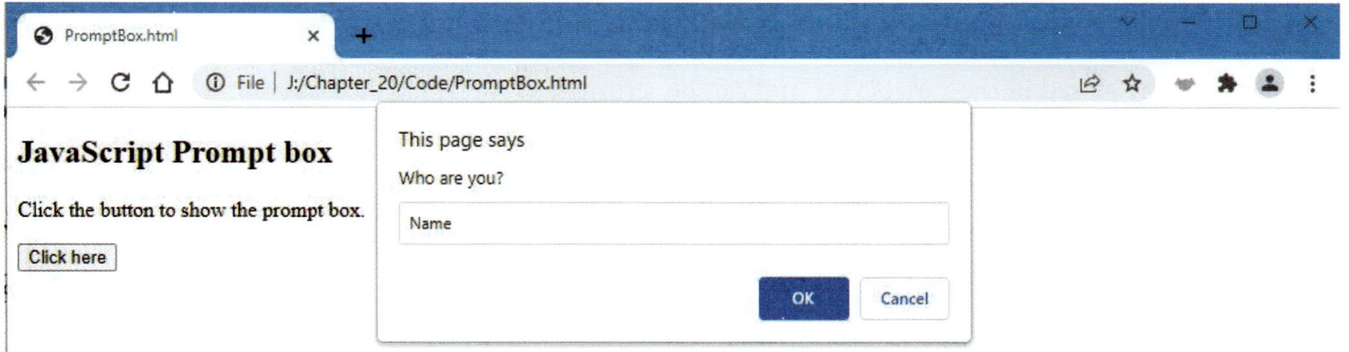

The returned values are stored in the variable and passed to the HTML code for
display on the web page.

Activity 21e

Write a script to create a prompt box that asks **Who are you?** and collects
the name that is typed in the box. When **OK** is pressed, the page displays the
message **Good day** followed by the name and also displays the question **Is this
really you?**. The button message should be **Please click here**.

Confirm()

The **confirm()** statement creates an interactive dialogue box for users, but
the only actions the user can take are to click **OK** or to click **Cancel**. The **OK**
button returns true and the cancel button returns false. The most common
use is for asking the user to accept (or confirm) an action or choice or to reject
it. A close (**X**) button may also be provided and this can also act as a **Cancel**

button to return false. A message can also be specified to explain what the user is asked to do:

```
confirm("Press a button!");
```

The **confirm()** statement is usually combined with **logical operators** in a function to determine the action to be taken when the **OK** or **Cancel** buttons are clicked. Messages to the user can also be specified depending on which button is clicked, as in this code:

```
function confirmbox() {

    let txt = "Click the OK or the Cancel button.";

    if (confirm(txt) == true) {

        txt = "You clicked the OK button.";

    } else {

        txt = "You clicked the Cancel button.";

    }
```

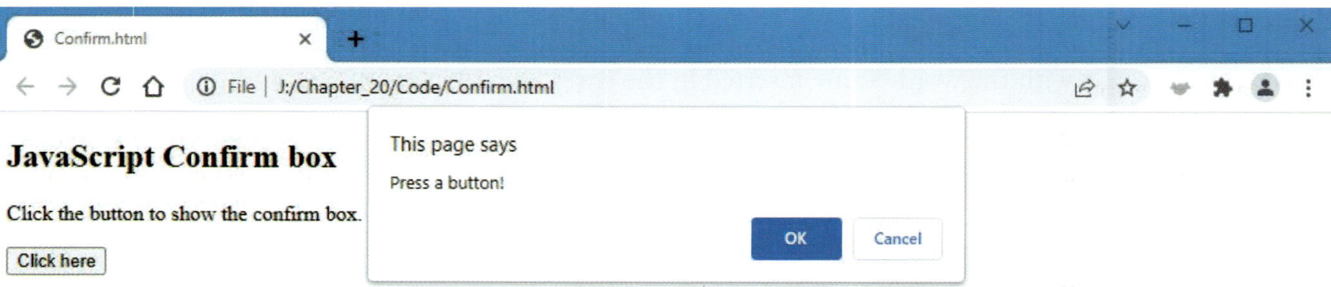

The messages are stored in the variable **txt** and can be returned to the HTML code in the web page to show to the user.

> ### Activity 21f
> Write a script to create a confirm box when a button is clicked. The button message on the web page should be **Click here**. The confirm box messages should be **Click a button** and on a new line **Click either the OK or Cancel button**. When a button is clicked, the web page should tell the viewer which button was clicked.

Like the **alert()** and **prompt()** statements, the user is forced to focus attention on, and interact with, the dialogue box until the box is closed. No other user interaction with the page can occur and other codes may stop working until the box is closed. Also, some dialogue box actions are not supported by all web browsers, so some of these may not work.

21.1.9 Structure and syntax of JavaScript code

JavaScript statements

Each line of JavaScript is called a statement. Statements end with a **;** character (the semi-colon) and are executed by the web browser. It is usual, for clarity, to put each **JavaScript statement** on a new line, but more than one statement can be on a single line provided they are separated by a semi-colon. While JavaScript statements do not need to be indented, it is usual to indent them. JavaScript ignores multiple white spaces and tabs so these can also be included for clarity when reading the script.

JavaScript statements consist of values, operators, expressions, keywords and comments. Specific sequences of statements of JavaScript code, called blocks, can be grouped together as named functions to carry out a specific task. A function is executed when it is called, or invoked, by name from another statement.

Values can be either fixed or variable. A fixed value is called a literal. A variable value can be changed during the execution of the code and is called a **variable**. For example:

```
let b = 21;
```

In the statement above, **21** is a literal value because it is always 21 and cannot be changed (it is said to be immutable), but **b** is a variable value (or simply called a variable) because it can be changed, for example:

```
let b = 22;
```

and so **b** is now 22, but **21** is still 21 and always will be 21.

Arithmetic operators (addition, subtraction, multiplication, division and more) work out values, and **assignment operators** (=, +=, *= and more) assign values.

Expressions are combinations of values (literals and variable) and operators that produce a value. The process carried out by expressions to return a value is called an evaluation.

Keywords are reserved words that are for use by JavaScript only and specify the action that is to be carried out. Keywords cannot be used as names for variables or functions. Examples of JavaScript keywords include **if**, **let** and **var**. There is a large number of keywords and more are being added as JavaScript develops, so programmers must keep their knowledge up to date.

Comments are ignored by JavaScript engines so are not executed. They are inserted to make the code more readable. They can be used to explain how the code works or what it is doing. Comments start with // on a single line and anything following on that line will not be executed. Comments that go over several lines are inserted by opening with /* and ending with */ so anything between these is ignored.

Data types

Data types are used to tell JavaScript statements how to deal with values when evaluating expressions.

Several data types are recognised in JavaScript:

» **Number** is used for any kind of number. Numbers are not distinguished into integer or decimal and are stored as floating-point numbers.
» **String** is used for storing any character.
» **Boolean** is used for true or false. True or false is the only value that a Boolean can have.
» **Array** is used for storing more than one value.
» **Object** is used for storing complex data structures.
» **Undefined** is used when a variable is empty and does not have a value.

The same variable in JavaScript can store any data type and the type can be changed during the running of the code, for example:

```
let b = 22;      // Here b is a number
let b = "22";    // This changes b into a string
let b;           // And here b is undefined
```

Type conversions

Variables can be converted into a different data type or into a new variable using JavaScript keywords. JavaScript can also be used to do the conversion automatically.

Conversion	Keyword/reserved word to use	Notes
Numbers into strings	String() toString()	Can be used on any type of numbers or values (literal and variable) and on expressions.
Strings into numbers	Number()	Strings are converted into numbers. Empty strings become zero, for example Number("") returns 0.

If the incorrect data type for a value is used during a JavaScript evaluation, it will automatically convert the value to the correct type. This may, however, give unexpected results. For example: adding a number and a string always produces a string, so "10" + 2 will return 102 because the 2 is converted to the string "2" during the evaluation. But adding two strings, for example "10" + "2", will return 12 because here the "10" + "2" become the numbers 10 and 2. Unexpected outputs may also be seen when assigning the values to variables for calculation. It is very important to check your code by thorough testing to ensure that the code is acting as expected.

When objects, variables, Booleans or numbers are output from JavaScript, it automatically uses toString() to convert the output to a string so it can be displayed on screen or made into hardcopy.

Variables

A variable is a container in which values are stored. Variables are given names so that they can be identified within the code. When a variable is given a name, it is said to be 'declared'. Each variable within a script must have a unique name. The name must not be a key or reserved word. Variables are declared with var, let or const. It is good practice to use const to declare variables that do not change and to use let if you think the variable may change. Older browsers may only support var. It is also possible to declare a variable with none of these. For example, the syntax is:

```
const buyprice = 10;
const = 15;
let profit = sellprice - buyprice;
var buyprice = 10;
```

The variables buyprice and sellprice are fixed and cannot be changed by the script (but they can be changed by manually retyping them), whereas profit can be changed. If the variables are not declared, they do not exist until the statement in which they appear is run and they cannot be referenced from elsewhere.

Variables that are declared outside a function are global. Other scripts and functions on a web page can use global variables. Variables declared within a function are local and are not available for use outside the function. Care must be taken when declaring values to ensure that unique names are used for variables so that there is no confusion between global and local variables.

Arrays

Arrays are special variables because they contain more than one value. A list of different types of transport, such as bus, train, car, bicycle, motorcycle, horse and scooter, can be stored in one variable called **transport**, instead of using a single variable for each type. Starting at 0, each value in an array has a number, or index, which denotes its position and can be used to access the values in the variable.

If the types of transport were stored in separate variables, the code could be:

```
let transport1 = "bus";

let transport2 = "train";

let transport3 = "car";

let transport4 = "bicycle";

let transport5 = "motorcycle"

let transport6 = "horse";

let transport7 = "scooter";
```

Using an array, it becomes:

```
const transport = ["bus","train","car","bicycle",
"motorcycle","horse","scooter"];
```

In arrays, line breaks and spaces are irrelevant and are ignored, so the code could be written as:

```
const transport = [
"bus",
"train",
"car",
"bicycle",
"motorcycle",
"horse",
"scooter"
];
```

The index of the value **"bus"** is 0 and the index of the value **"motorcycle"** is 4. The index can be used to access the individual values in the array, for instance accessing **transport(3)** would return **bicycle** as the value.

Objects

JavaScript is object-based. An object can be considered as a 'container' or 'box' storing a collection of data values. The value can be anything and the name is a string of characters used to identify the value. A JavaScript object is a type of variable that has many values. Each of the values can, but need not, have properties. A property is a value with a name.

The data values are not held in any particular order, but they are related and are stored in the form:

```
key:value
```

The **key:** can be a variable that is a property or a function that is a method when referring to an object.

A real object such as a textbook has properties that include its subject, title, name of author, number of pages and number of glossary pages. If coded in JavaScript, these properties of the object **textbook** could be named as **subject**, **title**, **author**, **pages** and **glossary**, each with its own value. The code would look something like:

```
const textbook = {
    subject: "IT",
    title: "Cambridge International A Level Information Technology",
    author: "Gillinder",
    pages: "464",
    glossary: "13"
    }
```

The code could also be written as:

```
const textbook = {subject: "IT", title: "Cambridge International
A Level Information Technology", author: "Gillinder", pages: "464",
glossary: "13"};
```

The properties of an object can be viewed with a **for/in loop**, as described later in this section.

There are also actions that can be carried out on a textbook. For example, the title can be changed, more pages added or the glossary amended. All textbooks can have these actions carried out on them. Actions carried out on JavaScript objects are called methods. For JavaScript objects, methods are functions that are stored as properties, for example:

```
titlechange()
addpages()
changerefs()
```

JavaScript operators

Arithmetic operators

Arithmetic on numbers is carried out by using the arithmetic operators. The numbers in an arithmetic calculation are called operands and these can be literals or variable values. The calculation to be carried out is determined by the operator. The operators are shown in Table 21.3.

▼ **Table 21.3** Arithmetic operators

To carry out this action	Example	Arithmetic operator used
Addition	Add 2 and 3	+
Subtraction	Subtract 3 from 5	–
Multiplication	Multiply 2 by 3	*
Division	Divide 6 by 3	/
Exponentiation (raise to the power of)	5 to the power of 2, i.e. 5^2	**
Modulus (give the remainder of a division)	The modulus of dividing 5 by 2 is 1	%
Increment (add one to an operand)	Incrementing 5 gives 6	++
Decrement (remove one from an operand)	Decrementing 6 gives 5	--

Operator precedence

JavaScript expressions are always evaluated from left to right. For example, evaluating:

```
let publication = 'text' + 'book';
```

will result in the variable **publication** being **textbook**, but operators may not always follow this rule. This is because arithmetic operators are actioned in a specific order, which is as follows:

» Brackets or parentheses, ( ), are actioned first so can be used to group together other operators
» Exponentiation is carried out before any other arithmetic operator
» Multiplication before division, addition and subtraction
» Division before addition and subtraction
» Addition before subtraction
» Subtraction.

For example: 6 + 4 * 5 – 3 evaluates to 23 because 4 * 5 is 20, which is added to 6 to give 26 and then 3 subtracted to give 23. Parentheses can be used to alter the order of the arithmetic because these are actioned first. (6 + 4) * (5 – 3) evaluates to 20 because the parentheses force 6 + 4 to be calculated to 10 and 5 – 3 to 2. 10 and 2 are then multiplied to give the result of 20.

It is important when writing JavaScript expressions to carefully consider the order of precedence when including arithmetic on values.

Assignment operators

Assignment operators assign values in the right-hand operand to the left-hand operand when an expression is evaluated. For example:

```
let b = 'text';
```

assigns the value **text**, the right-hand operand, to the variable **b**, the left-hand operand. Assignment operators are used as shorthand versions of other operators, for example:

```
b *= 20;          // This is shorthand for b = b * 20;
```

Activity 21g

Look up and create a table to list the assignment operators available in JavaScript.

String operators

Strings can be manipulated with an assignment operator. Strings can be added together using the + operator. This is called 'concatenation' and can also be carried out with +=, for example:

```
let a = "text";
let b = "book";
let c = a + b;
```

This results in the variable **c** storing **textbook**.

```
let a = "text";
a += "book";
```

This results in the variable **a** storing **textbook**.

Comparison operators

Comparison operators are used in statements to compare the condition of values or variables. The difference or equality is always returned as true or false and the result can be used to determine actions to be taken.

If, for example, the variable **A** is given the value of 1, then Table 21.4 shows the details of the conditions that can be tested.

▼ **Table 21.4** Comparison operators to compare the condition of values or variables

Description	Operator syntax	Comparison against A = 1	Return
equal to	==	A == 1	True
		A == 2	False
		A == "1"	True
greater than	>	A > 1	False
less than	<	A < 1	False
greater than or equal to	>=	A >= 1	True
less than or equal to	<=	A <= 1	True
equal value and equal type	===	A === 1	True
		A === "1"	False
not equal value or not equal type	!==	A !== 1	False
		A !== 2	True
		A !== "1"	True

Logical operators

The logic between values and variables is determined by the logical operators AND, OR and NOT.

If, for example, the variable **A** is given the value of 1 and **B** is given the value of 2, then Table 21.5 shows the details of the conditions that can be tested.

▼ **Table 21.5** The conditions that can be tested by the logical operators

Description	Operator syntax	Examples of comparison	Return
AND	&&	(A < 5 && B > 1)	True
		(A = 1 && B = 3)	False
OR	||	(A == 2 || B == 3)	False
		(A == 2 || B = 2)	True
		(A == 1 || B = 3)	True
NOT	!	!(A == B)	True
		!(A > B)	True

Conditional statements with operators

Conditional statements are used to carry out different actions depending on the outcome of decisions. The statements use comparison operators to test a condition. A condition is tested to determine whether it is true or false and, depending on the outcome, different blocks of code are executed. The conditional statements available are:

▼ **Table 21.6** Conditional statements

Conditional statement	Description
if	Used to specify a block of code if the condition is true
else	Used to specify a block of code if the condition is false
else if	Used to specify a new test if the first condition is false
switch	Used to test multiple conditions

For example:

```
if (X>Y) {           // Tests whether X>Y is true or false
    dothis();        // If true, this is executed
} else {
    dothisinstead();  // If false, this is executed
}
```

When another test is required when the outcome of the test of **X>Y** is false, **else if** is used:

```
if (X>Y) {           // Tests whether X>Y is true or false
    dothis();        // If true, this is executed
} else if (X>Z) {    // If false, this test is carried out
    dothisinstead();  // If true, this is executed
} else {
    dothisnow();     // If false, this is executed
}
```

An important benefit of using **else…if** is that more **else…if** statements with correspondingly different actions can be added.

It should be noted that while it is possible to construct code using the logical operators, the use of **if…else** or **else…if** makes the code much more readable and easier to understand.

The **if** operator does not need **else** or **else if** operators. If these are omitted, the specified code runs only if the outcome is true; otherwise it is skipped over and the next statement is executed.

The **switch**, and the associated **case**, statements are used to test multiple conditions and carry out different actions depending on the outcome of the tests. The **switch** expression is executed only once and its value is compared with the values of the following **case** statements. As soon as a match is found, the corresponding code block is executed. A **break** keyword is used to break out and stop the execution of a **switch** code block. The **default** keyword specifies the code to be executed if no match is found in the **case** statements. An example of the use of **switch** is below. It assigns a value to the variable **text** depending on the condition of the variable **flower**.

```javascript
switch flower = "marigold" {
// the variable with the value marigold is to be tested

    case "daisy":

        text = "I grow daisies";
// this case is false so JavaScript moves to the next case

        break;

    case "rose":

        text = "I grow roses";

        break;

    case "marigold":
// this case is true so that the next statement is executed

        text = "I grow marigolds";
// executed because the case is true

        break;
// included to exit the switch block

    case "hydrangea":

        text = "I grow hydrangeas";

        break;

    default:
// executed if no case is true

        text = "I don't grow flowers";
// default action

        break;

}
```

The **break** statement should be included because, if it is omitted, the next case will be executed after the match regardless of whether the following match conditions are met. In this example, **case "hydrangea":** and the default would also be executed. Note that the default case does not have to be at the end of the block but, if it is not, it is especially important to include **break**.

Code blocks can be used more than once by **switch** so that different matches can use the same code. In the following example, three values (marigold, daisy and rose) of the variable **flower** produce the text **I grow this flower** and the other two (hydrangea and lily) produce the text **I do not grow this flower**.

```javascript
let flower = "marigold";
switch (flower) {
    case "daisy":
    case "rose":
    case "marigold":
        text = "I grow this flower";
        break;
    case "hydrangea":
    case "lily"
        text = "I do not grow this flower";
        break;
    default:
        text = "I don't grow any flowers";
        break;
}
```

There are several important features of the **switch** statement to be aware of. The order of the cases is important because it executes only the first matching case that is found; any subsequent case matches are ignored and never executed. If no default case is included, JavaScript continues to the next statement *after* the switch block of code. The most important feature to note is that switch uses strict comparison when matching cases. Strict comparison means that there must be an exact match in the value and the type of value; it is the equivalent of using the operator === for the comparison.

Ternary operator

The **ternary operator** is written in JavaScript code as **?**, the question mark. It has three parts: a condition followed by the **?**, an expression that is executed if the condition is true, followed by a colon, and an expression to be executed if the condition is false. True is any value that can be converted to true and false is any value that can be converted to false. Expressions such as **null**, **NaN**, **0**, the empty string (**""**), and **undefined** are also deemed to be false. The use of the ternary operator can be considered to be equivalent to **If…Then…Else**.

The syntax of the use of the ternary operator is:

```javascript
condition ? dothisiftrue : dothisiffalse
```

An example of the use of the ternary operator is:

```javascript
let driver = (age <17) ? "Not old enough to drive" :
"Old enough to drive"
```

If the variable **age** is below 17, then the value of **driver** becomes **"Not old enough to drive"**, else it becomes **"Old enough to drive"**.

Loops

Loops are used so that the same blocks of code can be executed over and over again. There are several types of loop available to carry out JavaScript operations.

▲ **Table 21.7** Types of loop available to carry out JavaScript operations

Operation required	Type(s) of loop used
Looping through a block of code a specified number of times	for
Looping through the properties of an object	for/in
Looping through the values of an iterable object	for/of
Looping through a block of code while a specified condition is true	while do/while

For loop

The **for loop** has three parameters in its syntax:

```
    for (statement_1; statement_2; statement_3) {
 // The statements are separated by ;

 // The code to be executed is placed here inside the { } brackets

    }
```

In the for loop, statement_1 is executed once before the block of code and is used to initialise variables. More than one variable can be initialised here. This is the initialisation of the loop.

statement_2 sets out the condition to be tested, evaluates statement_1 and returns true or false. This is the test statement. The loop starts again if true is returned but the loop ends if it is false.

statement_3 is executed after the block of code has been executed. statement_3 is executed every time the loop comes around. This is the iteration statement. It is used to change the value(s) of the variable(s) in statement_1, often by adding to, or subtracting from, from the values.

There is no requirement to include the three statements inside the brackets. The initial variables can be set up before the for loop code. The other statements can be inside the code block of the loop. Leaving out statement_2 will cause the loop to run forever and the browser may crash, but inserting a break in the loop can avoid this. statement_3 can be omitted from the brackets if the changes are carried out elsewhere in the loop.

For example, the for loop uses the value of x to step through the array list and send each value to the console in the browser.

```
let list = [1, 2, 3, 4, 5];   // Initialises the array

    for (x = 0; x < 5; x++)    // Sets up the loop

    {

    console.log(list[x]);      // Outputs to the console

    }
```

The first statement in the brackets (x = 0) declares a variable, x, and the second statement (x < 5) is the condition that is checked as the loop goes around. The third statement (x++) uses the ++ operator to increment x by 1 each time the loop iterates (goes round the loop). The three statements are separated by ;. In this loop, the output is sent to the web browser console.

The for loop could have its statements outside of the brackets and still work:

```javascript
let list = [1, 2, 3, 4, 5];          // Initialises the array
let x = 0;                           // Initialises the condition
for (; ;) {                          // Sets up the loop but ; still required
   if (x >= 5)                       // The condition needed to stop the loop
   break;                           // Required to end the loop
   console.log(list[x]);            // Outputs to the console
   x++;                             // Increments the condition
}
```

Both versions produce the same output in the console: 1, 2, 3, 4 and 5.

For/in loop

For/in loops are used to iterate through the properties of an object. They are useful for checking the properties of an object when looking for errors. Each iteration over an object returns a key that is used to access the value of the key. In this example, the properties of the object **textbook** can be discovered by the for/in loop shown in blue:

```javascript
const textbook = {subject:"IT", title:"Cambridge International A Level
Information Technology", author:"Gillinder", pages:464, glossary:13};
let book = "";
for (let x in textbook) {
   book += textbook[x];
}
```

The properties are extracted at each iteration and added to the variable **book**, which can be passed to the HTML of the web page for display or used in other JavaScript code.

While for/in can be used with arrays, it is not recommended that these are used to discover the values in an array because the array values might not be accessed in the expected order. For/of loops should be used with arrays.

For/of loop

The for/of loop is used to iterate through objects, arrays and strings and assign the values of the properties to a variable in the same order as they are found. In the example above, **for...in** is replaced with **for...of**:

```javascript
for (let x of textbook)
```

While loop

While loops iterate through a block of code as long as a condition is true. The syntax of a while loop is:

```javascript
while (condition) {
        // The block of code to be executed goes here
}
```

condition is the specified condition that has to be true for the block of code to be executed. As soon as the condition is false, the iteration stops and the code ceases to be executed. If the condition is false at the start, the code is not executed at all.

Do/while loop

Do/while loops also iterate through a block of code as long as a condition is true. The syntax of a do/while loop is:

```
do {

        // The block of code to be executed goes here

}

while (condition);
```

condition is the specified condition that has to be true for the block of code to be executed. As in the while loop, as soon as the condition is false, the iteration stops and the code ceases to be executed. However, it is always executed at least once because the code is executed before the condition is tested. This happens even if the condition is false.

Break

A **break** statement is used to exit a loop. The **break** statement is also used to exit a **switch()** statement. When JavaScript encounters a **break** statement in a loop, it jumps out of the loop and continues with the statement immediately after the loop. If an exit from a loop is required during a specific iteration, the **continue** statement can be used. Using **continue** breaks a loop if a specific condition is met but then, unlike **break**, the loop carries on with the next iteration.

Functions

Functions are blocks of code that perform a specific task. The syntax of a function is to use the keyword **function()** with its parameters, separated by commas, inside the brackets and the code of the function placed between paired curly brackets { }:

```
function functionname(parameter1,parameter2,etc…) {

        // The code of the function goes here

}
```

The block of code in the function is said to be 'declared', or 'defined', and is stored in memory and executed only when it is invoked, or called, in a script. The line that declares a function is not an executable statement so it does not need to end in a semi-colon, for example:

```
function myaddingupfunction(X, Y) {
// Declaration, no need for semi-colon

    return X + Y;
// Performs the action, must have a semi-colon

}
```

The **return** statement sends the result back to the statement that called the function.

Functions can be called, or 'invoked', in several ways. JavaScript code can invoke a function or the function can be invoked when an event occurs. An event is a user action or an action by the web browser, for instance the click of a button by a user or an HTML page loading into the browser. Functions can also be 'self-invoked'. Self-invoked functions do not have a name and are invoked immediately after being defined. They execute once only and any variables declared in a self-invoking function are not available outside the function.

JavaScript timing events

Execution interval methods

The web browser window is represented by a window object, and any JavaScript objects, variables and functions are automatically part of this object. The window object allows JavaScript to be **executed at intervals** and timing events are part of the interaction between the web browser, HTML and JavaScript.

Timing events are used to specify the code block, for instance a function, to be executed and when it is executed. The timing can be set to execute the function after a set interval with **window.setTimeout()** or set to repeat the execution of the function at specified intervals with **window.setInterval()**. Both can be used with or without **window** at the front.

SetTimeout()

The **setTimeout()** statement takes two parameters:

```
setTimeout(name of function, wait time in milliseconds)
```

The function to be executed is called by its name, and the number of milliseconds to wait before it is executed is stated in numbers. Examples are:

```
setTimeout(mywaitingtime01, 4000)

setTimeout(mywaitingtime02, 8000)

setTimeout(mywaitingtime03, 10000)

setTimeout(mywaitingtime04, 14000)
```

Here, the functions **mywaitingtime01, 02, 03** and **04** are called and executed after 4, 8, 10 and 14 seconds respectively.

The execution of the function set in a **setTimeout()** statement can be stopped at any moment during the waiting time by using:

```
clearTimeout(name of function)
```

The named function is not executed if, within the waiting time, the **clearTimeout()** statement is called. It could be called by user action, such as a mouse click on a button.

In summary, **setTimeout()** starts a named function after a predetermined waiting time but the function can be prevented from being executed if, during the waiting time, **clearTimeout()** is used.

setInterval()

The **setInterval()** statement takes two parameters:

```
setInterval(name of function, time interval between
repeats in milliseconds)
```

The function to be executed is called by its name, and the number of milliseconds to wait before it is executed again is stated in numbers. A function started with **setInterval()** can be stopped from repeating by using **clearInterval()** with the same function name.

Errors and error handling

Errors in JavaScript code can occur for several reasons. Errors are reported to the browser and can be viewed in the console of the browser. Programmers can make mistakes when entering code, and errors can occur when values are input. There are three main types of errors that can occur in programming:

» Syntax errors are mistakes in the 'grammar' or spelling of the language used. Syntax errors are also called 'parsing errors', for example:

```
const transport = ["bus","train","car");
```

» This would cause an error because the closing bracket is not correct. Syntax errors show up when the code is run, when the JavaScript engine is trying to interpret, or parse, the statement.
» Runtime errors occur when the code is run; that is, when it is executed. Runtime errors are also called 'exceptions'. The syntax of the code may be correct but there may be a mistake in, for example, the JavaScript method used on an object or a non-existent function is called.
» Logic errors occur when there is an error in the logic of the processes or calculations in the script and unexpected results are obtained. Logic errors are very difficult to track down because they cannot be seen or spotted in the code but must be found by thorough testing and checking. A simple example is that a script may require a while loop to continue as long as a value is less than 10, but in the script this is written as >10 instead of <10. The error might not be noticed until runtime because it is not obvious, but the loop would not perform as expected.

Errors can also occur when input values are not what the script is expecting. For example, entering a name when a number is required produces, or 'throws', a type error; numbers that are out of a specified range, for example entering 232 years for your age on an application form, could throw a range error.

When an error occurs, the code block in which the error occurred is affected and the rest of the code could possibly carry on. However, if the remaining code is dependent on the statement or block, then the whole script will stop. Exceptions may affect only the block in which they occur but they can cause the interpretation of the script to stop at that point. Usually, errors cause a script to stop and a web page can appear to just stop and do nothing, or 'hang', as it is often called.

Also, whenever an error occurs with the loading of a script, the **onerror(msg,URL,line)** method can be used to extract information about the error. This method provides three pieces of information about the error:

» msg: the error message
» URL: the file in which the error occurred
» line: the line number where the error occurred.

These details can be used in customised error reporting.

To try to prevent errors from causing a total failure of the script and a web page crash, code can be written to try to catch errors and deal with them in a suitable way. Runtime errors (exceptions) can be caught using **try/catch/finally**, which has the code block to be executed, and the code block to run if there is an exception, followed by a code block to run regardless of the result from these blocks. Note that syntax errors cannot be caught like this. The code block syntax for **try/catch/finally** is:

```
try {

        // Code to be run goes here

}
catch ( myerror ) {

        // If an exception occurs this code is run and the value in
        myerror can be used or displayed

}
finally {

        // The code here is always executed regardless of an exception
        occurring

}
```

This can be used when testing a block of code, such as when a programmer has written some code and wants to try it and catch any errors. Also, **throw()** can be used to create or customise exceptions so the programmer can see what happens when testing code. For example, a **throw()** exception could be placed within the **try { }** block to produce an exception and the results tested.

JavaScript includes an error object that provides information about an error when one occurs. It provides the name of the error and returns a string containing an error message. Errors include a number being out of range or when syntax or type errors occur. Information about six different errors can be returned from the error object, which can be used in customised **error handling**.

Activity 21h

Dealing with errors

Writing code in JavaScript that works perfectly first time and in every possible situation is extremely difficult.

1 Describe two causes of errors that occur in JavaScript.
2 Describe what happens when JavaScript code encounters an error.
3 Research the errors that may occur during the execution of JavaScript code.

Programmers try to make their JavaScript code deal with errors so their code runs smoothly and can carry on. JavaScript has a number of methods for dealing with errors. Error information can be provided by the **error** object. Dealing with errors is called 'error handling'.

4 Research and try out the different methods that JavaScript provides for handling errors.

Activity 21i

Create a table of the **error name** values.

Practice questions

1 JavaScript code is used to provide interactivity between a user and a web page.

   a Explain how JavaScript can be used to make a user accept or reject an option. **(4)**

   b Explain how JavaScript can be used to make a user input a value in response to a question. **(6)**

2 Describe the three types of programming error that can occur in JavaScript. **(6)**

3 Comparison operators are used in statements to compare the condition of values or variables. **True** or **false** is returned as the result of a comparison.

Copy and complete the table below to state the result, **true** or **false**, of comparisons. The first is done for you. **(4)**

Comparison against X = 1	Return
X == 1	True
X !== "1"	
X <= 1	
X == "2"	
X === 1	

4 a Explain the purpose of conditional statements in JavaScript. **(2)**

   b State the different conditional statements used in JavaScript and describe their use. **(8)**

Glossary

2.4 gigahertz (2.4 GHz) and 5 GHz bands Parts of the electromagnetic spectrum available for public use. Wi-Fi works in these bands, dividing them into radio channels for data transmission.

2D and 3D animation 2D objects in animations are drawn with the dimensions of height and width around the *x*- and *y*-axes and are viewed from only one angle. 3D objects in animations are drawn with the dimensions of height, width and depth around the *x*-, *y*- and *z*-axes, allowing them to appear to be able to rotate though 360° and to be lit and viewed from different camera angles.

3D printing A type of printing that uses sets of instructions sent to a 3D printer to add successive layers of materials in order to create three-dimensional objects. *See* Additive manufacturing.

3G mobile communications Third-generation system for mobile communications. A global standard for digital cellular networks including the general packet radio service (GPRS), which is a standard for exchanging packets of data.

4G mobile communications Fourth-generation system for mobile communications, providing more simultaneous connections and a higher quality of voice call than 3G. 4G networks are based on packet-switching techniques and can implement IP networking to provide internet access.

5G mobile communications Fifth-generation system for mobile communications. 5G can provide greater bandwidth and more simultaneous connections than 4G. 5G has a very low latency. Devices can be interconnected using 5G to create an Internet of Things (IoT).

Acceptance criteria A list of requirements set out in the project plan that sets the conditions that must be met before a project can be considered completed.

Access rights/permissions Permissions that control who can and who cannot access files.

Active File Transfer Protocol (FTP) mode In this FTP mode, an FTP server responds to a client by asking the client for the data port to be used for exchanging data. The data connection is started by the server and is made on the port requested by the client.

Activity In project management, activities are carried out to complete a task. *See* Task.

Activity-on-arrow diagram Puts the milestones at the nodes in PERT charts and uses arrows to represent the tasks to be carried out to move from one milestone, shown as a circle, to the next.

Activity-on-node diagram Puts the tasks on the nodes in boxes in PERT charts, and the timings, along with other task information, inside the boxes.

Ad hoc mode Wi-Fi networking mode in which data is directly transferred between devices.

Adaptive maintenance Maintenance carried out to amend, modify or update a system.

Additive manufacturing Alternative name for 3D printing.

Address Resolution Protocol (ARP) A network protocol used to discover the MAC address of a NIC from the IP address on ethernet networks. Replaced by Neighbour Discovery Protocol (NDP, ND) on IPv6. Usually listed as working at the Data Link layer of the OSI model and the Link layer of TCP/IP, but it works with the Internet layer protocols when mapping addresses.

Agile method of software development A method of software development, based upon full involvement with the software developers by clients and other stakeholders, which focuses on the ability to adapt and change the product during development.

Alert box An on-screen box used when data or information, such as a message, is to be shown to a user.

alert() The JavaScript method used to produce an alert box.

Alignment tools Tools that make the precise positioning of elements, objects and layers in images and animations easier and much more accurate. Alignment can be with another layer, object or element or with the whole image.

Alpha testing Testing on new software, a product or system, as it nears completion and before it is released to end users, that is carried out by testers who are employed by the company.

Analysis stage The stage of the system life cycle where information is gathered by carrying out fact-finding investigations that are used during the design stage. The information is gathered by systems analysts from individuals or groups using questionnaires, interviews, observation and document analysis.

Animation variable Used to control the positions of objects in an animation.

Anti-malware software Software that scans every file for malware as it is accessed, or when a user instructs it to scan.

Anti-spyware software Software that scans every file for spyware as it is accessed, or when a user instructs it to scan.

Anti-virus software Software that scans every file for viruses as it is accessed, or when a user instructs it to scan.

Application layer This layer of the TCP/IP suite and OSI model contains all the protocols necessary for exchanging data or for providing services for users. Software applications use protocols at the Application layer.

Applications (or application) server A type of network server that uses software to provide and deliver applications to clients. Applications servers provide the interface for the client and the resources on the server to run the applications.

Arithmetic operator Addition, subtraction, multiplication, division (and more) used in calculations to work out values.

Array A data type in JavaScript used for storing more than one value. Arrays are special variables because they contain more than one value.

Artificial intelligence (AI) The ability of computer systems, or computer-based machines, to carry out tasks that would 'normally' only be possible by intelligent humans. AI mimics human thinking processes.

Assignment operator =, +=, *= (and more) used to assign values.

Augmented reality (AR) The enhancement of a real-world experience by overlaying computer-generated objects onto real objects.

Automatic number plate recognition (ANPR) A monitoring system that uses optical character recognition to read vehicle licence plates and creates data that can show the vehicle's location every time it is photographed by an ANPR camera.

Autonomous Transport System A vehicle or system that moves people or resources from one place to another with little or no input or control by a human operator or driver.

Backup strategy A strategy stating the cost of backup systems, how often backups are to be carried out, who is to be responsible for carrying out the backup procedures and where the backups are to be stored. It includes what data is to be backed up and how long backups should be stored.

Bandwidth The rate of transfer of data over a communication channel. Bandwidth is usually quoted in terms of the number of bits per second.

Beacon packet Wi-Fi frames containing data packets sent from a WAP to announce its presence, its name (service set identifier, SSID) and details of the Wi-Fi network, e.g. the frequencies in use.

Beta testing Testing that is carried out by the client, or by end users who are not employees of the company developing the system, on a version of new software, a product or system that is released to end users.

Biometrics Measurements of human body characteristics and the calculations that can be carried out using the measurements. Biometric measurements used for security must be sufficiently unique to an individual to identify a person among many others.

Bit rate The number of bits (binary digits) that are transferred along a communication channel in a single unit of time. The unit of time is usually one (1) second, therefore bit rate is bits per second or bit/sec.

Bitcoin A decentralised cryptocurrency using blockchains as a distributed ledger.

Bitmap graphics The creation of images made up of picture elements called pixels.

Black box testing High-level testing that tests how the new system functions and behaves without the need to know how it works.

Blending The blend setting of a layer determines how the layer appears when viewed with the other layers.

Blockchain A distributed ledger with 'blocks' that store information electronically and are secured by encryption.

Blockchain technology The use of cryptography to create data blocks containing details of the previous block, a timestamp, and a record of transactions. Blockchain technology uses a decentralised record or ledger of transactions on a peer-to-peer network. Used in cryptocurrency and for smart contracts.

Blog A social-network platform for posting content using software tools. Those who create blogs and post their content are called bloggers.

Bluetooth A wireless method of exchanging data over short ranges using radio waves in the 2.4 GHz band. Bluetooth is a separate networking system from the Internet Protocol suite and has its own protocol stack. *See* 2.4 gigahertz (2.4 GHz) and 5 GHz bands.

Blur Used to change the focus, give an appearance of motion or add artistic effects, e.g. to a digital photograph.

Boolean A data type in JavaScript used for true or false. True or false is the only value that a Boolean can have.

Border Gateway Protocol (BGP) Network protocols used by routers to communicate with each other and choose the paths for IP packets.

Botnet A number of networked devices that are running software, called a bot, carrying out repetitive tasks that can be used to deny access to services, send spam emails, steal files and data, and allow an attacker to remotely control devices by issuing commands that appear to come from the device itself.

Break A statement used to exit a JavaScript loop.

Bridge Connects two or more networks together. Frames are transferred to and from each network on to the others.

Brute force attack The systematic submission of many passwords or pass keys in the hope of correctly finding the actual password.

Buffer An area of RAM or other read/write data storage that is used as a temporary store for data that is being moved between components or devices. The term is also used to describe the area between two structures that is designed to keep them apart, e.g. in the construction of cables.

Buffering The temporary storage of data before it is used. Used when transferring data between devices or between components in devices. In video or audio streaming, buffering is used to avoid jerky video, lagging of audio, slow buffering, freezing of the data stream or complete loss of the stream.

Calculated field In mail merge, using and manipulating the data from different fields to create the content in another merge field.

Canvas The area in which a digital image or animation is drawn. It can be set by specifying the number of pixels on, or the length of, the *x*- and *y*-axes.

Cel animation A traditional animation technique of manually drawing and painting images onto transparent celluloid sheets.

Centralised banking system Financial system with a central regulator that controls financial activities.

CGI Computer-generated imagery that creates two-dimensional (2D) and three-dimensional (3D) objects that appear to move on screen.

Chat room An online space that is used to share information with other users using a protocol called Internet Relay Chat (IRC).

Circuit switching A continuous, direct communications link between two nodes or devices.

Cleaning data The process of removing incorrect, incomplete, irrelevant, improperly formatted, replicated or corrupt data from a data set.

Client–server network Devices, the clients, are connected to servers that provide the services requested by the clients. Clients can be personal computers, laptops or smartphones that request data, files, applications and websites from servers.

Client-sided The process in which JavaScript, or other scripts, are executed on a computing device by a JavaScript engine in a web browser.

Clone tool A software tool used to copy an area of an image and place it elsewhere in the image.

Cloud computing The sharing of data storage, server resources, files, databases, computational power and other computing resources on a LAN, WAN or the internet.

CMYK A colour system using four colours: Cyan, Magenta, Yellow and Key (or Black) to produce different colours.

Coaxial copper cable Cables that have a central copper core surrounded by an insulating material (the dielectric), usually made of plastic, and then a shield of braided copper.

Collaborative project management All project members are involved in the planning, monitoring and execution of the project from start to finish.

Collaborative working *See* Collaborative project management.

Colour depth The number of bits used to hold information about the intensity of each of the primary colours of a pixel. Also known as the 'bit depth'.

Colour fill Filling an object with colour; either a solid block of colour or a gradient fill.

Colour-management system (CMS) Uses uses a standard set of reference colours to create colour profiles for each device, screen or printer.

Colour-picker tool A software tool in image manipulation or animation software that allow colours to be chosen and set by conveniently clicking on the required colour from a colour chart or wheel.

Comment Statements that are ignored by JavaScript engines so are not executed.

Communication channel A method or means of conveying data between devices, applications or users. It is a pathway along a transmission medium over which data flows. A channel can be a real connection, e.g. a wire, or a logical connection, e.g. a connection across a packet-switched network.

Comparison operator Operators used in statements to compare the condition of values or variables.

Compression The process of encoding information in fewer bits of data than are used in the original representation of the information.

Computer-assisted translation (CAT) Translating one human language into another with the assistance of computer software.

Computer-based training (CBT) An interactive learning method that can also be called computer-assisted learning or computer-based instruction.

Conditional statement Statements used to carry out different actions depending on the outcome of decisions.

confirm() The JavaScript method used to create an interactive dialogue box for users. The only actions the user can take are to click OK or to click Cancel.

Connection-less mode A message, or data, is sent from one node to another node without the sender knowing whether or not the receiver is ready to accept the message or even whether it actually exists. There is no connection between the nodes and all that is known by the sender is the address of the recipient.

Connection-oriented mode A fixed link is made between devices over a network. Each device knows where the other device is located. Over a packet-switched network, a virtual circuit can be created between nodes and all the data packets travel over the same route during the exchange of data.

Console A log kept by a web browser that allows users to check on JavaScript for debugging and diagnostic purposes.

console.log() A method used by JavaScript to write data, such as values, variables and arrays, into a web browser console.

Contactless payment Secure payments by debit or credit cards, smart cards, smart phones or other devices, using NFC or RFID.

Cookies Text files placed on client devices, e.g. by web servers, to identify the device. The files contain data that the server can use when the device re-connects. Cookies can be used for managing a session, personalising the connection, e.g. personalised advertising or naming, and for tracking the activities of users.

Corrective maintenance Maintenance carried out to identify and correct problems or errors so that the system is returned to its fully functioning condition.

Crackers People who gain access to networks, or software applications, by taking advantage of security flaws, e.g. to use the resources or software without paying for the use.

Credit card Card issued to enable payments in return for services or goods. The payment is made by the issuing bank, which invoices the cardholder at intervals for repayment.

Critical path method (CPM) Used by project managers to discover and highlight important deadlines, the longest path through a project and to calculate the shortest time that a project could take to complete.

Cropping The removal of objects or parts of the image to leave only those that the user wishes to be seen.

Cross-industry standard process for data mining (CRISP-DM) reference model The standard method for carrying out data mining. It has six phases: business understanding, data understanding, data preparation, data modelling, evaluation and deployment.

Cryptocurrency A type of digital currency that uses cryptography technology to keep transactions authentic and secure.

Cut The process of selecting and removing a section out of a digital image.

Data centre Specialised buildings designed to provide all the power, network and telecommunication connections, and the physical storage space needed to hold, and maintain, clusters of servers in 'server farms'.

Data consolidation Cleaning and combining data from different sources.

Data destruction The process of making sure that it is impossible for data to be recovered. Data may be destroyed as a result of a malware attack.

Data flow diagram (DFD) A diagram that shows how the data flows through a system. DFDs start at Level 0 and progress through levels that show more and more detail of the data flow.

Data integration Combining data from different sources.

Data Link layer OSI model only: The protocols deal with the mapping of IP addresses to the MAC addresses of the hardware (the NIC) in devices.

Data manipulation The changing of the format of data so it is displayed in different ways. This can be visual presentation, e.g. presenting data in a table or in a chart.

Data mining The extraction of useful information from very large repositories of data, called data sets.

Data modification Changing the actual data, e.g. a value of 9 is changed to 8.

Data set A collection of data. A data set can be one or more tables in a database.

Data streaming A method of communication that allows users to access content immediately without having to wait for it all to be downloaded.

Data transformation The changing of data from one format, structure or value into another.

Data transmission The physical transfer of data over communication channels.

Data type A parameter used to tell JavaScript statements how to deal with values when evaluating expressions.

Debit card A card issued to enable payments in return for services or goods. The payment is taken almost immediately from the cardholder's account.

Decentralised system Currency systems that are not controlled by any central authority.

Deliverable An output from a project. The most important deliverable from a project is the intended final product.

Demilitarized zone (DMZ) A logical section of a local area network that can be accessed from outside the LAN while the remainder of the LAN is secure behind a firewall.

Denial of Service (DoS) A DoS attack attempts to make a computing resource unavailable to its legitimate users.

Dependencies The way that tasks in a project relate to each other.

Design documentation Documentation that includes the design specifications, decisions about the designs and other considerations, e.g. the end users and the product features.

Design specification Documentation that sets out how the system will work and look and how end users will interact with it. It is intended for the developer, or developers, to follow in order to create the new system exactly as required.

Design stage The third stage of the system life cycle, where the flow of work through the system and the processing of the data as it flows through the system from input to output is decided.

Development The stage of the system life cycle when the work begins on creating the product using the design documents from the previous stages.

Diary-management software Software that manages calendars and appointments to allow scheduling and re-scheduling of meetings, automatic sending of reminders of meetings and alerts of potential clashes in appointments.

Digital base money Digital currency set up and managed by a government.

Digital currency Money that is recorded electronically and does not exist in a physical form like banknotes and coins.

Digital graphics The processing, storage and display of digital images.

Digital wallet Encrypted software application or service that stores user details on a computer device.

Direct changeover A method of system changeover where the current system is completely replaced with the new system all at once.

Directory mail merge A mail merge that creates a list or document using specified fields from a data source, such as a spreadsheet or database.

Disaster recovery The process of recovering IT services or systems from a disaster in order to return them to the state before the disaster occurred.

Disaster-recovery management Preparing for disasters to reduce the effects of a disaster and planning for the restarting of the IT services in the shortest time possible.

Discussion board Alternative name for an online forum.

Distributed ledger A database of transactions spread across several computing devices (nodes) on a peer-to-peer network.

Distribution Describes how elements, objects and layers in images and animations are spaced and placed around the image canvas.

Do/while loop A loop in JavaScript that iterates through a block of code as long as a condition is true. It is always executed at least once before the condition is tested. This happens even if the condition is false.

Document analysis Reading and analysing documentation to discover how data moves through a system.

Documentation Documents that record, describe and explain how a new system was created and developed and how to use it.

Downlink The link from a communications satellite to a ground station.

Dynamic Host Configuration Protocol (DHCP) A network protocol used to assign IP addresses to devices and inform them of network parameters. It works at the Application layer of networking models and uses UDP to send data.

Dynamic routing table A routing table created when a router starts up by interrogating other routers and is updated by the frequent exchange of routing data between routers. Dynamic routing allows the routing of packets to adapt to any changes or conditions on a network. Routers use routing protocols to exchange messages about network conditions and dynamic tables are updated accordingly. Dynamic routing tables are not stored by the router and are lost when the power is turned off.

E-business Doing business over the internet.

E-commerce Commercial activities conducted electronically over the internet.

Electronic point of sale (EPOS) Self-contained computer system used at checkouts or sales points in stores to allow customers to pay using credit or debit cards. Carries out all of the tasks that a checkout person would do manually.

Element An individual component of a whole, e.g. a component of an image, animation or web page.

Emoticon A pictorial representation of facial expressions.

End point The start points and end points for tasks show when a particular set of activities has begun or ended.

Envelope An object that is used to distort or change the shape of other, selected objects.

Error handling The process of dealing with errors and exceptions in JavaScript code.

Ethernet A networking technology that divides data that is to be transmitted into small sections called frames.

Evaluation stage The stage of the system life cycle that determines whether or not the new system meets the specifications set out by the analysts and designers.

Evolutionary prototyping The development of a working prototype that has only the basic functionality. As development progresses, more functions and features are added to the basic prototype version and tested until all functions have been added.

E-waste Electronic waste is discarded electronic devices.

Execution interval method JavaScript methods used to control the timing of the execution of JavaScript code.

Expressions in JavaScript Combinations of values (literal and variable) and operators in JavaScript that produce a value.

Extensible Business Reporting Language (XBRL) A mark-up language based on XML used for business reports. It enables traders to sort and search the public reports of companies more easily.

Extensible Messaging and Presence Protocol (XMPP) A communication protocol over TCP used for sending XML elements and presence information in messaging systems.

Exterior Gateway Protocol (EGP) Network protocols used by routers to exchange their data about paths and links between LANs over the internet. Now known as the Exterior Border Gateway Protocol (EBGP).

External script A script file referenced and inserted into HTML when needed.

Facial recognition The identification of individuals by the unique features of their face. Software can be used to match images or match facial measurements with those stored in databases.

Feasibility study A study carried out to assess whether a new system can actually be created.

Feature creep Where more features are added during the project than were requested or decided during planning.

Fibre optic A data transmission method that uses light to carry data at high speeds.

File server A type of network server that stores files and file folders for access by users over the network.

File Transfer Protocol (FTP) A network protocol used for transferring files over a network between an application on a device, the client, and an FTP server.

File Transfer Protocol (FTP) server A type of network server that stores files and allows access to them using the File Transfer Protocol (FTP).

Fill Sets the colour within the boundaries of an object. The fill tool allows the colours of a selected object to be changed as required, e.g. a solid block of colour or a gradient fill.

Firewall A barrier between one network and other networks. Firewalls can also be set up on individual host computer devices. The firewall barrier is used to monitor and control incoming and outgoing network traffic.

Fitness tracker A wearable device or application that monitors, or tracks, physical or other fitness-related activities or health measurements, e.g. breathing rates, heart beats.

Flattening Combining, or merging, all the layers on a digital image into a single layer.

Flip book The illusion of movement is created using a booklet of still images that differ slightly from each other and are flipped or flicked so that each page is viewed quickly in sequence.

Float The amount of time allowed for a task to end later than originally planned. *See* Slack.

Floating point number The representation in binary of a number with a decimal point.

For loop Code that loops through JavaScript code. The for loop has three parameters in its syntax: the initialisation of variables, the condition to be tested and its evaluation, and the iteration statement.

For/in loop A loop in JavaScript used to iterate through the properties of an object.

Form control Methods used in input forms to force users to input data in a specific manner. Form controls include dropdown boxes, radio buttons and check boxes.

Fourth-generation optical discs *See* Holographic (and fourth-generation optical) data storage.

Frame One individual, complete image or drawing within a series of complete images in an animation, video or film.

Free float The amount of time that a project can be delayed without affecting the next task that is dependent on it.

Function JavaScript blocks of code that perform a specific task.

Fuzzy select tool *See* Magic wand select tool.

Gantt chart A chart that visually shows a project schedule as a linear time chart. Project tasks are placed on the vertical axis and the start times and end times of the tasks on the horizontal axis.

Gateway Connects two separate networks together. Unlike switches and routers, a gateway can use the different protocols on each network to allow network traffic to travel between the two networks.

Global mapping system A system that uses satellites orbiting the Earth to take photographs, monitor the weather, monitor the seas, and provide much information for producing accurate maps.

Global Positioning System (GPS) A positioning system that uses a constellation of satellites to transmit data that receivers can use to calculate their location.

Graphics Interchange Format (GIF) A bitmap image file format used mainly for basic images that are displayed on the internet, e.g. in web browsers. It supports 256 colours.

Grid An overlay of squares on an image or animation canvas used to provide reference points to align elements accurately horizontally and vertically.

Ground station The station that prepares a data signal from a control centre and transmits it to a satellite in orbit and receives data signals from a satellite and sends data to a control centre.

Grouping Merging objects, elements and layers to allow them to be treated as a single entity.

Guidelines Lines created from rulers to temporarily place lines across the image canvas to allow easier placement when moving or creating elements.

High-level task schedule An overview of the whole of a project, which can act as a project master schedule showing the major milestones and the most important deliverables.

Hologram A 3D image created by interference patterns using light beams, often from lasers.

Holographic (and fourth-generation optical) data storage High-capacity data-storage systems using holographic imaging. *See* Holographic imaging.

Holographic imaging The use of computer science, electrical engineering and optics to create three-dimensional images called holograms.

HSL Hue, saturation and lightness colour system.

HTML element A component of an HTML document.

HTML event A change that occurs as a result of an action by a web browser or a user of the browser, e.g. the page has finished loading or the user has pressed a key.

HTML styles The parameters of HTML elements, such as background colour and text colour, font, size and alignment.

Hubs Repeaters with many connections or ports. They work at the Physical layer of the OSI model as there is no processing of the data inside frames.

Hypertext Transfer Protocol (HTTP) A network protocol used to send requests from, and media content to, web browsers and web servers. It works at the Application layer of networking models. It uses TCP to provide the reliability and connections on ports 80 or 8080 and IP provides for the addressing and routing of the requests and responses.

Hypertext Transfer Protocol Secure (HTTPS) A network protocol based on HTTP that encrypts exchanges using SSL/TLS when used to send requests from, and media content to, web browsers and web servers. It works at the Application layer of networking models.

Identity theft The deliberate use of another person's identity. There is usually a financial or social loss to the individual whose identity has been stolen.

Image manipulation software Software that provides the tools for creating and editing digital graphics.

Implementation The stage of the system life cycle in which the new system is prepared and placed in its working environment in the real world. Parallel running, direct changeover, phased implementation and pilot implementation are methods of system changeover.

Inbetweening The process of creating the 'in between' frames between key frames is called 'tweening'. In animation software this function creates all the frames between two key frames.

Incremental method of software development A method of software development in which a new system or software application is divided into sections, each of which represents a required function or feature and is developed in the same way as a whole new product would be. As each function or feature is completed, it is added to the next version of the software until all the requirements are met and in place.

Incremental prototyping Development process in which a new software application is divided into its component parts and each part is developed into a prototype. Each part of the new application is fully tested as a prototype before being combined into the whole.

Information overload A situation where there is too much data and information to understand easily and quickly.

Infrared (IR) Electromagnetic radiation with wavelengths longer than those of visible light, extending from the red end of the visible spectrum. IR is used to exchange data over short distances between devices.

Infrastructure mode Wi-Fi networking mode in which a network is created. A service set identifier (SSID) is assigned that all devices on the Wi-Fi network must use in order to connect.

Input form Online forms used to enable the accurate and easy entry of data. Data is verified and validated on entry.

Instant messaging (IM) A real-time text-based communication over the internet between two users known to each other.

Interior Gateway Protocol (IGP) Network protocols that work within networks and are used by routers to exchange information and select paths on LANs. IGPs work at the Link layer of the network models. Messages are exchanged between routers in UDP packets and are used to update routing tables.

International Organization for Standardization (ISO) The international organisation that sets and manages technical and manufacturing standards.

Internet Control Message Protocol (ICMP) A network protocol used for carrying error messages between devices. It does not carry user data. It works at the Network layer of the OSI model and the Transport layer of TCP/IP model.

Internet forum Alternative name for online forum.

Internet layer TCP/IP suite only: Protocols at this layer form the rules that enable the internet to work and set out how packets should be constructed.

Internet Message Access Protocol (IMAP) A network protocol used for the retrieval of email from email servers. When a message is retrieved using IMAP, it is not deleted from the server. It works at the Application layer of networking models.

Internet of Things (IoT) The interconnection of devices to exchange data over any communications network. The network is usually the internet but does not have to be. The devices can be any object that can connect and exchange data, such as laptops, smartphones, smart watches, smart television sets and home appliances.

Internet Protocol (IP) IP specifies the addressing system for labelling and routing network datagrams, or packets, and how the packets are structured.

Internet Protocol Security (IPSec) A network protocol used to authenticate and encrypt packets of data. It provides authentication at the start of a session and establishes the encryption keys to be used during the session. When used to encrypt and authenticate an entire IP packet that is placed within a new packet and sent over a network, this is tunneling and creates a VPN. It works at the Internet layer of the TCP/IP model and the Network layer of the OSI model.

Internet Relay Chat (IRC) A protocol used by chat rooms.

Interview Gathering information from people by verbally asking questions.

Internet (TCP/IP) protocol suite *See* Transmission Control Protocol/Internet Protocol (TCP/IP) suite.

Intrusion detection Used at the point of entry to networks to detect unwanted, unusual or suspicious activity by devices or users.

Inverse Address Resolution Protocol (InARP) A network protocol used to discover the IP address from the MAC address on networks.

IP address A unique 32- (IPv4) or 64- (IPv6) bit address for identifying each NIC (in a device) in a network. IP addresses are used for locating a destination NIC and for finding a route from the sending device to the destination device, even if the destination is on another network or if the location of the destination device is not known to the sending device.

Iterative method of software development A method of software development that repeats the stages over and over, using informed feedback, to perfect an end product.

Iterative review A review that is repeated until all issues discovered by the review have been addressed.

JavaScript A scripted programming language for use on the web.

JavaScript statement Each line of JavaScript is called a statement. A statement ends with ;

Joint Photographic Experts Group (JPEG) image file format A bitmap image file format that uses lossy compression to reduce the file size. There are several file extensions associated with JPEG images, e.g. jpg, jpeg and jif.

Kanban board Uses cards or boxes to represent tasks or activities and allows work to be managed at an individual or group level.

Key frame A frame containing the start and end points of an element, movement or action in an animation.

Keyword Reserved words that are only for use by JavaScript.

Knowledge discovery Data mining in databases.

Laser Used for sending light down fibre-optic cables for data transfer. Also used for transmitting data across free space.

Lasso tool A free selection tool that is used to mark out an area by freehand drawing around the area that is to be edited.

Layer In digital graphics and animation, a layer holds separate image elements laid or stacked on top of each other. In networking, protocol stacks are visualised in layers.

Layer 2 Tunneling Protocol (L2TP) A network protocol used to send private data between devices on different networks across public telecommunication networks. It works at the Link layer of networking models.

Link layer TCP/IP suite only: The Link layer defines the rules for sending data between the two hosts over the connection between them.

Litecoin A cryptocurrency derived from Bitcoin that also uses blockchains as a distributed ledger. Transactions are faster and there is a higher maximum number of coins.

Local area network (LAN) An internal network that is restricted to a specific area or organisation.

Logical operator Operator used to determine the logic between values and variables. The logical operators are AND, OR and NOT.

Logical topology Shows the ways that data travels in the network.

Looping Repeating a section of an animation several times in succession.

Loops Used so that the same blocks of JavaScript code can be executed over and over again.

Lossless compression A method of compression of images that retains all the image data so that, when it is decompressed, the image is perfectly restored with no loss or change of detail or colour from the original.

Lossy compression A method of compression of images that takes into account the perception of detail and colour by humans and removes data without affecting the overall visual perception of the image.

Machine intelligence *See* Artificial intelligence (AI).

Machine learning The use of algorithms and data as part of artificial intelligence to copy how humans learn.

Machine translation Translating languages by computer without human input.

Magic wand select tool A selection tool that uses the similarity of contiguous colours to mark out areas of the active layer. *Also called* the fuzzy select tool.

Mail merge The automatic combining of source materials, e.g. text and data, from different sources. Mail merge automates the creation of personalised documents from a standard document. The standard document is the template and is called the master document.

Mail server A type of network server that is responsible for receiving, storing and forwarding email messages between mail clients. Mail servers and their clients work in the client–server model of networking.

Maintenance stage The stage of the system life cycle that involves checking that the system is functioning correctly and ensures that it keeps functioning properly.

Malicious actors People, also called threat actors, who perform acts intended to harm the computer networks, systems and devices of individuals or organisations.

Malware Malicious software created to inflict damage on computer systems, applications, files or data.

Masking tool A software tool used to cover, or mask, part of an image.

Massive open online course (MOOC) A MOOC can have almost an unlimited number of students enrolled on the course at the same time. The course is delivered using the internet and its purpose is to teach a specific topic.

Master document A standard document used as a template in mail merge. It can be a word-processed document or an email template.

Maximum Transmission Unit (MTU) The limit to the size of a unit of data that can be transmitted on any particular network.

Media access control (MAC) address A unique identifying address used when sending and receiving frames on a network. MAC addresses assigned to NICs during manufacture cannot be changed, but can be overridden in some circumstances.

Media access control (MAC) software Software responsible for the interaction between the transmission medium and the hardware in the network interface card (NIC). MAC software works at the Data Link layer of the network models.

Merge field In mail merge, the details in the master document that link a position in the document to the data in the source file.

Message switching A whole, complete message is sent over a communications channel in one go.

Microblog A small blog.

Microwaves Radio waves, electromagnetic radiation, with short wavelengths from about 1 mm to 10 cm. Microwaves are used for data transmission in wireless networking technologies.

Milestone A significant and important event that marks the progress through a project.

Mind map A visual representation of an idea. The central idea or topic has sub-topics radiating out from the centre, branching out and eventually ending in 'twigs' that are an indivisible activity.

Mining for cryptocurrency The process of creating new digital currency by the use of a user's computer resources as a reward for checking transactions.

Mobile communications Technology that allows data connections between portable devices, e.g. smartphones, so they can access internet services using the mobile telephone networks.

Mobile electronic wallet Software application or app installed on a mobile device that stores details of the user's credit or debit cards, loyalty cards and other payment options.

Mobile hotspot A Wi-Fi hotspot for devices using 4G or 5G communications systems to connect to the internet.

Mobile, or cellular (cell), networks Networks working from base stations, located on poles, tall buildings or towers, that cover a small geographic area called a cell and allow devices to connect to networks while users move about.

Molecular data storage The storage of data using chemical molecules as the storage medium instead of using electronic circuitry.

Morphing Changing the shape of an object by making one shape appear to transition smoothly, dissolve or distort into another in an animation.

Most likely time A realistic estimate of the time a task will take to complete.

Multipurpose Internet Mail Extensions (MIME) Used to extend the basic email format to allow more character sets to be used and to allow attachments. Email messages with MIME formatting are transmitted with the usual email protocols of SMTP, POP3 and IMAP.

Near Field Communication (NFC) Wireless communications system working at 13.56 MHz in the radio spectrum. NFC has a range of only a few centimetres so devices have to be very close to each other, usually under 4 cm.

Network architecture The design of a network and the way network devices and services are arranged and connected together, along with the rules by which they can communicate.

Network-attached storage (NAS) A device or node specifically designed to provide access from a LAN to stored files. A NAS system consists of the secondary storage devices and a computing device specifically designed to allow users to connect from the network for access to the files.

Network components The hardware items required to create a network and make it function.

Network interface card (NIC) A network interface controller (also called a network adapter) connects computing devices to a wired or wireless transmission medium.

Network layer OSI model only: This layer provides protocols for the addressing and routing of packets.

Network policies Documentation used to set the rules on how networks are used.

Network protocol Agreed set of rules and conventions that determines how computer devices can communicate with each other.

Network security Includes all the procedures and policies that must be carried out and used to protect stored data and files from security threats.

Network server A node on a network that provides services, data or applications to other nodes (host computers), locally on the network or globally by using the internet. The servers respond to requests from hosts (the clients) in a client–server relationship using the request–response model of networking.

Network stack A representation of the layers in the TCP/IP suite and OSI model of protocol layers. *Also known as* the protocol stack.

Network switch A device with the same role as a network hub with the added ability to manage the network traffic by examining the contents of frames.

Networked courses A type of computer-based training used in schools and colleges as well as in businesses. Networked courses are taken online with access to course materials and assessments provided by a school or business.

Node In computer networks, a device or host. In PERT charts, symbols that hold and show information about the tasks, activities, events and milestones that make up a project. In vector graphics, a control point.

Number A data type in JavaScript used for any kind of number. Numbers are not distinguished into integer or decimal, and are stored as floating point numbers.

Object A data type in JavaScript used for storing complex data structures.

Observation Watching in person the workings and requirements of the current system.

On-demand streaming The distribution of previously recorded content to users as and when they wish to access it.

One-time code, token or password (OTP) A code sent to a user when logging in. The OTP is sent via another form of communication, e.g. an OTP sent via SMS when logging into a website over the internet. Used in two-factor authentication.

Online bulletin board Alternative name for online forum.

Online forum A website, or part of a website, that allows users to communicate with other users by posting messages.

Online meeting The use of IT systems, e.g. video conferencing or an online chat system, to represent a face-to-face meeting but with the participants at remote locations. It enables the sharing of documentation, allows progress to be discussed and team members to be updated even if they are distributed around the world.

Online tutorial A course studied online as an activity, carried out by the students as self-study and with a specific outcome for their learning.

Online wallet Online storage, usually encrypted, e.g. on an online store users store their online shopping information such as login, password, credit card details and shipping address.

Opacity/transparency A measure of how little, or how much, of the layer can be seen through. An inverse relationship exists between opacity and transparency.

Open Systems Interconnection (OSI) model A networking model that defines protocols to be used when devices communicate with each other. It is a conceptual model visualised as being made up of seven layers: Application, Presentation, Session, Transport, Network, Data link and Physical.

Operator precedence The order in which operators are always carried out: brackets or parentheses ( ) and exponentiation are carried out before any other arithmetic operator, followed by multiplication, division, addition and lastly subtraction.

Optimistic time The minimum possible time a task will take to complete.

Packet header The section of a data packet that contains the destination and source addresses, sequence numbers, error-checking fields and flags indicating the status of the packet. The contents of the packet headers differ between types of data packet.

Packet switching On IP networks, messages or files are divided into data packets ready for sending. Individual Internet Protocol (IP) packets are addressed and routed, or switched, across networks to their destination. Packets may travel along different routes to get to the destination.

Parallel running A method of system changeover that has both the current system and the new system running at the same time until the new system is shown to be working as expected.

Passive File Transfer Protocol (FTP) mode In this FTP mode, the client starts a data connection and makes it to a data port that the server has already opened and is available for use.

Path In vector graphics, a line determined by the number and position of nodes.

Peer-to-peer network Nodes are connected together in a network on an equal basis. Peers can share resources but there is no central server to control or co-ordinate the sharing of resources, e.g. processing power, data storage systems or bandwidth.

Perfective maintenance Maintenance carried out to improve the performance of a new system during its intended lifespan.

Performance Evaluation and Review Technique (PERT) A tool that represents the tasks and events needed in a project.

Performance Evaluation and Review Technique (PERT) chart A chart used to show, and analyse, the tasks, activities and timescales involved in a project. A PERT chart shows which tasks are dependent on others, which tasks run concurrently and the links between them.

Perpetrator analysis The analysis of cybercrime perpetrators to assess the threats to IT services, systems and data from cybercrime.

Perspective Giving flat, two-dimensional images the impression of being three-dimensional by making near objects appear larger than those further away from the viewing point.

Pessimistic time The longest possible time a task will take to complete.

Phased implementation A method of system changeover that replaces the parts or modules in a current system with new parts or modules in stages.

Physical layer OSI model only: At this layer, the collections of binary data that make up the frames are converted into signals that are required for the transmission medium.

Physical topology Shows the physical layout and connections of the transmission medium and how they connect to each node in a network.

Pilot implementation A method of system changeover that involves replacing the current system in one part of a company or business but leaving the rest of the company using the current, old system.

Placeholder In mail merge, a merge field.

Planning stage The stage in the system life cycle where resources, costs, timings and other requirements are considered.

Portable Network Graphics (PNG) A bitmap image file format that supports lossless compression. The PNG format supports transparency but not animation.

Post Office Protocol (POP3) A network protocol used for the retrieval of email from email servers. When a message is retrieved using POP3, it is deleted from the server. It works at the Application layer of networking models.

Presentation layer Layer 6 of the OSI networking model, responsible for converting data from the Application layer into the correct format and applying any required encryption.

Preventative maintenance Planned maintenance carried out to try to ensure that minor faults and errors ('latent faults') do not develop into more significant issues or problems.

Print server A type of network server that collects print jobs from computing devices (clients) on a network and sends them to printers.

Product breakdown structure (PBS) analysis The determination of what has to be created and delivered at the end of the project.

Project closure The formal end of a project when the contracts have been fulfilled. The responsibility for the deliverables passes to aftersales departments.

Project execution The stage of a project where the project charter is accepted, the resources are allocated to tasks and the project deliverables are produced.

Project initiation The conception and start of a project. It includes brainstorming, setting of objectives, SWOT analysis, project scoping and the creation of a project charter.

Project life cycle The complete set of stages in a project. The main stages in a project are the initiation, planning, executing, monitoring and control stages and the project closure stage.

Project management software (PMS) Software application(s) that provide the tools to carry out, organise and document a project. It also enables collaboration between the project team members.

Project monitoring and control The stage in a project that includes keeping track of where in the project the team has got to (i.e. what tasks are completed and what are still to be done) identifying the work done and problems and ensuring that steps are taken to solve them. Testing of the deliverables ensures an acceptable end product is created. It includes acceptance testing.

Project planning The stage in a project where a detailed plan is made that includes descriptions of the tasks and the costs, timings and resources needed to carry out the project.

prompt() The JavaScript method used to open a box that requires a user to input a value.

Property key frames Key frames that hold the settings for specific properties of the objects in the frame.

Prosthetic limb An artificial limb.

Protocol stack *See* Network stack.

Prototype An early version of a new software application that can be used and tested. A prototype is usually functional but may not have all the functions and features of the final product.

Prototyping The process of creating a prototype that can be used during the development of a new product, system or software application.

Proxy servers A type of network server that is logically located on networks between clients making requests and the responding servers. A proxy server acts as an intermediary between clients and servers on a network.

Push notifications Messages, notifications, alerts or requests for a transaction sent from a server to a client device instead of being initiated by the client.

Questionnaire Hardcopy or online questionnaires gather information from many people by asking written questions.

Radio wave All wireless technology uses radio waves, which are all wavelengths that are longer (lower in frequency) than infrared (IR).

Ransomware Malware that blocks access to, or encrypts, user data and then requires users to pay a ransom for the encryption keys. It works 'best' when targeting network, or cloud, storage systems.

Rapid application development (RAD) A method of software development in which the requirements are set out, prototypes are quickly created, reviewed using repeated feedback from clients and end users, and the application quickly completed and released.

Rapid prototyping Development process that creates a working version of a new software application in a very short time.

Real-time streaming The capturing, encoding and distribution of live events, e.g. with video and audio content, over the internet.

Recovery Point Objective (RPO) The minimum time between backups that is acceptable to business operations and how much data loss can be acceptable.

Recovery Time Actual (RTA) The actual recovery time from a disaster. It may be shorter or longer than the RTO and depends on the nature of the disaster.

Recovery Time Objective (RTO) A target time limit for recovery from a disaster. The maximum time allowed for recovery from the disaster to the data before the work of the business is seriously harmed.

Recycling The recovery and re-use of complete devices or systems or their component parts or materials.

Red-eye removal The process, in digital image manipulation, of replacing the red colour produced by light reflecting off the back of the retina with a more natural-looking dark colour.

Regular expression A sequence of one or more characters that makes up the pattern to be replaced or searched in JavaScript code.

Rendering In animation, the conversion and saving of an animation into a format that can be played on different computer platforms without the need for the specific software that created it. The result is a standalone animation.

Repeater Used to extend networks by connecting several network sections together into a bigger network to increase the network range. They work at the Physical layer of the OSI model as there is no processing of the data inside frames.

Replication Replicating elements or objects to produce multiple copies of the original.

Requirements documentation Documentation that contains all the system and user requirements.

Resizing an image Resizing an image changes its actual, physical size.

Resolution A measure of how much detail an image has. It depends on the number of pixels present in the image.

RFID Radio-frequency identification. It uses electromagnetic fields to identify and track tags on objects.

RGB An additive colour system produced by combining, or adding, varying amounts of red, green and blue.

Risk matrix A matrix in which threat activities are placed according to the assessment of their likelihood of occurring and the impact they could have.

Risk testing The process of establishing the extent to which threats pose a risk to IT services, files and data.

Robot A machine with a computer system, mechanical structure or frame and a power supply.

Robotics The study and science of the creation of machines that replace or copy human actions.

Rotate Moving an image, layer or element in a circular motion around a central fixed point.

Router Connects sections of networks, and different networks, together. A router may connect networks with different transmission media by containing the appropriate network interface cards for each type of medium. Routers use the store-and-forward method of transmission of data frames and are responsible for sending, and directing, data frames across a network.

Routing The directing of datagrams (packets) along a path through and between networks. Packets are said to be 'routed' through networks.

Routing protocols Network protocols that make the rules and conventions for routing Internet Protocol (IP) packets through networks. Routing protocols work at the Network layer of the OSI model and the Internet layer of the TCP/IP model.

Rulers Used for alignment in images and animations and are usually placed on the left side and on the top of the canvas.

Sandbox A computing environment that isolates malware and restricts the resources it can access.

Satellite communications system The use of satellites to distribute data, TV and radio content around the world.

SatNav A receiver that uses the data in signals from GPS satellites to make calculations to determine and display the location to users as longitude and latitude coordinates or a point on a map.

Scalable Vector Graphics (SVG) An open standard markup language for storing vector graphics that describes two-dimensional images in an XML-based text format.

Scaling an image Scaling changes the viewed size of an image but may not change its physical size.

Script kiddie An unskilled individual who writes scripts and program code, or uses other people's scripts or codes, to attack computer networks and systems.

Secure Shell (SSH) A network protocol used to create a connection to a remote host in the same manner as telnet but uses cryptography to make the connection secure. It works at the Application layer of networking models.

Selection tools Software tools used to choose, select or highlight areas, e.g. of images and animations, to allow them to be worked on.

Server farms A collection of servers connected and networked to appear to users as one server. The servers work together and provide services and functions.

Server-side scripting The process in which JavaScript, or other scripts, are executed on a server.

Service set identifier (SSID) The name of a Wi-Fi network.

Session layer OSI model only: This layer controls the connections between hosts. It adds more ways of checking and recovering data compared to the Transport layer of the TCP/IP suite.

setInterval() A JavaScript statement that repeatedly executes a function. It specifies, in milliseconds, the delay before the specified function is repeated. The statement takes two parameters: setInterval(name of function, time interval between repeats in milliseconds).

setTimeout() A JavaScript statement that specifies, in milliseconds, the delay before a specified function is executed. The statement has two parameters: setTimeout(name of function, wait time in milliseconds).

Sharpen The process that increases the contrast at the edges of shapes so they appear more prominent to the viewer.

Shear Applies a horizontal or vertical slant to an image.

Short Message Service (SMS) Text messaging system commonly used on mobile (cell) phones to allow devices to exchange text messages over communication networks.

Signature-based threat detection A process that looks for known patterns of suspicious activity in order to detect threats to network security.

Simple Mail Transfer Protocol (SMTP) A network protocol used for sending email messages from email clients to mail servers and between mail servers. It works at the Application layer of networking models.

Simulation A computer simulation is an imitation of a real-world process.

Skew The amount of slant applied to an image.

Slack The amount of time allowed for a task to begin later than was originally planned. *See* Float.

Smart television sets TV sets connected to the internet so they can access online content and streaming services. Some can be controlled by voice commands.

Smartglasses Eye glasses or devices worn on the head that provide network or internet connectivity so that the user can be provided with visual data and information.

Smartphone A portable computing device with both computing and mobile telephone capabilities.

Smudge tool A software tool that produces a smearing effect similar to wiping a finger across an area of real paint that is not fully dry.

Snapping The automatic alignment of image or animation elements to a grid.

Social media Applications and websites that have facilities for people to post and share content in real time.

Social networking The use of social media to connect with friends, family and other people with shared interests.

Social-networking service/platform Online service/platform that allows people to socially interact and build a network of contacts.

Source file In mail merge, the source file is used to provide the data to be inserted into the master document.

Specification Documentation that sets out the exact details of a piece of work.

Stakeholder Anyone who is involved in a project, i.e. project managers, the team working for the managers, the clients who asked for the product to be developed and initiated the project, and the potential end users.

Static routing table A manually configured table used to ensure that a router always uses the same paths when sending on packets. The tables are stored in the non-volatile RAM (NVRAM) of the routers.

Stop motion An animation technique that involves physically altering or moving real objects very slightly between individual photographs of the objects.

Store-and-forward method A whole message, part of a message or a data packet is stored before it is forwarded to the next point in its travel to its destination.

Stored-value card Card issued to represent a fixed amount of money that can be withdrawn or used to buy goods or services.

String A store of a collection of any type and number of characters, including none at all, inside quotes in JavaScript code.

Stroke Stroke sets the colour of the line that surrounds an object.

Structured query language (SQL) Language used for programming and for managing data in relational databases. SQL is used to access, return and present the results of the access to a user.

Subtractive manufacturing The creation of 3D objects by removing surplus materials, as in sculpting or wood-carving. Traditional manufacturing, e.g. of car parts, cuts them out of solid metals.

Success criteria The measurable outcomes required from a project by the client, stakeholders and end users.

Syntax The rules that govern the way that characters and symbols are combined in JavaScript.

System An information system is one that is able to gather information, and process, store and distribute the information. A computer system can be the collection of hardware or software or it can be a combination of both hardware and software.

System flowchart A chart that shows how an entire system operates by showing how the work flows between all the operations or processes in the system and showing the relationships between all the inputs, processing and outputs.

System life cycle The development cycle of a new system, consisting of planning, analysis, design, development and testing, implementation, evaluation and maintenance stages.

Tagged Image File Format (TIFF) A bitmap image file format that can carry extra image information data. Some versions support lossless compression and some may support lossy compression.

Task A set of activities in a project necessary to complete the project. The terms 'task' and 'activity' in project management are often used interchangeably, but a task is a major job that needs to be done in a project. Tasks are divided into smaller activities, or assignments, that are required to complete the task.

TCP/IP suite *See* Transmission Control Protocol/Internet Protocol (TCP/IP) suite.

Technical documentation Documentation that details and explains every part of a system so that advanced users, technicians, future analysts and future developers can know and understand exactly how the system was created, how it works, how to maintain it and how to improve it.

Technology-enhanced learning The use of IT to support learning.

Telepresence The appearance of being present by using technology to connect and interact remotely.

Telesurgery The use of remote robotic systems to carry out surgery. The robotic systems receive their instruction in real time over networks including the internet.

Television match official (TMO) A match official, using video technology and wireless communications, who assists the referee on the field in making decisions that affect the play in a game.

Telnet A network protocol used to create a connection between terminal software and remote hosts or servers. It grants almost total, unrestricted access to the host server and there is little or no security. It works at the Application layer of networking models.

Ternary operator This is written in JavaScript code as ? and has three parts: a condition to be tested, what to do if the condition is true, and what to do if the condition is false.

Test data Data that is used when carrying out tests on software or systems.

Test plan Documentation, a guide or a set of instructions on how the testing of a system is to be carried out to ensure that the requirements set out in the designs are met.

Testing The stage of the system life cycle that checks that a product, software component or module does what it is supposed to do.

Text message Messages using the Short Message Service (SMS) for text only or the Multimedia Messaging Service (MMS) if other media are included, e.g. images, video, audio.

Threat control The appropriate and proper use of security measures and procedures to reduce, or mitigate, the effect of threats.

Threat detection The process of detecting threats to IT systems, files and data in order to mitigate the effects of disasters or to prevent disasters from happening.

Throwaway prototyping Development process that creates a working version of a new software application for testing and review. When this is completed, the prototype is discarded and a new one created with all the corrections required. This is repeated until the final product is error-free and complete.

Time lapse A technique where photographs or frames in a film or video are captured at a lower frequency, or rate, than the rate at which they are subsequently played back or displayed.

Timeline A linear representation of the passage of time, e.g. in project management or in an animation.

Timings Project and task timings include: the expected duration (t_e), the earliest possible start time (E_S), the latest possible start time (L_S), the earliest possible finish time (E_F) and the latest possible finish time (L_F).

Topology Network topology refers to how the devices and the links, i.e. the connections, in a network are arranged. *See* Physical topology *and* Logical topology.

Total float The amount of time that a project can be delayed without affecting the finish time of the overall project.

Transformation tools Software tools used to change the size, shape or orientation of an element or object in an image or animation.

Transmission Control Protocol/Internet Protocol (TCP/IP) suite Commonly known as the Internet Protocol suite, a collection of protocols that control the transmission (TCP) and the make-up (IP) of packets. It is a conceptual model of networking protocols and is visualised as being made up of four layers: Application, Transport, Internet and Link.

Transparency/opacity *See* Opacity/transparency.

Transponder A radio/microwave receiver that receives signals, converts them to another frequency and retransmits them. Commonly used in satellite communications systems where uplinked signals are converted and retransmitted to receivers on the ground.

Transport Control Protocol (TCP) A protocol responsible for passing data between applications and the IP software. It controls the flow of the data by setting up and maintaining the logical connection between devices.

Transport layer TCP/IP suite and OSI model: The protocols at this layer are concerned with the type, set-up and control of the connection between the network hosts where software is exchanging data.

Transport Layer Security/Secure Socket Layer (TLS/SSL) A network protocol used to provide security for data on networks. Transport Layer Security (TLS) has superseded Secure Socket Layer (SSL), which is now part of TLS. TLS is used with other protocols to secure data and is the basis of the digital certificate system. It works at the Transport and Application layers of networking models.

Tunneling The use of protocols (tunneling protocols) that encapsulate, i.e. wrap, data so that it can be moved between private networks over public telecommunications systems.

Twisted pair cabling Cables that consist of two cables, each with a copper core encased in an insulating plastic sheath, twisted around each other. Cables with no shielding are called unshielded twisted pair (UTP) cables and those with shielding are shielded twisted pair (STP) cables.

Two-factor authentication (two-step verification) Use of an OTP, e.g. when logging into a system.

Unauthorised access When files or data are read, viewed or otherwise accessed by people, or by software, without the required permissions to do so.

Undefined A data type in JavaScript used when a variable is empty, i.e. does not have a value.

Uplink The link from a ground station to a communications satellite.

User Datagram Protocol (UDP) A network protocol used to send packets on networks by the connection-less method. It is used where error-checking of packets is not necessary and so reduces the response time. It works at the Transport layer of networking models.

User documentation Documentation that describes how end users can install a new system and carry out the tasks for which the system was created.

User requirements specification Documentation that sets out the details of an agreement between the user and the development team about what a new system will do.

Variable A data type in JavaScript that is a container in which values are stored.

Vector graphics The creation of images from data that specifies geometric shapes.

Video-assistant referee (VAR) A match official, using video technology and wireless communications, who assists the referee on the field in making decisions that affect the play in a game.

Video conferencing The use of multiple cameras and screens so that the participants can all see each other, a set of microphones and speakers to exchange the spoken words, and video-conferencing software to exchange the video and audio over the internet.

Virtual currency Currency where the amount of money is stored only in digital form.

Virtual private network (VPN) Nodes are connected to private LANs using public telecommunications systems so users can use it to exchange data as if they were connected directly to the private LAN.

Virtual reality (VR) A simulation of the real world using computing technology.

Virtual servers A software instance of a server running on the same physical hardware as other virtual servers.

Vision enhancement The use of technology and devices to improve the quality of life of people with vision impairment.

Waterfall method A method of software development with a series of stages that follow on from each other: full analysis, design and production of the specifications, development work, testing and review by the client and end users, followed by deployment and maintenance.

Waypoint A point on the timeline of an animation. Most animation software has more than one way to adjust parameters. 'Waypoints' and 'graphs' can be used in Synfig to adjust the opacity of layers, for example. Waypoints can be shown by selecting the parameter and displaying its graph (by selecting the graph icon) on the timeline.

Wearable computing Also known as 'body-worn computing', the wearing or attaching of computing devices on the body to provide useful information or services to the wearer, e.g. fitness trackers.

Web forum Alternative name for online forum.

Web server A type of network server that stores websites and enables access to the contents of the websites.

Weblog Alternative name for blog.

While loop A loop in JavaScript that iterates through a block of code as long as a condition is true. As soon as the condition is false, the iteration stops and the code ceases to be executed. If the condition is false at the start, the code is not executed at all.

White box testing Low-level testing that checks the internal functioning of a system. It tests every line of code, every branching of the program flow, every path and every condition in the system.

Wide area network (WAN) A computer network that covers a much larger area than a LAN. It makes use of public and private telecommunications systems to connect nodes from distant locations.

Wi-Fi A set of protocols specifying how data is exchanged by wireless networking. Wi-Fi works at the Link layers of the network protocol models.

Wi-Fi hotspot A wireless access point that provides connections to the internet for mobile devices.

Wi-Fi Protected Access (WPA) A method of securing Wi-Fi frames. A passphrase entered by a user is used along with the SSID to generate a unique key for the encryption of the frames. WPA uses the TKIP (Temporal Key Integrity Protocol) methods and has additional checks on the data packets to ensure that they have not been changed during transmission. It has replaced WEP as the standard method of securing Wi-Fi connections and WEP should not be used.

Windows Bitmap (BMP) A bitmap image file format created by Microsoft.

Wired Equivalent Privacy (WEP) An optional method used for the encryption of Wi-Fi data frames. The key entered by the user when a device joins a Wi-Fi network is used for authentication and to generate the encryption keys. WEP is considered insecure and WPA should be used instead.

Wireless access point (WAP) A device used to connect devices with wireless capability into a wired network.

Wireless network interface card (WNIC) Used by a device to connect to and exchange data over a radio-based network instead of a wired network.

Wireless security protocols Protocols that provide the security and confidentiality of data to the same standards as those found on wired networks, e.g. WEP and WPA.

Wireless technology Communications technology uses radio waves, part of the electromagnetic spectrum, to carry data.

Wireless transmission of power A method of transmitting energy without a physical connection. DC electrical energy is captured from electromagnetic waves, including those that carry Wi-Fi signals.

Work breakdown structure (WBS) Includes every job, work, activity and task needed to produce the deliverables from a project and can be used to create Gantt and PERT charts. A WBS is also a deliverable and can be developed early in a project by mind mapping.

Workflow The structured and repeatable set, or pattern, of tasks or activities in a project, consisting of the steps required to complete an activity, the resources needed to complete the activity and the team members involved.

Working level schedule A schedule of activities and tasks showing what needs to be done and when in a project.

Zero-login The use of authentication techniques that enable the user to be recognised without the need for usernames or passwords. The authentication is based on a probability or risk score of the user not being who they say they are. Only the risk score of the identification being acceptable is sent from the client to the server.

Index